Organic Chemistry I Workbook FOR DUMMIES®

Organic Chemistry I Workbook FOR DUMMIES®

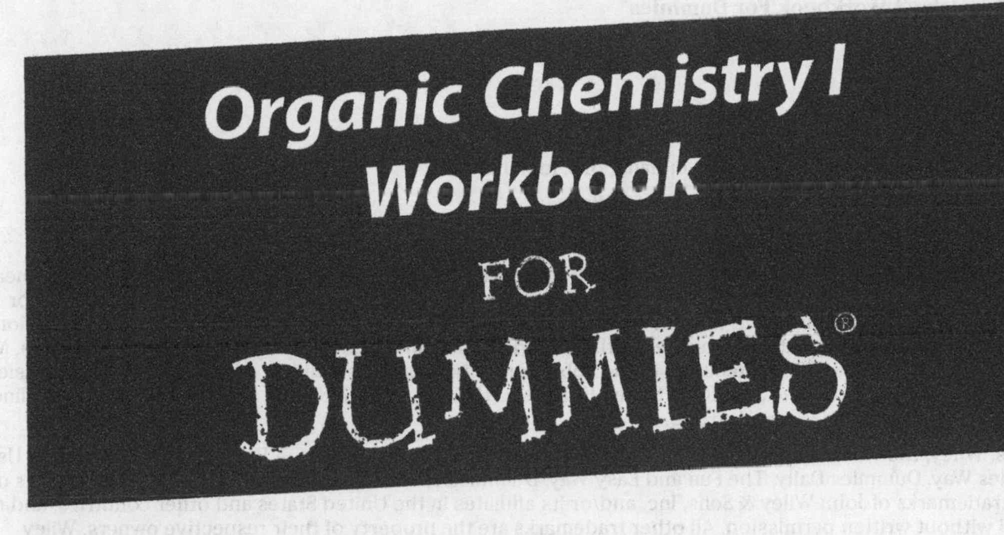

by Arthur Winter, PhD
Creator, Organic Chemistry Help!
Web site at chemhelper.com

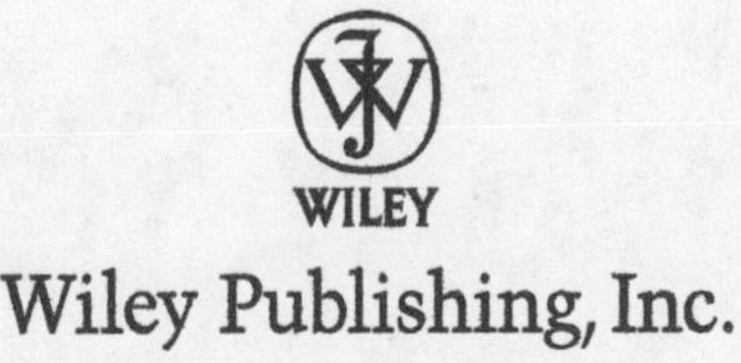

Wiley Publishing, Inc.

Organic Chemistry I Workbook For Dummies®

Published by
Wiley Publishing, Inc.
111 River St.
Hoboken, NJ 07030-5774
www.wiley.com

Published by Wiley Publishing, Inc., Indianapolis, Indiana

Published simultaneously in Canada

For general information on our other products and services, please contact our Customer Care Department within the U.S. at 877-762-2974, outside the U.S. at 317-572-3993, or fax 317-572-4002.

For technical support, please visit www.wiley.com/techsupport.

Wiley also publishes its books in a variety of electronic formats and by print-on-demand. Some content that appears in standard print versions of this book may not be available in other formats. For more information about Wiley products, visit us at www.wiley.com.

Library of Congress Control Number: 2008927913

ISBN: 978-0-470-25151-5

ISBN 978-0-470-25151-5 (pbk); ISBN 978-0-470-40524-6 (ebk)

Manufactured in the United States of America

SKY10028511_080921

About the Author

Arthur Winter received his PhD in chemistry from the University of Maryland. He is the creator of the popular Organic Chemistry Help! Web site at `chemhelper.com` and is the author of *Organic Chemistry I For Dummies* (Wiley). His two major research interests involve exploiting photochemistry to solve challenging problems in medicine and using high-powered lasers to start small laboratory fires. He is currently a post-doctoral student at Ohio State University.

Dedication

For Dan Falvey: Best. Advisor. Ever.

Author's Acknowledgments

I thank the good folks at Wiley for making this workbook possible. First, I thank Lindsay Lefevere and Kathy Cox for getting the ball rolling on this project and keeping it on a steady course. I also thank Chad Sievers, Danielle Voirol, and Alissa Schwipps for their dedication and patience with the editing aspects of this book. For their friendship, I thank Jonathan, Julian, Katie, and Suzanne Winter. On a personal level, I am also grateful to Becky Veiera, Brian Borak, Raffaele Perrotta, Owen McDonough, Kostas Gerasopoulis, Alex Tzannes, and Mike Hughes. I also thank Philip DeShong, Steve Rokita, Jeff Davis, Dan Falvey, and Christopher Hadad for their support.

Publisher's Acknowledgments

We're proud of this book; please send us your comments through our Dummies online registration form located at www.dummies.com/register/.

Some of the people who helped bring this book to market include the following:

Acquisitions, Editorial, and Media Development

Project Editor: Chad R. Sievers

Acquisitions Editor: Lindsay Lefevere

Senior Copy Editor: Danielle Voirol

Editorial Program Coordinator: Erin Calligan Mooney

Technical Editor: Joe C. Burnell, PhD

Editorial Manager: Michelle Hacker

Editorial Assistants: Joe Niesen, Jennette ElNaggar

Cover Photos: © cosmin4000/istockphoto

Cartoons: Rich Tennant (www.the5thwave.com)

Composition Services

Project Coordinator: Erin Smith

Layout and Graphics: Carl Byers, Carrie A. Foster, Stephanie D. Jumper

Proofreaders: Jessica Kramer, Arielle Carole Mennelle

Indexer: Broccoli Information Management

Special Help

Alissa Schwipps, Carrie Burchfield

Publishing and Editorial for Consumer Dummies

Kathleen Nebenhaus, Vice President and Executive Publisher

Kristin Ferguson-Wagstaffe, Product Development Director

Ensley Eikenburg, Associate Publisher, Travel

Kelly Regan, Editorial Director, Travel

Publishing for Technology Dummies

Andy Cummings, Vice President and Publisher

Composition Services

Debbie Stailey, Director of Composition Services

Contents at a Glance

Table of Contents

Introduction

Organic chemistry is a subject that blends basic chemistry, logic problems, 3-D puzzles, and stick-figure art that looks like something you may find in a prehistoric cave. To say that organic chemistry covers a pretty large amount of material is a bit like saying that oxygen is pretty important for human survival. You're probably somewhat familiar with an organic chemistry textbook if you're reading this workbook. I'd be proud to catch a fish that weighed as much! Organic chemistry does cover a lot of material, so much that you can't possibly hope to memorize it all.

But good news! You don't need to memorize the vast majority of the material if you understand the basic concepts at a fundamental level, and indeed, memorization beyond the basic rules and conventions is even frowned upon. The catch is that to really understand the concepts, you have to practice at it by working problems. Lots of problems. Lots. Did I mention the whole working problems thing? Mastering organic chemistry without working problems is impossible — kind of like becoming an architect without bothering to draw up any plans.

This workbook is for getting hands-on experience. I've heard that organic exams are a lot like a gunfight. You act out of instinct only if you've drilled the material you need to know. Classmates who haven't worked the problems will see the problems gunning at them on an exam and spook. They'll come down with a bad case of exam-block and start sucking their thumbs and crying for Momma. You, on the other hand, having been to boot camp and practiced by drilling the problems every day, will stare the exam down like a cool-headed soldier and get down to the serious business of whooping it up until it begs for its life.

About This Book

Ideally, you should use this book in conjunction with some other reference book, such as a good introductory organic textbook or *Organic Chemistry I For Dummies*. This book doesn't cover the material in great detail; for each section, I give a brief overview of the topic followed by problems that apply the material.

The organization of this book follows the *For Dummies* text, which in turn is organized to follow most organic texts fairly closely. The basic layout of this workbook is to give you straightforward problems for each section to really drill the concepts and build your confidence — before spicing things up with a mischievous humdinger or two at the end of each section to make you don the old thinking cap.

For added convenience, the book is modular, meaning that you can jump around to different chapters without having to have read or worked problems in other chapters. If you need to know some other concepts to get you up to speed, just follow the cross-references.

Conventions Used in This Book

As with all *For Dummies* books, I've tried to write the answers in a simple conversational style, just as if you and I were having a one-on-one tutoring session, coffee in hand. Here are some other conventions I've followed concerning the problems:

- At the beginning of each section, I present one or two example problems to show you the thought process involved in working that problem type before you take a stab at similar problems. You can refer back to the example while you're working the other problems in that section if you get stuck.
- Short answers appear in bold in the Answer Key, followed by a detailed breakdown of how I solved each problem. This includes my personal thought process of how to solve a particular problem type, such as where to start and how to proceed. Although other thought processes may lead to the same answer, my explanation can at least give you a guide for problems on which you get stuck.
- Sometimes, I discuss common mistakes that people make with a certain problem type. My basic philosophy is that I'd rather over-explain than give too little explanation.
- In naming molecules, I've used official nomenclature of the International Union of Pure and Applied Chemistry (IUPAC).

Foolish Assumptions

When writing this book, I made a few general assumptions about you, the reader. You probably meet at least one of these assumptions:

- You have a background in general chemistry, and ideally, you've taken a one- or two-semester course in introductory chemistry.
- You're in the midst of or are getting ready to enter your organic chemistry I class in college, and you need some extra help practicing the concepts.
- You took organic chemistry a few years ago, and you want to review what you know.

No matter where you stand, this book provides multiple chances to practice organic chemistry problems in an easy-to-understand (and dare I say fun) way.

How This Book Is Organized

I divide this workbook into five parts that cover the most important topics in first-semester organic chemistry. Here's an overview.

Part I: The Fundamentals of Organic Chemistry

Here's where you first practice speaking the words of the organic chemist. You put charges on structures, work with resonance, and draw structures using the various drawing schemes — all the skills that you just gotta know to do well in the class. You also work with the functional groups and do a bit of magic with acid and base chemistry, because these concepts are so important when you work with organic reactions a little later in the course.

Part II: The Bones of Organic Molecules: Hydrocarbons

In this part, you enter the cemetery of the organic chemist and take a look at the hydrocarbons. These are the bones of organic molecules that bind organic structures together, and they consist of just hydrogen and carbon atoms. You first practice working with alkanes, the sturdy carbon backbones that hold all the reactive centers on organic molecules in place and keep things nice and stable. When you're finally straight with these organic molecules, I take you into the third dimension through stereochemistry. Stereochemistry is the way that atoms can orient in space, and here you get to practice your 3-D visualization skills. You also see how organic molecules can bend, flex, and pretzel themselves to form different conformations, and you see how to predict the various energies of these conformations. Finally, you get a first appetizer of organic reactions in the discussion of alkenes and alkynes, molecules containing carbon-carbon double and triple bonds.

Part III: Functional Groups and Their Reactions

This is the part where you get the full entrée of organic reactions: the discussion of various functional groups and their reactions, spiced up with a few healthy helpings of nomenclature. Included are the alkyl halides, aromatic rings, and — my favorite! — the alcohols (of which there are thousands more than the alcohol you find cheering up the local spirits and inspiring karaoke singers in your favorite watering hole).

Part IV: Detective Work: Spectroscopy and Spectrometry

In this part, you put on your overcoat and fedora and break out the magnifying glass and dusting powder. You practice your detective work in solving for unknown structures using spectroscopy and spectrometry, instrumental techniques that let you nail down a structure of an unknown molecule. You work on extracting the various parts of *spectra* (the data plots coming out of these instruments) for clues to the identity of your molecule and then put all the clues together, just as if you were in a cornball TV murder mystery trying to figure out whodunit. So go get 'em, Sherlock.

Part V: The Part of Tens

In this part, I give you some tips on how to ace orgo exams. As an added bonus, I've included the long-lost Ten Commandments of Organic Chemistry, which help you avoid committing the common sins that lead organic chemistry students into the abyss. Disobey these commandments at your own peril!

Icons Used in This Book

This book uses icons to direct you to important info. Here's your key to these icons:

The Tip icon highlights orgo info that can save you time and cut down on the frustration factor.

This symbol points out especially important concepts that you need to keep in mind as you work problems.

The Warning icon helps you steer clear of organic chemistry pitfalls.

This icon directs you to the examples at the beginning of each set of problems.

Where to Go from Here

Organic chemistry builds on the concepts you picked up in general chemistry, so I strongly suggest starting with Chapter 1. I know, I know, you've already taken a class in introductory chemistry and have stuffed yourself silly with all that basic general-chemistry goodness — and that's all in the past, man, and you're now looking to move on to bigger and better things. However, winter breaks and days spent at the beach during summer vacations have a cruel tendency to swish the eraser around the old bean, particularly across the places that contain your vast, vast stores of chemistry knowledge. That's why I suggest that you start with Chapter 1 for a quick refresher and that you at least breeze through the rest of Part I. In a sense, Part I is the most important part of the book, because if you can get the hang of drawing structures and interpreting what they mean, you've reached the first major milestone. Getting versed in these fundamental skills can keep you out of organic purgatory.

Of course, this book is designed to be modular, so you're free to jump to whatever section you're having trouble with, without having to have done the problems in a previous chapter as reference. Feel free to flip through the Table of Contents or the Index to find the topic that most interests you.

Part I
The Fundamentals of Organic Chemistry

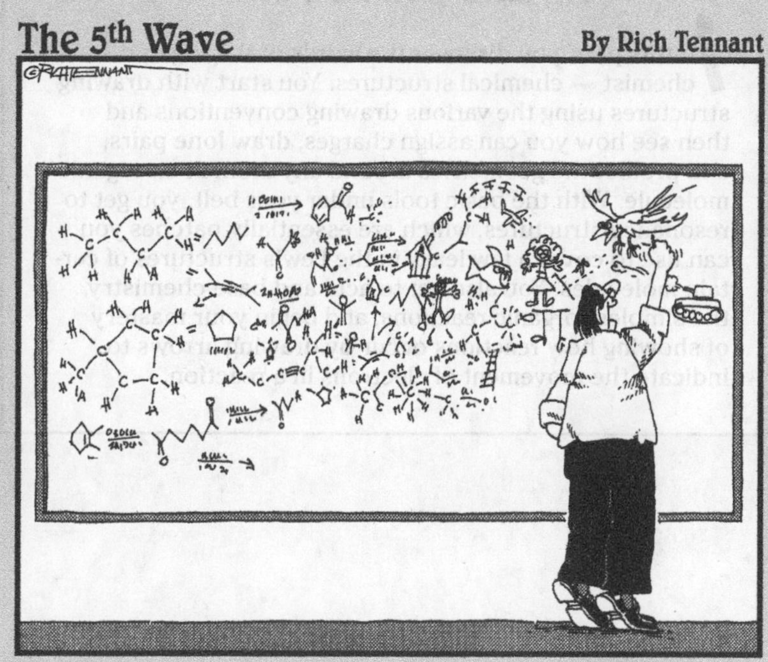

In this part . . .

In this part, you discover the words of the organic chemist — chemical structures. You start with drawing structures using the various drawing conventions and then see how you can assign charges, draw lone pairs, and predict the geometries around any atom in an organic molecule. With the basic tools under your belt, you get to resonance structures, which are essentially patches you can use to cover a few leaks in the Lewis structures of certain molecules. You also get to acid and base chemistry, the simplest organic reactions, and begin your mastery of showing how reactions occur by drawing arrows to indicate the movement of electrons in a reaction.

Chapter 1

Working with Models and Molecules

In This Chapter

- Diagramming Lewis structures
- Predicting bond dipoles and dipole moments of molecules
- Seeing atom hybridizations and geometries
- Discovering orbital diagrams

Organic chemists use models to describe molecules because atoms are tiny creatures with some very unusual behaviors, and models are a convenient way to describe on paper how the atoms in a molecule are bonded to each other. Models are also useful for helping you understand how reactions occur.

In this chapter, you use the Lewis structure, the most commonly used model for representing molecules in organic chemistry. You also practice applying the concept of atom hybridizations to construct orbital diagrams of molecules, explaining where electrons are distributed in simple organic structures. Along the way, you see how to determine dipoles for bonds and for molecules — an extremely useful tool for predicting solubility and reactivity of organic molecules.

Constructing Lewis Structures

The *Lewis structure* is the basic word of the organic chemist; these structures show which atoms in a molecule are bonded to each other and also show how many electrons are shared in each bond. You need to become a whiz at working with these structures so you can begin speaking the language of organic chemistry.

To draw a Lewis structure, follow four basic steps:

1. **Determine the connectivity of the atoms in the molecule.**

   Figure out how the atoms are attached to each other. Here are some guidelines:

   - In general, the central atom in the molecule is the least electronegative element. (Electronegativity decreases as you go down and to the left on the periodic table.)
   - Hydrogen atoms and halide atoms (such as F, Cl, Br, and I) are almost always peripheral atoms (not the central atom) because these atoms usually form only one bond.

2. **Determine the total number of valence electrons (electrons in the outermost shell).**

   Add the valence electrons for each of the individual atoms in the molecule to obtain the total number of valence electrons in the molecule. If the molecule is charged, add one electron to this total for each negative charge or subtract one electron for each positive charge.

3. **Add the valence electrons to the molecule.**

   Follow these guidelines:

   - Start adding the electrons by making a bond between the central atom and each peripheral atom; subtract two valence electrons from your total for each bond you form.
   - Assign the remaining electrons by giving lone pairs of electrons to the peripheral atoms until each peripheral atom has a filled octet of electrons.
   - If electrons are left over after filling the octets of all peripheral atoms, then assign them to the central atom.

4. **Attempt to fill each atom's octet.**

   If you've completed Step 3 and the central atom doesn't have a full octet of electrons, you can share the electrons from one or more of the peripheral atoms with the central atom by forming double or triple bonds.

You can't break the octet rule for second-row atoms; in other words, the sum of the bonds plus lone pairs around an atom can't exceed four.

Q. Draw the Lewis structure of CO_3^{2-}.

A.

$$\left[\ \ddot{O}{=}C(\ddot{\underset{\cdot\cdot}{O}}{:})_2\ \right]^{2-}$$

Most often, the *least* electronegative atom is the central atom. In this case, carbon is less electronegative than oxygen, so carbon is the central atom and the connectivity is the following:

```
    O
    |
    C
  /   \
 O     O
```

Carbon has four valence electrons because it's an atom in the fourth column of the periodic table, and oxygen has six valence electrons because it's in the sixth column. Therefore, the total number of valence electrons in the molecule is 4 + 6(3) + 2 = 24 valence electrons. You add the additional two electrons because the molecule has a charge of –2 (if the molecule were to have a charge of –3, you'd add three electrons; if –4, you'd add four; and so forth).

Start by forming a bond between the central carbon atom and each of the three peripheral oxygen atoms. This accounts for six of the electrons (two per bond). Then assign the remaining 18 electrons to the oxygens as lone pairs until their octets are filled. This gives you the following configuration:

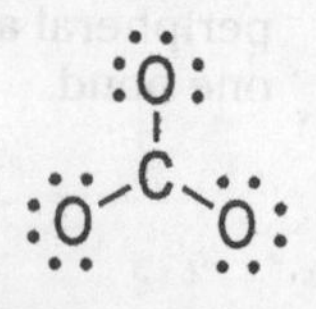

```
   :Ö:
    |
    C
  /   \
:Ö:   :Ö:
```

The result of the preceding step leaves all the oxygen atoms happy because they each have a full octet of electrons, but the central carbon atom remains unsatisfied because this atom is still two electrons short of completing its octet. To remedy this situation, you move a lone pair from one of the oxygens toward the carbon to form a carbon-oxygen double bond. Because the oxygens are identical, which oxygen you take the lone pair from doesn't matter. In the final structure, the charge is also shown:

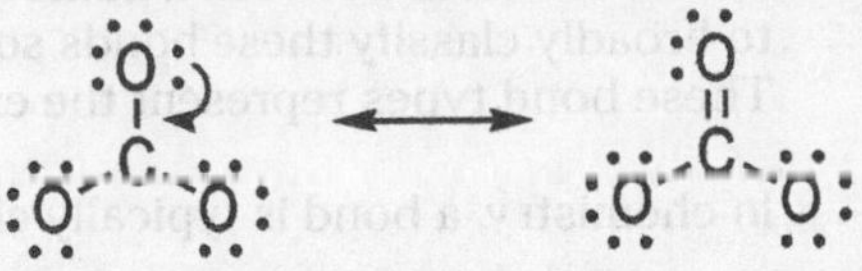

1. Draw the Lewis structure of BF_4^-.

Solve It

2. Draw the Lewis structure of H_2CO.

Solve It

3. Draw the Lewis structure of NO_2^-.

Solve It

Predicting Bond Types

Bonds can form between a number of different atoms in organic molecules, but chemists like to broadly classify these bonds so they can get a rough feel for the reactivity of that bond. These bond types represent the extremes in bonding.

In chemistry, a bond is typically classified as one of three types:

- **Purely covalent:** The bonding electrons are shared equally between the two bonding atoms.
- **Polar covalent:** The electrons are shared between the two bonding atoms, but unequally, with the electrons spending more time around the more electronegative atom.
- **Ionic:** The electrons aren't shared. Instead, the more electronegative atom of the two bonding atoms selfishly grabs the two electrons for itself, giving this more electronegative atom a formally negative charge and leaving the other atom with a formal positive charge. The bond in an ionic bond is an attraction of opposite charges.

You can often determine whether a bond is ionic or covalent by looking at the difference in electronegativity between the two atoms. The general rules are as follows:

- If the electronegativity difference between the two atoms is 0.0, the bond is purely covalent.
- If the electronegativity difference is between 0.0 and 2.0, the bond is considered polar covalent.
- If the electronegativity difference is greater than 2.0, the bond is considered ionic.

Figure 1-2 shows the electronegativity values.

H 2.1						
Li 1.0	Be 1.5	B 2.0	C 2.5	N 3.0	O 3.5	F 4.0
Na 0.9	Mg 1.2	Al 1.5	Si 1.8	P 2.1	S 2.5	Cl 3.0
K 0.8	Ca 1.0					Br 2.8
						I 2.5

Figure 1-2: Electronegativity values for common atoms.

Q. Using the following figure, classify the bonds in potassium amide as purely covalent, polar covalent, or ionic.

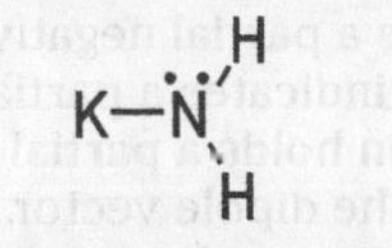

potassium amide

A. **You classify the N-H bonds as polar covalent and the N-K bond as ionic.**

To determine the bond type, take the electronegativity difference between the two atoms in each bond. For the nitrogen-potassium (N-K) bond, the electronegativity value is 3.0 for nitrogen and 0.8 for potassium, giving an electronegativity difference of 2.2. Therefore, this bond is considered ionic. For the N-H bonds, the nitrogen has an electronegativity value of 3.0 and hydrogen has an electronegativity value of 2.2, so the electronegativity difference is 0.8. Therefore, the N-H bonds are classified as polar covalent.

4. Classify the bond in NaF as purely covalent, polar covalent, or ionic.

Solve It

5. Using the following figure, classify the bonds in hexachloroethane as purely covalent, polar covalent, or ionic.

hexachloroethane

Solve It

Determining Bond Dipoles

Most bonds in organic molecules are of the polar covalent variety. Consequently, although the electrons in a polar covalent bond are shared, on average they spend more time around the more electronegative atom of the two bonding atoms. This unequal sharing of the bonding electrons creates a separation of charge in the bond called a *bond dipole*.

Bond dipoles are used all the time to predict and explain the reactivity of organic molecules, so you need to understand what they mean and how to show them on paper. You represent this separation of charge on paper with a funny-looking arrow called the *dipole vector.* The head of the dipole vector points in the direction of the partially negatively charged atom (the more electronegative atom) and the tail (which looks like a + sign) points toward the partially positive atom of the bond (the less electronegative atom).

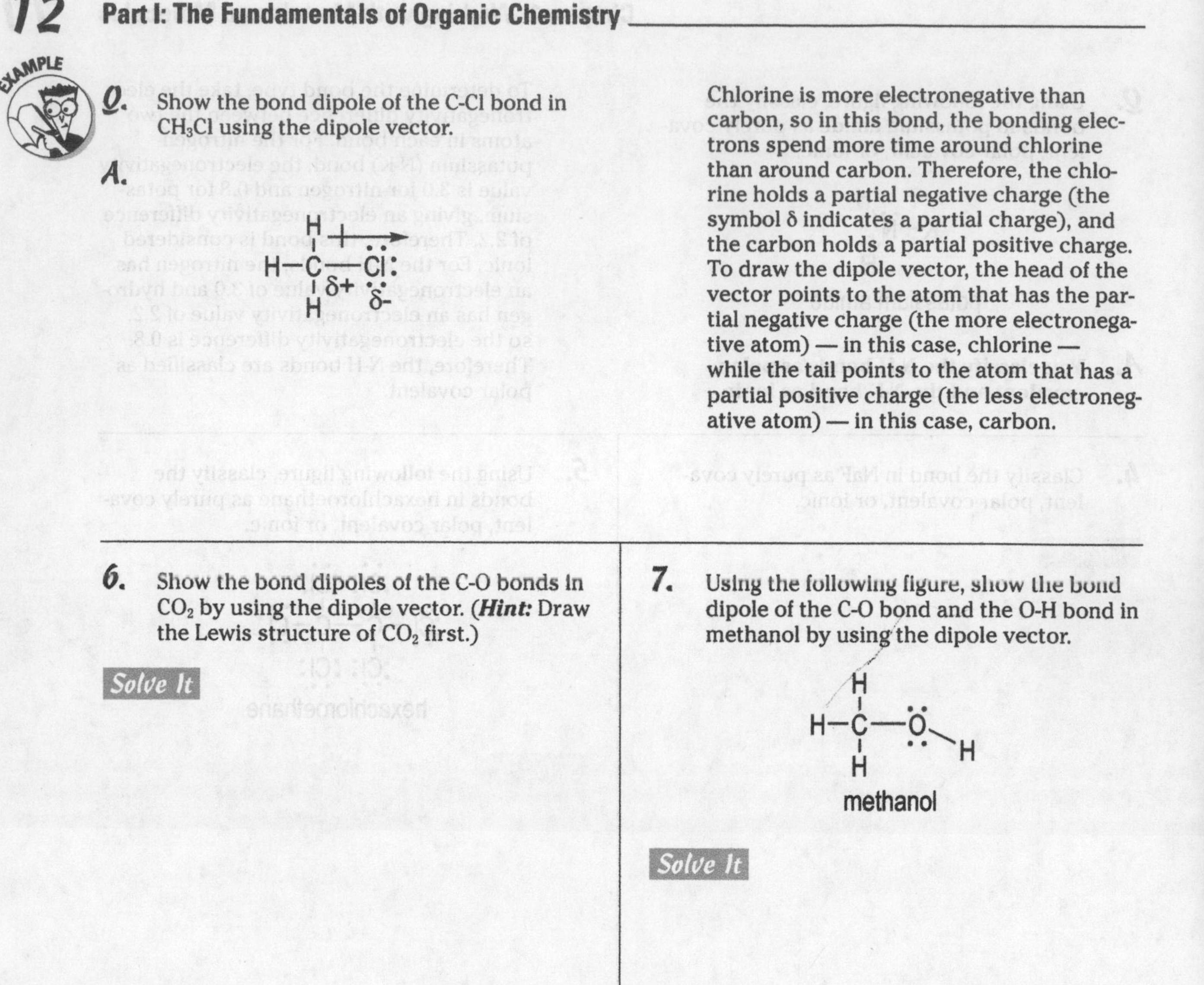

Q. Show the bond dipole of the C-Cl bond in CH_3Cl using the dipole vector.

A.

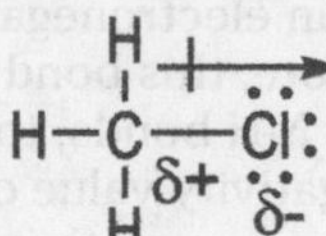

Chlorine is more electronegative than carbon, so in this bond, the bonding electrons spend more time around chlorine than around carbon. Therefore, the chlorine holds a partial negative charge (the symbol δ indicates a partial charge), and the carbon holds a partial positive charge. To draw the dipole vector, the head of the vector points to the atom that has the partial negative charge (the more electronegative atom) — in this case, chlorine — while the tail points to the atom that has a partial positive charge (the less electronegative atom) — in this case, carbon.

6. Show the bond dipoles of the C-O bonds in CO_2 by using the dipole vector. (***Hint:*** Draw the Lewis structure of CO_2 first.)

Solve It

7. Using the following figure, show the bond dipole of the C-O bond and the O-H bond in methanol by using the dipole vector.

H
H—C—O—H
H

methanol

Solve It

Determining Dipole Moments for Molecules

The sum of all the bond dipoles on a molecule is referred to as the molecule's *dipole moment.* Molecule dipole moments are useful in predicting the solubility of organic molecules. For example, by using dipole moments, you can predict that oil and water won't mix and will be insoluble in each other, whereas water and alcohol will mix. Solubilities are important for practical organic chemistry because it's hard to get a reaction between two molecules that don't dissolve in the same solvent.

To determine the dipole moment of a molecule, follow these steps:

1. Draw the bond dipole vector for each of the bonds in the molecule.

2. Add the individual bond dipole vectors using mathematical vector addition to obtain the molecule's overall dipole moment.

A simple method to add vectors is to line them up head to tail and then draw a new vector that connects the tail of the first vector with the head of the second one.

You can generally ignore contributions to the molecular dipole moment from C-H bonds because the electronegativity difference between carbon and hydrogen is so small that the C-H bond dipoles don't contribute in any significant way to the overall molecule dipole moment.

Q. Using the following figure, determine the dipole moment of *cis*-1,2-dichloroethene.

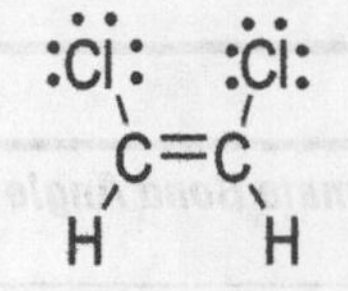

cis-1,2-dichloroethene

A. First draw the bond dipoles for each of the C-Cl bonds. You can ignore the bond dipoles from the other bonds in the molecule because C-H bonds have such small bond dipoles that you can ignore them and because C-C bonds have no bond dipole. After you draw the two C-Cl bond dipoles (labeled *a* and *b*), you add the vectors to give a third vector (labeled *c*). This new vector *(c)* is the molecule's overall dipole moment vector.

individual bond dipoles | vector addition | molecular dipole

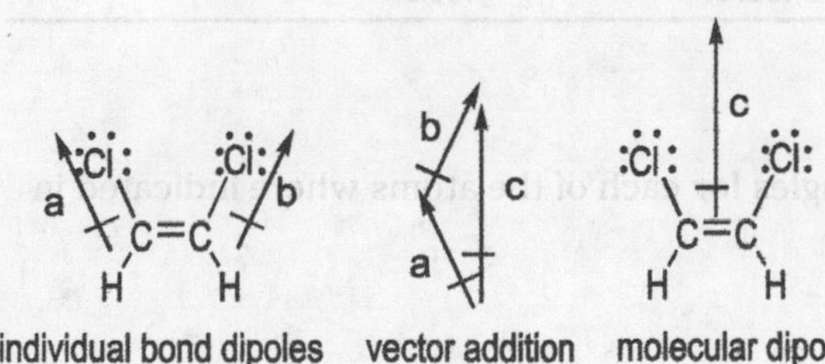

8. Determine the dipole moment of dichloromethane, CH_2Cl_2, shown here. For this problem, you can assume that the molecule is flat as drawn.

Solve It

9. Determine the dipole moment of *trans*-1,2-dichloroethene shown here.

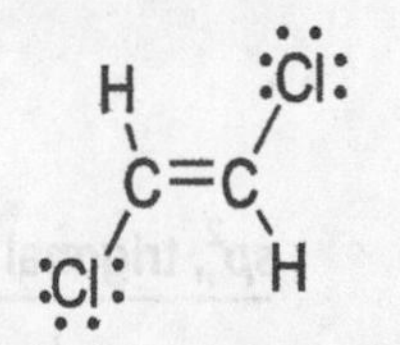

trans-1,2-dichloroethene

Solve It

Predicting Atom Hybridizations and Geometries

Organic molecules often have atoms stretched out into three-dimensional space. Organic chemists care about how a molecule arranges itself in 3-D space because the geometry of a molecule often influences the molecule's physical properties (such as melting point, boiling point, and so on) and its reactivity. The 3-D shape of molecules also plays a large role in a molecule's biological activity, which is important if you want to make a drug, for example. To predict the geometry around an atom, you first need to determine the hybridization of that atom.

You can often predict the *hybridization* of an atom simply by counting the number of atoms to which that atom is bonded (plus the number of lone pairs on that atom). Table 1-1 breaks down this information for you.

Table 1-1 The Hybridization of an Atom

Number of Attached Atoms Plus Lone Pairs	*Hybridization*	*Geometry*	*Approximate Bond Angle*
2	*sp*	Linear	180°
3	sp^2	Trigonal planar	120°
4	sp^3	Tetrahedral	109.5°

Q. Predict the hybridizations, geometries, and bond angles for each of the atoms where indicated in the shown molecule.

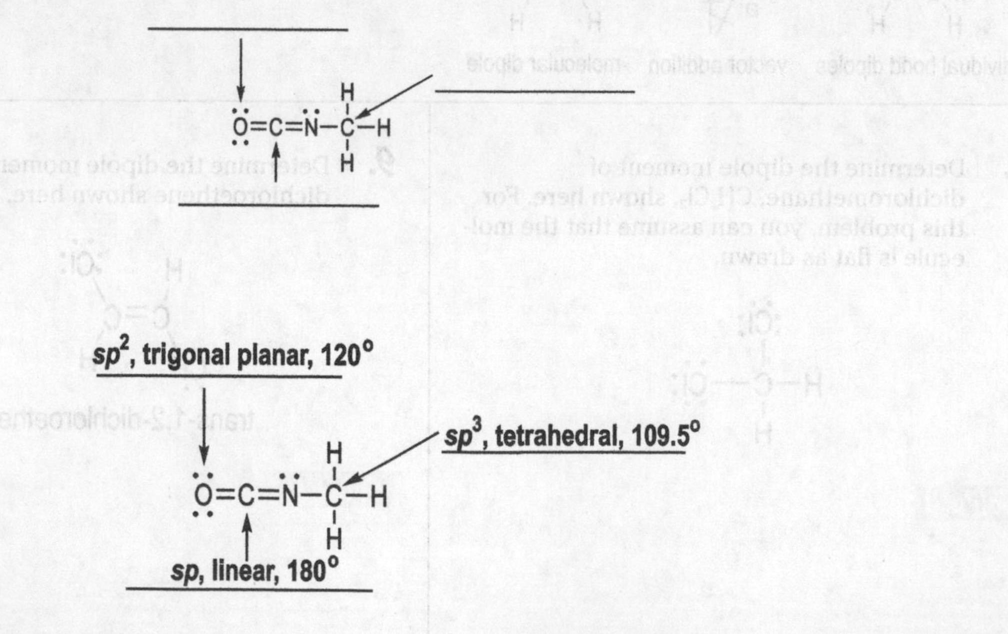

$O=C=N-CH_3$ (arrows point to N, C of the CH_3 group, and the central C, each with a blank: ______)

A.

$O=C=N-CH_3$

N: **sp^2, trigonal planar, 120°**

CH_3 carbon: **sp^3, tetrahedral, 109.5°**

Central C: ***sp*, linear, 180°**

The oxygen has three attachments from the adjacent carbon plus the two lone pairs, making this atom sp^2 hybridized. Atoms that are sp^2-hybridized have a trigonal planar geometry and bond angles of 120° separating the three attachments. ***Note:*** Don't take the oxygen's double bond into account; rather, simply count the number of attached atoms plus lone pairs. The carbon has two attachments and so is *sp* hybridized with a linear geometry and 180° bond angles between the attachments. And the right-most carbon, with four attachments, is sp^3 hybridized with a tetrahedral arrangement between the four attachments and bond angles of 109.5°.

10. Predict the hybridizations, geometries, and bond angles for each of the atoms where indicated in the shown molecule.

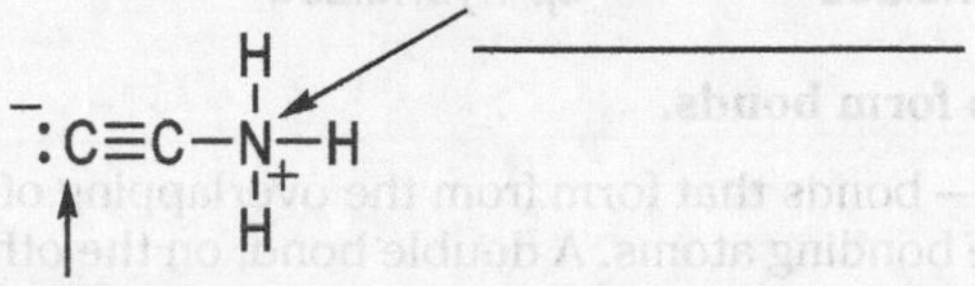

Solve It

11. Predict the hybridizations, geometries, and bond angles for each of the atoms where indicated in the shown molecule.

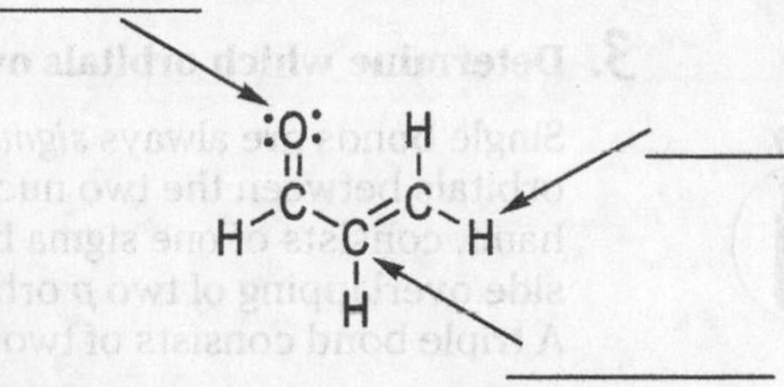

Solve It

Making Orbital Diagrams

An *orbital diagram* expands on a Lewis structure (check out the "Constructing Lewis Structures" section earlier in this chapter) by explicitly showing which orbitals on atoms overlap to form the bonds in a molecule. Organic chemists use such orbital diagrams extensively to explain the reactivity of certain bonds in a molecule, and the diagrams also do a better job than Lewis structures of showing exactly where electrons are distributed in a molecule. Follow these three steps to draw an orbital diagram:

1. **Determine the hybridization for each atom in the molecule.**

   Check out the preceding section for help on this step.

2. **Draw all the valence orbitals for each atom.**

   Sp^3-hybridized atoms have four valence sp^3 orbitals; sp^2-hybridized atoms have three sp^2-hybridized orbitals and one *p* orbital; and *sp*-hybridized atoms have two *sp*-hybridized orbitals and two *p* orbitals. You may find the following templates helpful for constructing your orbital diagrams (where A represents the hybridized atom):

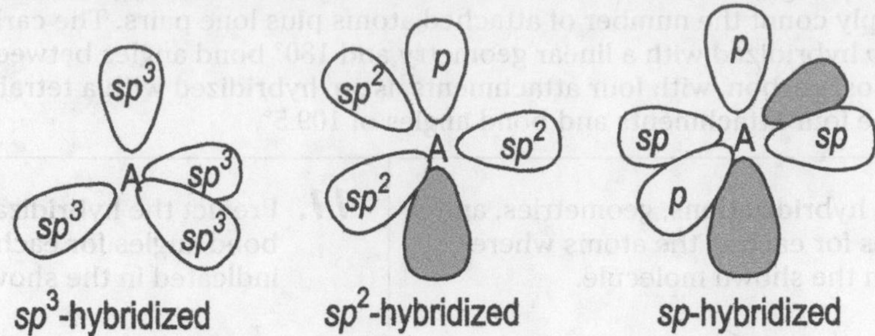

3. Determine which orbitals overlap to form bonds.

Single bonds are always *sigma bonds* — bonds that form from the overlapping of orbitals between the two nuclei of the bonding atoms. A double bond, on the other hand, consists of one sigma bond and one pi bond. A *pi bond* is formed from the side-by-side overlapping of two *p* orbitals above and below the nuclei of the two bonding atoms. A triple bond consists of two pi bonds and one sigma bond.

Q. Referring to the following figure, draw the orbital diagram of acetylene.

H—C≡C—H

acetylene

A.

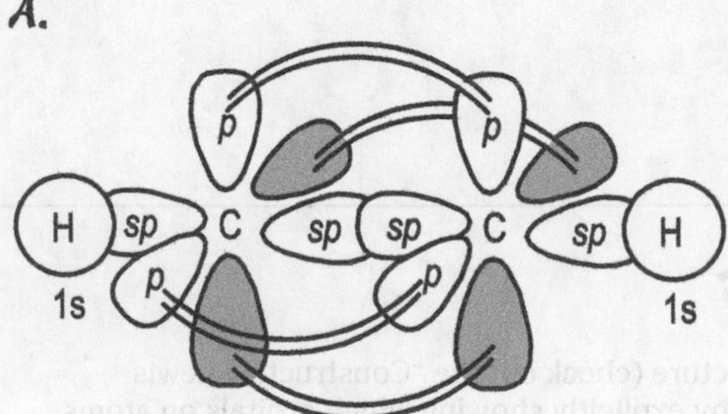

This problem is daunting, but you can tackle it step by step. The first thing to do is determine the hybridizations for all the atoms. The two carbons are *sp* hybridized. The hydrogens, having only one electron, remain unhybridized (hydrogen is the only atom that doesn't rehybridize in organic molecules):

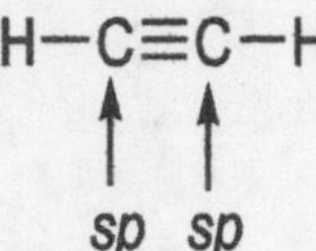

Next, draw the valence orbitals as shown here. Hydrogen has only the 1*s* orbital, and you can use the earlier template for *sp*-hybridized atoms for each of the carbons.

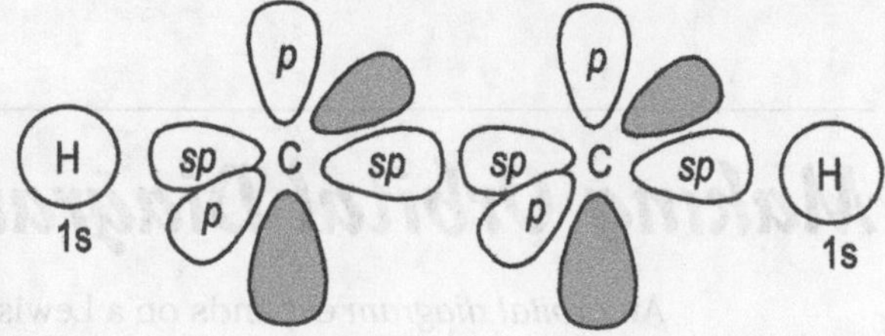

Next, you need to figure out which orbitals overlap to give rise to the bonds in acetylene. The C-H bonds form from overlap of the hydrogen 1*s* orbitals with the *sp* orbitals on carbon. Triple bonds consist of two pi bonds and one sigma bond. The one sigma bond comes from overlap of the two carbon *sp* orbitals. The two pi bonds come from overlap of the two *p* orbitals on each carbon, giving you the final answer shown earlier.

12. Draw the orbital diagram for methane, CH_4.

Solve It

13. Draw the orbital diagram of formaldehyde, H_2CO. (***Hint:*** Draw the full Lewis structure first.)

Solve It

14. Use the following figure to draw the orbital diagram for allene (very challenging).

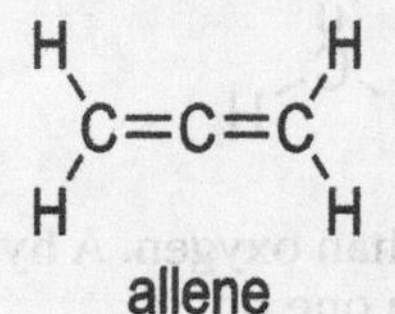

allene

Solve It

Answer Key

The following are the answers to the practice questions presented in this chapter.

1

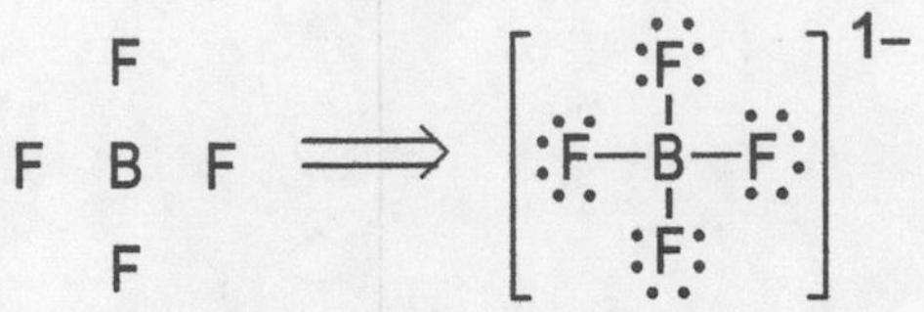

Boron is the central atom because it's less electronegative than fluorine (and in any case, halogens such as F almost never form more than one single bond). Boron has three valence electrons, fluorine has seven, and the charge on the molecule is –1, so the total number of valence electrons in this molecule is 3 + 7(4) + 1 = 32. Adding single bonds from boron to each of the four fluorines (for a total of eight electrons, two per bond) and adding the remaining 24 electrons to the fluorines as lone pairs gives the Lewis structure shown. Each atom is happy because it has a full octet of electrons, so there's no need to make multiple bonds.

2

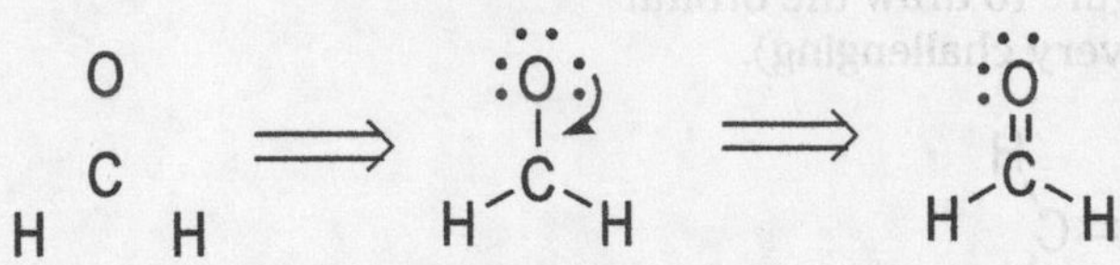

Carbon is the central atom because it's less electronegative than oxygen. A hydrogen can never be the central atom because hydrogens don't form more than one bond.

Hydrogen has one valence electron, carbon has four valence electrons, and oxygen has six valence electrons, so the total number of valence electrons is 2(1) + 4 + 6 = 12 valence electrons.

Adding a single bond from carbon to each of the two hydrogens and a single bond to the oxygen and peppering the remaining lone pairs onto the oxygen gives you the structure in the middle. Although oxygen is happy because it has a full octet of electrons, carbon isn't faring as well because it's two electrons short of its octet. Therefore, you push down one of the lone pairs from oxygen to form a double bond from oxygen to carbon. After that move is complete, all the atoms are happy because each atom has a full octet of electrons. ***Note:*** You can't give any lone pairs to hydrogen because with one bond already, hydrogen has satisfied its valence shell with two electrons (recall that the first shell holds only two electrons, and then it's eight in the second shell).

3

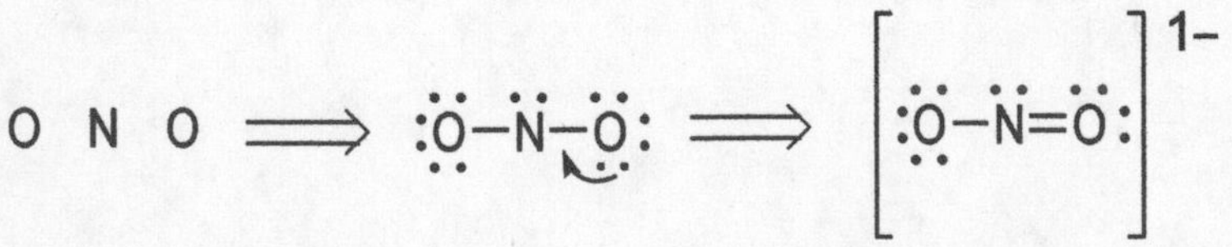

Nitrogen is the central atom in NO_2^- because nitrogen is less electronegative than oxygen.

Nitrogen has five valence electrons, oxygen has six, and the charge on the molecule is –1, so the molecule has 5 + 2(6) + 1 = 18 valence electrons.

Making single bonds from N to both oxygens (for a total of four electrons, two per bond) leaves 14 electrons. Adding these electrons onto the oxygens until both oxygens have completed their octet still leaves two electrons left over. Place these two electrons on the central nitrogen. Examining this structure reveals that both oxygens have a complete octet, but nitrogen is still shy two electrons. So a lone pair on one of the oxygens is pushed onto the nitrogen to form a nitrogen-oxygen double bond. Last, add the charge to complete the final structure.

4 **Ionic.** Fluorine has an electronegativity of 4.0, and sodium has an electronegativity of 0.9, so the electonegativity difference is 3.1, making this bond an ionic bond.

5 **The C-C bonds are purely covalent; the C-Cl bonds are polar covalent.** The C-C bond in hexachloroethane is purely covalent because there's 0.0 electronegativity difference between the two atoms (because they're the same). The C-Cl bonds are all polar covalent because the electronegativity difference between chlorine (3.0) and carbon (2.5) is 0.5.

6

Oxygen is more electronegative than carbon, so oxygen is partially negatively charged and carbon is partially positively charged. Therefore, the bond dipole vectors point toward the oxygens.

7

methanol

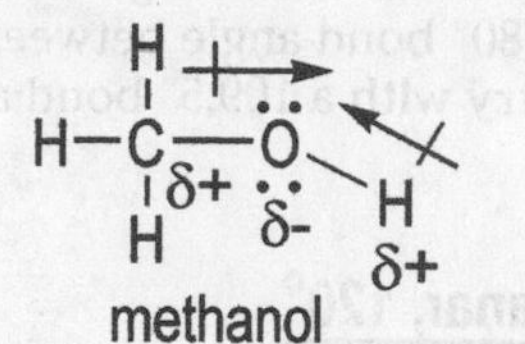

In methanol, the oxygen is more electronegative than either carbon or hydrogen. Therefore, the oxygen is partially negative charged and the carbon and hydrogen are partially positively charged. As a result, both bond dipole vectors point toward the oxygen.

8

individual bond dipoles vector addition molecular dipole

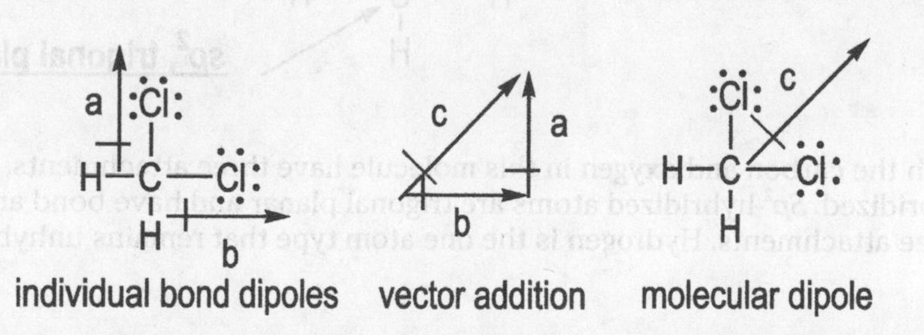

The two C-Cl bonds have dipole vectors pointing toward the chlorine (because chlorine is more electronegative than carbon). Summing these two vectors gives the dipole moment vector (vector *c*) for the molecule, which points between the two carbon-chlorine bonds.

9

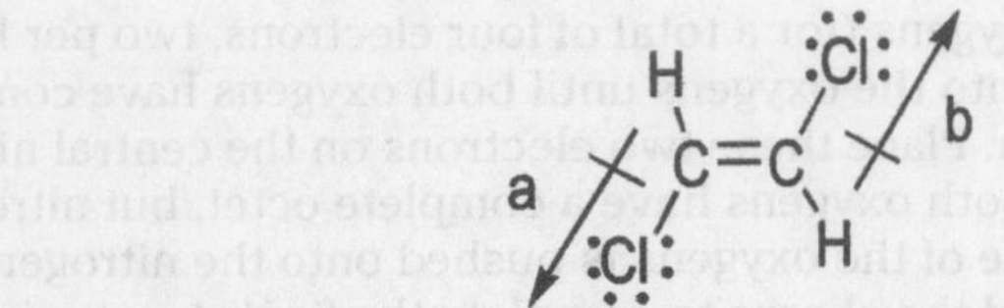

trans-1,2-dichloroethene

Both C-Cl bond vectors point toward the chlorine because chlorine is more electronegative than carbon. However, summing up the two vectors gives a net dipole moment of 0.0 — the two individual bond dipole vectors cancel each other out. Therefore, although the individual C-Cl bonds do have bond dipoles, the molecule has no net dipole moment.

10

sp^3, tetrahedral, 109.5°

$^-$:C≡C—N$^+$H$_3$

sp, linear, 180°

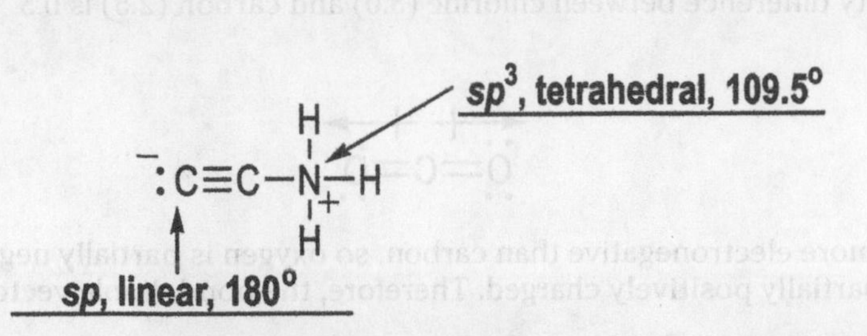

The carbon has two attachments (one being the lone pair), making this atom *sp* hybridized. The nitrogen has four attachments, making this atom *sp*^{3} hybridized. *Sp*-hybridizied atoms have a linear geometry with a 180° bond angle between the two attachments. *Sp*^{3}-hybridized atoms have a tetrahedral geometry with a 109.5° bond angle between the four attachments.

11

sp^2, trigonal planar, 120°

unhybridized

sp^2, trigonal planar, 120°

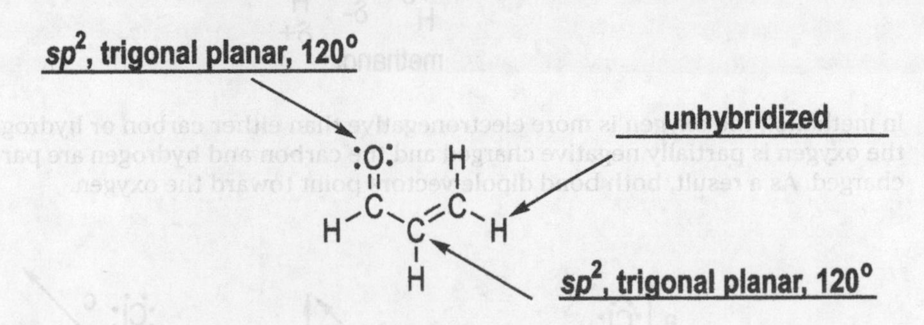

Both the carbon and oxygen in this molecule have three attachments, so both atoms are *sp*^{2} hybridized. *Sp*^{2}-hybridized atoms are trigonal planar and have bond angles of 120° between the three attachments. Hydrogen is the one atom type that remains unhybridized.

12

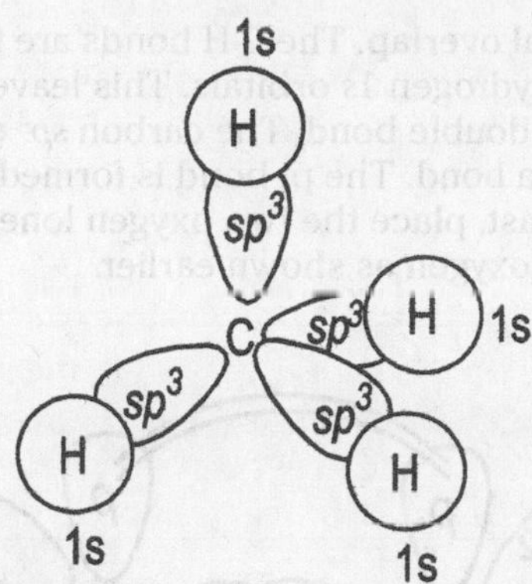

The carbon has four attachments, so this atom is *sp*^{3}-hybridized, with four *sp*^{3} orbitals to bond with the four hydrogen 1*s* orbitals.

13

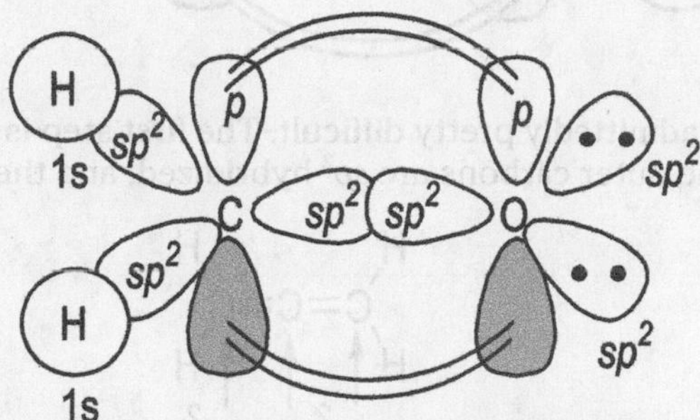

First drawing the Lewis structure of formaldehyde and then assigning the hybridizations shows that both the carbon and the oxygen are *sp*^{2} hybridized.

$$\underset{\substack{\uparrow \\ sp^2}}{\overset{H}{\underset{H}{C}}}\!=\!\underset{\substack{\uparrow \\ sp^2}}{\ddot{O}}$$

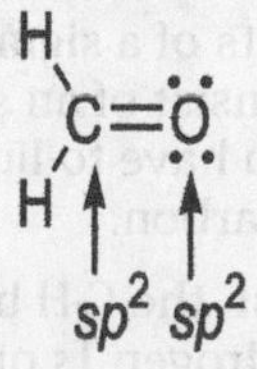

Next, drawing out all the valence orbitals for the atoms gives the following (using the templates here may help to speed up this process).

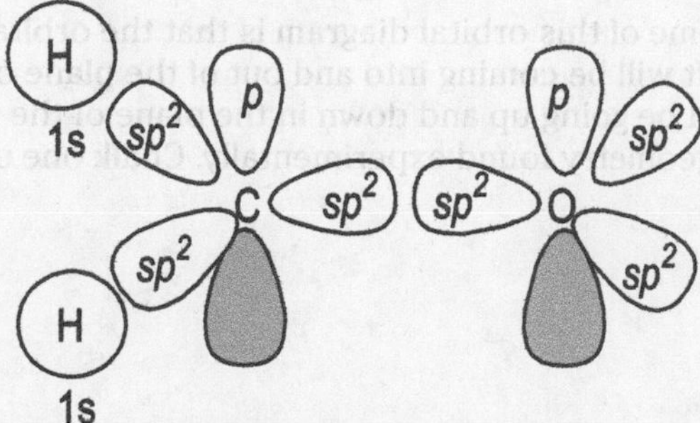

Finally, show the orbital overlap. The C-H bonds are formed from overlap of two carbon sp^2 orbitals with the two hydrogen 1*s* orbitals. This leaves one carbon sp^2 orbital and one carbon *p* orbital for forming the double bond. The carbon sp^2 orbital and one of the oxygen sp^2 orbitals overlap to form a sigma bond. The pi bond is formed from overlap of the carbon *p* orbital and the oxygen *p* orbital. Last, place the two oxygen lone pairs into the remaining unoccupied sp^2 hybridized orbitals on oxygen as shown earlier.

14

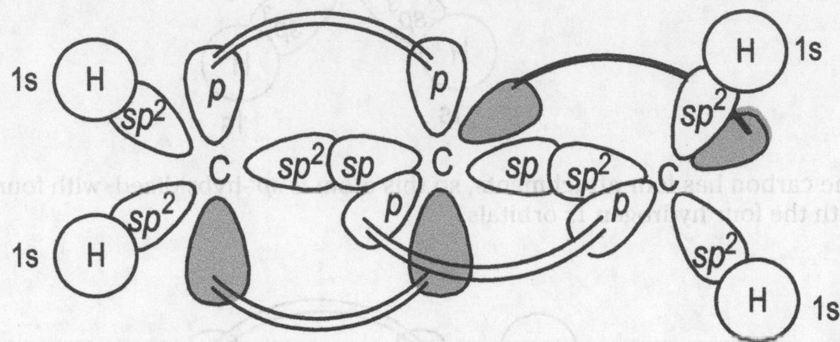

This problem is admittedly pretty difficult. The first step is assigning the hybridizations of each of the atoms. The outer carbons are sp^2 hybridized, and the inner carbon is *sp* hybridized.

H H
C=C=C
H H
sp^2 sp sp^2

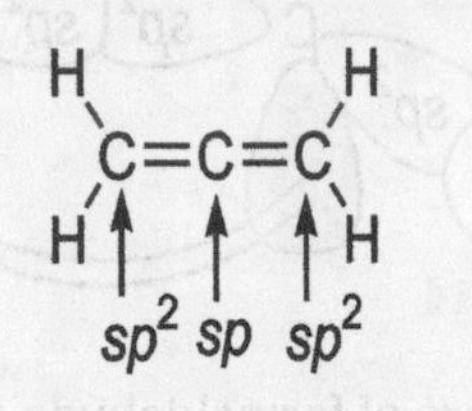

Next, show all the valence orbitals on each of the atoms. The tricky part is lining up the orbitals from the middle carbon to the outer carbons so the orbitals can overlap to form one double bond each. Each double bond consists of a sigma bond and a pi bond. Therefore, each of the carbon-carbon sigma bonds must consist of an sp^2-*sp* orbital overlap. Pi bonds are formed from the *p* orbital overlaps. Therefore, you have to line up the *p* orbitals so it's possible for the orbitals to overlap with the central carbon.

Finally, show the orbital overlap. First, the C-H bonds are formed from the overlap between the outer carbon sp^2 orbitals and the hydrogen 1*s* orbitals. The sigma bonds in the two double bonds are formed in both cases from the overlap between the central carbon *sp* orbital and the two outer carbon sp^2 orbitals. The pi bonds are then formed from the overlap of the two *p* orbitals on the central carbon and the lone *p* orbitals on the outer carbons.

An interesting outcome of this orbital diagram is that the orbital diagram predicts that the two hydrogens on the left will be coming into and out of the plane of the paper, while the two hydrogens on the right will be going up and down in the plane of the paper. As a matter of fact, this turns out to be the geometry found experimentally. Chalk one up to orbital diagrams!

Chapter 2

Speaking Organic Chemistry: Drawing and Abbreviating Lewis Structures

In This Chapter

- Figuring out how to assign formal charges
- Sketching condensed structures and line-bond structures
- Taking a look at lone pairs and hydrogens

The language of chemistry isn't a spoken language or a written language but a language of pictures. Lewis structures are the pictorial words of the organic chemist, much like hieroglyphics were the pictorial words of the ancient Egyptians. Organic chemists currently use a number of different methods for drawing structures. You may already be familiar with the full Lewis structure (if not, see Chapter 1), but organic chemists often like to abbreviate Lewis structures by using simpler drawings to make speaking the language of organic chemistry faster and easier, much like you abbreviate words when text messaging your friends.

Two abbreviations to Lewis structures that you should become familiar with are the condensed structure and the line-bond structure, because you see these two structural abbreviations again and again throughout organic chemistry. This chapter familiarizes you with drawing and interpreting these structural abbreviations (condensed and line-bond structures) and helps you understand what the structural abbreviations mean. Before you get down to the dirty business of drawing structures, you practice determining formal charges and the number of lone pairs on atoms in a structure, two skills that are essential to mastering organic structures.

Assigning Formal Charges

The following equation shows a down-'n'-dirty method of calculating the formal charge on an atom. The dots are the non-bonding electrons assigned to an atom, and the sticks are the total number of bonds attached to an atom (a single bond counts as one stick, a double bond counts as two sticks, a triple, three):

Formal charge of an atom = number of valence electrons – dots – sticks

Of course, the secret is that no organic chemist actually calculates the charges on every atom in every molecule he or she looks at. Instead, the chemist becomes familiar with the structures of atoms with charges. For instance, neutral carbon has four bonds, but carbon is positively charged when it has three bonds and is negatively charged when it has three bonds and a lone pair.

Figure 2-1 shows the patterns of charges for common atoms. Do yourself a big favor and memorize this table — or at the very least, master how many bonds and lone pairs are found on common neutral atoms. For example, when neutral, carbon has four bonds, nitrogen has three bonds and a lone pair, oxygen has two bonds and two lone pairs, and so on.

Atom	Neutral	Cationic	Anionic
C	—C— (four bonds)	—C+ (three bonds)	—C:⁻ (three bonds, one lone pair)
N	—N: (three bonds, one lone pair)	—N+— (four bonds)	N⁻ (two bonds, two lone pairs)
O	O (two bonds, two lone pairs)	—O:+ (three bonds, one lone pair)	—O:⁻ (one bond, three lone pairs)
X	—X: (one bond, three lone pairs)	—X+— (two bonds, two lone pairs)	X⁻ (four lone pairs)

X = F, Cl, Br, I

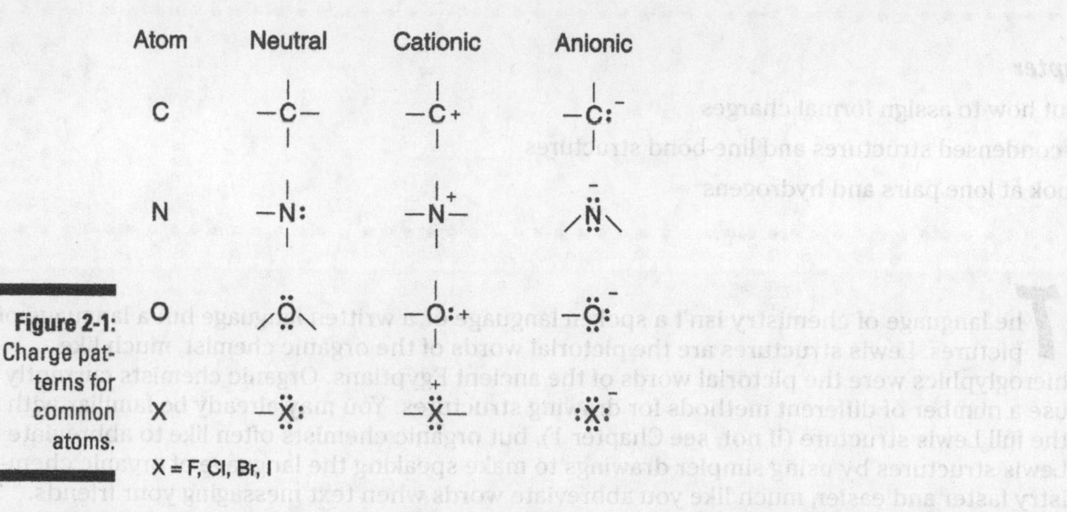

Figure 2-1: Charge patterns for common atoms.

Q. Calculate the formal charges on each of the indicated atoms in the following molecule:

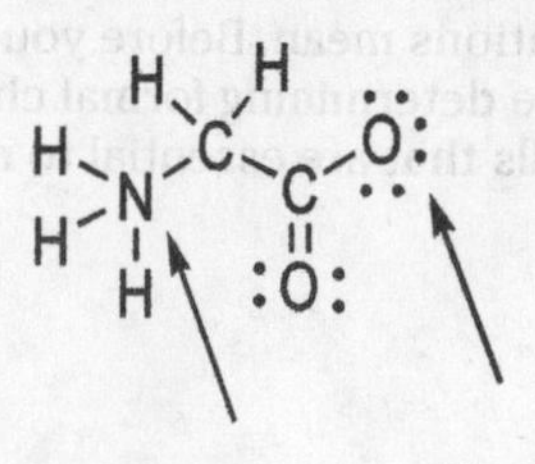

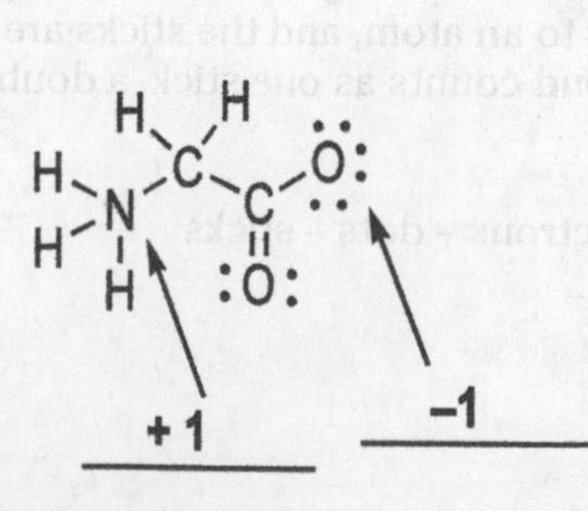

______ ______

A. To calculate the formal charge for the nitrogen, you plug the values for valence electrons, dots, and sticks into the equation. Nitrogen is in the fifth column of the periodic table, so it has five valence electrons. This atom has no dots because it doesn't have any non-bonding electrons, but it has four sticks (one stick for each of the four single bonds). Plugging these values into the equation produces 5 – 0 – 4 = +1.

The oxygen has six valence electrons (sixth column of the periodic table), six dots, and one stick. Plugging these values into the equation gives 6 – 6 – 1 = –1.

+1 –1

1. Calculate the formal charge for the indicated atoms in the following structure:

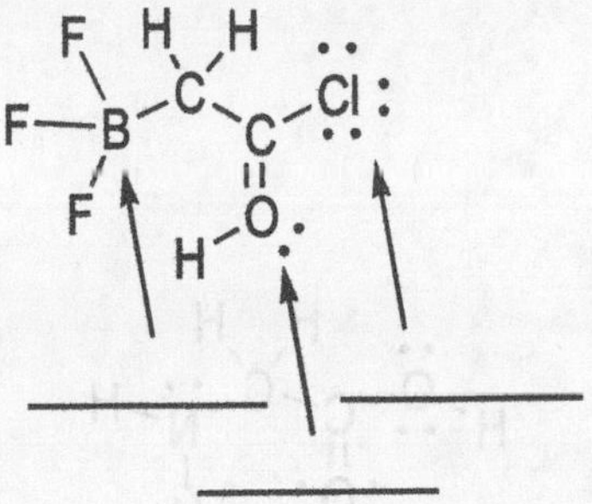

Solve It

2. Calculate the formal charge for the indicated atoms in the following structure:

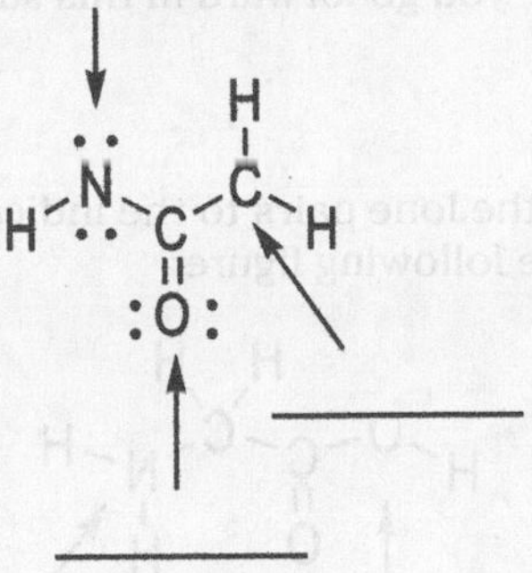

Solve It

3. Without actually calculating any of the charges, refer to the common patterns in Figure 2-1 and then scan the following structure for any charged atoms. Add formal charge designations (+ or –) to any charged atoms you see here.

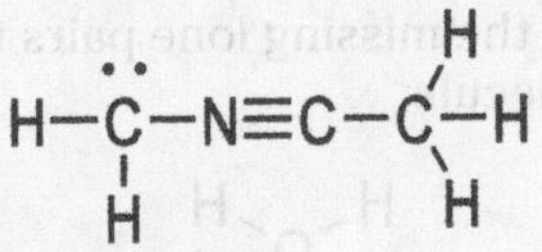

Solve It

4. Without actually calculating any of the charges, refer to the common patterns in Figure 2-1 and then scan the following structure for any charged atoms. Add formal charge designations (+ or –) to any charged atoms you see here.

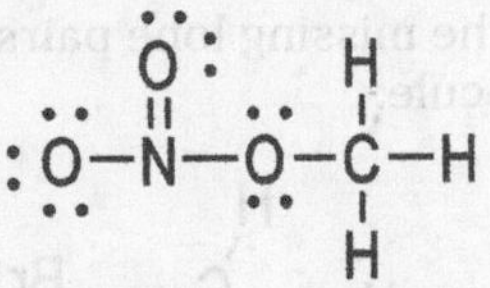

Solve It

Determining Lone Pairs on Atoms

Chemists are often lazy creatures (I would know; I am one!), and so many times in Lewis structures, the lone pairs of electrons on atoms aren't drawn because it's assumed that if the charge is specified, you can figure out for yourself how many lone pairs are on a given atom. Of course, to figure out how many lone pairs an atom owns, you can always plug the values into the rearranged formal charge equation I provide in the earlier "Assigning Formal Charges" section:

Dots = valence electrons – formal charge – sticks

But just as with formal charges, no organic chemist actually calculates the lone pairs on every atom in a structure where they aren't shown (did I mention that chemists are lazy?). Instead, after some time working with these structures, organic chemists master all the different patterns of lone pairs on atoms. For example, a negatively charged carbon has one lone pair of electrons (two dots); an oxygen that's neutral has two lone pairs (four dots); and so forth (refer to Figure 2-1).

Chemists can instinctively look at a molecule and know how many lone pairs are on each atom simply because they've memorized the patterns. And that's exactly what you need to do, because having this skill makes your life much easier and gives you a big advantage as you go forward in this subject.

Q. Add the lone pairs to the indicated atoms in the following figure:

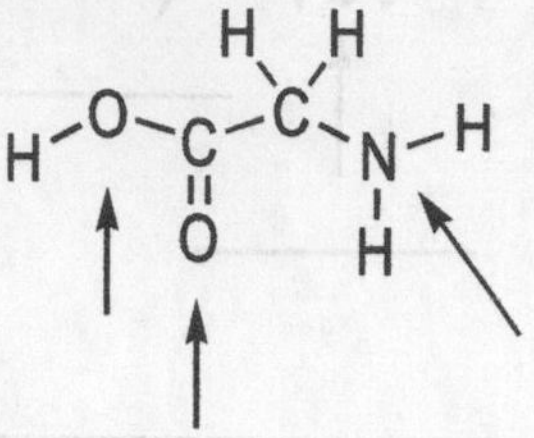

A.

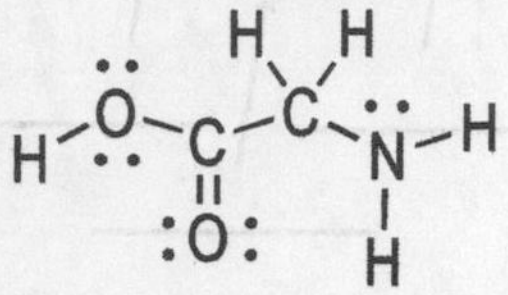

Although you can calculate the number of lone pairs on each atom by using the dots equation, I recommend that you try to remember the common patterns. All the atoms here are neutral, and neutral oxygen has two bonds and two lone pairs. Neutral nitrogen atoms have three bonds and a lone pair.

5. Add all the missing lone pairs to the following molecule:

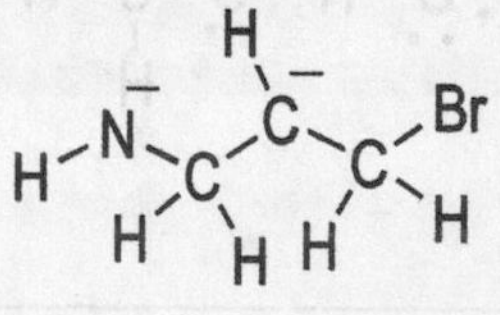

Solve It

6. Add all the missing lone pairs to the following molecule:

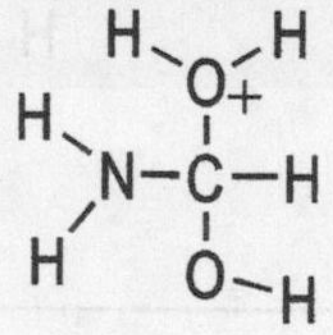

Solve It

Abbreviating Lewis Structures with Condensed Structures

One of the most common structural abbreviations to Lewis structures is called the condensed structure. In a *condensed structure,* bonds to hydrogen from second- or third-row atoms aren't explicitly drawn; instead, these atoms are grouped into clusters (such as CH_2 or CH_3 clusters), and the clusters are written in a chain to show the connectivity. The rules (and some additional quirks you should be familiar with) for drawing condensed structures include the following:

- The bonds between the clustered atoms in a condensed structure can be shown, but they're often omitted. The top condensed structure explicitly shows the bonds between the clusters; in the bottom condensed structure, they're omitted.

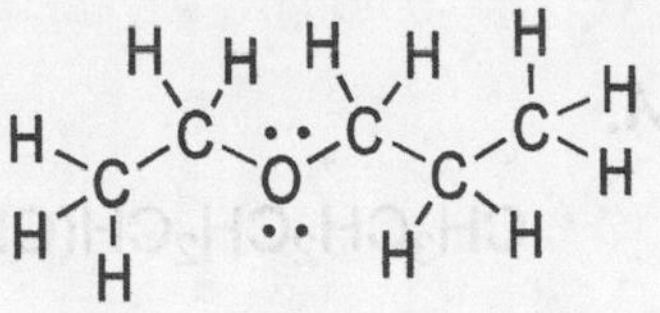

$CH_3\text{-}CH_2\text{-}O\text{-}CH_2\text{-}CH_2\text{-}CH_3$

or

$CH_3CH_2OCH_2CH_2CH_3$

- Condensed structures are most commonly used to abbreviate the structure of simple organic molecules connected in a linear chain. However, condensed structures can also abbreviate more complicated molecules (such as complex molecules containing branches or rings) if the connectivity of the atoms is made clear by explicitly drawing the bonds between the clusters.

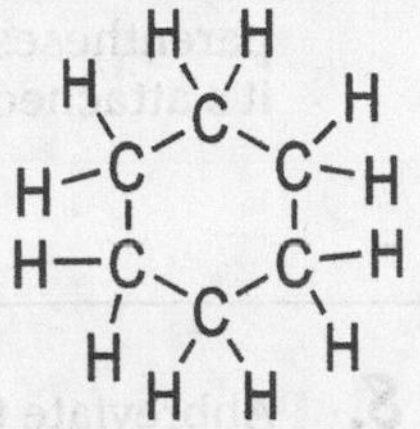

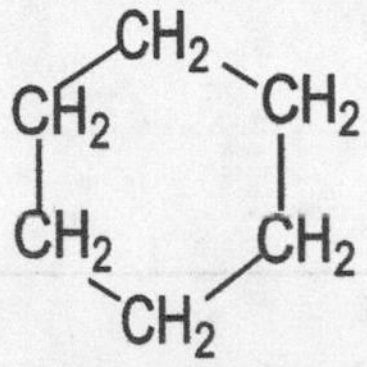

- You can show simple substituents off a linear chain by enclosing each substituent in parentheses *after* you've written the atom to which it's attached. However, if the substituent is just a single atom, such as a halide (like F, Br, and Cl), or is in a carbonyl group (C=O group), you don't need to put the substituent in parentheses; just place this single atom after the atom to which it's attached in the chain.

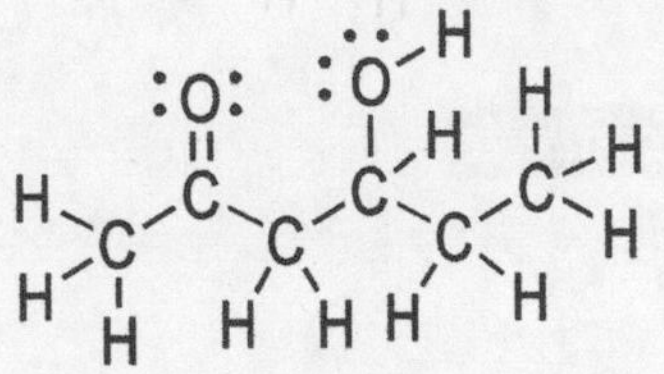

$CH_3COCH_2CH(OH)CH_2CH_3$

- Clusters that repeat in a straight chain (usually CH_2 clusters) can be abbreviated further by placing the repeating cluster in parentheses and using a subscript to show how many times the cluster repeats itself down a linear chain.

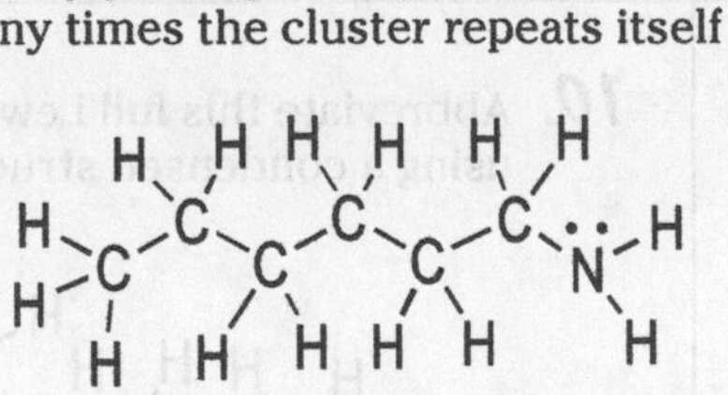

$CH_3CH_2CH_2CH_2CH_2NH_2$

or

$CH_3(CH_2)_4NH_2$

- When two or more identical clusters connect to a single atom, you can use parentheses along with a subscript to show the number of identical clusters attached to that atom.

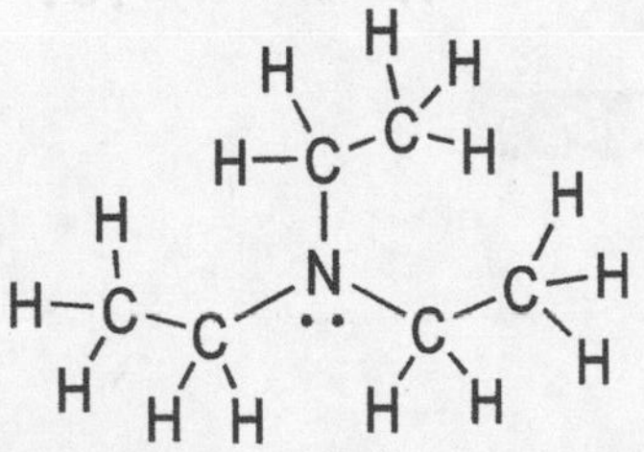

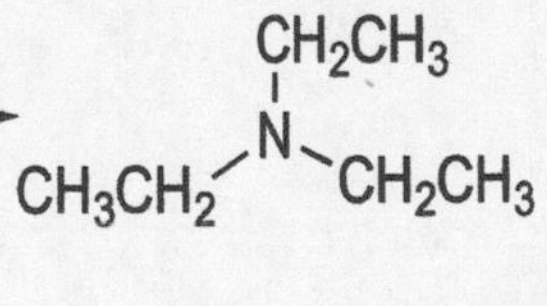

or $(CH_3CH_2)_3N$

Q. Draw the condensed structure of the following molecule:

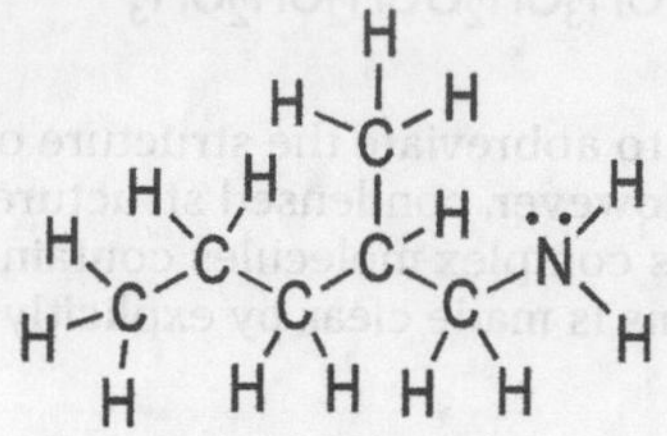

A. $CH_3CH_2CH_2CH(CH_3)CH_2NH_2$

The CH_2 clusters are attached in a straight line, and you can show the bonds between the clusters, or you can omit them — your choice (I've omitted them). One CH_3 cluster sticks off the main chain. You can indicate this branching by placing this cluster in parentheses *after* the CH cluster to which it's attached.

7. Abbreviate this full Lewis structure by using a condensed structure.

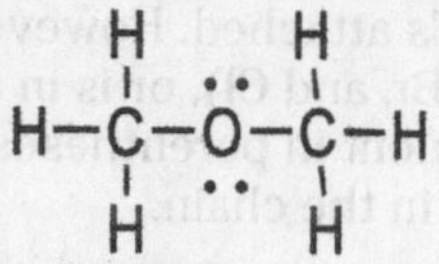

Solve It

8. Abbreviate this full Lewis structure by using a condensed structure.

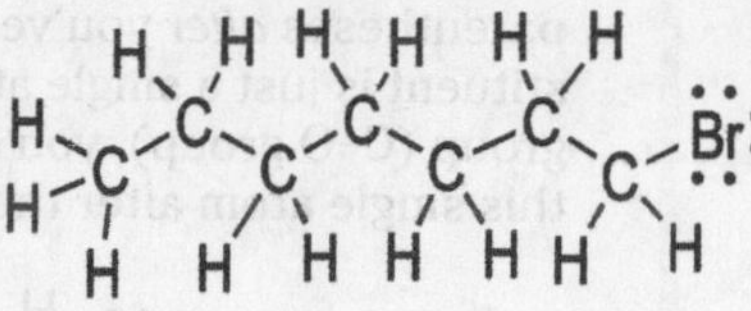

Solve It

9. Abbreviate this full Lewis structure by using a condensed structure.

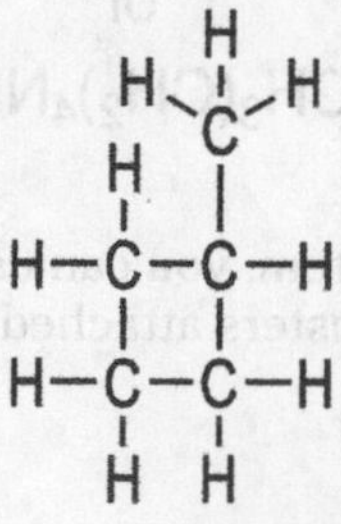

Solve It

10. Abbreviate this full Lewis structure by using a condensed structure:

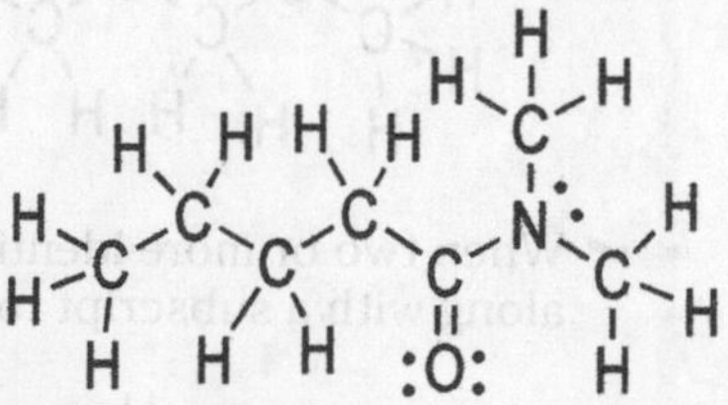

Solve It

Drawing Line-Bond Structures

The most common structural abbreviations of the full Lewis structure is the *line-bond structure* (also called a *Kekulé structure* or a *line-angle structure*). In a line-bond structure, carbons and hydrogen atoms aren't explicitly shown; instead, a chain of carbons is represented by a jagged line, with each point (or node) on the jagged line representing a carbon atom. All charges are assumed to be neutral unless a charge is shown, so it's understood that you can mentally figure out how many hydrogens a carbon has by making the total number of bonds to carbon equal to four (you practice this concept in the "Determining Hydrogens on Line-Bond Structures" section later in this chapter). Non-carbon atoms, however, must be explicitly shown, and any hydrogens attached to atoms other than carbon are also explicitly drawn.

Q. Draw the line-bond representation for the following Lewis structure:

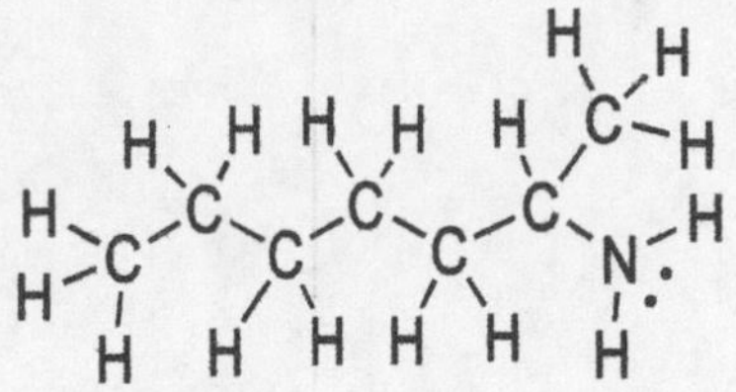

A.

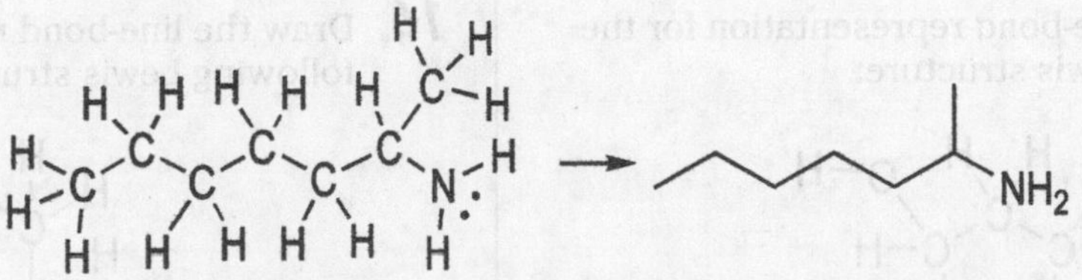

$\rightarrow$ NH_2

As in all line-bond drawings, the carbon chain is represented by a jagged line. ***Note:*** The ends of the lines (or tips of the lines) represent carbon atoms. The nitrogen and the hydrogens attached to the nitrogen are explicitly shown in the line-bond structure because the only atoms in a line-bond structure that aren't shown are carbon atoms and the hydrogen atoms that are attached to carbons.

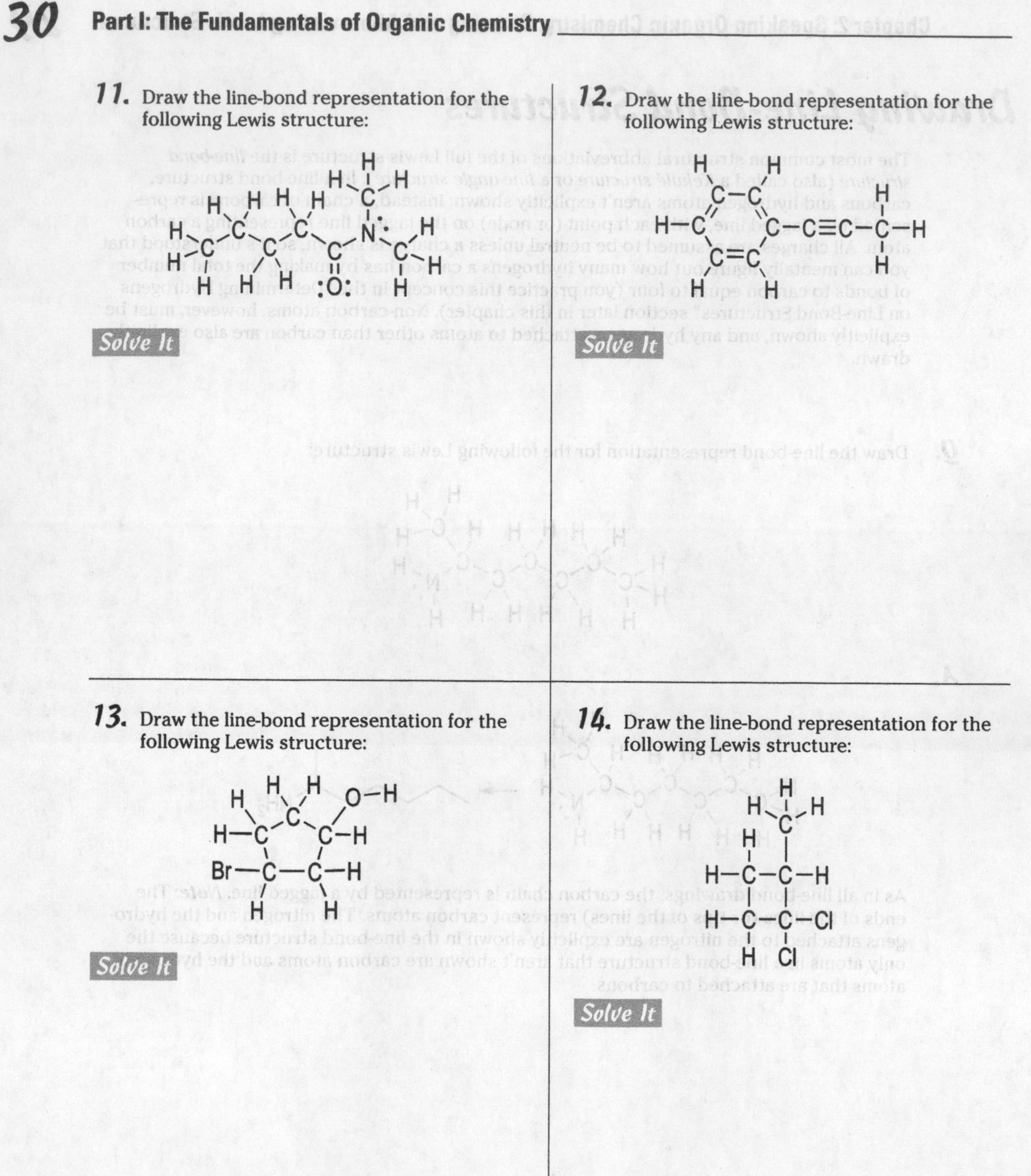

11. Draw the line-bond representation for the following Lewis structure:

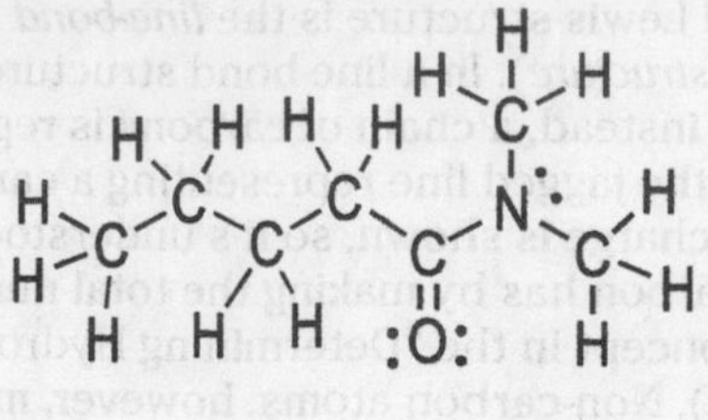

Solve It

12. Draw the line-bond representation for the following Lewis structure:

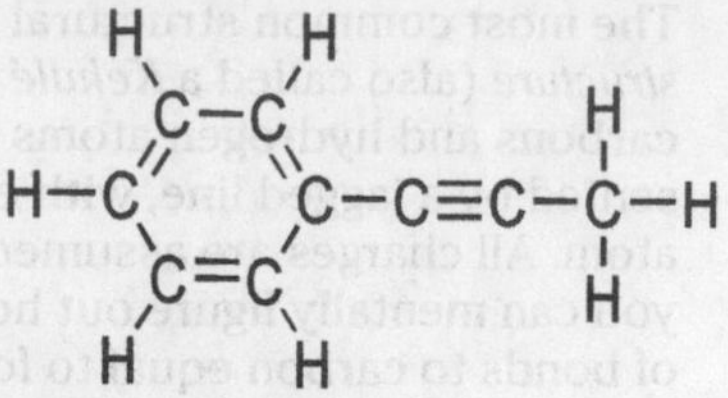

Solve It

13. Draw the line-bond representation for the following Lewis structure:

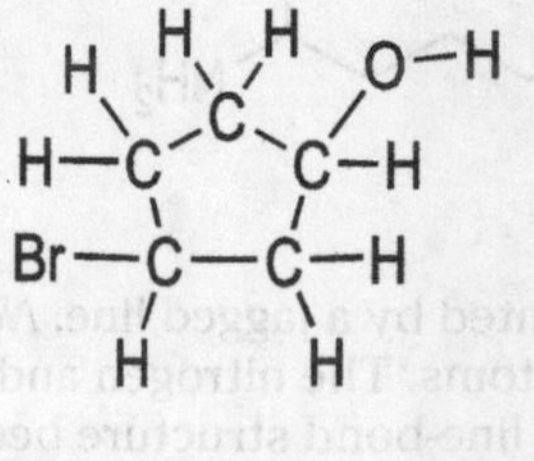

Solve It

14. Draw the line-bond representation for the following Lewis structure:

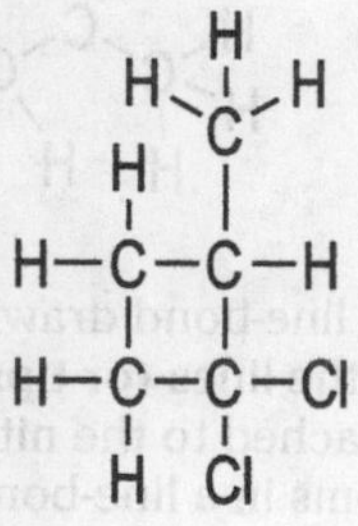

Solve It

15. Convert the following line-bond structure into the full Lewis structure:

O

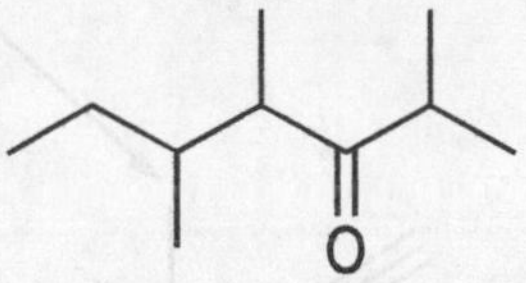

Solve It

16. Convert the following line-bond structure into the full Lewis structure:

O

N—

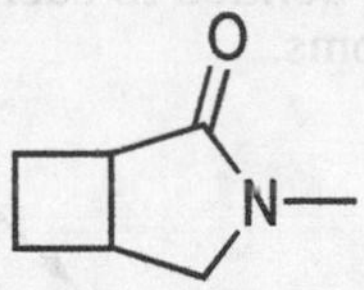

Solve It

Determining Hydrogens on Line-Bond Structures

Line-bond structures assume you can mentally supply the number of hydrogens to each carbon. You can accomplish this by adding hydrogens to a carbon until the carbon's total number of bonds equals four. Although this process is straightforward, you should still practice a few of these problems so you get the hang of it.

Q. Determine how many hydrogens that aren't shown are bonded to each of the indicated carbon atoms.

A.

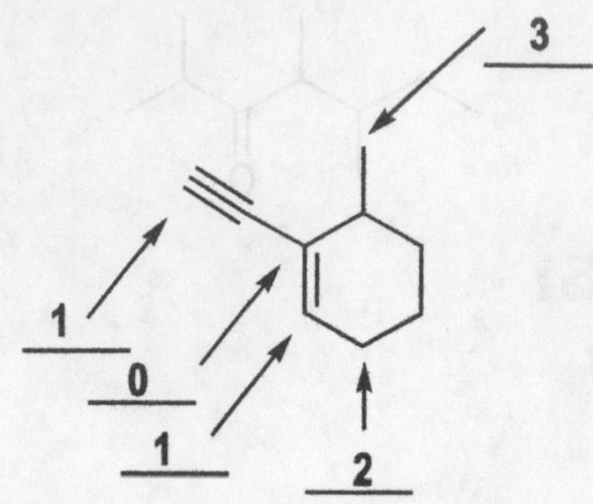

The key is to determine how many bonds a carbon already has and then see how many hydrogens you need to add so the number of bonds to carbon becomes four. Of course, if the carbon already has four bonds, then that carbon doesn't have any hydrogens.

17. Determine how many unshown hydrogens are bonded to each of the indicated carbon atoms.

C
Cl

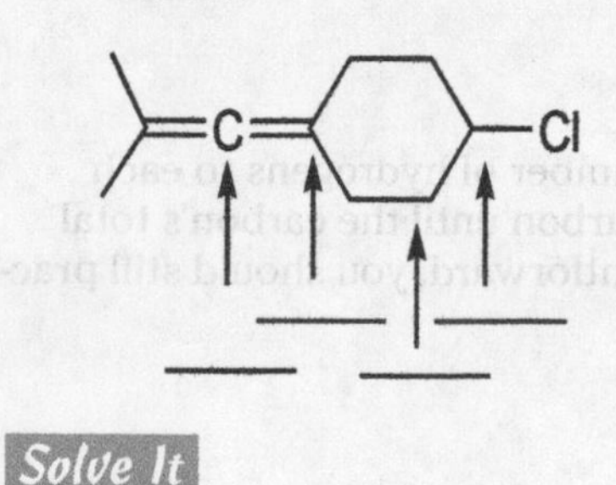

Solve It

18. Determine how many unshown hydrogens are bonded to each of the indicated carbon atoms.

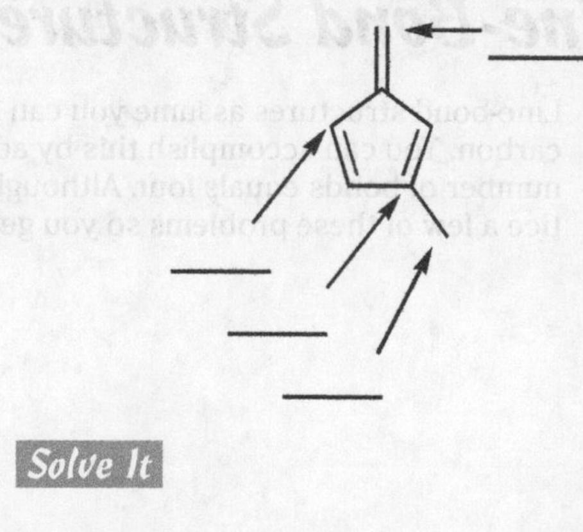

Solve It

Answer Key

The following are the answers to the practice questions presented earlier in this chapter.

1 **The formal charges are as follows: boron –1, oxygen +1, and chlorine 0.**

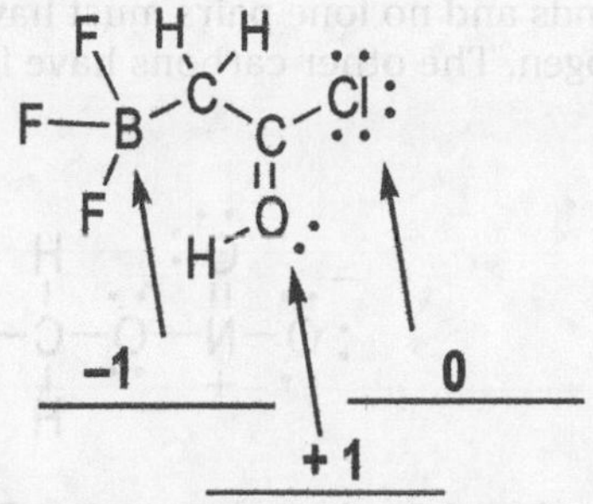

Boron has three valence electrons, no dots (no non-bonding electron pairs), and four sticks (four single bonds), so plugging these values into the equation produces the following: Formal charge on B = 3 – 0 – 4 = –1.

Oxygen has six valence electrons, two dots, and three sticks: Formal charge on O = 6 – 2 – 3 = +1.

Chlorine has seven valence electrons, six dots, and one stick: Formal charge on Cl = 7 – 6 – 1 = 0.

2 **The formal charges are as follows: nitrogen –1, oxygen 0, and carbon +1.**

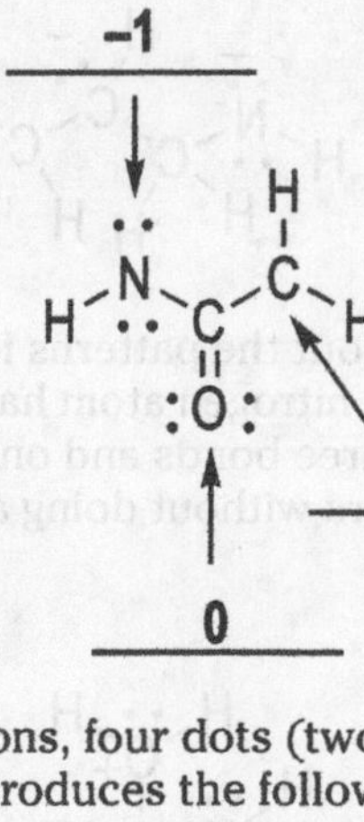

Nitrogen has five valence electrons, four dots (two lone pairs), and two sticks, so plugging these values into the equation produces the following: Formal charge on N = 5 – 4 – 2 = –1.

Oxygen has six valence electrons, four dots, and two sticks: Formal charge on O = 6 – 4 – 2 = 0.

Carbon has four valence electrons, no dots, and three sticks: Formal charge on C = 4 – 0 – 3 = +1.

3

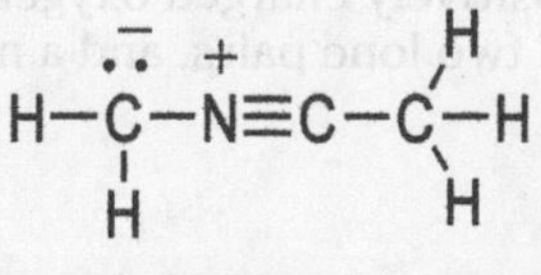

Using the scanning method, look for atoms that have a number of bonds different from neutral form. For example, carbon has four bonds when neutral, so look for carbons that have anything other than four bonds. The left-most carbon has three bonds and a lone-pair, so this atom has a charge. Looking at the patterns shown in Figure 2-1, this configuration is for a negatively-charged carbon.

For nitrogen, the neutral configuration is three bonds and a lone pair. Therefore, the nitrogen shown that has four bonds and no lone pairs must have a charge. This configuration is for a positively-charged nitrogen. The other carbons have four bonds, so these atoms are neutral.

4

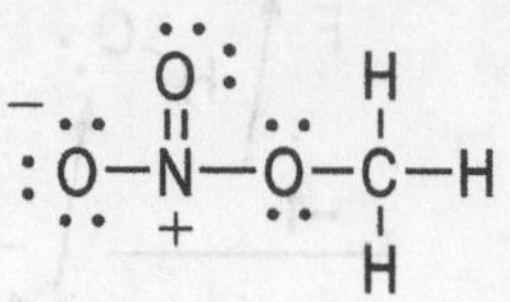

A neutral oxygen has two bonds and two lone pairs. Therefore, the right-most oxygen and the oxygen double-bonded to the nitrogen are neutral because both of these atoms have that configuration. However, the left-most oxygen has one bond and three lone pairs, the configuration for a negatively charged oxygen. Nitrogen's neutral valency is three bonds and a lone pair, but the nitrogen in this structure has four bonds, a configuration matching a positively-charged nitrogen (if you're ever in doubt, plug the values into the equation to make sure). The carbon atom has four bonds, so it's neutral.

5

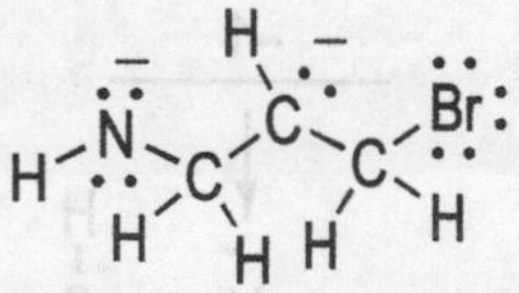

For determining lone pairs, check out the patterns for the common charged atoms in Figure 2-1. For instance, a negatively charged nitrogen atom has two bonds and two lone pairs. A negatively charged carbon atom has three bonds and one lone pair. Knowing these configurations allows you to quickly add lone pairs without doing any math.

6

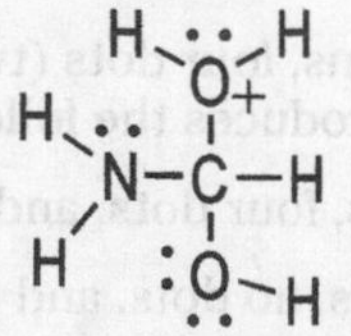

Knowing the common configurations in Figure 2-1 can help you supply the number of lone pairs to all the atoms. Here, a positively charged oxygen has three bonds and one lone pair. A neutral oxygen has two bonds and two lone pairs, and a neutral nitrogen has three bonds and one lone pair.

7

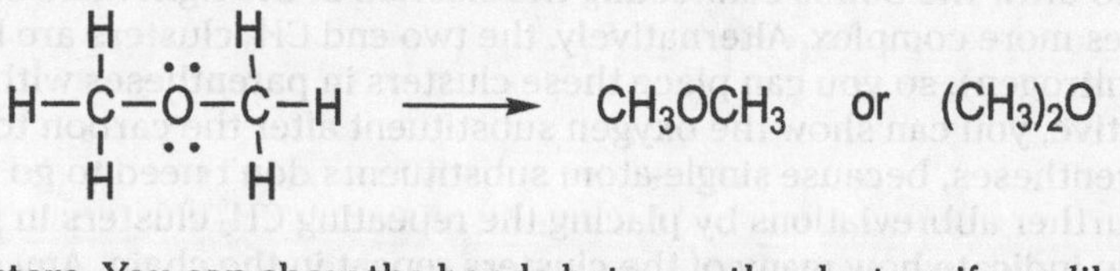

CH_3OCH_3 or $(CH_3)_2O$

Line up the clusters. You can show the bonds between the clusters if you like, although I've omitted them. Because two identical groups (CH_3 groups) are attached to the oxygen, you can make a further abbreviation by placing the CH_3 groups in parentheses with the subscript 2.

8

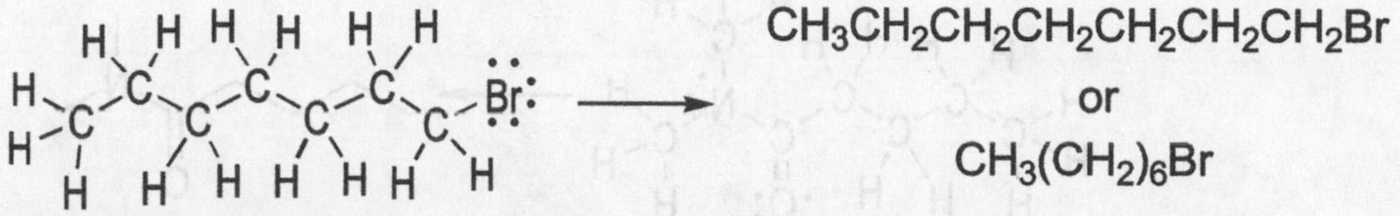

$CH_3CH_2CH_2CH_2CH_2CH_2CH_2Br$

or

$CH_3(CH_2)_6Br$

One way of condensing this structure is to simply line the clusters into a chain. Because a number of repeating clusters exist, you can draw an even more condensed form that places the repeating clusters in parentheses with a subscript to show how many clusters are repeated down the chain.

9

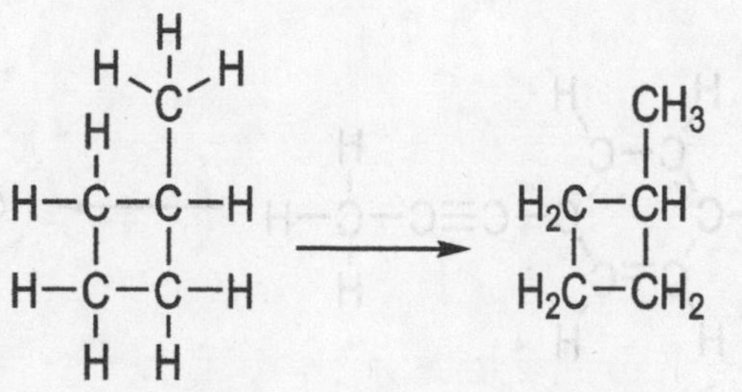

This structure is complex because it contains a ring and doesn't consist merely of a linear chain of atoms. Therefore, you should explicitly draw the bonds between the clusters to make the connectivity clear.

10

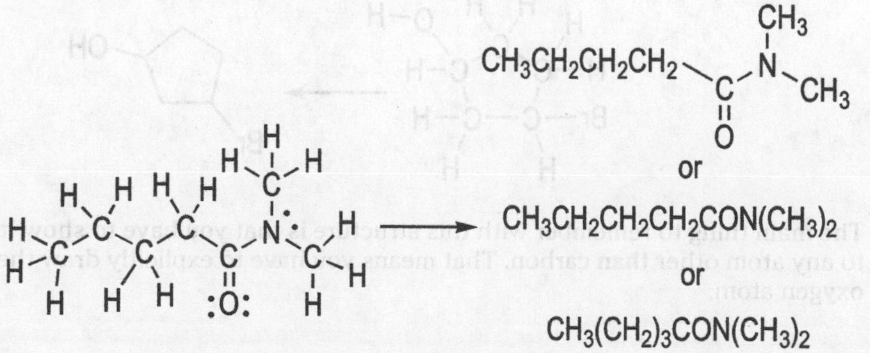

$CH_3CH_2CH_2CH_2CON(CH_3)_2$

or

$CH_3(CH_2)_3CON(CH_3)_2$

This molecule has a number of acceptable condensed structures. Perhaps the most obvious one is to draw the bonds connecting the clusters at the right-hand side where the structure becomes more complex. Alternatively, the two end CH_3 clusters are both attached to the same atom (nitrogen), so you can place these clusters in parentheses with a subscript. As another alternative, you can show the oxygen substituent after the carbon to which it's connected without parentheses, because single-atom substituents don't need to go in parentheses. You can make further abbreviations by placing the repeating CH_2 clusters in parentheses with a subscript to indicate how many of the clusters repeat in the chain. Any of these are acceptable condensed structures.

11

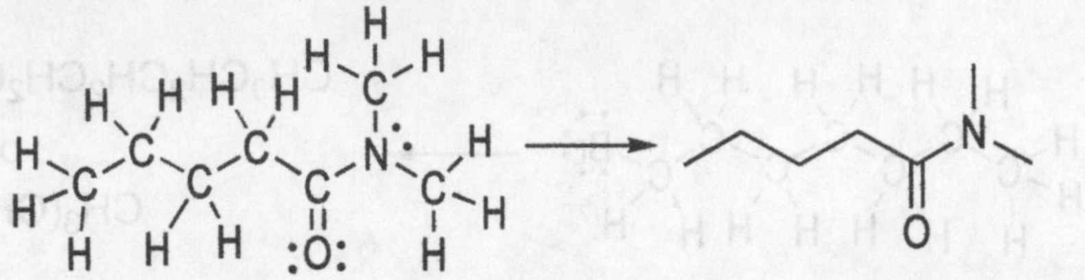

Use a jagged line to represent the carbon chain, with each point representing a carbon atom. Double and triple bonds, such as the carbon-oxygen double bond, must be explicitly shown. Additionally, you should explicitly show atoms that aren't hydrogen or carbon, such as nitrogen and oxygen.

12

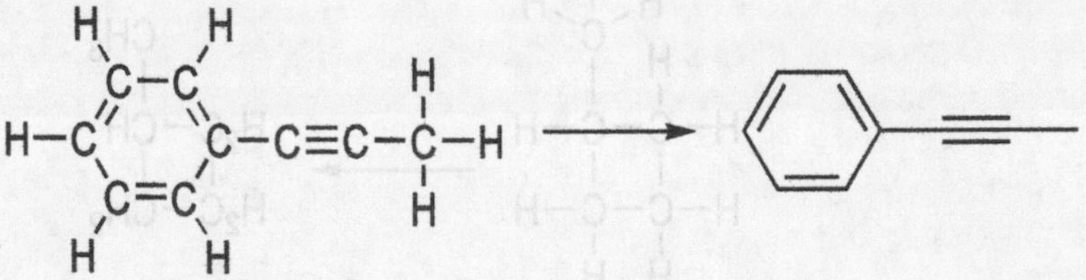

You should explicitly draw all double bonds and triple bonds. Additionally, by convention, you should draw all triple bonds in a straight line to show that these triple bonds adopt a linear geometry.

13

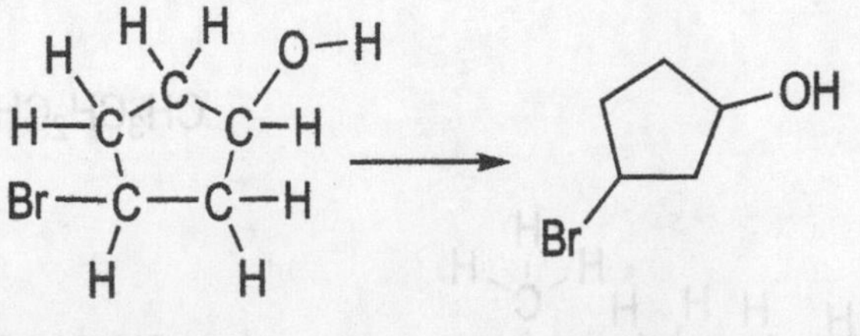

The main thing to remember with this structure is that you have to show hydrogens attached to any atom other than carbon. That means you have to explicitly draw the H attached to the oxygen atom.

14

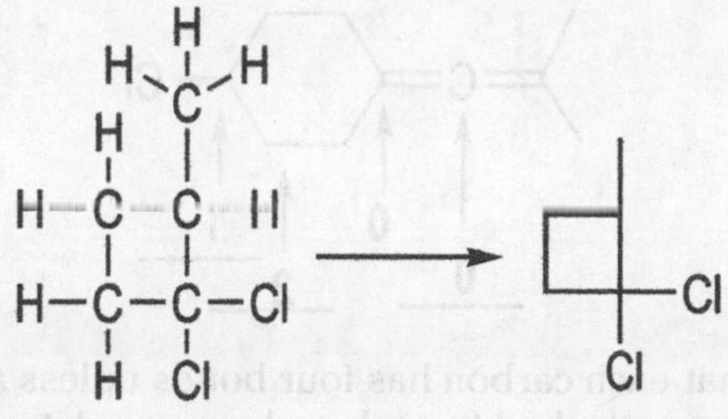

Here, to represent a four-membered ring, you draw a square. Moreover, you should explicitly draw out the chlorine atoms.

15

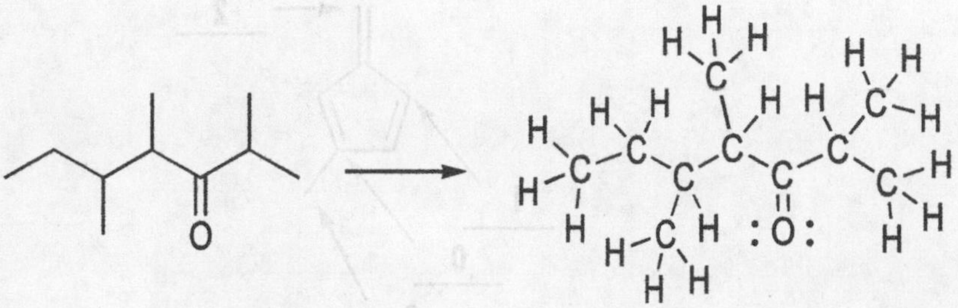

When drawing out a full Lewis structure from a condensed structure, make sure you don't forget that the ends of the lines (tips) represent carbon atoms. After you've drawn the carbon skeleton, add hydrogens to the carbons until you've made the total number of bonds to each carbon equal to four. You may add any lone pairs that aren't shown if you like.

16

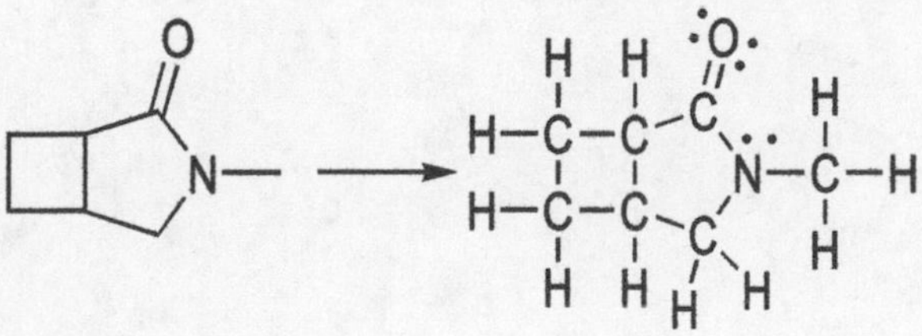

Just as in question 15, after you draw the carbon skeleton, add hydrogens to the carbons until you've made the total number of bonds to each carbon equal to four. You can then add any lone pairs not shown if you like.

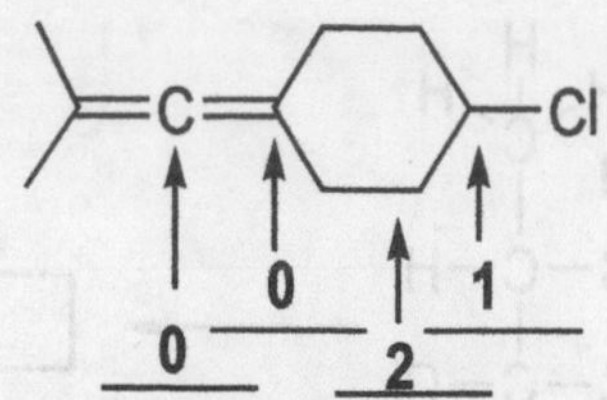

You can assume that each carbon has four bonds unless a charge is shown; therefore, to get the number of hydrogens attached to each carbon, you determine how many bonds to carbon are shown — the difference is the number of hydrogens that aren't shown but are assumed to be attached.

18

2

1

0

3

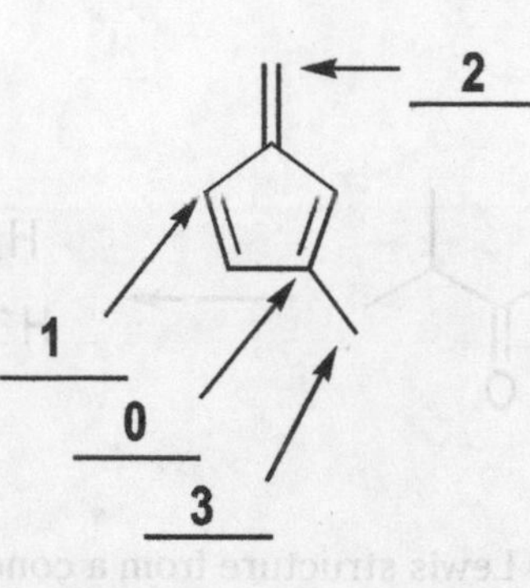

Just as in question 17, you find the difference between the number of shown bonds and the number four to get the number of hydrogens not shown attached to each carbon atom.

Chapter 3

Drawing Resonance Structures

In This Chapter

- Coming to grips with resonance structures of cations
- Moving lone pairs next to double and triple bonds
- Becoming like Robin Hood in using arrows

The Lewis structure is really quite a simple model for such a very complicated thing as a molecule, and sometimes this simple model gets it wrong. So what do chemists do? Instead of adopting a more complicated model to describe a molecule (such as the complex *molecular orbital theory model* I discuss in Chapter 14), chemists decided to patch the Lewis structure so that it *does* describe the molecule better in the few instances that it fails — much like a plumber may prefer to patch a few leaks in a cheap drain instead of installing a more expensive one. These patches to the Lewis structure are called *resonance structures.*

Lewis structures do a good job of representing single bonds. Where Lewis structures make their boo-boo is in molecules that contain lone pairs and double or triple bonds, because a single Lewis structure often doesn't accurately describe their electron distribution. For these molecules, resonance structures give a better representation of where the electrons are located in the molecule.

Keep the following rules for resonance structures in mind as you work through this chapter:

- **Don't move single bonds.** Because Lewis structures describe single bonds correctly, you don't move single bonds, only lone pairs or double or triple bonds.
- **Ensure that all resonance structures are valid Lewis structures.** That means that you can't break the *octet rule* for any of the resonance structures (the sum of lone pairs and bonds around a second-row atom can't exceed four).
- **Don't move atoms.** In resonance structures, the only thing that changes is the distribution of pi electrons (double bonds, triple bonds, and lone-pair electrons).
- **Remember charges.** In many cases, resonance structures have formal charges. Just remember that the sum of the charges must be the same for each resonance structure because each one describes the same molecule — you can't lose any electrons in going from one resonance structure to the next. Of course, the total *number* of charges may change from one resonance structure to another, but the *sum* of the charges must remain the same.

Keeping track of charges can be a bit tricky in this chapter because I switch between full Lewis structures, condensed structures, and line-bond structures to give you further experience with each of these drawing styles (refer to Chapter 2). If you're still a little shaky when using these different drawings and determining formal charges, don't worry; with enough practice manipulating and interpreting all three, it should soon become second nature to you.

Seeing Cations Next to a Double Bond, Triple Bond, or Lone Pair

One pattern that indicates that more than one resonance structure exists for a molecule is a cation situated next to a double bond, a triple bond, or a lone pair. A *cation* is organic-speak for a positively charged atom; when a cation is located on a carbon atom, the charge is often specifically referred to as a *carbocation*.

To draw alternative resonance structures matching this pattern, follow these steps:

1. **Push the electrons from the double bond, triple bond, or lone pair toward the charged atom.**

   Draw an arrow from the electrons you're moving toward the positively charged atom.

2. **Move the arrows correctly.**

   Arrows represent the movement of two electrons from where they started to where they're going, so make sure the arrow starts in one of these places:

   • At the double or triple bond, to indicate the movement of these pi electrons

   • From the lone pair of electrons

   Then make the arrow point between the positively charged atom and the closest adjacent atom to indicate that these two electrons are being moved here in order to form a new double bond.

3. **Draw the resonance structure showing the new double bond, and check your charges.**

   The positive charge should've moved either to one of the original double-bond or triple-bond atoms (if you moved a double or triple bond) or to the atom that originally contained the lone pair (if you moved a lone pair).

Q. **Draw the other resonance structures for the following cationic molecule by arrow-pushing. After you've drawn all the resonance structures, draw what the actual structure of the molecule looks like, including charges.**

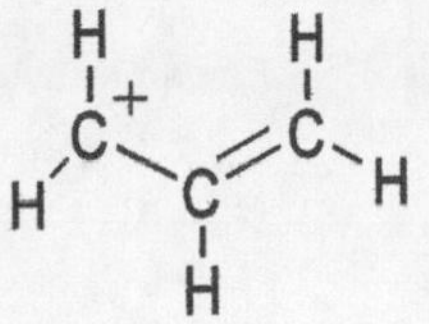

A. **Move the double-bond electrons and push them to make a new double bond on the other side. To do this, you start by drawing an arrow originating from the double bond (the electrons you're going to move) and point the arrow to the middle of the adjacent C-C bond (where the electrons are moving to). This change in the electron distribution moves the positive charge from the left carbon to the right carbon. The actual structure of the molecule looks like the hybrid of the two resonance structures. Thus, each C-C bond is expected to be roughly 1.5 bonds (partial bonds are indicated with the dotted line), and the positive charge is expected to be shared equally between the two terminal carbons (+½ each).**

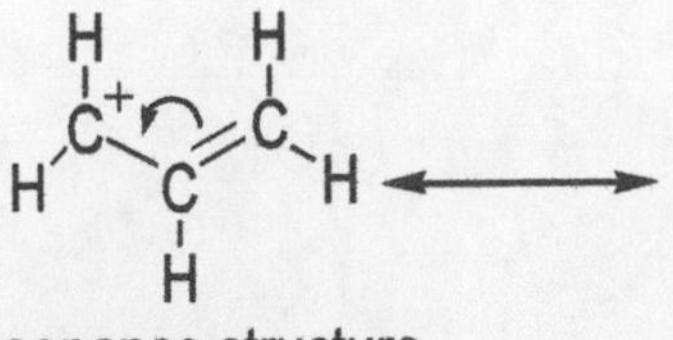

resonance structure

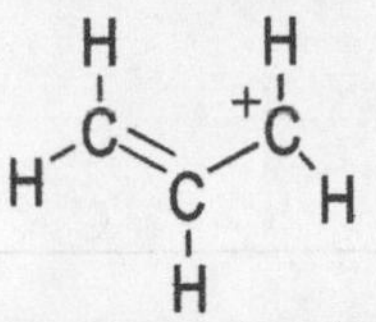

resonance structure

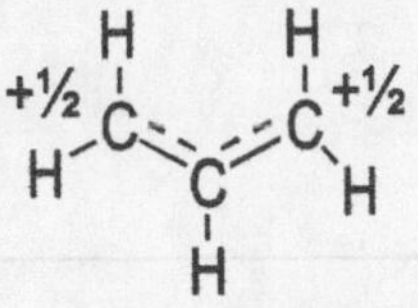

actual structure

1. **Draw the resonance structure for the following molecule by using arrow-pushing. After you've drawn the resonance structure, sketch what the actual structure looks like, including charges.**

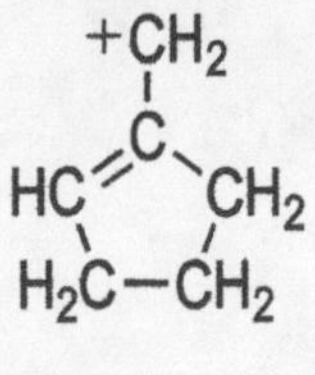

Solve It

2. **Draw the resonance structure for the following molecule by using arrow-pushing. After you've drawn the resonance structure, sketch what the actual structure looks like, including charges.**

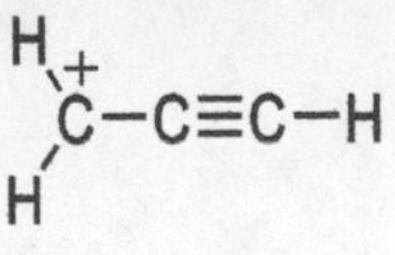

Solve It

3. Draw the resonance structures for the following molecule by using arrow-pushing. Then draw what the actual structure looks like, including charges. Two hints:

- There are *two* additional resonance structures for this one.
- Be careful about where you put the charges because they aren't as easy to calculate using the line-bond notation. If you get confused, draw the full Lewis structures showing all the atoms to help you keep the charges straight.

Solve It

4. Draw the resonance structure for the following structure.

Solve It

5. Draw all the other resonance structures for the following structure.

Solve It

6. Draw all the other resonance structures for the following structure.

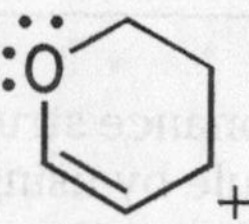

Solve It

Pushing Lone Pairs Next to a Double or Triple Bond

Another pattern you see that indicates that a molecule can be described by more than one Lewis structure is an atom with a lone pair directly attached to a double bond or triple bond.

To draw alternative resonance structures matching this pattern, follow these steps:

1. **Push the electrons from the lone pair toward the double bond or triple bond.**

   To accomplish this electron pushing, draw an arrow from the lone-pair electrons pointing toward the center of the adjacent bond. This move indicates that you're using the lone pair to form a new double bond.

2. **Move the double- or triple-bond pi electrons onto the farthest atom as a lone pair.**

   Draw an arrow starting from the middle of the double or triple bond and move the electrons onto the farthest atom in the double or triple bond by pointing the head of the arrow to this atom.

3. **Draw the resonance structure and check your charges.**

   The atom with the initial lone pair becomes one charge more positive in the resonance structure than what it started with, and the atom that received the double- or triple-bond electrons as a lone pair becomes one charge more negative than what it started with.

Q. Draw the other resonance structures for the following molecule by arrow-pushing.

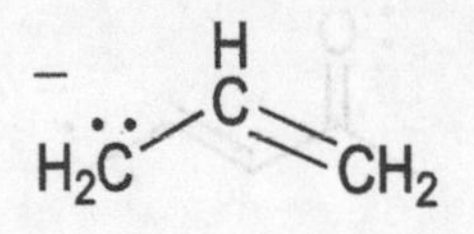

$H_2\ddot{C}^{-}$–CH=CH_2

A. $H_2\ddot{C}^{-}$–CH=CH_2 ⟷ H_2C=CH–$\ddot{C}H_2^{-}$

Start by moving the lone-pair electrons to make a double bond that goes from the left-most carbon to the middle carbon. You always draw the arrow from the electrons to where they're going, so to do this, you draw an arrow starting from the lone pair of electrons, with the head of the arrow pointing to the C-C single bond (not pointing to the carbon atom because the electrons are going to form the C=C double bond). After you've drawn this arrow, you can't stop here, or the central carbon will have five bonds and be *pentavalent* (five bonds times two electrons each would give ten electrons around the central carbon — a big no-no). Therefore, you have to simultaneously move the existing double-bond electrons and place them as a lone pair on the right-most carbon. You do this by drawing an arrow originating from the double bond and pointing it to the right-most carbon. This move has the effect of moving the negative charge to the right-most carbon.

7. **Draw the resonance structure for the following molecule by using arrow-pushing.**

:NH_2

Solve It

8. **Draw the resonance structure for the following ion by using arrow-pushing.**

H–C(:⁻)(H)–C≡C–H

Solve It

9. **Draw the resonance structures for the following ion by using arrow-pushing. (*Hint:* There may be more than one additional resonance structure.)**

Solve It

10. **Draw the resonance structures for the following ion by using arrow-pushing.**

Solve It

Pushing Double or Triple Bonds Containing an Electronegative Atom

If a double or triple bond contains an electronegative atom, such as oxygen or nitrogen, you can draw a resonance structure in which the double bond moves onto the electronegative atom as a lone pair. For neutral molecules, this leads to a resonance structure in which the electronegative atom has a negative charge and the less electronegative atom is left positively charged.

To draw alternative resonance structures matching this pattern, follow these steps:

1. **Push the pi bond electrons from the double or triple bond onto the electronegative atom as a lone pair.**

   Start with an arrow from the middle of the double or triple bond and point it toward the electronegative atom, showing how the pi bond electrons are reassigned as a lone pair onto the more electronegative atom.

2. **Draw the resonance structure and check the charges.**

   The electronegative atom, having accepted electrons as a lone pair, should've become one charge more negative in the resonance structure than it began. Thus, if the electronegative atom started positive, it should become neutral in the resonance structure. Likewise, the more electropositive atom should've become one charge unit more positive in the resonance structure than it began.

Q. Draw the other resonance structure for the following molecule by arrow-pushing.

:Ö:
||
H_3C–C–CH_3

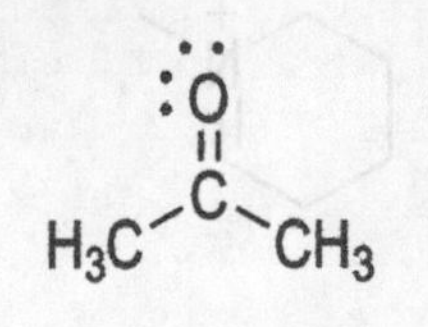

A. Oxygen is an electronegative atom involved in a double bond, so a resonance structure can be drawn for this molecule. To start, draw an arrow originating from the carbon-oxygen double bond and move the electrons onto the oxygen as a lone pair. This makes the double bond become a single bond, because two of the electrons have been moved. Additionally, the oxygen has become negatively charged and the carbon, positively charged. (Because the molecule started with a net neutral charge, the net charge in the resonance structure must remain neutral.)

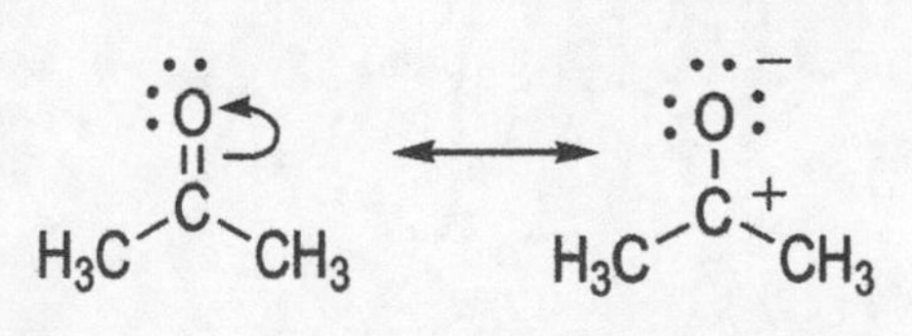

H_3C–C(=Ö:)–CH_3 ⟷ H_3C–C^+(–:Ö:$^-$)–CH_3

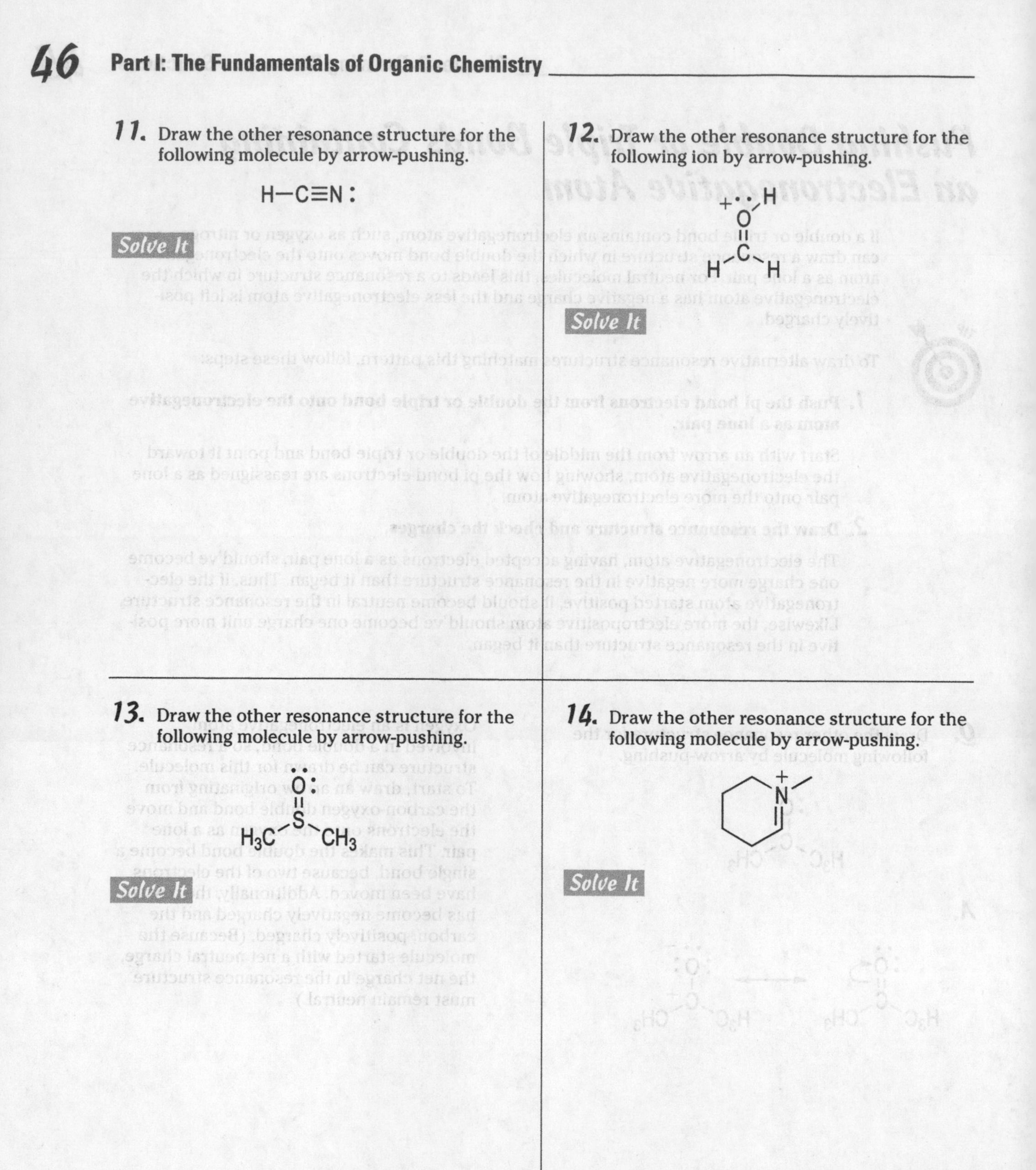

11. Draw the other resonance structure for the following molecule by arrow-pushing.

H—C≡N:

Solve It

12. Draw the other resonance structure for the following ion by arrow-pushing.

Solve It

13. Draw the other resonance structure for the following molecule by arrow-pushing.

Solve It

14. Draw the other resonance structure for the following molecule by arrow-pushing.

Solve It

Alternating Double Bonds around a Ring

A very interesting resonance structure can occur when you have a molecule that has alternating double bonds all the way around a ring. In such a molecule, you can make an alternative resonance structure by pushing each double bond one bond over all the way around the ring. Note that stopping at any step and not coming full circle leaves an atom that breaks the octet rule.

To draw these types of structures, follow these steps:

1. **Push the double-bond electrons all the way around the ring to reform each of the double bonds between different atoms.**

   Start by placing an arrow at the center of one of the double bonds and point it toward the adjacent bond (the direction around the ring doesn't matter). Continue this same procedure all the way around the ring until you're back where you started.

2. **Draw the resonance structure, making sure you don't break the octet rule.**

   When you draw the resonance structure, each of the double bonds is reassigned to a different location on the ring. Make sure you don't break the octet rule by stopping before you've pushed the double bonds all the way around the ring to end up at the atom that you started with.

Q. Draw the resonance structure for the following compound.

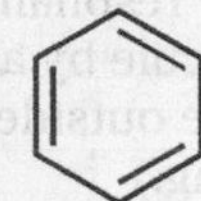

A. 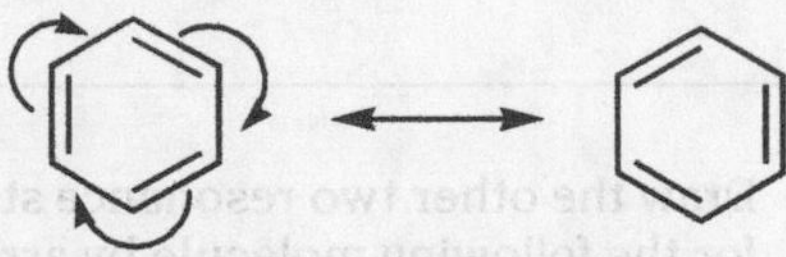

This molecule has double bonds that alternate all the way around the ring, so you can draw an alternative resonance structure as shown.

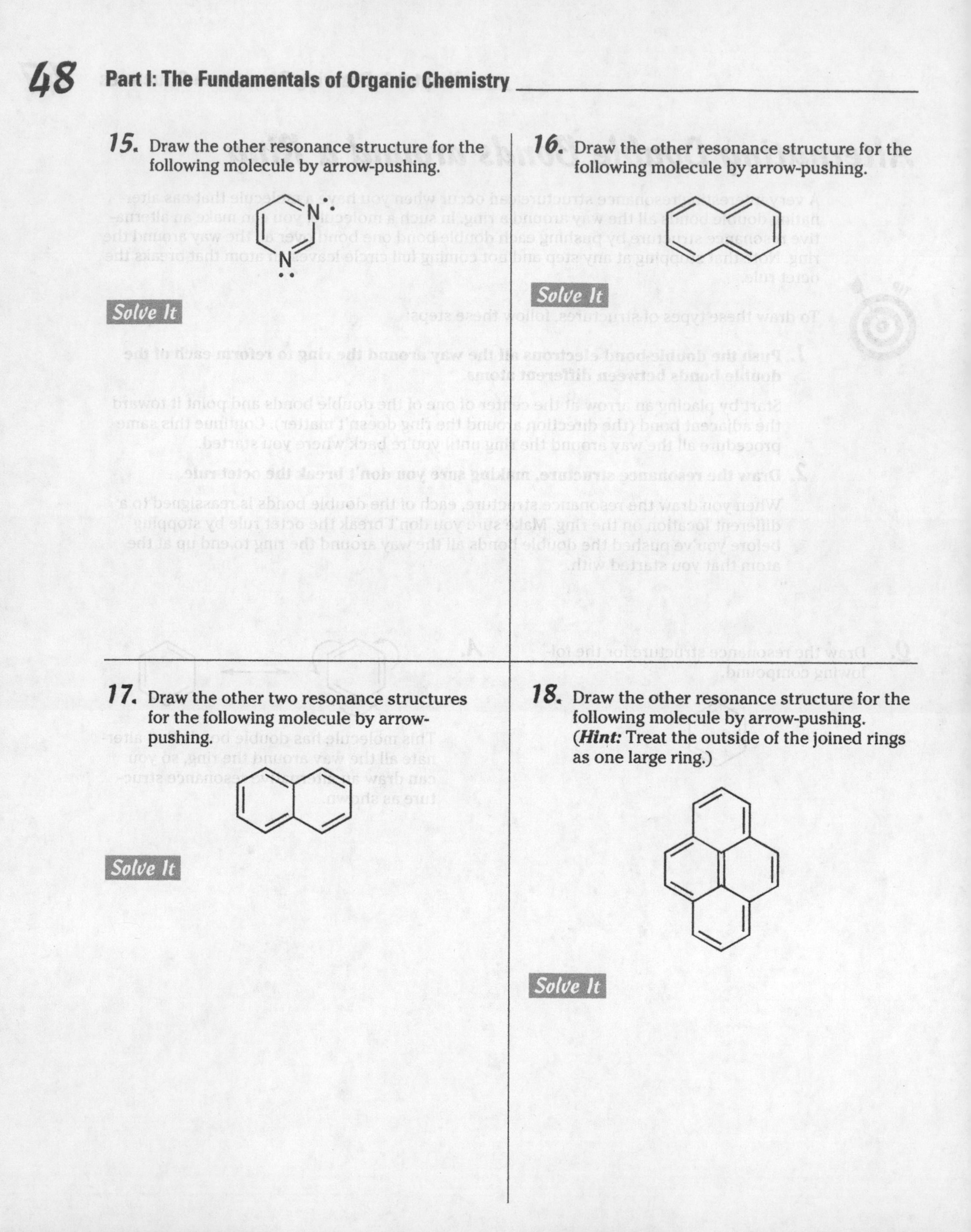

15. Draw the other resonance structure for the following molecule by arrow-pushing.

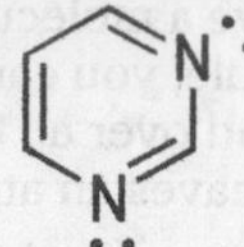

Solve It

16. Draw the other resonance structure for the following molecule by arrow-pushing.

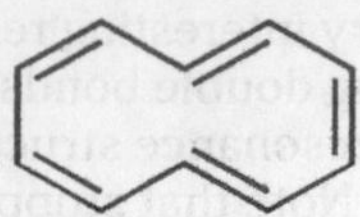

Solve It

17. Draw the other two resonance structures for the following molecule by arrow-pushing.

Solve It

18. Draw the other resonance structure for the following molecule by arrow-pushing. (***Hint:*** Treat the outside of the joined rings as one large ring.)

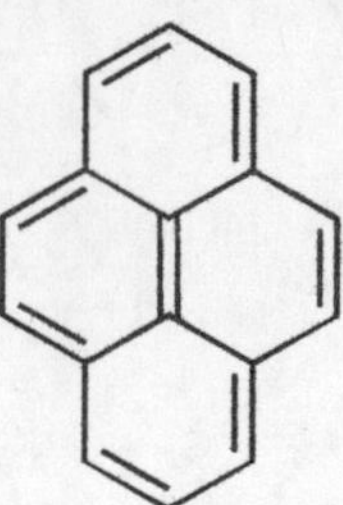

Solve It

Drawing Multiple Resonance Structures

Sometimes a compound has a number of resonance structures. Practicing on a few of these molecules that need to be described with multiple resonance structures is a good idea so you can get some more practice at spotting molecules that have resonance structures.

To draw multiple resonance structures, follow these steps:

1. **Look for one of the patterns to find a first alternative resonance structure.**

   The four patterns are as follows:

   - Cations next to a double bond, triple bond, or lone pair
   - Lone pairs next to a double or triple bond
   - Double or triple bonds containing an electronegative atom
   - Double bonds alternating around a ring

2. **After drawing the alternative resonance structure, look for additional patterns in the new resonance structure.**

   Often, drawing a resonance structure leads to a structure that meets one of the preceding patterns, in which case an additional resonance structure can be drawn. Repeat this step until you can't find any more alternative resonance structures.

Q. Draw all the resonance structures for the following molecule by arrow-pushing.

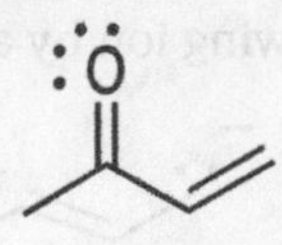

A. The first pattern is a double bond containing an electronegative element (oxygen). To draw the first resonance structure, you push the double-bond electrons on the oxygen as a lone pair. This leads to another resonance structure, which has another pattern — a cation next to a double bond. From this resonance structure, you can push the double-bond electrons over one bond and reform the cation on the right-most carbon. In this way, you can draw all three resonance structures for this molecule.

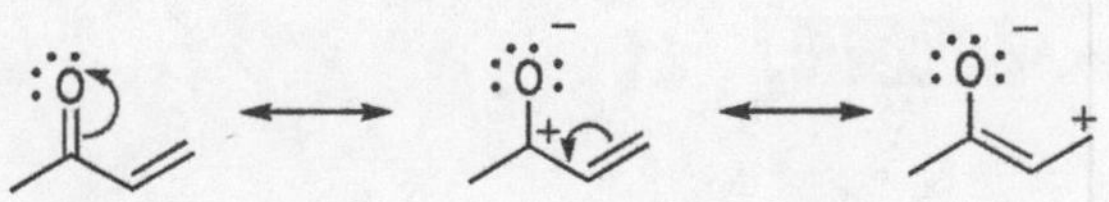

19. Draw the other resonance structures for the following molecule by arrow-pushing.

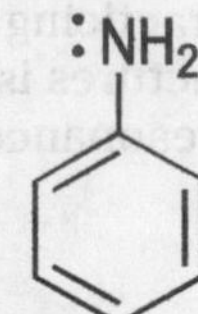

Solve It

20. Draw the other resonance structures for the following ion by arrow-pushing.

Solve It

21. Draw the other resonance structures for the following ion by arrow-pushing.

Solve It

22. Draw the other resonance structures for the following ion by arrow-pushing.

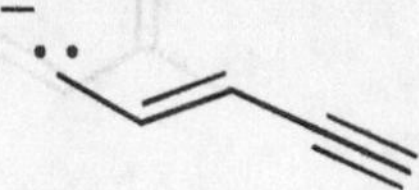

Solve It

Assigning Importance to Resonance Structures

A molecule with more than one resonance structure looks like the hybrid of all the resonance structures — but not all resonance structures necessarily contribute equally to the overall hybrid. In other words, if a molecule has two resonance structures, the actual structure may look more like one resonance structure than the other.

Generally, resonance structures that correspond to stable structures contribute more to the overall hybrid than do resonance structures that are less stable. So how do you determine stability? Here are some general guidelines:

- **Resonance structures with fewer charges are more stable than resonance structures with more charges.** For example, a resonance structure that has no charges contributes more to the overall hybrid than a resonance structure that has two. This rule comes as a result of the energy cost of separating charges.
- **For those molecules that are charged, the most stable resonance structures are the ones that place the charge on the best atom.** Negative charges prefer to rest on electronegative atoms (like oxygen and nitrogen), and positive charges prefer to rest on electropositive atoms (like carbon).
- **Resonance structures that have all atoms with complete octets of electrons are more stable than resonance structures that have atoms with incomplete octets.** This rule comes as a result of the desire of atoms to own a complete octet of electrons.
- **The desire of a molecule to have all atoms with complete octets trumps the desire to put the charge on the best atoms.** (The following example demonstrates this rule.)

Q. Predict which resonance structure of the following molecule contributes more to the overall hybrid.

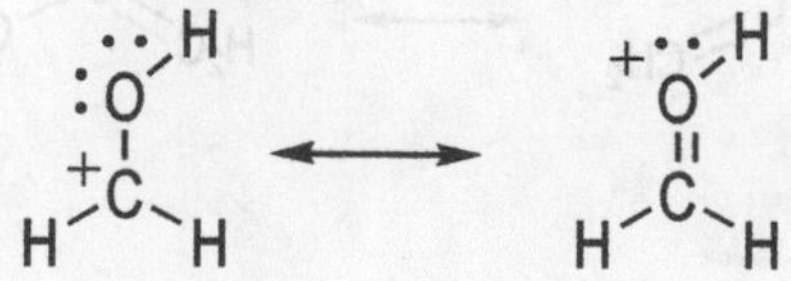

A. In this example, the right-hand resonance structure is more stable than the one on the left. In the right resonance structure, all atoms own complete octets of electrons, whereas in the left structure, the carbon is two electrons short of an octet. Thus, even though a positive charge prefers to rest on carbon rather than oxygen (because carbon is more electropositive than oxygen), the desire of a molecule to have filled octets trumps the desire to put the charge on the best atoms. Therefore, the actual structure is a mixture of the two resonance structures but looks more like the resonance structure on the right than the one on the left.

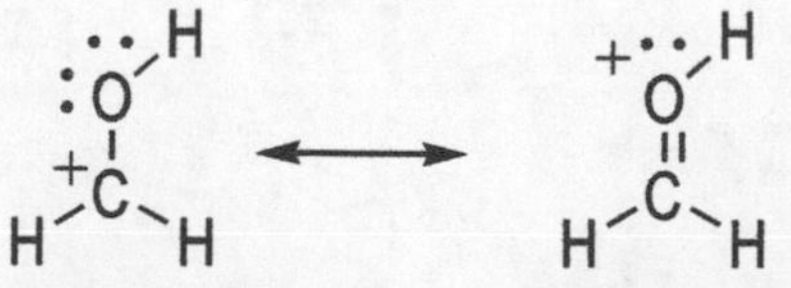

less stable (carbon owns only 6 e$^-$)

more stable (all atoms have full octets)

23. **Predict which resonance structure of the following molecule contributes more to the overall hybrid.**

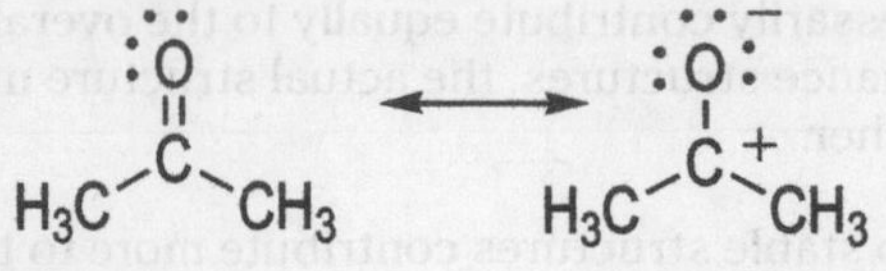

Solve It

24. **Predict which resonance structure of the following ion contributes more to the overall hybrid.**

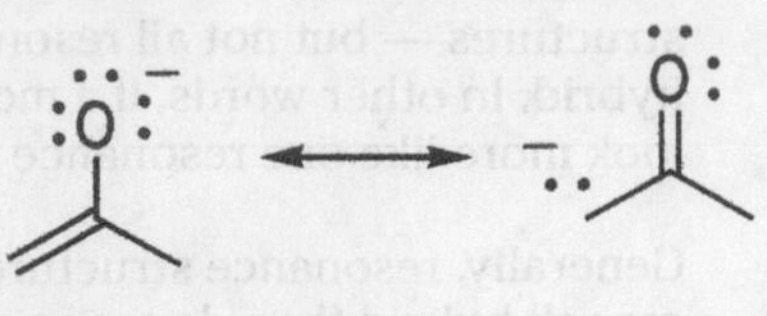

Solve It

25. **Predict which resonance structure of the following ion contributes more to the overall hybrid.**

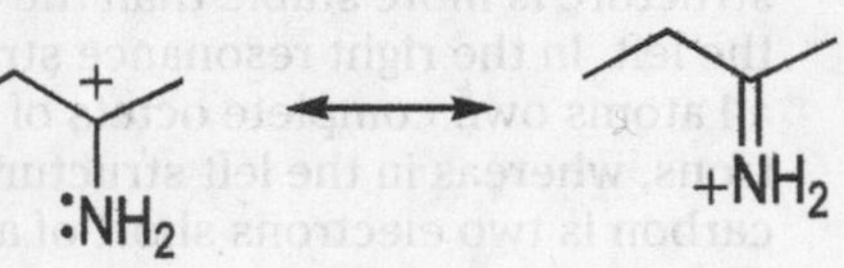

Solve It

26. **Predict which resonance structure of the following ion contributes more to the overall hybrid.**

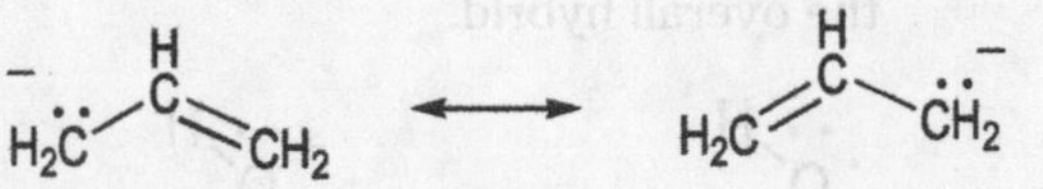

Solve It

Answer Key

The following are the answers to the practice questions presented in this chapter.

1

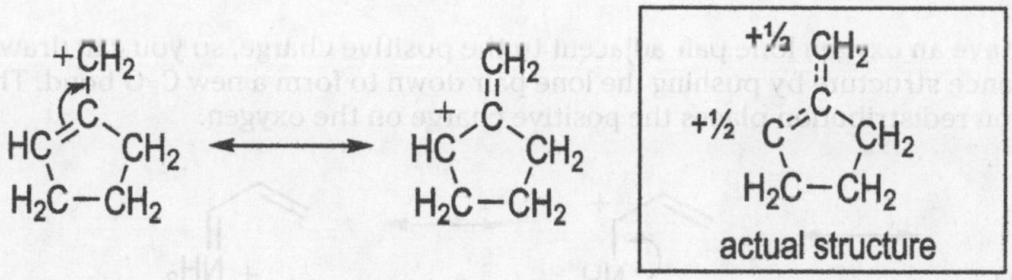

Start by moving the double-bond electrons to the adjacent bond. Doing so moves the positive charge. The actual structure is a mixture of the two structures, with approximately half of the positive charge residing on each of the two carbons that were charged in the resonance structures.

2

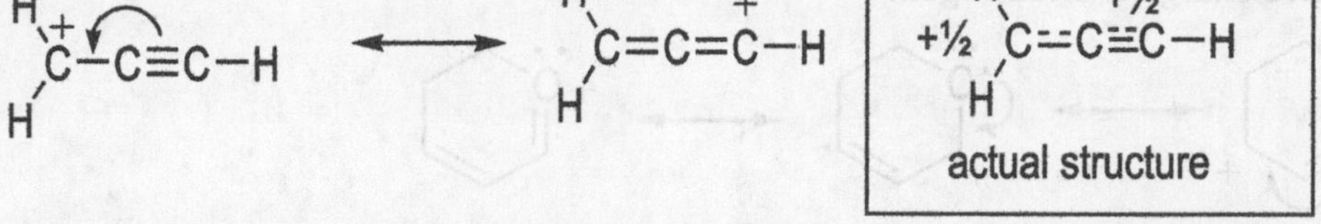

Start by moving the triple-bond pi electrons onto the adjacent bond, which results in the charge being moved to what was once a triple-bond carbon. The actual structure looks like a hybrid of the two resonance structures, with roughly half of the charge residing on each of the two carbons shown having a charge in the resonance structures.

3

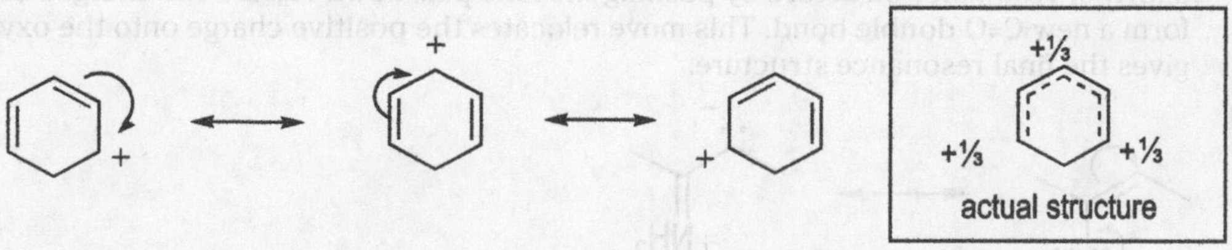

Draw the first resonance structure by moving the double bond adjacent to the cation over one bond in the direction of the charge. This results in the formation of a new C=C double bond and the cation moving two carbons to the left. You can draw a further resonance structure in a similar fashion starting from this new structure. By moving the adjacent double bond to the cation in the direction of the charge and redrawing the charge another two carbons to the left, you determine the final resonance structure. The actual structure looks like the average of the three resonance structures, with roughly a third of the charge occupying each of the carbons that held charge in the resonance structures.

4

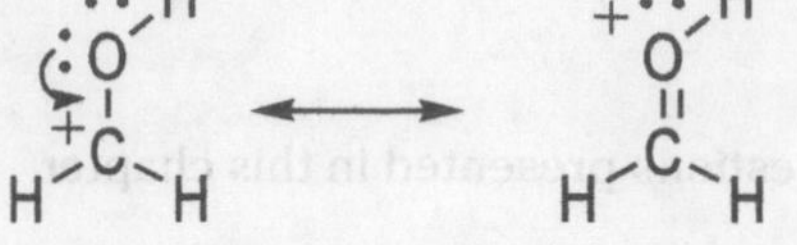

Here you have an oxygen lone pair adjacent to the positive charge, so you can draw an alternative resonance structure by pushing the lone pair down to form a new C=O bond. The result of this electron redistribution places the positive charge on the oxygen.

5

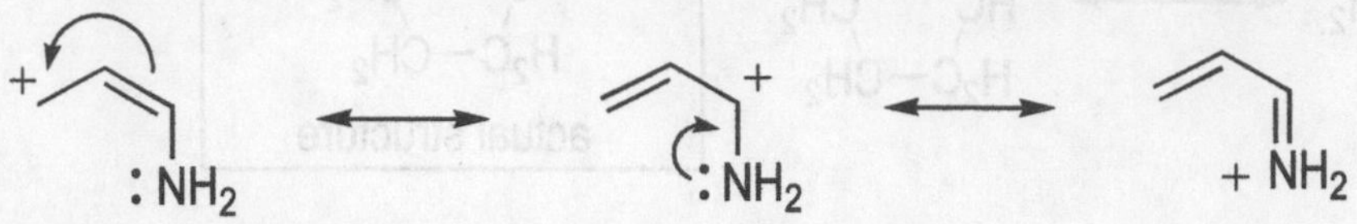

Here you have a double bond adjacent to a positive charge, so you can draw a resonance structure by pushing the double-bond electrons toward the positive charge to form a new double bond. The result of this electron redistribution moves the charge two carbons to the right. In this new resonance structure, you have a lone pair adjacent to a positive charge, so you can draw another resonance structure by pushing the lone pair down toward the charged carbon and forming a new C=N double bond. Doing so yields the final resonance structure, in which the positive charge is on nitrogen.

6

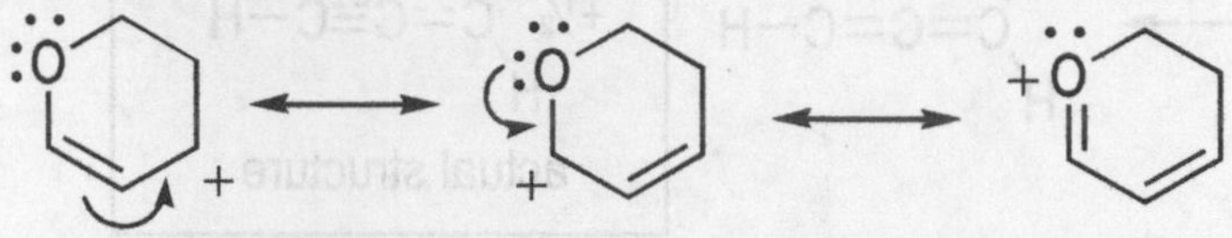

Because the double bond is adjacent to the positive charge, you can draw an alternative resonance structure by moving the double bond in the direction of the charge and reforming the C=C double bond. The result of this move is that the cation moves two carbons to the left. This leads to a new structure in which a lone pair is adjacent to the positive charge, so you can draw a further resonance structure by pushing the lone pair down toward the charged carbon to form a new C=O double bond. This move relocates the positive charge onto the oxygen and gives the final resonance structure.

7

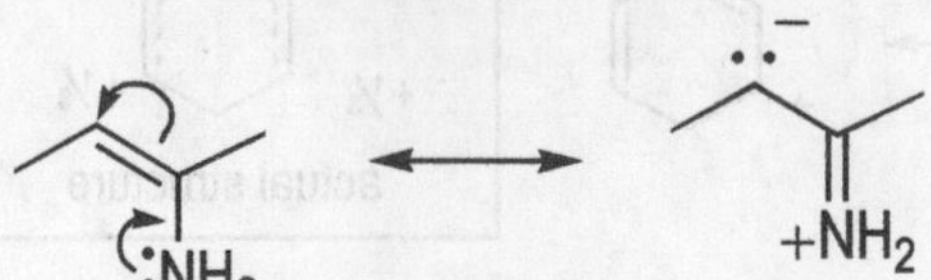

Here you have a lone pair adjacent to a double bond, so you start by moving the lone pair down from nitrogen to form a C=N double bond. You can't stop at this step, however, or the carbon attached to the nitrogen becomes pentavalent, so you simultaneously move the double bond as a lone pair to the adjacent carbon.

8

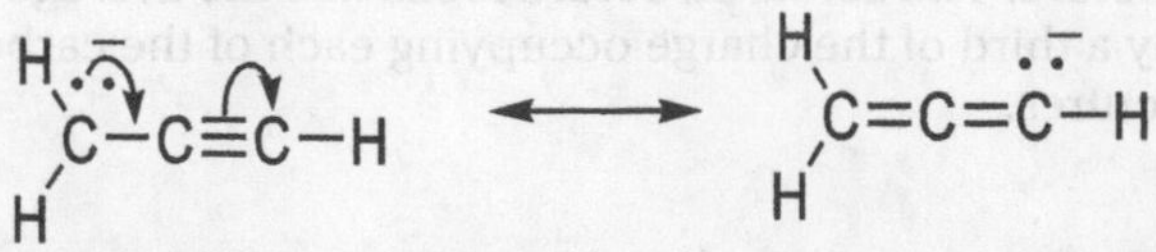

Start by moving the lone pair to form a double bond. Then move electrons from one bond of the triple bond onto the adjacent carbon. Because two electrons move from the triple bond, it becomes a double bond, and the negative charge moves to the right-most carbon.

9

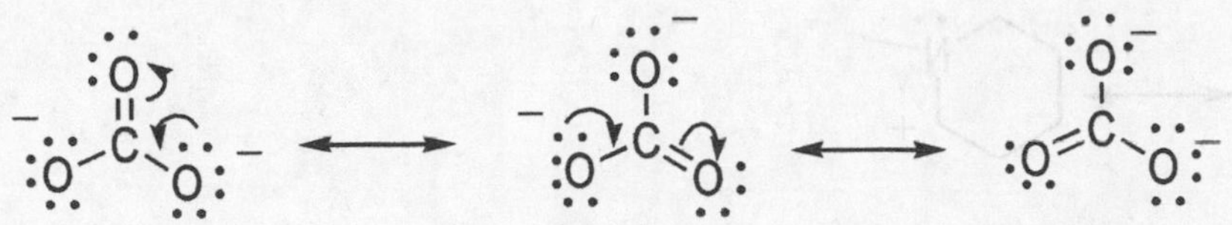

This problem has two atoms with lone pairs adjacent to the double bond, so you have two additional resonance structures. It doesn't matter which lone pair you start with, but move one of them down off the negatively-charged oxygen to form a C=O double bond, and move the electrons from the existing C=O double bond to the oxygen as a lone pair. The oxygen atom that received the lone pair becomes negatively charged. You can then repeat this process for the other oxygen atom to obtain the other resonance structure.

10

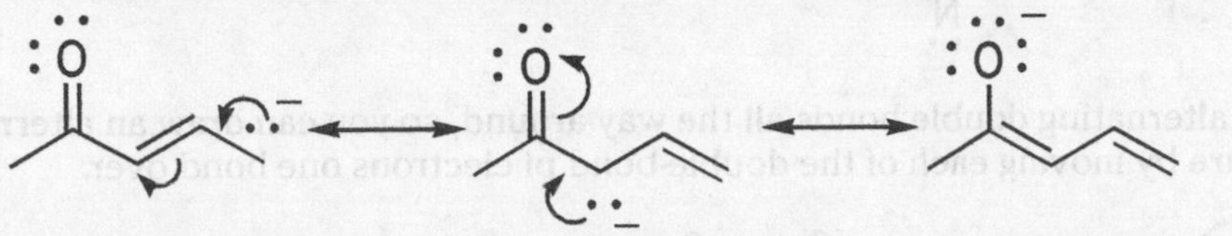

You can move the lone pair adjacent to the double bond in the same way as with the other problems in this section by forming a new double bond and pushing the existing double-bond electrons as a lone pair onto the adjacent carbon. The new structure has a lone pair adjacent to a double bond (the C=O double bond). Therefore, you can draw another resonance structure in similar fashion, forming a new C=C double bond and pushing the C=O double-bond electrons onto the oxygen, leaving the oxygen negatively charged.

11

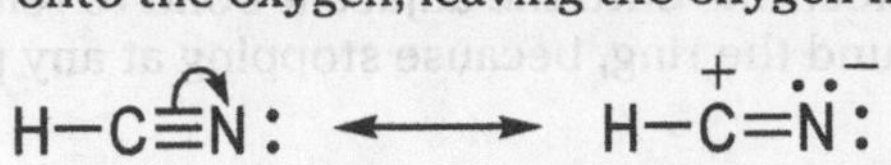

H–C≡N: ⟷ H–C⁺=N:⁻

Nitrogen is an electronegative element, so you can draw a resonance structure in which one of the triple bonds moves as a lone pair onto nitrogen. This move makes the carbon positively charged and the nitrogen negatively charged, which is okay because the net charge is still neutral (the net charge has to stay the same in all resonance structures, but the number of charges doesn't).

12

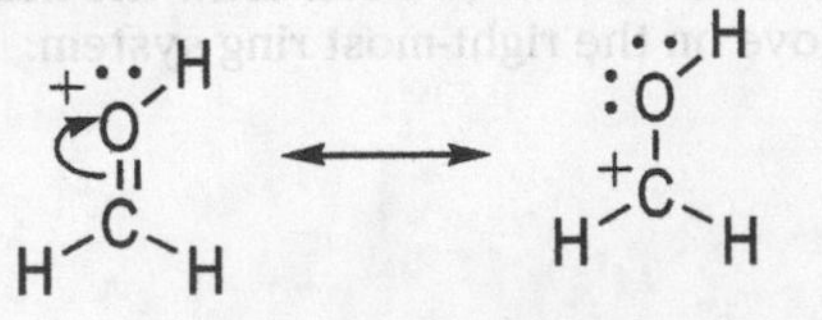

The double bond contains an electronegative atom, which is oxygen. Therefore, you can draw an alternative resonance structure by moving the double bond onto the oxygen as a lone pair. As a result of this electron redistribution, the positive charge moves onto the carbon in the alternative resonance structure.

13

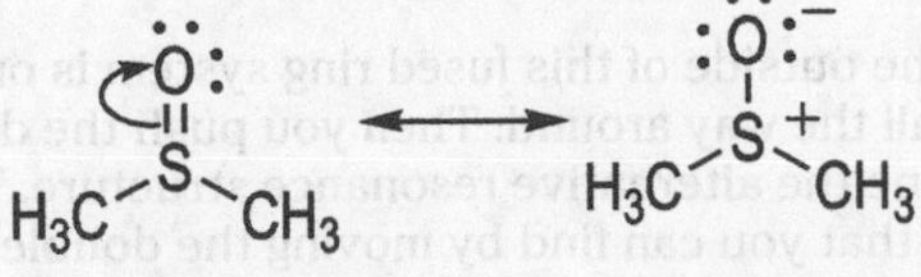

In a similar problem to question 12, the electronegative atom in the double bond is oxygen, so you can draw an alternative resonance structure by moving the electrons in one of the bonds of the double bond onto the oxygen as a lone pair. The move results in a negative charge on the oxygen and a positive charge on the sulfur atom in the resonance structure.

14

The nitrogen is the electronegative atom in the double bond, so you can draw a resonance structure by moving the double-bond electrons onto the nitrogen as a lone pair and leaving the carbon positively charged.

15

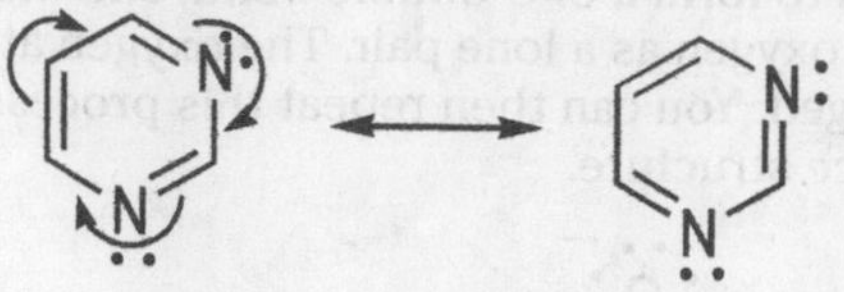

This ring has alternating double bonds all the way around, so you can draw an alternative resonance structure by moving each of the double-bond pi electrons one bond over.

16

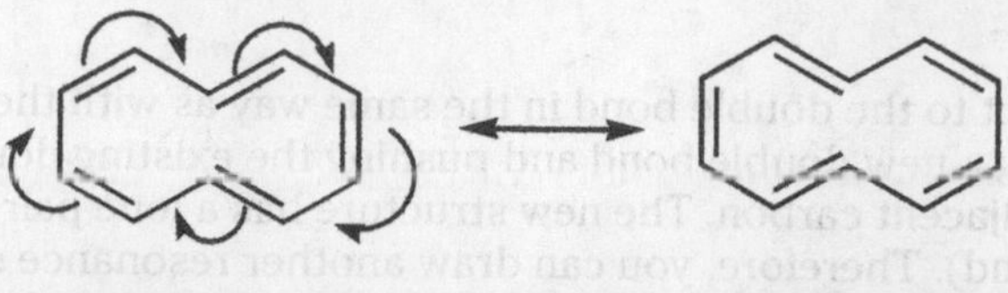

This one is a very large ring, but it works in the same fashion as problem 15. You start at one of the double bonds and push the double-bond electrons onto the adjacent bond to reform the double bond. You must do this all the way around the ring, because stopping at any point before then results in a pentavalent carbon.

17

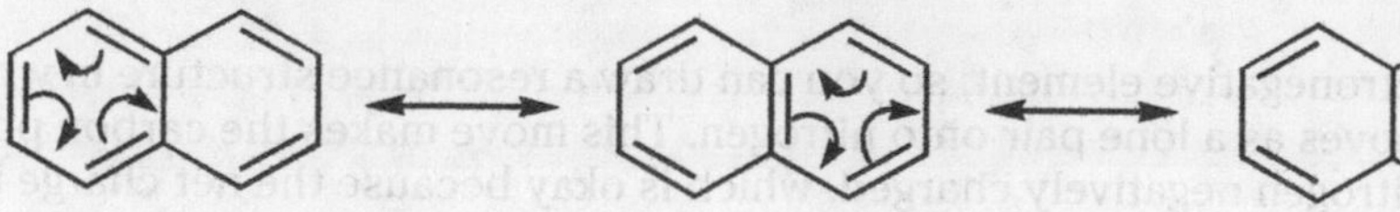

This fused ring system has two additional resonance structures. You can draw the first by moving the double bonds in the left-most ring one carbon over. You can draw the next by starting from this new structure and doing a similar move on the right-most ring system.

18

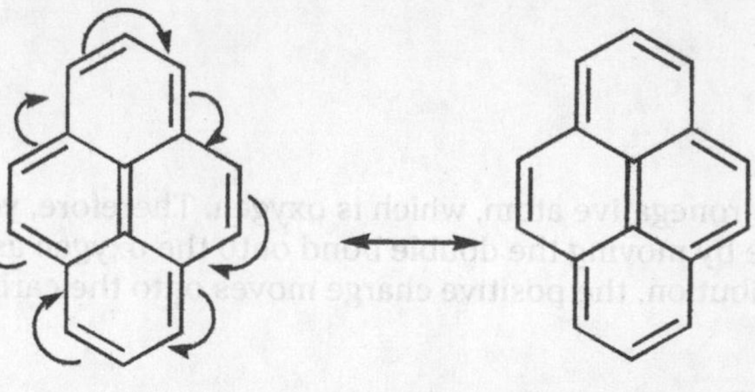

The trick to this problem is to pretend that the outside of this fused ring system is one giant ring that contains alternating double bonds all the way around. Then you push the double bonds all the way around the ring to determine the alternative resonance structure. There are also a number of other resonance structures that you can find by moving the double bonds around the individual six-membered rings.

19

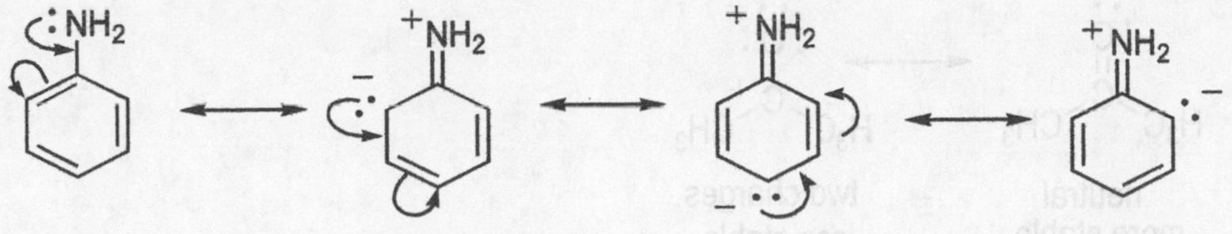

The first pattern to notice is a lone pair adjacent to a double bond. You can move this lone pair down to form a C=N double bond and push the other double bond as a lone pair onto the joining carbon. This process leads to a resonance structure that has another lone pair adjacent to a double bond, so you can draw another resonance structure, which leads to two more structures by a similar process.

20

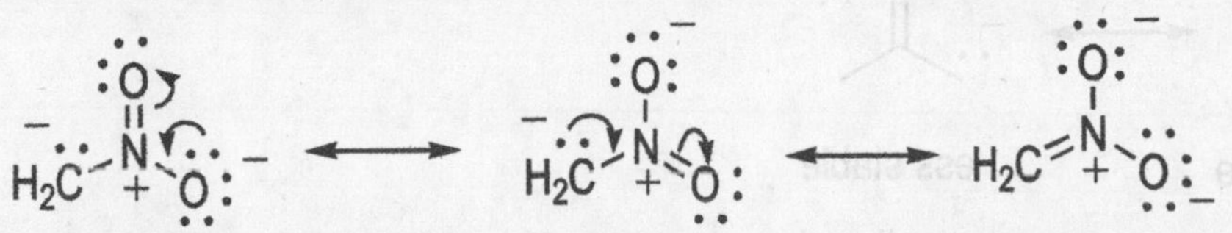

Here you have two adjacent lone pairs next to the N=O double bond, so there are two additional resonance structures. It doesn't matter which one you start with, because starting with one or the other ultimately leads to the same resonance structures. Keep in mind that these arrows are simply tools to help locate other resonance structures and don't have any physical meaning.

21

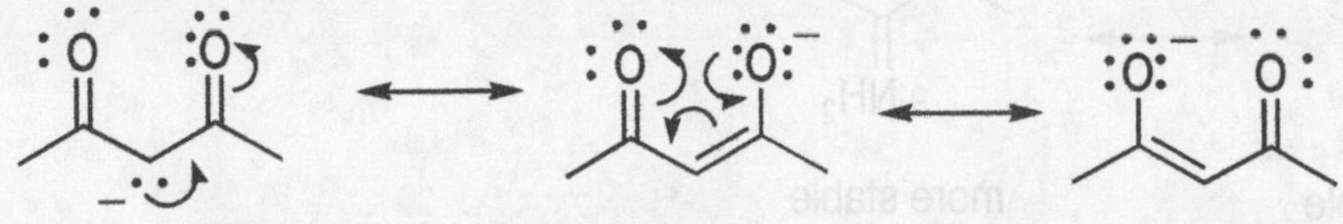

This one is a little tricky. The first step is the same: moving the carbon lone pair in the direction of one of the C=O double bonds. Because there are two adjacent double bonds, two possible resonance structures exist. You can get from the second resonance structure to the last by drawing three arrows.

22

You have a lone pair adjacent to a double bond, so you can draw an alternative resonance structure by moving the lone pair in the direction of the double bond and moving the double-bond electrons as a lone pair onto the carbon atom. This leads to an alternative resonance structure with a lone pair adjacent to a triple bond, so you can find a further resonance structure by pushing the lone pair toward the triple bond and moving one of the bonds in the triple bonds onto the farthest carbon as a lone pair.

23

neutral
more stable

two charges
less stable

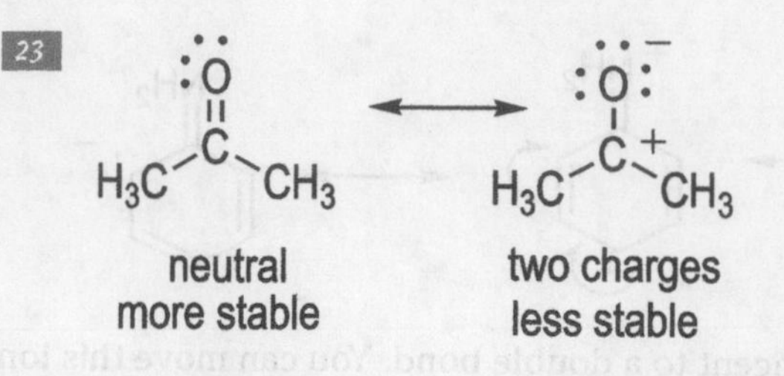

The neutral resonance structure contributes more to the overall hybrid (the actual structure) because separating charges costs energy.

24

more stable

negative charge
on oxygen

less stable

negative charge
on carbon

The left-hand resonance structure in which the negative charge rests on oxygen contributes more to the overall hybrid.

25

$:NH_2$

less stable
(carbon owns
only 6 e^-)

$+NH_2$

more stable
(all atoms have
full octets)

The resonance structure that contributes most to the overall hybrid is the right-most structure because this structure has all atoms with complete octets of electrons. Recall that the desire to have all atoms with complete octets trumps the desire to put the charges onto the most desirable atoms.

26

equally stable resonance structures
(both contribute equally to hybrid)

Trick question! Both resonance structures are equally stable, so both contribute equally to the overall hybrid.

Chapter 4

Working with Acids and Bases

In This Chapter

- Defining Lewis and Bronsted-Lowry acids and bases
- Seeing structural elements that affect acidity on molecules
- Predicting the direction of acid-base equilibria using pKa values

Most people — even those who've never taken a class in chemistry — have heard of acids. But when most people think of an acid, they often attach a negative image — the corrosive liquid found in car batteries, the stuff that causes your stomach to reflux after you eat sausage pizza, the vat of goo that disfigured the face of the Joker in *Batman,* and so on. But acids are in many other things, too. For example, acids make oranges, grapefruits, and lemons taste sour; acids make soft drinks last; and acids such as cream of tartar help cakes rise by reacting with baking soda (a base). Acids are even found in your genome (the *a* in DNA stands for *acid*).

And here's another little factoid you may find interesting: Nearly all the reactions that you see in organic chemistry can be classified as acid-base reactions. Therefore, it's essential that you understand acid-base chemistry right away, arming you to deal with the reactions of organic compounds that you see later on in organic chemistry.

In this chapter, you see the two most important classifications of acids and bases that you deal with in organic chemistry — the Bronsted-Lowry and Lewis definitions of acids and bases — and how these definitions differ. Additionally, you get the lowdown on how specific structural elements on a molecule affect its acidity; then you use that knowledge to compare the acidity of different acids. You work with a quantitative scale of acidity, called the pKa scale, which you can use to predict the direction of an acid-base reaction at equilibrium.

Defining Acids and Bases

The definitions of an acid and a base have evolved over the centuries, but the two definitions that organic chemists use widely today come in two flavors: the Bronsted-Lowry and the Lewis:

- **Bronsted-Lowry acids and bases:** A Bronsted-Lowry acid (or Bronsted acid for short) is a molecule or ion that donates a proton, and a Bronsted base is a molecule or ion that accepts a proton.

- **Lewis acids and bases:** A Lewis acid is a molecule or ion that accepts a pair of electrons to make a new covalent bond, and a Lewis base is a molecule or ion that donates a pair of electrons to make a new covalent bond.

This section takes a closer look at these two types and shows you how to solve some basic problems associated with them.

Bronsted-Lowry acids and bases

Chemists care deeply about the *mechanism* of a reaction, which is organic-speak for how a reaction occurs. The complete mechanism of a reaction shows all the bond-making and bond-breaking steps in a reaction, including the order in which those steps take place. You depict the mechanism on paper by using arrows to show the movement of the electrons in a reaction, taking the starting material into the product.

Figure 4-1 shows a schematic of the reaction between a Bronsted-Lowry acid and a Bronsted-Lowry base. Here, the Bronsted base (the proton acceptor) pulls the proton off the Bronsted acid (the proton donor).

$$\text{BASE}:^{-} \quad \text{H—ACID} \longrightarrow \text{BASE—H} \quad :\text{ACID}^{-}$$

conjugate acid　　conjugate base

Figure 4-1: Bronsted-Lowry acids and bases reacting.

Always draw the arrow starting from the electrons and then "push" the arrow to where the electrons are going. Acid-base reactions are the simplest organic reactions, so the mechanisms of these reactions are simple — a good place to start practicing your arrow-pushing skills. In the example in Figure 4-1, the lone pair of electrons then becomes the bonding electrons in the base-H bond.

Never draw an arrow originating from an H^+ in a mechanism (a common mistake), because an H^+ has no electrons.

Two pieces of organic-speak you should remember for Bronsted acids and bases: The deprotonated acid is referred to as the *conjugate base*, and the protonated base is called the *conjugate acid*.

Q. Label the acid and base in the following reaction. Then show the mechanism (arrow-pushing) of the deprotonation reaction. Designate the conjugate acid and the conjugate base of this reaction.

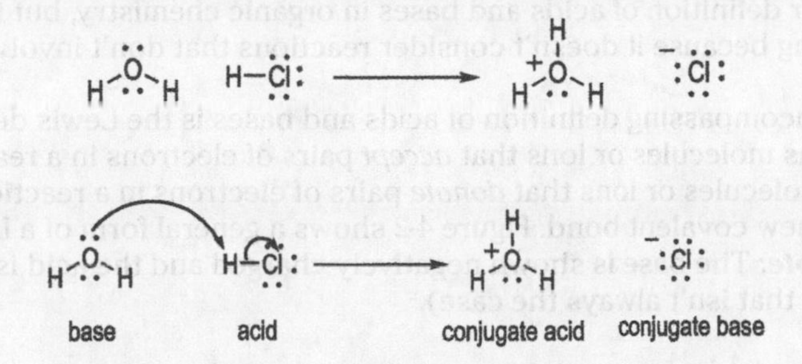

$$H_2\ddot{O} + H{-}\ddot{Cl}: \longrightarrow H_3O^+ + :\ddot{Cl}:^-$$

A.

$$\underset{\text{base}}{H_2\ddot{O}} + \underset{\text{acid}}{H{-}\ddot{Cl}:} \longrightarrow \underset{\text{conjugate acid}}{H_3O^+} + \underset{\text{conjugate base}}{:\ddot{Cl}:^-}$$

The proton donor in this reaction is HCl, so this molecule is the Bronsted acid in this reaction; the proton acceptor is water, H_2O, the Bronsted base. After water accepts the proton, it becomes the hydronium ion, H_3O^+, the conjugate acid. After HCl gives up its proton, it becomes the chloride ion, the conjugate base.

To draw the mechanism, start by moving the lone pair of electrons from the oxygen on water and pushing it to the proton on H-Cl. This push represents a new bond forming between the oxygen and the hydrogen. The bond between H-Cl is then broken (shown by another arrow) and those two bonding electrons are then assigned to the chlorine as a lone pair.

1. Label the acid and base in the following reaction. Then show the mechanism of the deprotonation reaction. Designate the conjugate acid and the conjugate base of this reaction.

$$H_3C{-}\ddot{O}{-}H + H{-}\ddot{I}: \longrightarrow H_3C{-}\overset{+}{O}H_2 + :\ddot{I}:^-$$

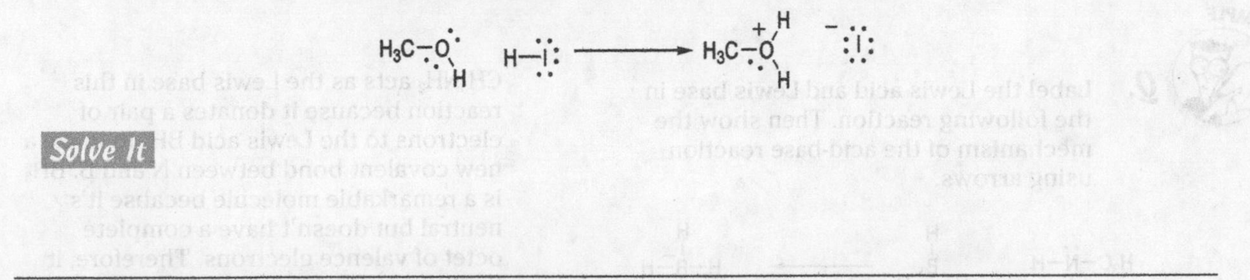

Solve It

2. Label the acid and base in the following reaction. Then show the mechanism of the deprotonation reaction. Designate the conjugate acid and the conjugate base of this reaction. (***Hint:*** Double-bond electrons can act as bases just like lone pairs).

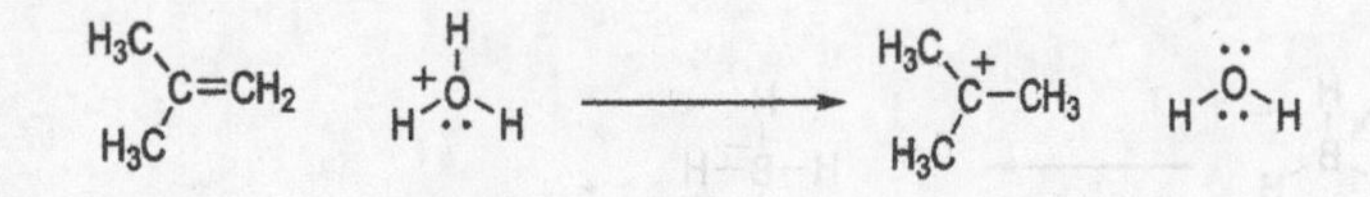

$$(H_3C)_2C{=}CH_2 + H_3O^+ \longrightarrow (H_3C)_2\overset{+}{C}{-}CH_3 + H_2\ddot{O}$$

Solve It

Lewis acids and bases

The Bronsted-Lowry definition of acids and bases (described in the preceding section) is the most popular definition of acids and bases in organic chemistry, but that definition isn't all-encompassing because it doesn't consider reactions that don't involve proton (H^+) transfers.

A more all-encompassing definition of acids and bases is the Lewis definition, in which acids are defined as molecules or ions that *accept* pairs of electrons in a reaction and bases are defined as molecules or ions that *donate* pairs of electrons in a reaction. The result is the formation of a new covalent bond. Figure 4-2 shows a general form of a Lewis acid and base reaction. (***Note:*** The base is shown negatively charged and the acid is shown positively charged, but that isn't always the case).

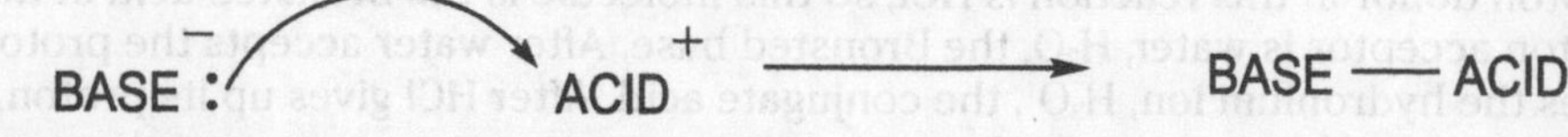

Figure 4-2: Lewis acids and bases.

The Lewis acid and base definitions encompass the Bronsted-Lowry definitions, because any molecule that pulls off a proton is necessarily donating electrons (a Lewis base) and any molecule that loses a proton (an H^+) is accepting electrons (a Lewis acid). Therefore, Bronsted acids and bases are also Lewis acids and bases. Chemists simply use the Bronsted-Lowry definition because it's easier to think of most acid-base reactions in terms of proton transfers rather than in terms of electron transfers.

Q. Label the Lewis acid and Lewis base in the following reaction. Then show the mechanism of the acid-base reaction using arrows.

$$H_3C-\ddot{N}H_2 + BH_3 \longrightarrow H_3C-\overset{+}{N}H_2-\overset{-}{B}H_3$$

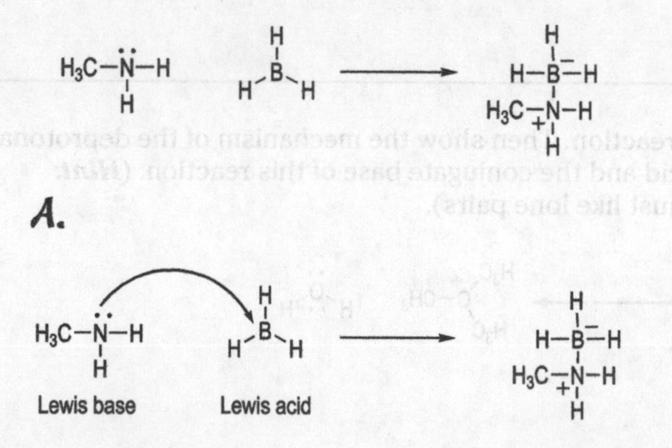

A. CH$_3$NH$_2$ acts as the Lewis base in this reaction because it donates a pair of electrons to the Lewis acid BH_3 to make a new covalent bond between N and B. BH_3 is a remarkable molecule because it's neutral but doesn't have a complete octet of valence electrons. Therefore, it badly wants to accept a pair of electrons to complete its octet, making this molecule a powerful Lewis acid (electron acceptor).

$$\underset{\text{Lewis base}}{H_3C-\ddot{N}H_2} + \underset{\text{Lewis acid}}{BH_3} \longrightarrow H_3C-\overset{+}{N}H_2-\overset{-}{B}H_3$$

3. Label the Lewis acid and Lewis base in the following reaction. Then show the mechanism of the acid-base reaction using arrows.

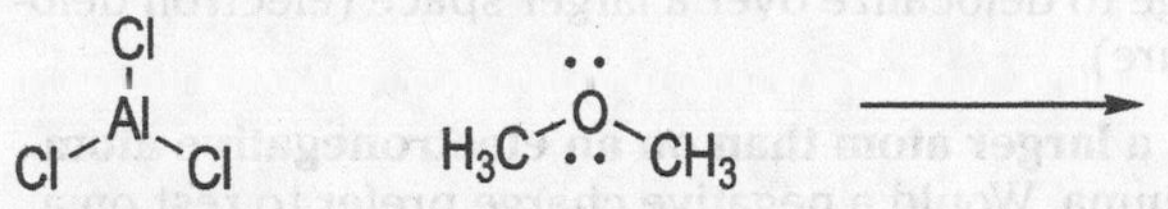

Solve It

4. Label the Lewis acid and Lewis base in the following reaction. Then show the mechanism of the acid-base reaction using arrows.

$$H^+ \quad :H^- \longrightarrow H\text{—}H$$

Solve It

Comparing Acidities of Organic Molecules

Whether an acid-base reaction occurs depends on the strength of the acid. Because acid and base reactions are so important in organic chemistry, being able to compare two acid structures so you can say which acid is stronger than another is a really important skill.

The secret to comparing the strength of two acids is this: *Strong acids have stable conjugate bases.* Therefore, the more stable the conjugate base of an acid, the stronger the acid, because an acid is more willing to give up a proton when doing so leads to a stable conjugate base. Conversely, an acid with an unstable conjugate base is less willing to give up its proton because doing so leads to an unstable conjugate base.

The question then becomes this: What features on a molecule stabilize a conjugate base? Because most acids are neutrally charged and upon deprotonation become negatively charged in the conjugate base form, any structural features that help to stabilize this negative charge in the conjugate base leads to a stronger acid.

Contrasting atom electronegativity, size, and hybridization

Charges are more stable on some atoms than others. Here are a few rules:

- **Negative charges are more stable on more electronegative atoms.** Recall that electronegativity increases as you go up and to the right on the periodic table. Therefore, any conjugate base that places the negative charge on a more electronegative atom is more stable than a conjugate base that places the negative charge on a more electropositive atom.

- **Negative charges are more stable on larger atoms.** The size of atoms increases as you go down the periodic table. Negative charges prefer to rest on larger atoms because larger atoms allow the negative charge to delocalize over a larger space (electron delocalization is always a stabilizing feature).
- **A negative charge is more stable on a larger atom than on an electronegative atom.** Of course, you sometimes face a dilemma. Would a negative charge prefer to rest on a more electronegative atom or a larger atom that isn't as electronegative? In these cases, atom size trumps electronegativity.
- **Negative charges prefer to rest on *sp*-hybridized atoms over sp^2-hybridized atoms, and they prefer sp^2 over sp^3 atoms.** Negative charges prefer to be placed in orbitals with more *s* character because *s* orbitals are closer to the atom nucleus. This means that a negative charge prefers to rest on an *sp*-hybridized atom over an sp^2 atom and on an sp^2 atom over an sp^3 atom (refer to Chapter 1 to brush up on determining atom hybridization).

Q. Which acid is stronger, HF or HI?

A. **HI is the stronger acid.** The strength of an acid depends on the stability of the conjugate base, so the first thing to do is to deprotonate these acids and see which conjugate base is more stable. The conjugate base of HF is F^-, and the conjugate base of HI is I^-. Fluorine is more electronegative than iodine, but iodine is a bigger atom because it's three rows down from fluorine on the periodic table. This presents a dilemma, because although negative charges prefer to rest on more electronegative atoms, they also prefer to rest on bigger atoms. However, atom size trumps electronegativity, so I^- is more stable than F^-, and HI is therefore a stronger acid than HF.

5. Which acid is stronger, HCl or HF?

Solve It

6. Which acid is stronger, CH_4 or NH_3?

Solve It

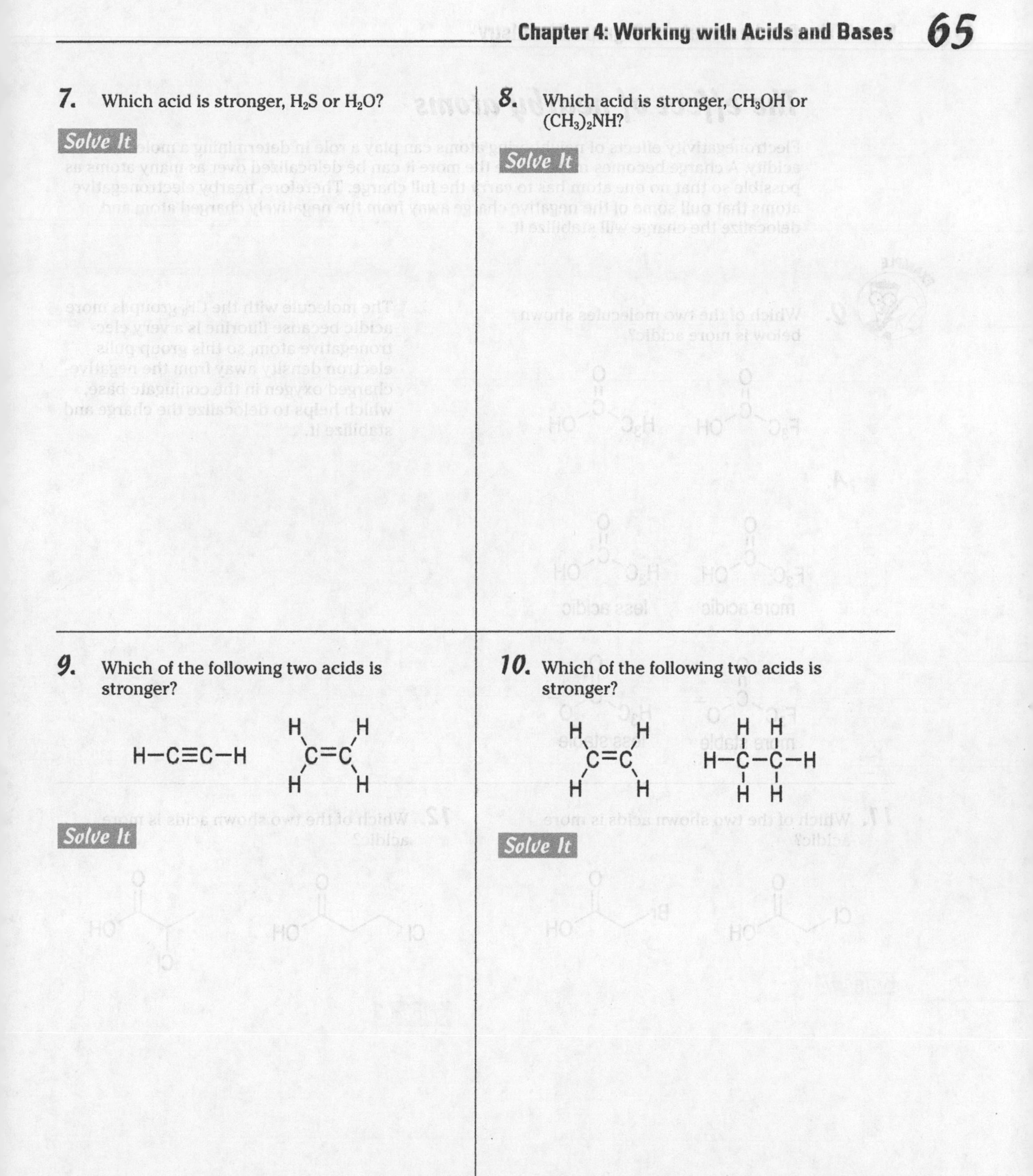

7. Which acid is stronger, H_2S or H_2O?

Solve It

8. Which acid is stronger, CH_3OH or $(CH_3)_2NH$?

Solve It

9. Which of the following two acids is stronger?

H–C≡C–H

$H_2C{=}CH_2$ (H₂C=CH₂ drawn with all H atoms)

Solve It

10. Which of the following two acids is stronger?

$H_2C{=}CH_2$

$H_3C{-}CH_3$

Solve It

The effect of nearby atoms

Electronegativity effects of neighboring atoms can play a role in determining a molecule's acidity. A charge becomes more stable the more it can be delocalized over as many atoms as possible so that no one atom has to carry the full charge. Therefore, nearby electronegative atoms that pull some of the negative charge away from the negatively charged atom and delocalize the charge will stabilize it.

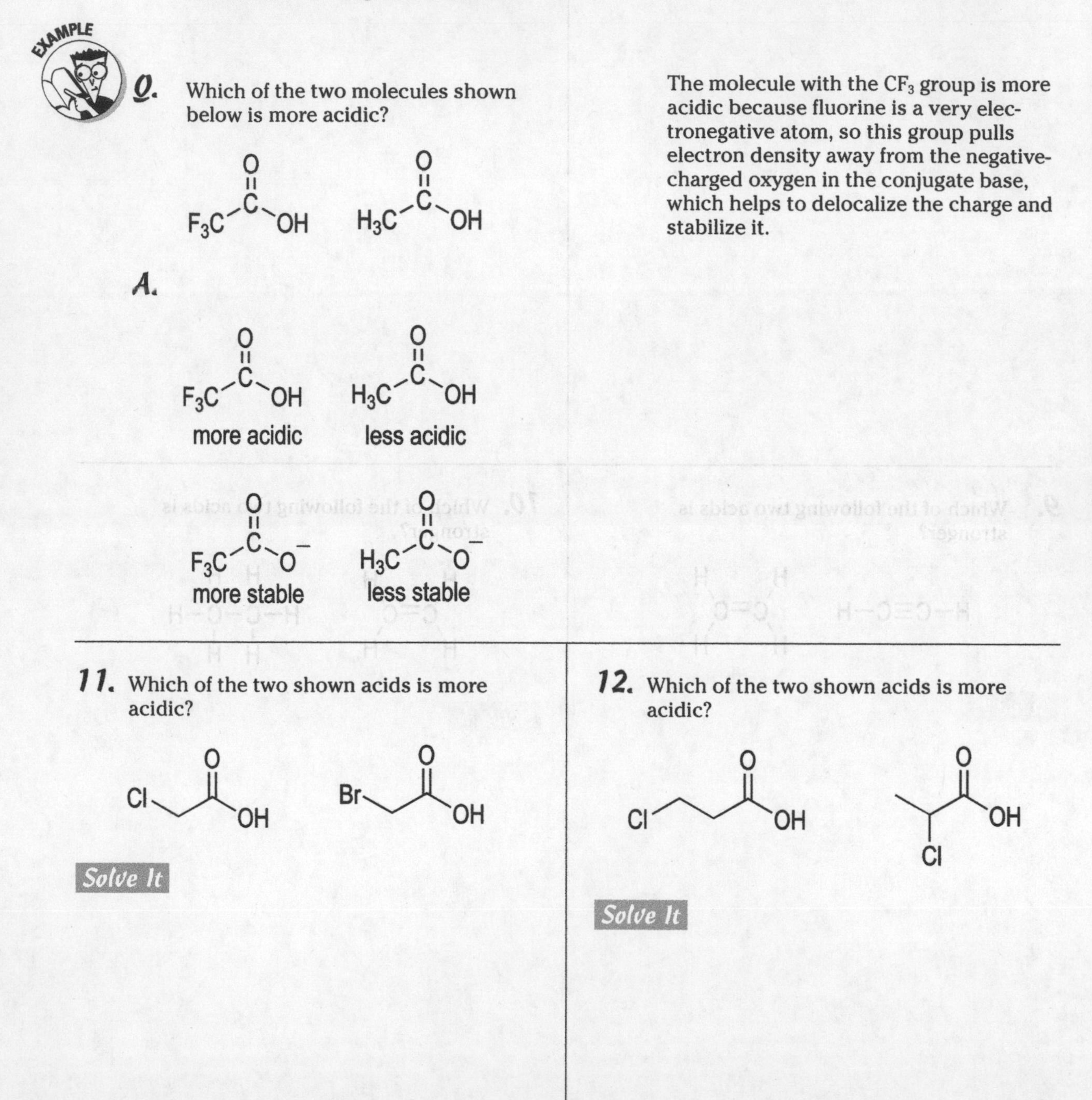

Q. Which of the two molecules shown below is more acidic?

$F_3C-C(=O)-OH$ $H_3C-C(=O)-OH$

A. The molecule with the CF_3 group is more acidic because fluorine is a very electronegative atom, so this group pulls electron density away from the negative-charged oxygen in the conjugate base, which helps to delocalize the charge and stabilize it.

$F_3C-C(=O)-OH$ more acidic $H_3C-C(=O)-OH$ less acidic

$F_3C-C(=O)-O^-$ more stable $H_3C-C(=O)-O^-$ less stable

11. Which of the two shown acids is more acidic?

$ClCH_2C(=O)OH$ $BrCH_2C(=O)OH$

Solve It

12. Which of the two shown acids is more acidic?

$ClCH_2CH_2C(=O)OH$ $CH_3CH(Cl)C(=O)OH$

Solve It

13. Which of the two shown acids is more acidic?

Solve It

14. Which of the two shown acids is more acidic?

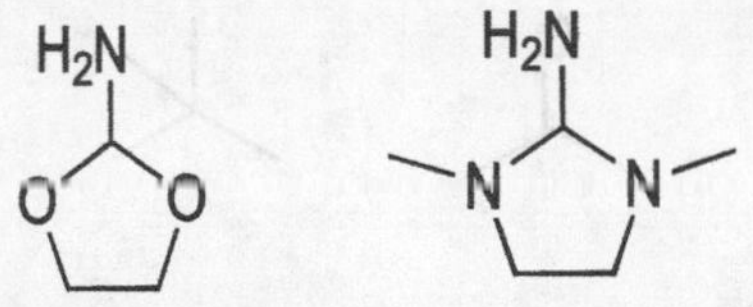

Solve It

Resonance effects

Acids with conjugate bases that can delocalize the negative charge to other atoms through resonance are more acidic than molecules lacking conjugate base resonance structures (for a refresher on resonance structures, see Chapter 3). Resonance structures stabilize charges because they allow a charge to delocalize over two or more atoms and don't require just a single atom to bear the full charge. As a general rule, the more resonance structures a molecule has, the more stable the structure.

Q. Which of the two shown acids is more acidic?

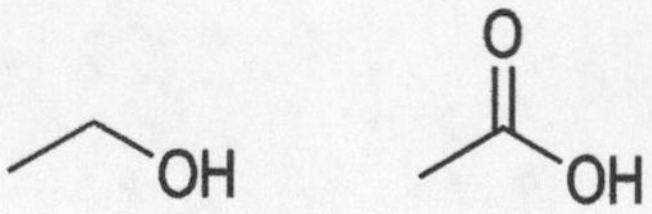

A.

less acidic

more acidic

localized charge

charge delocalized through resonance

First, draw the conjugate base of both acids and see which one is more stable. The left-most conjugate base in this case is less stable because this structure has no resonance structures, so the negative charge is localized on a single oxygen. In the right-hand structure, on the other hand, the charge can delocalize through resonance over two oxygens, stabilizing the conjugate base. Therefore, the right-hand acid is the more acidic because it has the more stable conjugate base. You may find it helpful when finding alternative resonance structures to draw out the full Lewis structures as I show here.

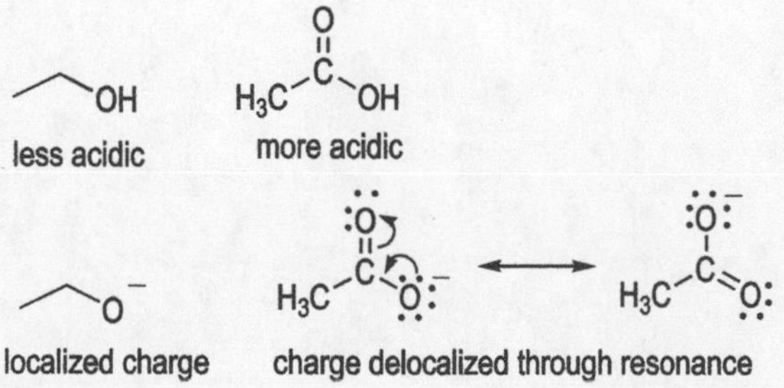

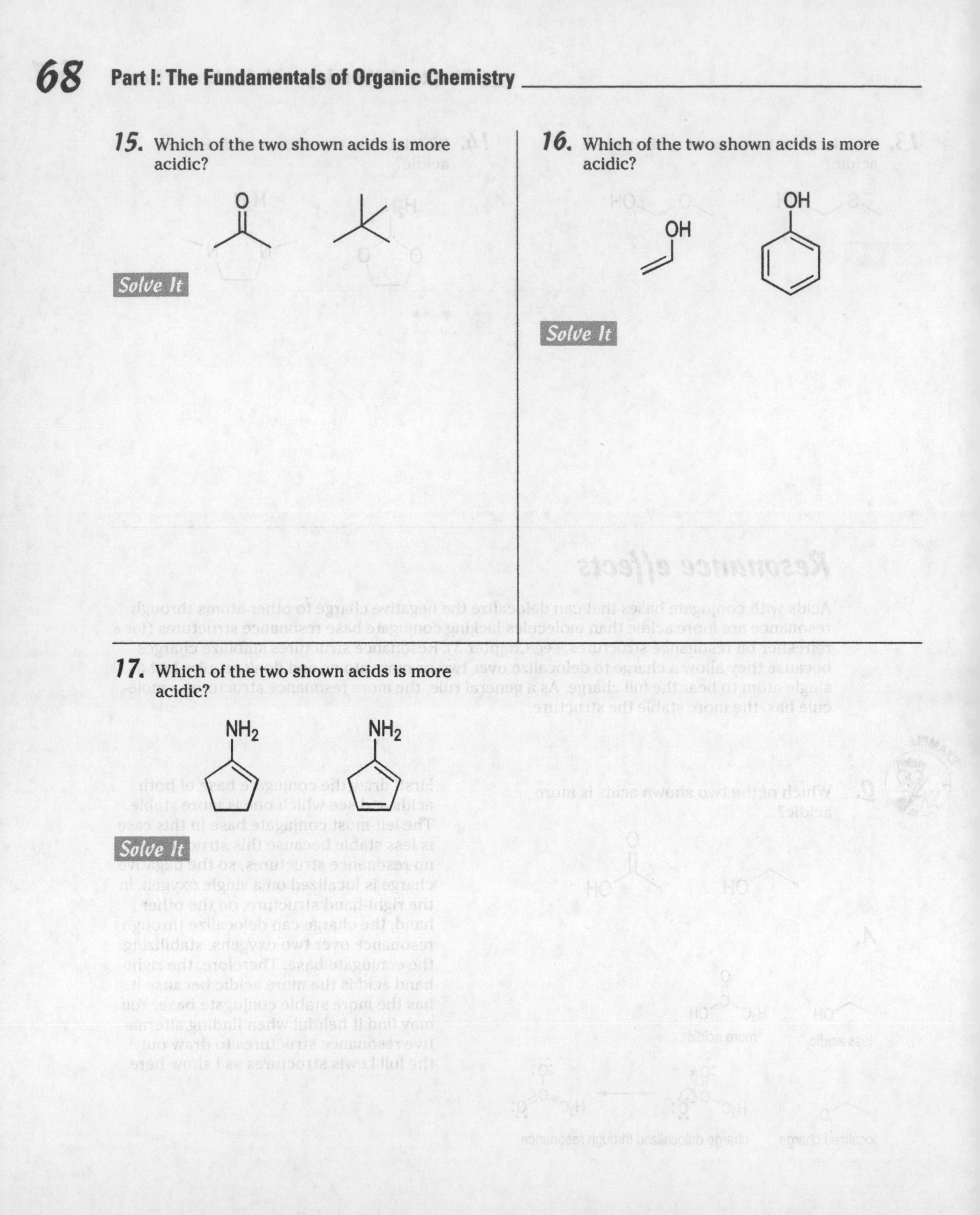

15. **Which of the two shown acids is more acidic?**

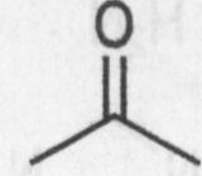

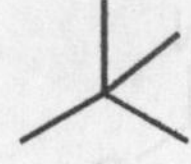

Solve It

16. **Which of the two shown acids is more acidic?**

OH

OH

Solve It

17. **Which of the two shown acids is more acidic?**

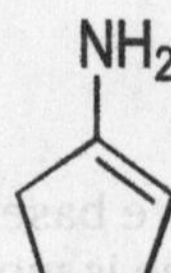

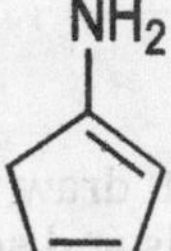

Solve It

Predicting Acid-Base Equilibria Using pKa Values

Other problems in this chapter show the ways that different structural elements affect the acidity of a molecule, but a quantitative scale of the acidity of a molecule is given by the molecule's pKa value. The *pKa value* is a logarithmic measure of acidity based on the acid's equilibrium constant for dissociation in water. The bottom line is that the more acidic a molecule is, the lower its pKa value. In general, a reaction equilibrium favors the side with the lower-energy molecules, and because strong acids and bases are high in energy, acid-base reactions favor the side with the weaker acids and bases.

For any acid-base reaction, if you know the pKa of the acid and the pKa of the conjugate acid, you can determine the direction of the equilibrium. The equilibrium lies in the direction of the side that has the weaker acid (that is, the acid with the higher pKa value).

I provide the pKa values in the following questions where needed, but keep in mind that if you're a student, your professor may require you to memorize a chart of these pKa values. You can find tables of pKa values in *Organic Chemistry I For Dummies* (Wiley) by yours truly or in most any introductory organic chemistry text.

Q. Predict the direction of the equilibrium in the following acid-base reaction.

O, OH (pKa = 5) + OH^- ⇌ O, O^- + H_2O (pKa = 16)

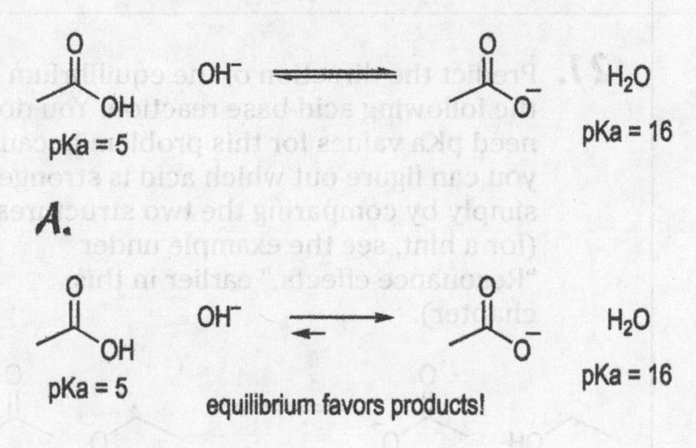

A. The equilibrium favors the products. The pKa of the acid and the conjugate acid are given to you. Because the pKa of the acid in this reaction is lower than the pKa of the conjugate acid, the equilibrium favors the products because the product side has the weaker acid.

O, OH (pKa = 5) + OH^- ⇌ O, O^- + H_2O (pKa = 16)

equilibrium favors products!

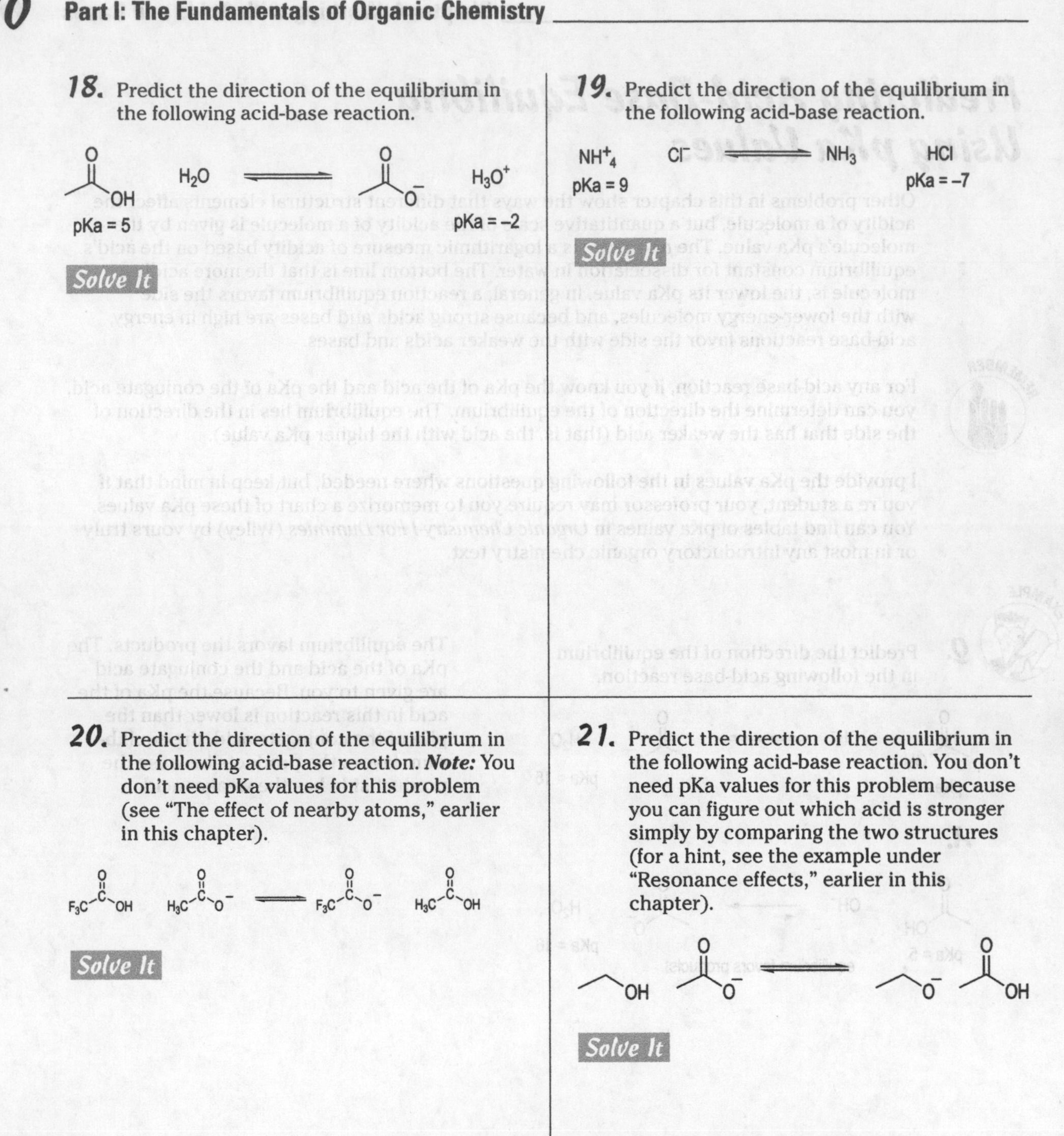

18. Predict the direction of the equilibrium in the following acid-base reaction.

CH_3COOH (pKa = 5) + H_2O ⇌ CH_3COO^- + H_3O^+ (pKa = −2)

Solve It

19. Predict the direction of the equilibrium in the following acid-base reaction.

NH_4^+ (pKa = 9) + Cl^- ⇌ NH_3 + HCl (pKa = −7)

Solve It

20. Predict the direction of the equilibrium in the following acid-base reaction. ***Note:*** You don't need pKa values for this problem (see "The effect of nearby atoms," earlier in this chapter).

$F_3C{-}C(=O){-}OH$ + $H_3C{-}C(=O){-}O^-$ ⇌ $F_3C{-}C(=O){-}O^-$ + $H_3C{-}C(=O){-}OH$

Solve It

21. Predict the direction of the equilibrium in the following acid-base reaction. You don't need pKa values for this problem because you can figure out which acid is stronger simply by comparing the two structures (for a hint, see the example under "Resonance effects," earlier in this chapter).

CH_3CH_2OH + CH_3COO^- ⇌ $CH_3CH_2O^-$ + CH_3COOH

Solve It

Answer Key

The following are the answers to the practice questions presented in this chapter.

1

$H_3C-\ddot{O}-H$ + $H-\ddot{I}:$ → $H_3C-\overset{+}{O}H_2$ + $:\ddot{I}:^-$

base acid conjugate acid conjugate base

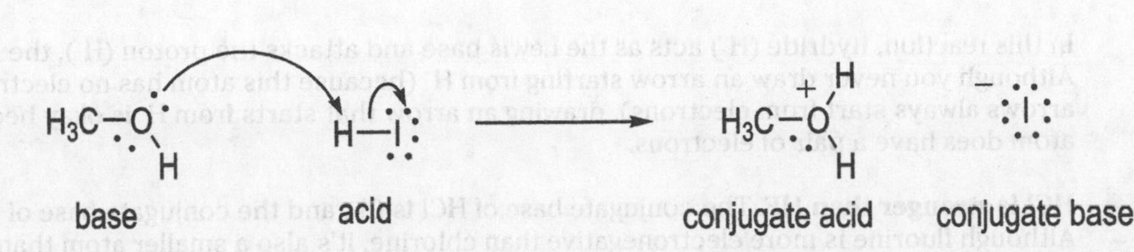

Methanol (CH_3OH) picks up the proton in this reaction, so this molecule is the Bronsted base; HI gives up its proton, so this molecule is the acid. Show the mechanism of this reaction by drawing an arrow from a lone pair on the oxygen and pushing it to the hydrogen on HI. The meaning of this arrow is that a new bond is forming between the oxygen and the hydrogen using the oxygen's lone-pair electrons. Then draw a second arrow from the H-I bond to the iodine to show the breaking of this bond and the reassignment of these electrons as a lone pair on iodine.

2

$(H_3C)_2C=CH_2$ + H_3O^+ → $(H_3C)_2\overset{+}{C}-CH_3$ + H_2O

base acid conjugate acid conjugate base

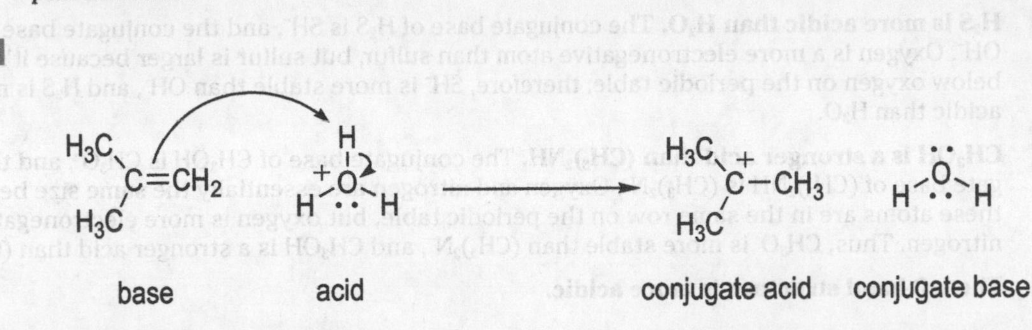

This reaction is slightly different because a double bond rather than a lone pair is protonated. However, the reaction works in the same way. Draw an arrow from the double bond to a hydrogen on the hydronium ion (H_3O^+). Then use a second arrow to show the hydronium ion O-H bond being broken, with the bonding electrons placed onto the oxygen as a new lone pair.

3

$:N\equiv C:^-$ + $H-C(=O)-CH_3$ → $N\equiv C-CH(O^-)-CH_3$

Lewis base Lewis acid

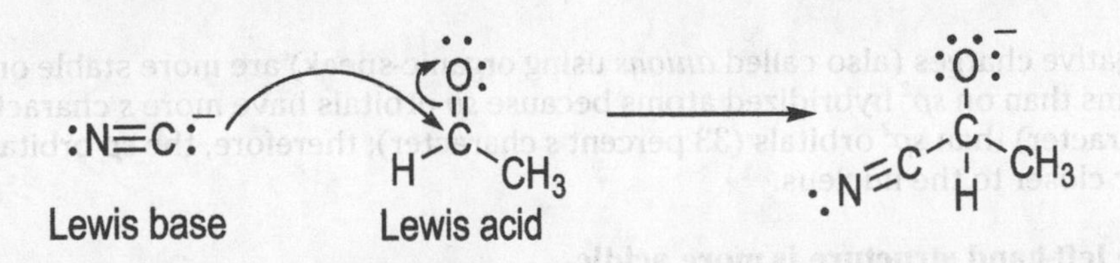

$AlCl_3$ is a powerful Lewis acid because aluminum doesn't have a complete octet of valence electrons. Therefore, a molecule with lone pairs of electrons (a Lewis base) readily adds to the aluminum. Draw an arrow from an oxygen lone pair and push the arrow to the aluminum to show the formation of an O-Al bond. Don't forget to compute the charges when you're done.

H^+ ← :H^- ⟶ H—H

Lewis acid Lewis base

In this reaction, hydride (H^-) acts as the Lewis base and attacks the proton (H^+), the Lewis acid. Although you never draw an arrow starting from H^+ (because this atom has no electrons, and arrows always start from electrons), drawing an arrow that starts from H^- is okay because this atom does have a pair of electrons.

5 **HCl is stronger than HF.** The conjugate base of HCl is Cl^-, and the conjugate base of HF is F^-. Although fluorine is more electronegative than chlorine, it's also a smaller atom than chlorine because chlorine is a row down from fluorine on the periodic table. Size trumps electronegativity, so Cl^- is more stable than F^-, and consequently, HCl is a stronger acid than HF.

6 **NH_3 is a stronger acid than CH_4.** The conjugate base of CH_4 is CH_3^-, and the conjugate base of NH_3 is NH_2^-. Nitrogen and carbon are roughly the same size because both atoms are in the same row of the periodic table, but nitrogen is more electronegative than carbon, so NH_2^- is more stable than CH_3^-. Consequently, NH_3 is a stronger acid than CH_4.

7 **H_2S is more acidic than H_2O.** The conjugate base of H_2S is SH^-, and the conjugate base of H_2O is OH^-. Oxygen is a more electronegative atom than sulfur, but sulfur is larger because it's a row below oxygen on the periodic table; therefore, SH^- is more stable than OH^-, and H_2S is more acidic than H_2O.

8 **CH_3OH is a stronger acid than $(CH_3)_2NH$.** The conjugate base of CH_3OH is CH_3O^-, and the conjugate base of $(CH_3)_2NH$ is $(CH_3)_2N^-$. Oxygen and nitrogen are essentially the same size because these atoms are in the same row on the periodic table, but oxygen is more electronegative than nitrogen. Thus, CH_3O^- is more stable than $(CH_3)_2N^-$, and CH_3OH is a stronger acid than $(CH_3)_2NH$.

9 **The left-hand structure is more acidic.**

H—C≡C:$^-$ (sp) H$_2$C=C$^-$H (sp^2)

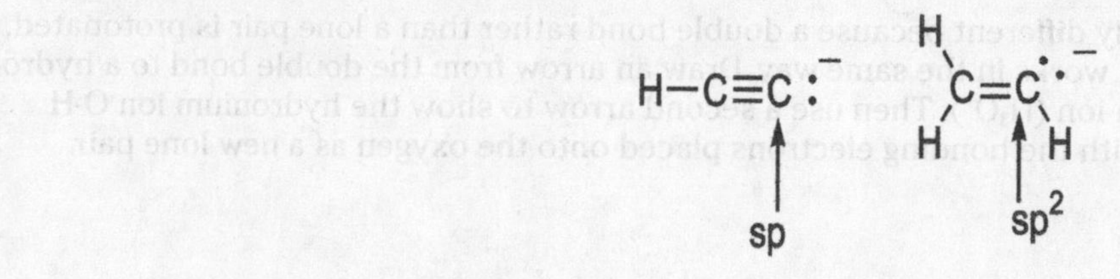

Negative charges (also called *anions* using organic-speak) are more stable on *sp*-hybridized atoms than on sp^2 hybridized atoms because *sp* orbitals have more *s* character (50 percent *s* character) than sp^2 orbitals (33 percent *s* character); therefore, the *sp* orbital places the lone pair closer to the nucleus.

10 **The left-hand structure is more acidic.**

H$_2$C=C$^-$H (sp^2) H$_3$C—C$^-$H$_2$ (sp^3)

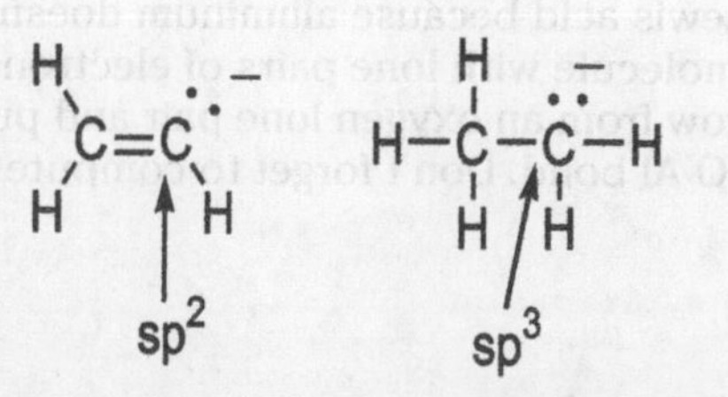

Anions are more stable on sp^2 hybridized atoms than on sp^3 hybridized atoms because sp^2 hybrid orbitals have more *s* character (33 percent *s* character) than sp^3 hybrid orbitals (25 percent *s* character).

11

Cl–CH2–C(=O)OH more acidic   Br–CH2–C(=O)OH less acidic

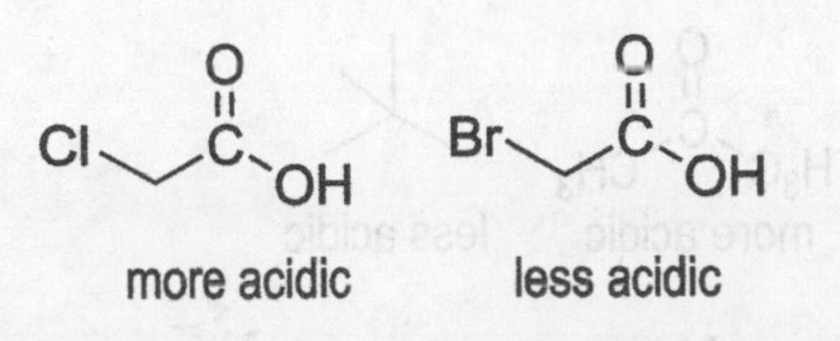

Chlorine is more electronegative than bromine, so the conjugate base on the molecule containing the chlorine is more stable because more of the negative charge can be pulled away and delocalized than that of the conjugate base with the bromine. Electronegativity increases as you go up and to the right on the periodic table.

12

less acidic   more acidic

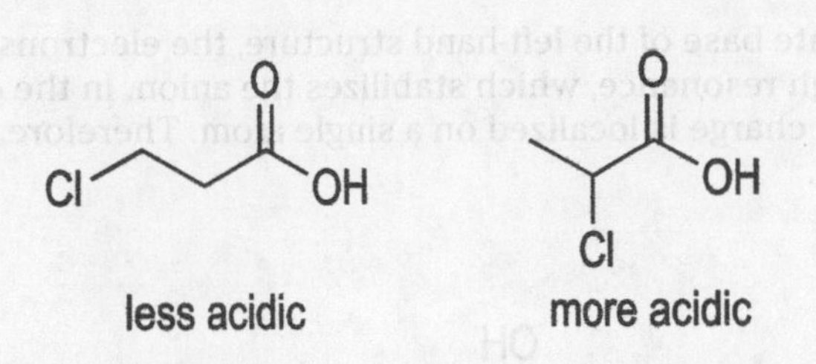

The difference between these two molecules is that the chlorine is closer to the negative charge in the conjugate base of the right structure than on the left. The influence of electronegative atoms diminishes the farther they're away from the acidic proton. Thus, the right-hand molecule with the closer chlorine is more acidic than the left-hand molecule.

13

less acidic   more acidic

Oxygen is more electronegative than sulfur, so oxygen is able to pull more of the electrons from the conjugate base to stabilize it. Keep in mind that the size of the neighboring atoms is unimportant in terms of affecting the acidity: Size of the atom matters only when the charge is on that particular atom. Therefore, the fact that sulfur is larger than oxygen isn't relevant here — only the fact that oxygen is more electronegative than sulfur.

14

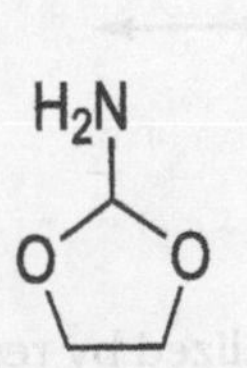

more acidic   less acidic

Oxygen is more electronegative than nitrogen and is able to pull more of the negative charge away from the conjugate base anion (negative charge) than nitrogen. Therefore, the left structure is more acidic than the right one.

15

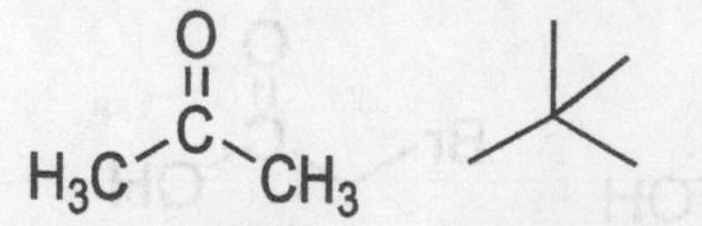

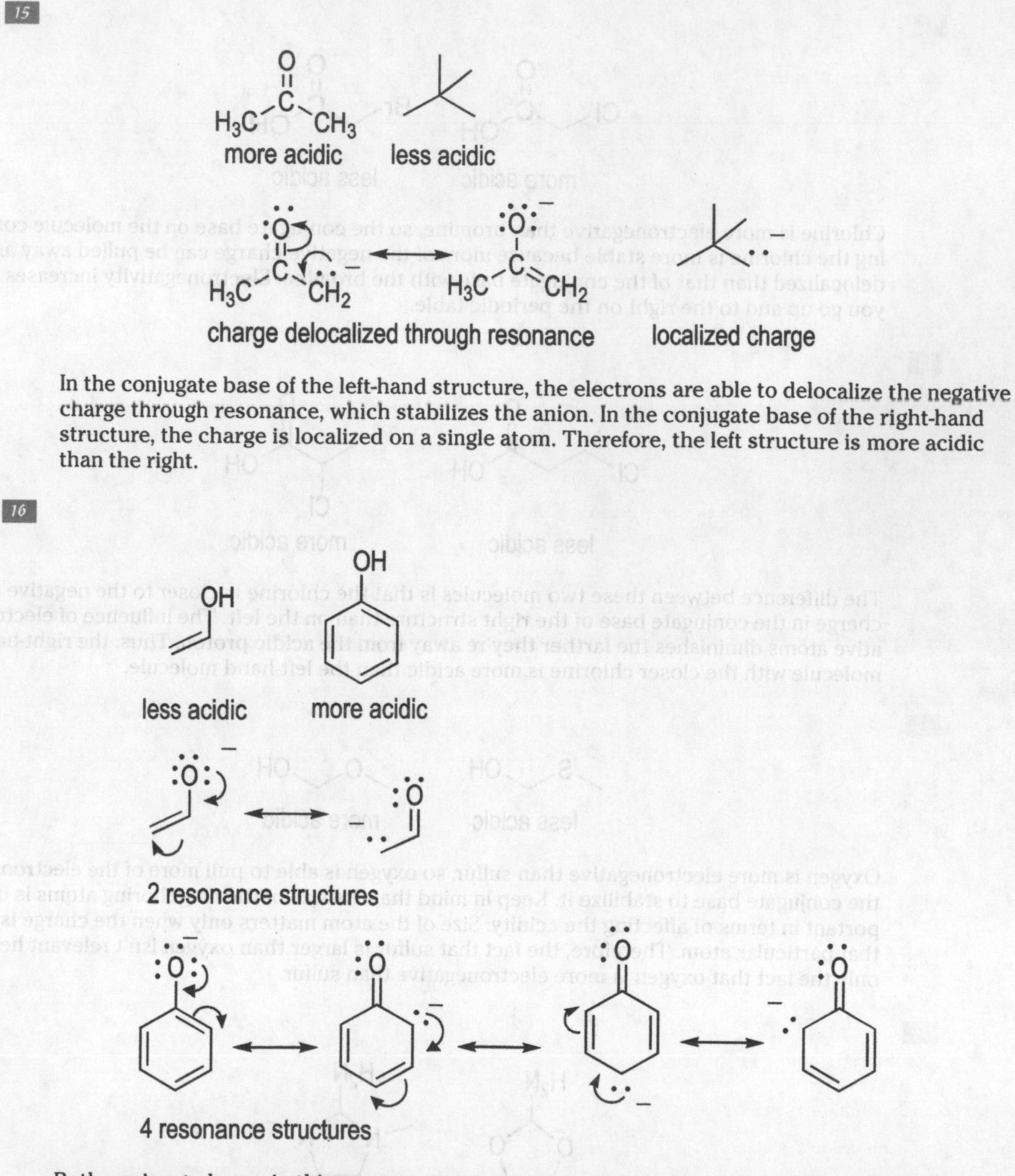

more acidic   less acidic

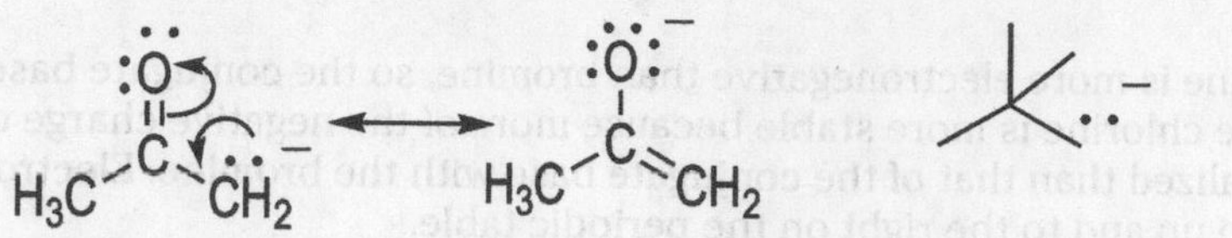

charge delocalized through resonance   localized charge

In the conjugate base of the left-hand structure, the electrons are able to delocalize the negative charge through resonance, which stabilizes the anion. In the conjugate base of the right-hand structure, the charge is localized on a single atom. Therefore, the left structure is more acidic than the right.

16

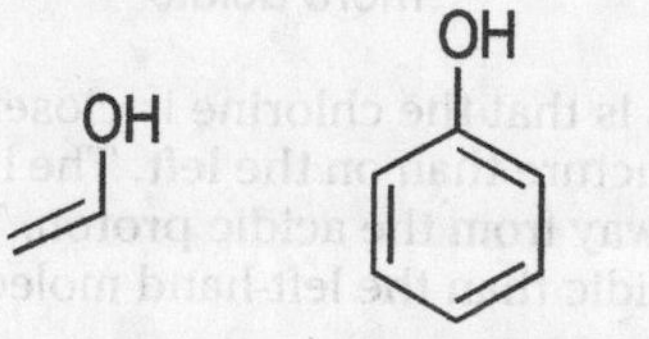

less acidic   more acidic

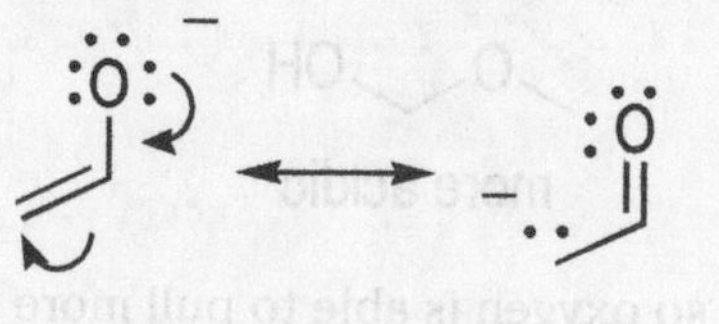

2 resonance structures

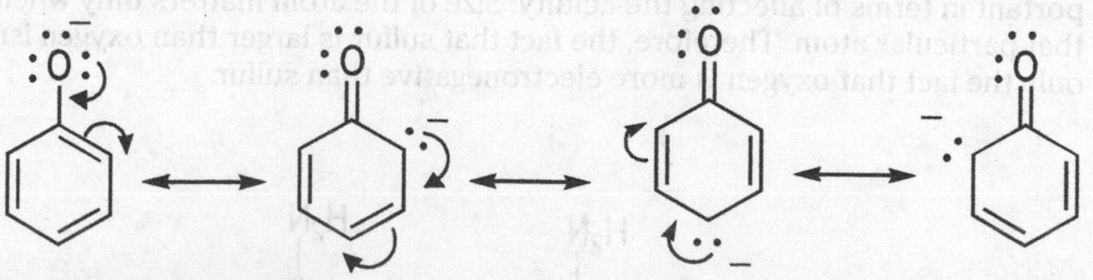

4 resonance structures

Both conjugate bases in this case are stabilized by resonance. However, the base of the left structure has only two resonance structures, whereas the base of the right-hand structure has four resonance structures. Typically, the more resonance structures a molecule or ion has, the more stable it is. Therefore, the right structure is more acidic than the left.

17

NH2 NH2

less acidic more acidic

2 resonance structures

3 resonance structures

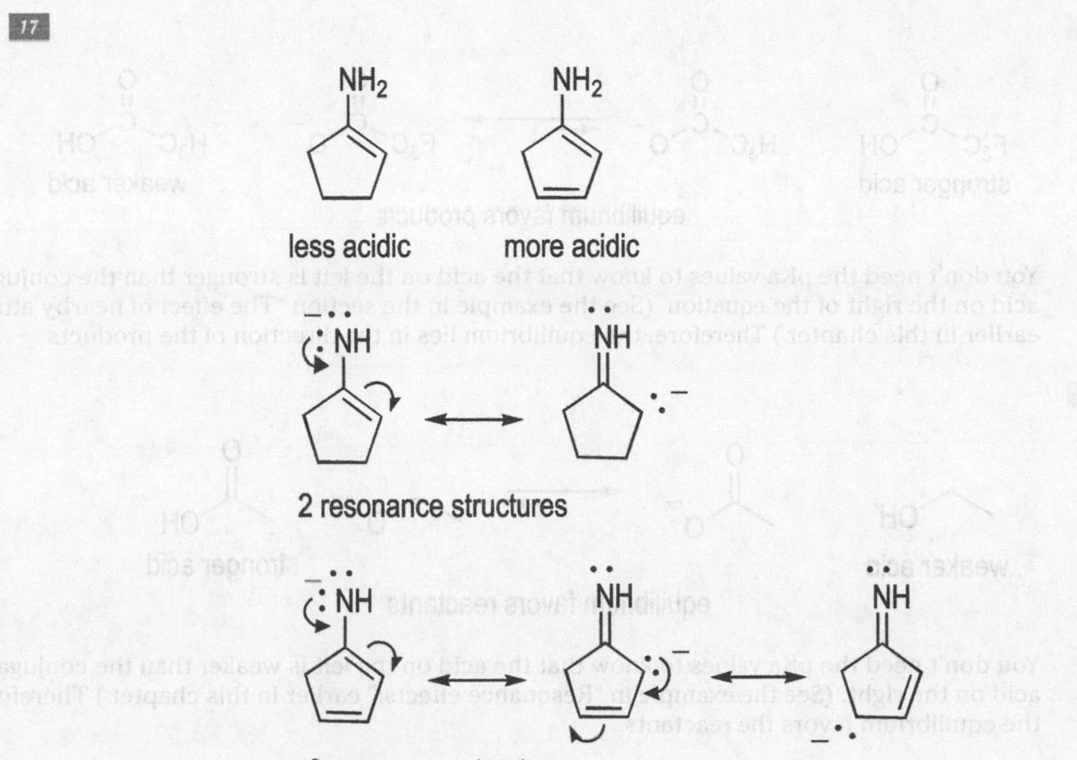

In this case, resonance stabilizes both conjugate bases. However, the left structure has only two resonance structures, whereas the right-hand structure has three. Because the more resonance structures a molecule has, the more stable it is, the right structure is more acidic than the left.

18

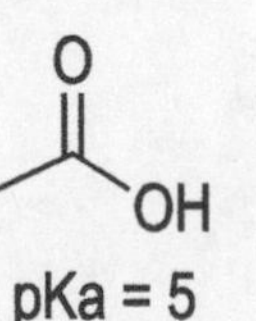

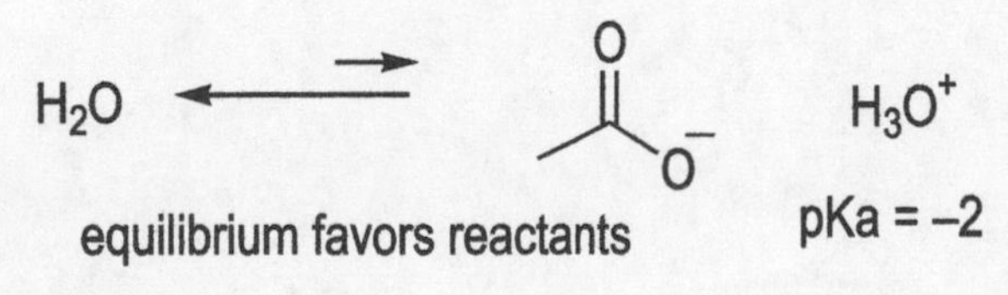

OH H_2O H_3O^+

pKa = 5 equilibrium favors reactants pKa = –2

The equilibrium favors the reactants because acetic acid has a higher pKa (5) than H_3O^+ (–2).

19

NH_4^+ Cl^- NH_3 HCl

pKa = 9 pKa = –7

equilibrium favors reactants

The equilibrium favors the reactants because ammonium (NH_4^+) has a higher pKa (9) than HCl (–7).

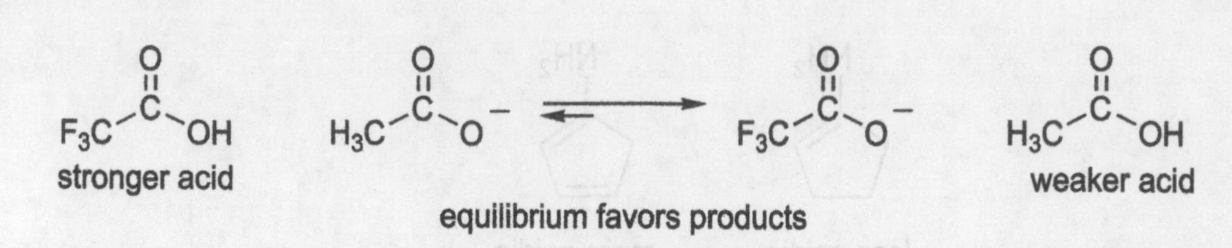

$$F_3C{-}COOH + H_3C{-}COO^- \rightleftharpoons F_3C{-}COO^- + H_3C{-}COOH$$

stronger acid (left) — weaker acid (right)

equilibrium favors products

You don't need the pKa values to know that the acid on the left is stronger than the conjugate acid on the right of the equation. (See the example in the section "The effect of nearby atoms," earlier in this chapter.) Therefore, the equilibrium lies in the direction of the products.

21

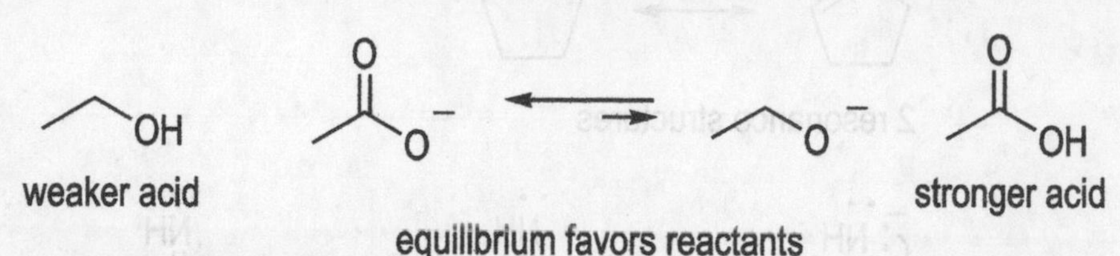

$$CH_3CH_2OH + CH_3COO^- \rightleftharpoons CH_3CH_2O^- + CH_3COOH$$

weaker acid (left) — stronger acid (right)

equilibrium favors reactants

You don't need the pKa values to know that the acid on the left is weaker than the conjugate acid on the right. (See the example in "Resonance effects," earlier in this chapter.) Therefore, the equilibrium favors the reactants.

Part II

The Bones of Organic Molecules: The Hydrocarbons

"I know it's a placebo effect, but they seem to help me work through the 3-D stereochemistry problems better than if I didn't wear them."

In this part . . .

In this part, you get down to business by working with the simplest organic structures, those molecules consisting of just hydrogen and carbon atoms, better known as *hydrocarbons.* These molecules form the bones of organic structures, so this part lays the brickwork that forms the foundation of your knowledge for the rest of the course. Here, you discover how to name hydrocarbons, study their 3-D properties, and get a close look at some introductory reactions with the reactions of *alkenes,* molecules containing carbon-carbon double bonds.

Chapter 5

Seeing Molecules in 3-D: Stereochemistry

In This Chapter

- Recognizing chiral centers and substituent priorities
- Configuring R & S stereochemistry
- Using Fischer projections
- Examining the relationship between stereoisomers and meso compounds

Think of your hands. Both have fingers connected in the same way, but your right and left hands aren't identical. They're mirror images of each other, and you can't superimpose one hand upon the other. In a similar way, two molecules can have the same connections of atoms but differ from each other by the orientation of those atoms in three-dimensional space. Such molecules are said to have different *stereochemistry,* or differing spatial arrangements of atoms.

The study of stereochemistry is important for understanding the molecular basis of biology. Stereochemistry has far-reaching consequences because nature often treats two molecules with different stereochemistry quite differently. For example, a molecule with one stereochemistry may smell like mint, while the same molecule with a different stereochemistry smells like cinnamon; the first molecule may treat a disease, while the other is poisonous.

So how does stereochemistry manifest itself in molecules? Every carbon that has four different attachments to it (called a *chiral center*) can have two different configurations, much like a hand has two possible configurations, right or left. But organic chemists have a thing for Latin, so they use the labels *R* and *S*, which stand for *rectus* (right) and *sinister* (left). The goal of this chapter is to familiarize you with working with molecules of differing stereochemistry. Part of that involves being able to assign whether a chiral center is of R or S configuration and also to distinguish the relationships between molecules with different stereochemistry, called *stereoisomers.* Also on the menu is working with a handy tool called a Fischer projection, which aids in quickly comparing the stereochemistry of molecules.

Identifying Chiral Centers and Assigning Substituent Priorities

Because stereochemistry can be conceptually challenging, I take a gradual approach in this chapter to assigning the configuration of chiral centers as R or S. The basic procedure for assigning the configuration of a chiral center is as follows:

1. **Locate a chiral center.**

   A *chiral center* is a carbon atom with four different attached atoms or groups.

   Often, spotting chiral centers by the process of elimination is easy: You just ignore atoms that *can't* be chiral centers. For example, all methyl groups (CH_3 groups), methylene groups (CH_2 groups), or carbons involved in a double or triple bond can't be chiral centers because these atoms can't have four different groups attached to them.

2. **Assign the priorities of each of the four groups attached to the chiral center using the Cahn-Ingold-Prelog rules.**

   Prioritize the groups from 1 to 4 based on the atomic number of the first atom attached to the chiral center. The highest priority (1) goes to the group whose first atom has the highest atomic number; the lowest priority (4) goes to the group whose first atom has the lowest atomic number (usually hydrogen).

   In the event of a tie — that is, two groups attach with the same atom type — continue down the chain of both groups until the tie is broken. The tie can break in two ways:

   - **You reach a higher-priority atom on one of the chains, giving that group the higher priority.** Thus, a carbon attached to an oxygen has priority over a carbon attached to another carbon because oxygen has a higher atomic number than carbon.
   - **You find that a larger number of identical atoms are attached to one group over another.** Thus, a carbon attached to two other carbons has a higher priority than a carbon attached to just one other carbon.

   However, atom type trumps the number of groups, so a carbon attached to an oxygen has higher priority than a carbon attached to two or three other carbons. Figure 5-1 shows some examples.

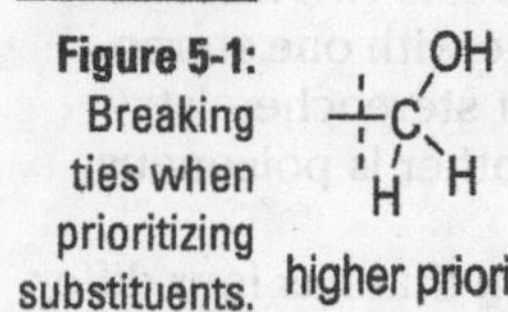

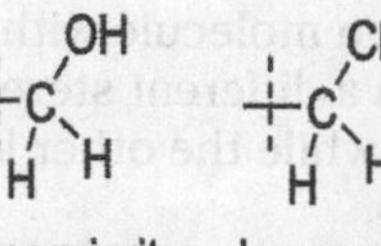

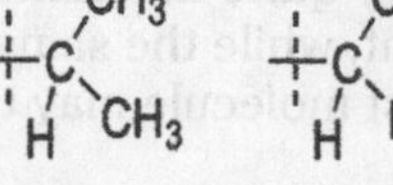

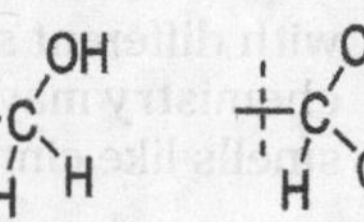

Figure 5-1: Breaking ties when prioritizing substituents.

   Double and triple bonds are treated somewhat strangely because they're expanded into an equivalent number of single bonds for each multiple bond. For example, you treat a carbon-nitrogen triple bond as though the carbon were single-bonded to three nitrogens, with each of the nitrogens in turn bonded to a carbon. You treat a carbon double-bonded to nitrogen as though the carbon were two single bonds to nitrogen, each of which would be bonded back to the carbon. The effect is that for identical atom types, triple bonds have higher priority than double bonds, and double bonds have higher priority than single bonds. Figure 5-2 shows an example.

3. **Use a procedure for assigning the R or S designation to that chiral center using the group priorities you assigned in the second step.**

   See the next section for more specifics.

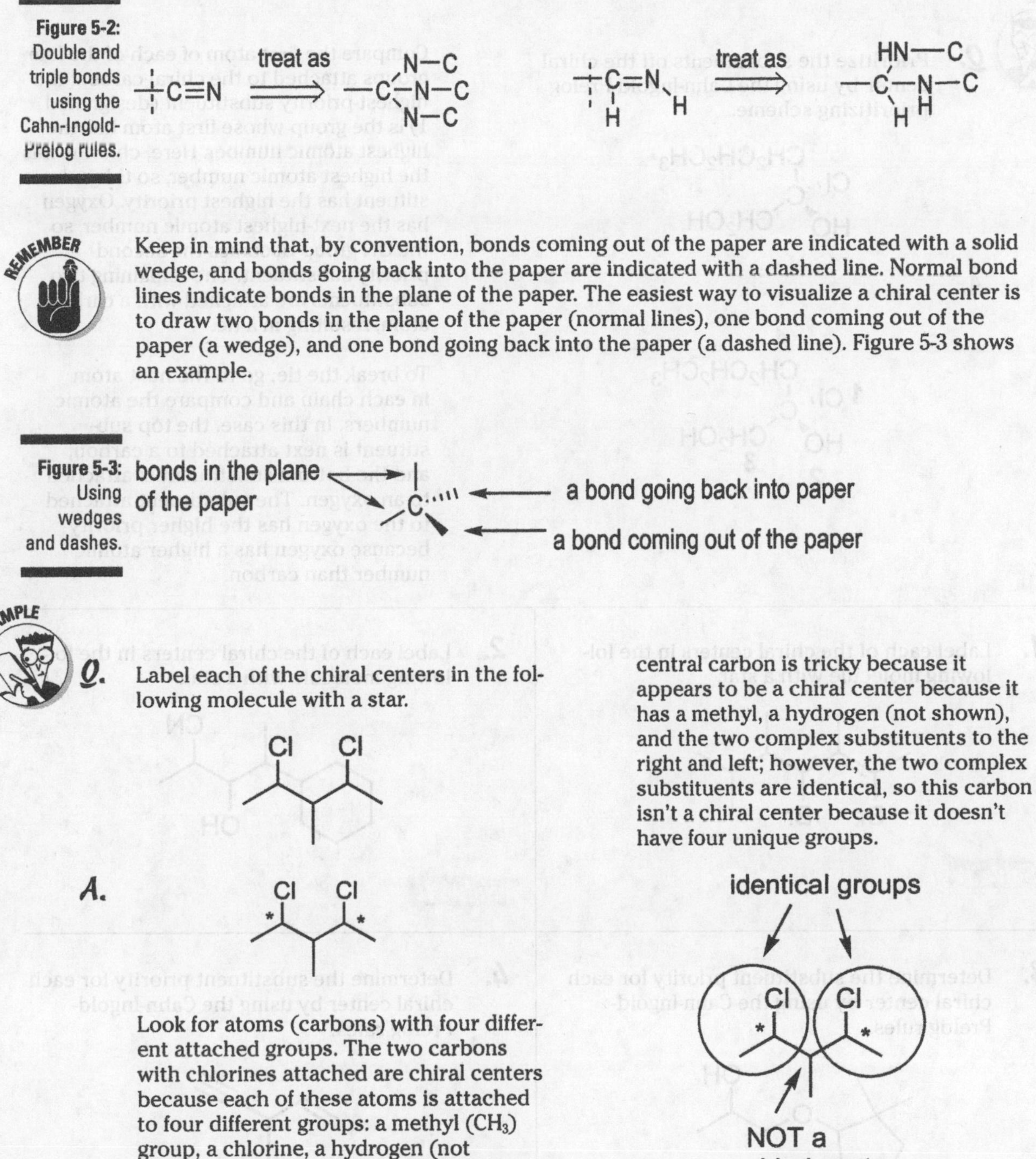

Figure 5-2: Double and triple bonds using the Cahn-Ingold-Prelog rules.

Keep in mind that, by convention, bonds coming out of the paper are indicated with a solid wedge, and bonds going back into the paper are indicated with a dashed line. Normal bond lines indicate bonds in the plane of the paper. The easiest way to visualize a chiral center is to draw two bonds in the plane of the paper (normal lines), one bond coming out of the paper (a wedge), and one bond going back into the paper (a dashed line). Figure 5-3 shows an example.

Figure 5-3: Using wedges and dashes.

Q. Label each of the chiral centers in the following molecule with a star.

A. Look for atoms (carbons) with four different attached groups. The two carbons with chlorines attached are chiral centers because each of these atoms is attached to four different groups: a methyl (CH_3) group, a chlorine, a hydrogen (not shown), and a complex substituent. The central carbon is tricky because it appears to be a chiral center because it has a methyl, a hydrogen (not shown), and the two complex substituents to the right and left; however, the two complex substituents are identical, so this carbon isn't a chiral center because it doesn't have four unique groups.

Q. Prioritize the substituents off the chiral center by using the Cahn-Ingold-Prelog prioritizing scheme.

$CH_2CH_2CH_3$, Cl, HO, CH_2OH on C

A.

4 $CH_2CH_2CH_3$, 1 Cl, 2 HO, 3 CH_2OH on C

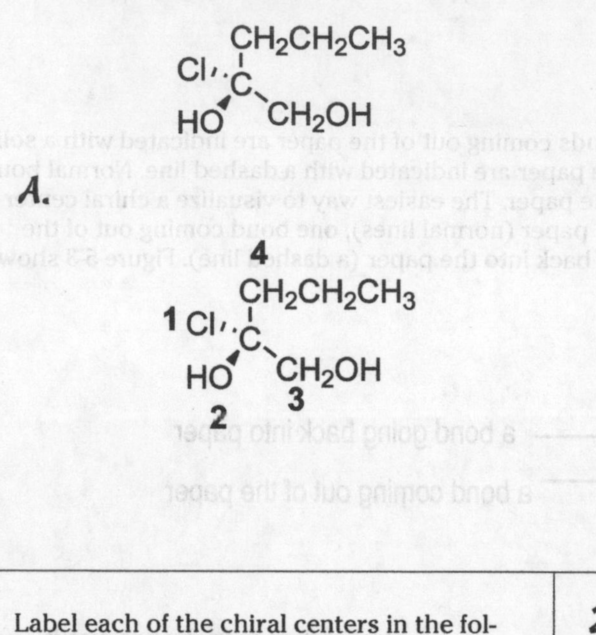

Compare the first atom of each of the four groups attached to the chiral carbon; the highest-priority substituent (designated 1) is the group whose first atom has the highest atomic number. Here, chlorine has the highest atomic number, so this substituent has the highest priority. Oxygen has the next-highest atomic number, so the OH group becomes the second-priority substituent. The remaining two substituents are attached with a carbon atom, resulting in a tie.

To break the tie, go to the next atom in each chain and compare the atomic numbers. In this case, the top substituent is next attached to a carbon, and the bottom substituent is attached to an oxygen. The substituent attached to the oxygen has the higher priority because oxygen has a higher atomic number than carbon.

1. Label each of the chiral centers in the following molecule with a star.

Br Br

Solve It

2. Label each of the chiral centers in the following molecule with a star.

CN OH

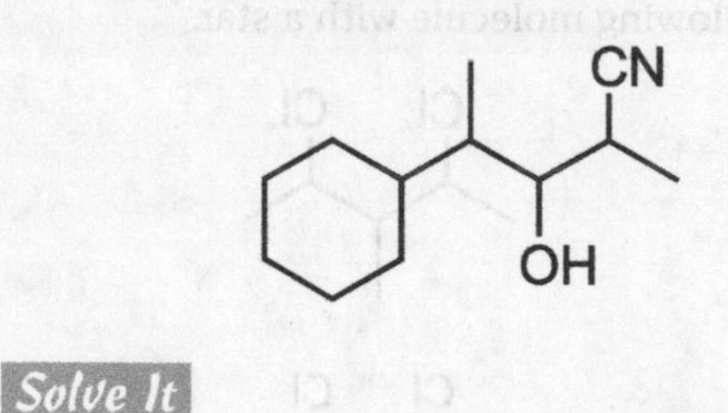

Solve It

3. Determine the substituent priority for each chiral center by using the Cahn-Ingold-Prelog rules.

OH O

Solve It

4. Determine the substituent priority for each chiral center by using the Cahn-Ingold-Prelog rules.

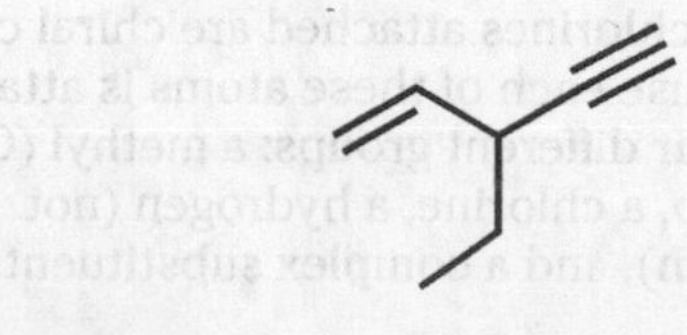

Solve It

Assigning R & S Configurations to Chiral Centers

After you spot a chiral center in a molecule and assign priority of each substituent from 1 to 4 (see the preceding section), you can determine the configuration of that chiral center as R or S in two steps:

1. **Rotate the molecule so the fourth-priority substituent is in the back (on a dashed bond sticking back into the paper).**

   If you have difficulty rotating the molecule in 3-D, use a molecular modeling kit (those plastic balls and sticks that represent atoms and bonds) until you get the feel of doing this kind of transformation.

2. **Draw a curve from the first- to second- to third-priority substituent.**

   If the curve is clockwise, the configuration is designated *R;* if the curve is counterclockwise, the configuration is *S. **Tip:*** Remember this step in terms of driving and picture the curve as a steering wheel: If you steer clockwise, the car goes right (R); turn the wheel counterclockwise, and the car goes left (S, which stands for *sinister*, or *left,* in Latin).

If you have difficulty visualizing in 3-D — or want to save some time on an exam — I have a trick you can use: Swapping any two groups inverts the stereochemistry (it makes the chiral center go from R to S or S to R). Therefore, if you swap two groups to put the fourth-priority substituent in the back, you can then continue in assigning the stereochemistry — so long as you remember that you'll get the *opposite* stereochemistry of the one you're trying to determine!

Q. Assign the chiral center in the following molecule as R or S.

Cl
H''C
F Br

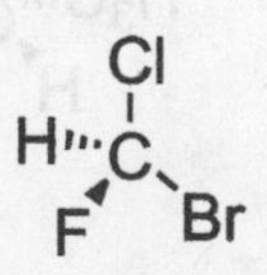

A. **The chiral center is S.** Bromine has the highest priority because this atom has the highest atomic number; chlorine is second, fluorine is third, and hydrogen (as always) is last, in order of descending atomic number.

4 Cl 2
H''C
F Br 1
3

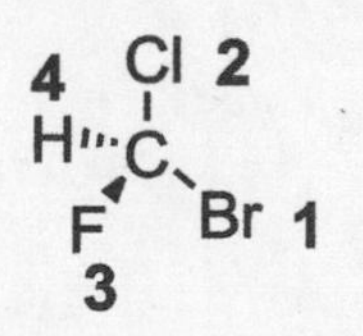

The next step is to point the fourth-priority substituent to the back. In this case, the hydrogen (four) is already in the back (it points back into the paper with a dashed bond). Draw an arrow connecting the first- to second- to third-priority substituents.

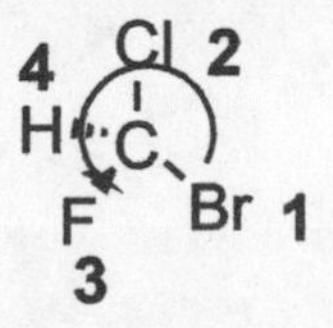

4 Cl 2
H''C
F Br 1
3

The curve is counterclockwise, so the configuration of this chiral center is S.

Q. Assign the chiral center in the following molecule as R or S.

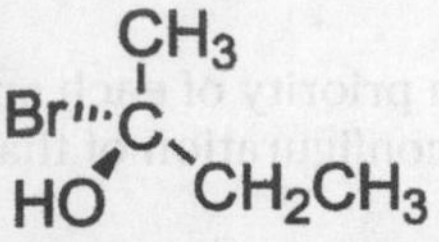

A. The chiral center is R. Prioritize the substituents on the chiral center. Bromine is the highest-priority substituent because it has the highest atomic number, and the OH is second because oxygen has a higher atomic number than carbon. Finally, the CH_2CH_3 group has higher priority than CH_3 because in the former group, the first carbon is attached to an additional carbon, breaking the tie.

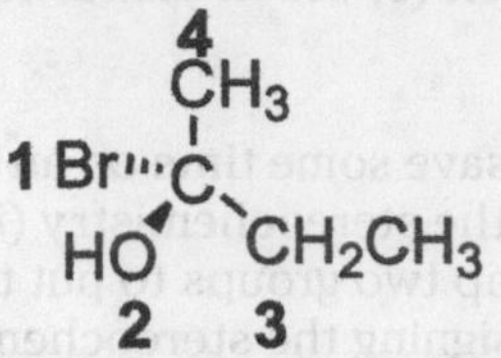

Rotate the molecule so the fourth-priority substituent (CH_3) sticks in the back. This step is the most difficult because it requires you to visualize the molecule in three dimensions.

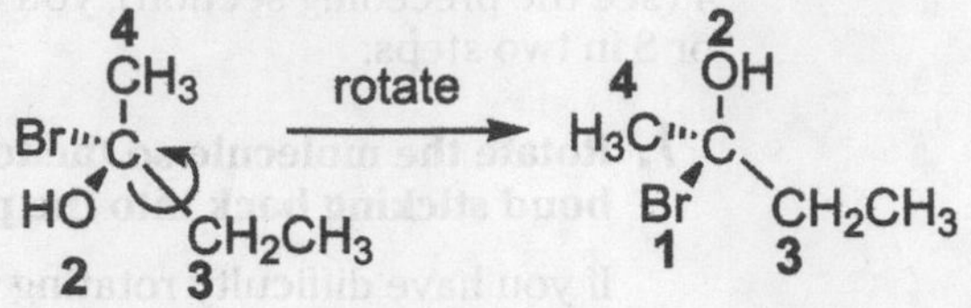

Finally, draw the curve from the first- to second- to the third-priority substituent. This curve is clockwise, so the configuration for this chiral center is R.

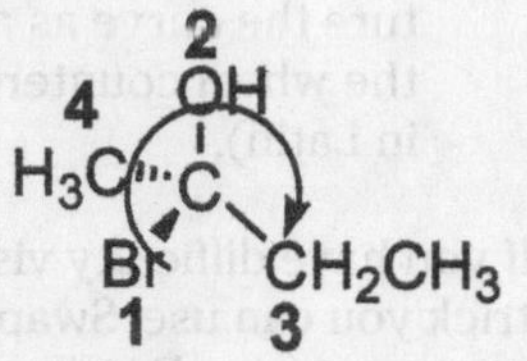

5. Identify any chiral centers in the following molecule and then label each chiral center as R or S.

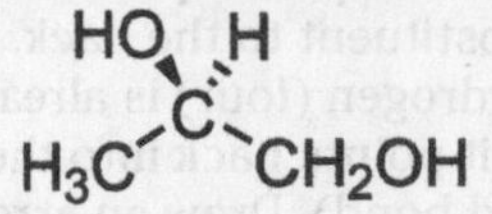

Solve It

6. Identify any chiral centers in the following molecule and the label each chiral center as R or S.

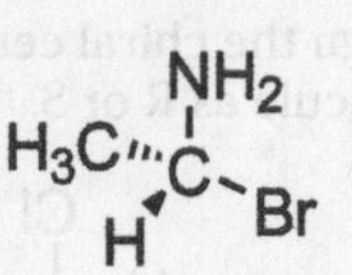

Solve It

7. Identify any chiral centers in the following molecule and then label each chiral center as R or S.

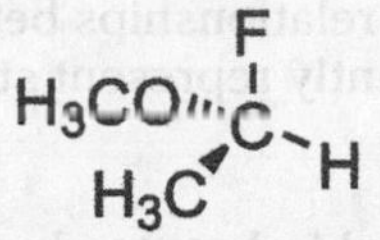

Solve It

8. Identify any chiral centers in the following molecule and then label each chiral center as R or S.

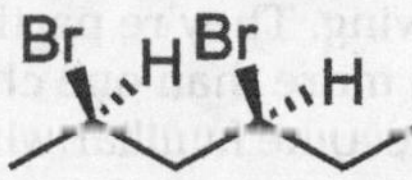

Solve It

9. Identify any chiral centers in the following molecule and then label each chiral center as R or S.

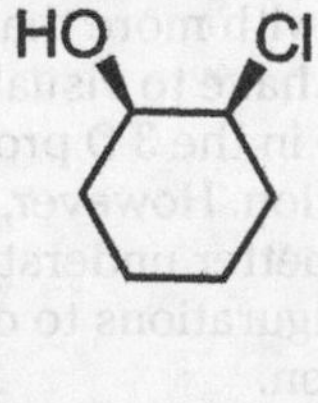

Solve It

10. Identify any chiral centers in the following molecule and then label each chiral center as R or S.

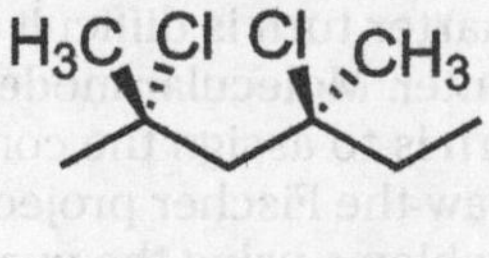

Solve It

Working with Fischer Projections

Fischer projections are a handy drawing method for representing 3-D chiral centers in a 2-D drawing. They're particularly convenient for comparing the relationships between molecules with more than one chiral center. Fischer projections frequently represent stereochemistry, so become familiar with these projections.

By convention, a cross on a Fischer projection represents a chiral center, but don't confuse a Fischer projection with a Lewis structure. Even though Fischer projections look flat, by convention, you assume that the horizontal bars indicate bonds coming out of the plane of the paper (like wedged bonds) and the vertical bars indicate bonds going back into the paper (like dashed bonds).

Figure 5-4 shows how to convert a 3-D projection (wedge and dash structure) into a Fischer projection. You rotate a quarter turn around one of the bonds so that two of the bonds on the chiral center are coming out of the plane of the paper and two bonds are going back into the paper. The bonds coming out of the plane become the horizontal bars in the Fischer projection, and the bonds going back into the plane become the vertical bars.

CH_3, Cl, H, CH_2CH_3 — 3-D projection — quarter turn → Cl—C(H)—: CH_3 / Cl —— H / CH_2CH_3 = Fischer projection

Figure 5-4: Converting a 3-D representation to a Fischer projection.

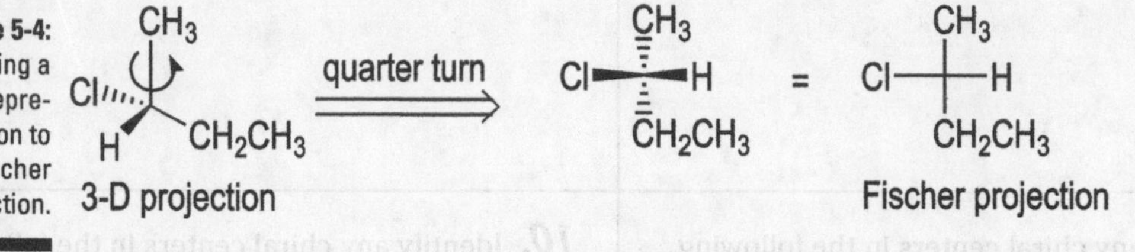

Many people have trouble converting 3-D projections into Fischer projections because the quarter turn is difficult to visualize, particularly in molecules with more than one chiral center. Molecular models can help, but one trick so you don't have to visualize the quarter turn is to assign the configuration of the chiral center (R or S) in the 3-D projection and then draw the Fischer projection so that it has the same configuration. However, trying a few problems using the quarter-turn method can help give you a better understanding of what the Fischer projection means. You can always assign the configurations to double-check your work, ensuring you didn't make mistakes during the conversion.

Here's how to determine the R or S stereochemistry of a chiral center on a Fischer projection:

1. **Assign the priorities to each of the substituents using the Cahn-Ingold-Prelog rules.**

   See "Identifying Chiral Centers and Assigning Substituent Priorities," earlier in this chapter, for the rules.

2. **Draw the curve from the first- to second- to third-priority substituents.**

   If the fourth-priority substituent is on a vertical line (either at 12 o'clock or 6 o'clock), a clockwise curve indicates an R configuration and a counterclockwise curve indicates an S configuration.

   However, if the fourth-priority substituent is on a horizontal bar (at 3 o'clock or 9 o'clock), just the opposite is true: A clockwise curve indicates an S configuration, and a counterclockwise curve indicates an R configuration.

Q. Assign R or S stereochemistry to the chiral center in the following Fischer projection.

H
Br—+—CH_3
OCH_3

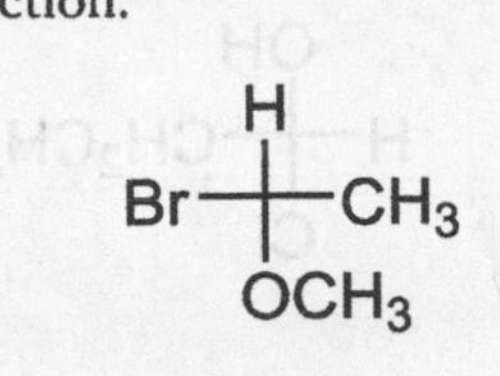

A. **The chiral center is S.** The first thing to do is to assign the priorities to each of the substituents as usual.

4 H
1 Br—+—$\overset{3}{C}H_3$
OCH_3
2

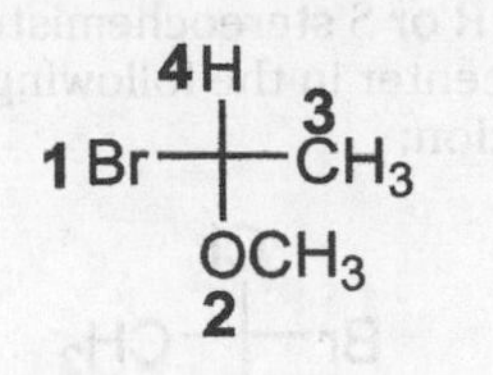

In this case, the fourth-priority substituent is on a vertical bar (at 12 o'clock) and the curve is counterclockwise, so the configuration is S.

4 H
1 Br—+—$\overset{3}{C}H_3$
OCH_3
2

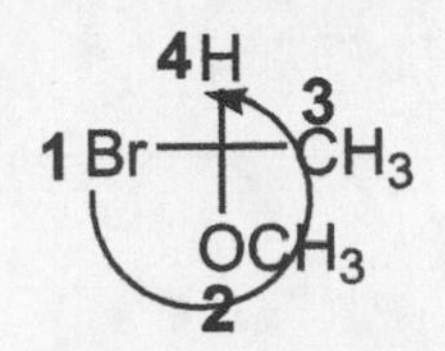

Q. Convert the following molecule into its Fischer projection.

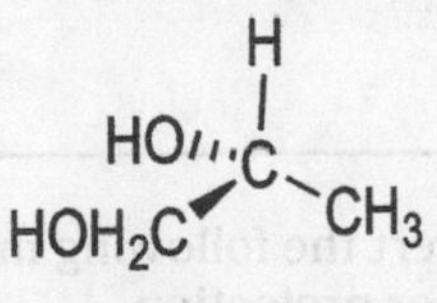

H, HO, C, HOH_2C, CH_3

A.

H
HO—+—CH_2OH
CH_3

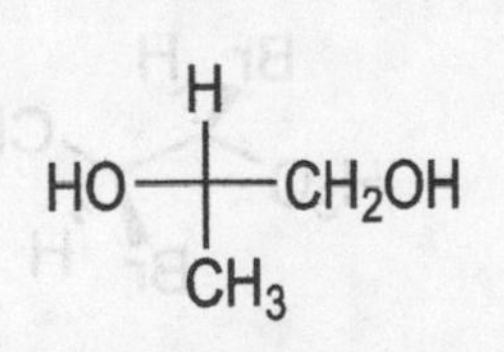

First, perform the quarter turn around one of the bonds so that two horizontal bonds are coming out of the plane of the paper and two vertical bonds are going back into the paper. This becomes the Fischer projection.

H, HO, C, HOH_2C, CH_3 —quarter turn→

HO—C—CH_2OH (H above, CH_3 below) = HO—+—CH_2OH (H above, CH_3 below)

Fischer projection

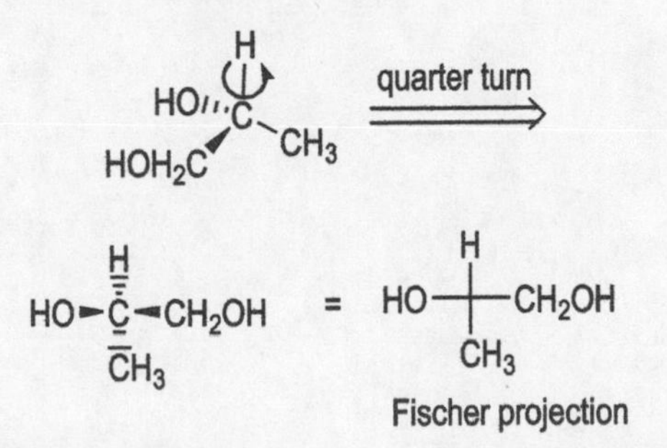

Keep in mind that your primary goal when performing this conversion into a Fischer projection is to make sure that the configuration of the chiral center stays the same. Thus, you can double-check your answer by assigning the configuration to the 3-D projection and the Fischer projection to make sure they're the same. In this case, they're both R, so you know you transferred the stereochemistry successfully.

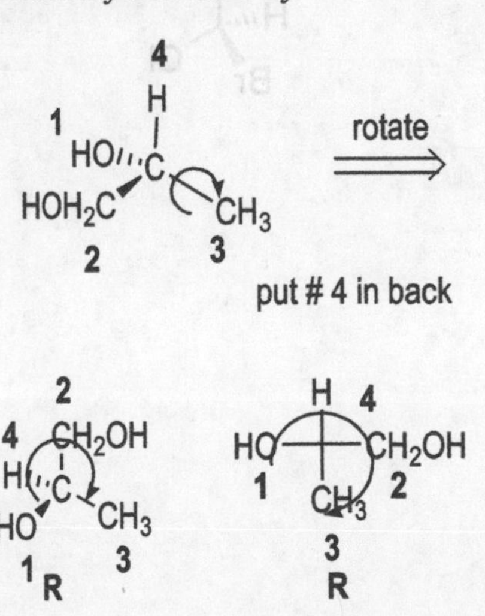

4 H, 1 HO, C, 2 HOH_2C, 3 CH_3 —rotate→

put # 4 in back

2 CH_2OH, 4 H, C, 1 HO, 3 CH_3 — R

H 4, HO 1, CH_2OH 2, CH_3 3 — R

11. Assign R or S stereochemistry to each chiral center in the following Fischer projection:

Cl
Br—|—CH3
H

Solve It

12. Assign R or S stereochemistry to each chiral center in the following Fischer projection:

OH
H—|—CH2CH3
Cl

Solve It

13. Convert the following molecule into its Fischer projection.

F
H
Cl
Br

Solve It

14. Convert the following molecule into its Fischer projection.

Br H
H3C CH3
Br H

Solve It

Comparing Relationships between Stereoisomers and Meso Compounds

Stereoisomers are molecules that have the same atom connectivity but differ in the configuration of their chiral centers. Just as the relationship of a right hand to a left hand is that of mirror images, a molecule with a chiral center of R configuration is the mirror image of the same molecule with the S configuration. Chemists call molecules that are mirror images of each other *enantiomers,* so in orgo-speak, your right hand is the enantiomer of your left.

The problem with this analogy is that molecules can have more than one chiral center. Now imagine an arm with two left hands attached (weird, I know). The mirror image of this arm (its enantiomer) is an arm with two right hands attached. Similarly, the enantiomer of a molecule with two R chiral centers is a molecule with two S chiral centers. The bottom line is that the enantiomer of a molecule is a molecule with all the configurations switched (all R configurations go to S, and all S configurations go to R).

When you have more than one chiral center on a molecule, getting stereoisomers that aren't mirror images is possible. If you were to have an arm with two right hands and another arm with one right hand and one left hand, the two arms would have a different relationship than that of mirror images. Stereoisomers that aren't mirror images of each other are in a *diastereomeric* relationship. One example of a pair of diastereomers is one molecule that has two R chiral centers and a molecule with one R and one S chiral center. In other words, a *diastereomer* is a stereoisomer that doesn't have each and every chiral center switched and is not a mirror image.

With one exception, any molecule that has a chiral center is a chiral molecule. A *chiral molecule* is a molecule that has a non-superimposable mirror image. What that means is that you can't superimpose the mirror image of the molecule onto the original molecule. A molecule with an R chiral center isn't superimposable on its mirror image, the enantiomer with an S chiral center — the atoms simply don't align, much like the fingers of your right hand don't align with the fingers on your left hand.

Molecules that have chiral centers are chiral molecules — with one exception. *Meso compounds* are molecules that have chiral centers but are *achiral* (not chiral) as a result of having a plane of symmetry in the molecule. A *plane of symmetry* is an imaginary line that you can draw in a molecule for which both halves are mirror images of each other. Unlike chiral molecules, meso compounds have mirror images that are superimposable on the original molecule (that is, the mirror images are identical to the original molecule). Thus, being able to spot planes of symmetry in molecules that have chiral centers is important for determining whether the molecule is chiral or achiral.

Watch for rotations around single bonds. Anytime you can rotate around a single bond to give a structure that has a plane of symmetry — in other words, when any conformation has a plane of symmetry — the molecule is meso and achiral.

Q. Identify the relationship between the two shown molecules as identical molecules, enantiomers, or diastereomers.

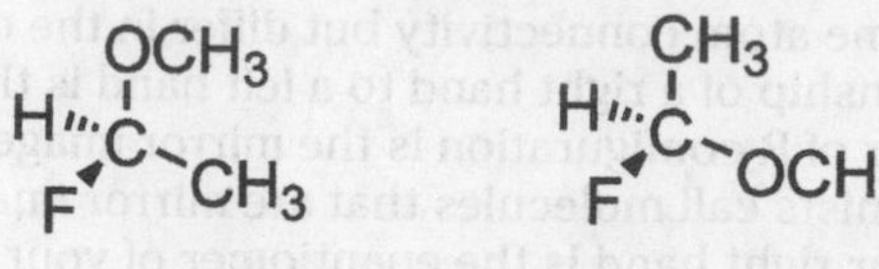

A. **The two molecules are enantiomers of each other.** The easiest way to determine the relationship between two stereoisomers is to assign and compare the configurations of the chiral centers in the two molecules. In this case, the configurations are opposite, with the left structure being R and the right structure being S. Therefore, these molecules are mirror images, or enantiomers, of each other.

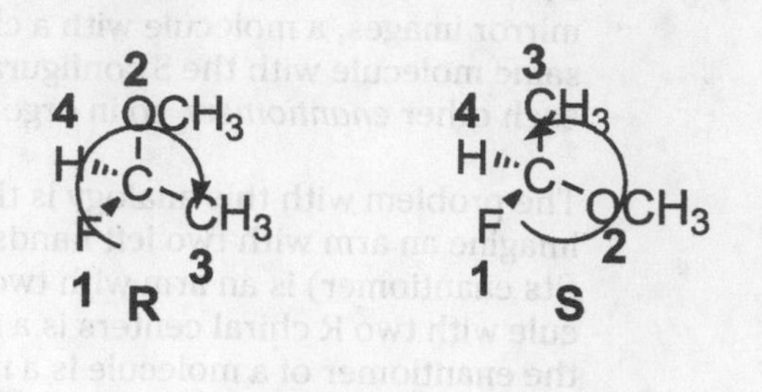

Q. State whether the following compound is chiral.

A. **This molecule is achiral and meso.** This molecule has two chiral centers, so you may expect it to be chiral. However, close inspection of this molecule shows that it has a plane of symmetry that cuts the molecule in half. Therefore, the mirror image of this compound is identical to itself, so the compound is classified as achiral and meso.

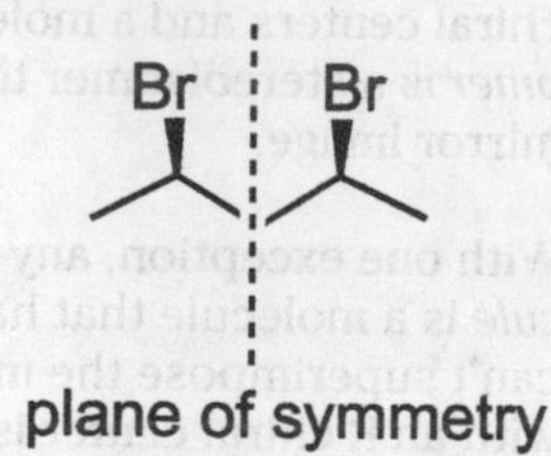

15. Identify the relationship between the two shown molecules as identical molecules, enantiomers, or diastereomers.

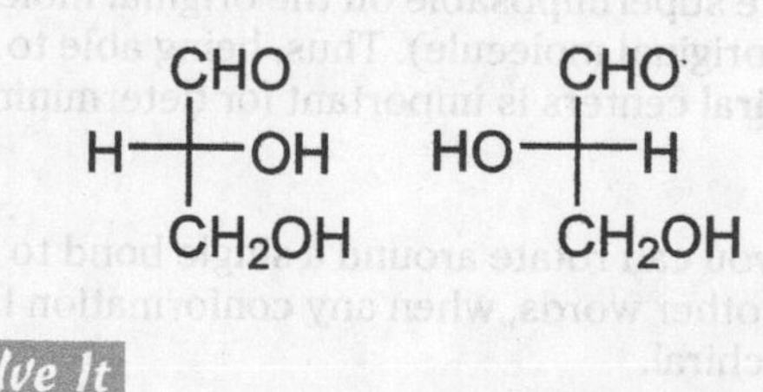

Solve It

16. Identify the relationship between the two shown molecules as identical molecules, enantiomers, or diastereomers.

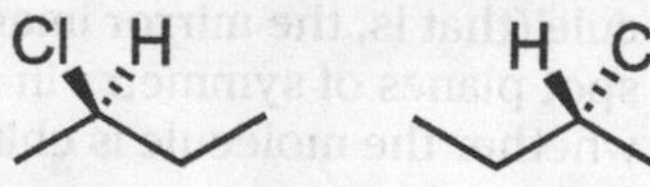

Solve It

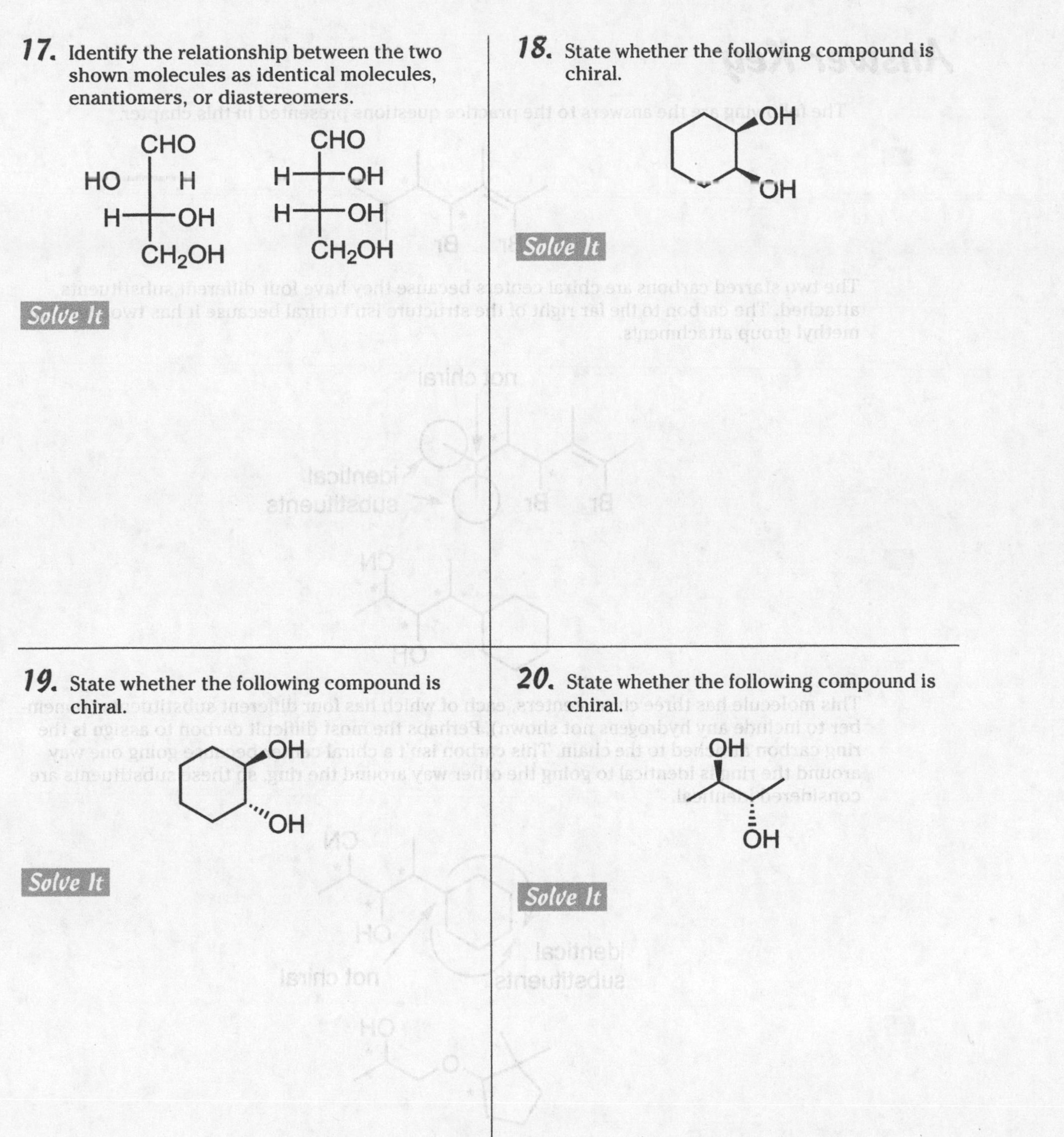

17. Identify the relationship between the two shown molecules as identical molecules, enantiomers, or diastereomers.

CHO / HO H / H—OH / CH_2OH CHO / H—OH / H—OH / CH_2OH

Solve It

18. State whether the following compound is chiral.

Solve It

19. State whether the following compound is chiral.

Solve It

20. State whether the following compound is chiral.

Solve It

Answer Key

The following are the answers to the practice questions presented in this chapter.

1

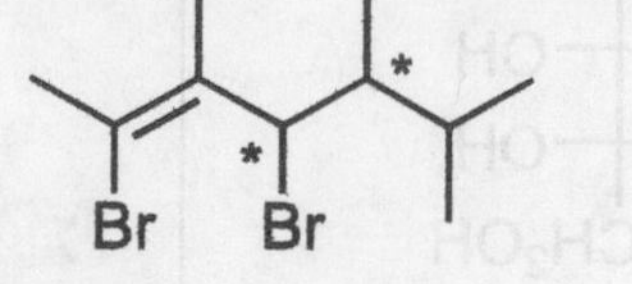

The two starred carbons are chiral centers because they have four different substituents attached. The carbon to the far right of the structure isn't chiral because it has two identical methyl group attachments.

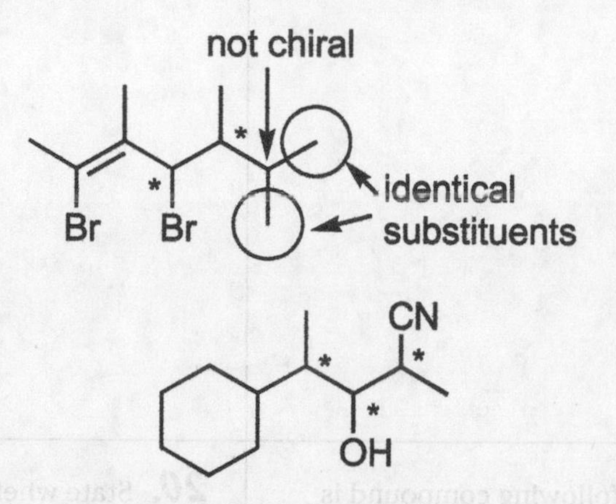

2

This molecule has three chiral centers, each of which has four different substituents (remember to include any hydrogens not shown). Perhaps the most difficult carbon to assign is the ring carbon attached to the chain. This carbon isn't a chiral center because going one way around the ring is identical to going the other way around the ring, so these substituents are considered identical.

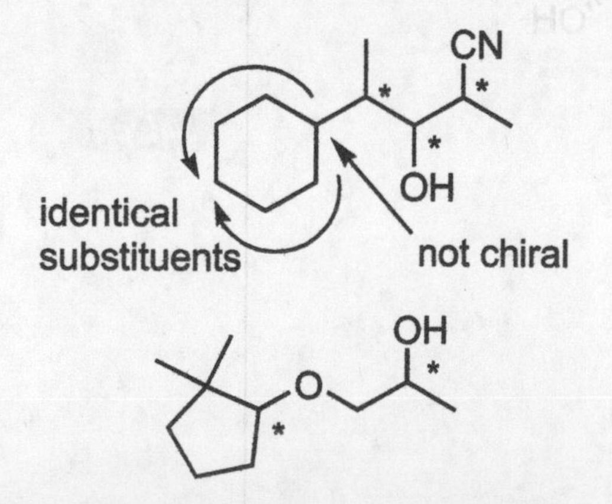

3

This molecule has two chiral centers. The ring carbon with the chain attached is a chiral center because the top side of the ring is different from the bottom side of the ring. Unlike in problem 2, these two sides of the ring are considered different substituents.

Starting with the right chiral center, prioritizing the substituents is reasonably straightforward. The oxygen gets the highest priority because this atom has the highest atomic number. The hydrogen gets the lowest priority because it has the lowest atomic number. Finally, the methyl group and the complex substituent to the left both have first-atom carbons, but the methyl gets a lower priority than the complex substituent because the complex substituent is attached to an oxygen, breaking the tie.

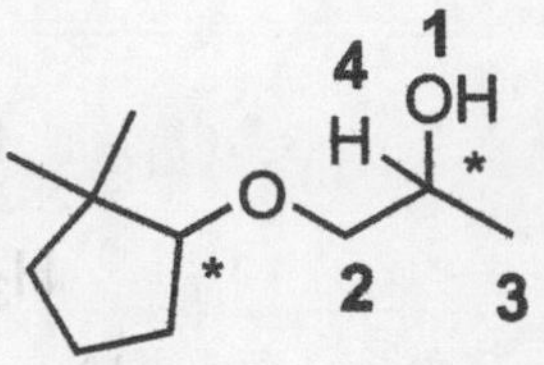

For the ring-carbon chiral center, the oxygen substituent has the highest priority and the hydrogen, the lowest, because these atoms have the highest and lowest atomic numbers, respectively. Both of the adjacent atoms in the ring are carbons. The top carbon in the ring is attached to three carbons and the bottom carbon in the ring is attached to just one other carbon, so the top ring carbon has the higher priority (note that the number of attached atoms matters only in the event of a tie).

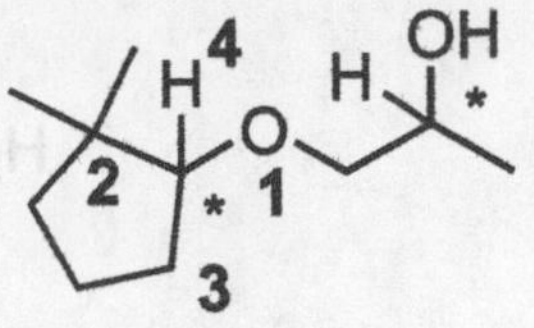

4

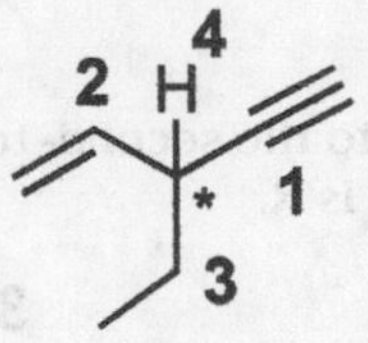

The atom has only one chiral center: the carbon bridging the double bond and the triple bond (no double-bond or triple-bond carbon can be chiral). Hydrogen has the lowest priority as usual. Under the Cahn-Ingold-Prelog prioritizing scheme, carbon-carbon double bonds are treated as though they represent two single bonds from the first carbon to the second carbon; therefore, one of the second carbons has two single bonds to the first atom. Triple bonds represent three single bonds from the first carbon to the second carbon, so one of the second carbons then has three single bonds back to the first atom. The basic idea to remember is that double bonds have higher priority over single bonds between the same atoms, and triple bonds have priority over double bonds between the same atoms. Thus, the carbon-carbon triple bond has priority over the carbon-carbon double bond, and the carbon-carbon double bond has priority over the carbon-carbon single bond.

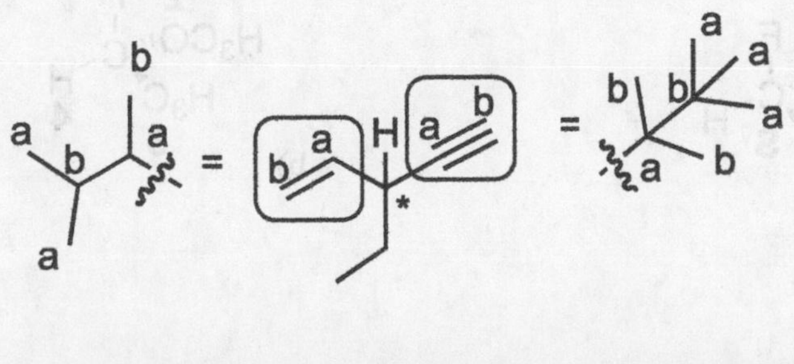

5 **The chiral center is R.** First, determine the substituent priorities. The OH is first because oxygen has the highest atomic number, and hydrogen is last. The CH_2OH group has higher priority than CH_3 because the carbon is attached to oxygen, breaking the tie (as in *a*). Because the fourth-priority substituent is already in the back, you can draw the curve from the first- to second- to third-priority substituent. The curve is clockwise, so the chiral center is designated R (as in *b*).

a) b)

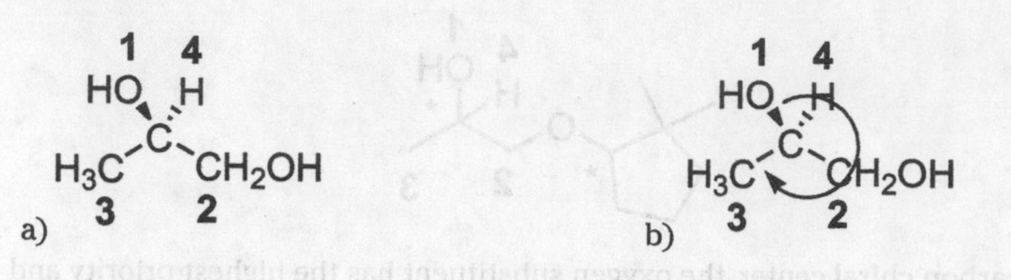

6 **The chiral center is R.** First, determine the substituent priorities. Based on atomic numbers, bromine is the highest, NH_2 is second, CH_3 is next, and hydrogen is last (as in *a*). Next, rotate the molecule so the fourth-priority substituent is in the back (as in *b*).

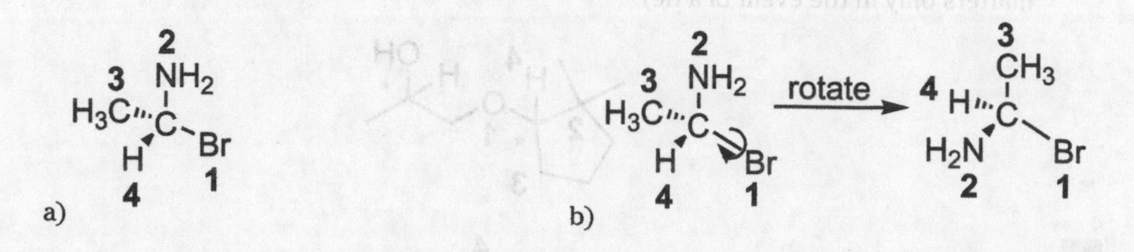

a) b) rotate

Draw the curve from the first- to the second- to the third-priority substituent. This curve is clockwise, so the chiral center is R.

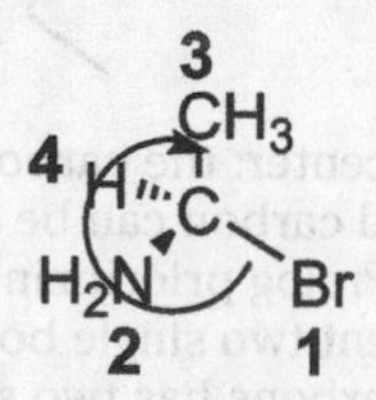

7 **The chiral center is S.** Determine the substituent priorities. Based on atomic numbers, fluorine is the first, OCH_3 is second, CH_3 is third, and hydrogen is fourth (as in *a*). Rotate the molecule so the fourth-priority substituent is in the back (as in *b*).

a) b)

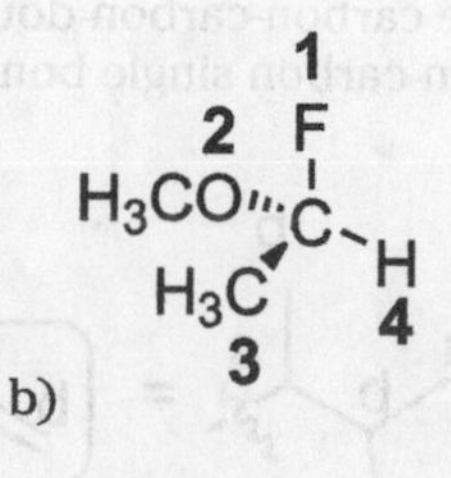

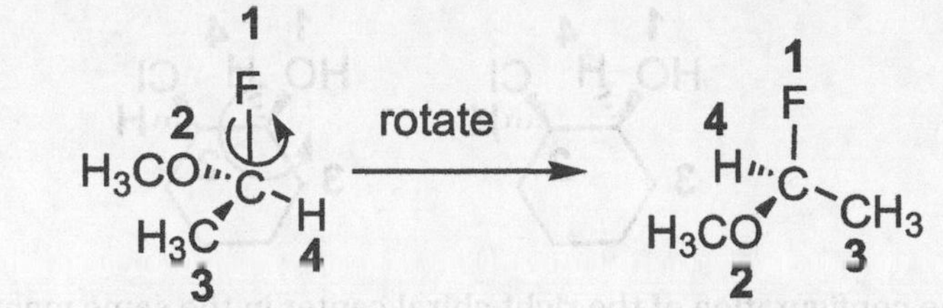

Draw the curve from the first-priority substituent to the second to the third. This curve is counterclockwise, so the configuration is S.

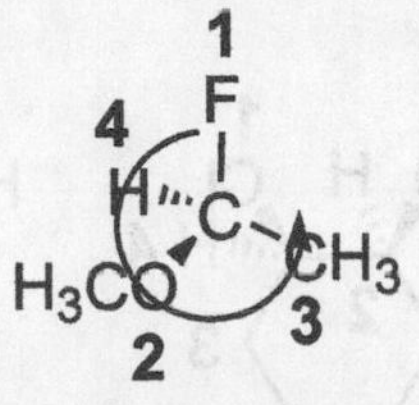

8

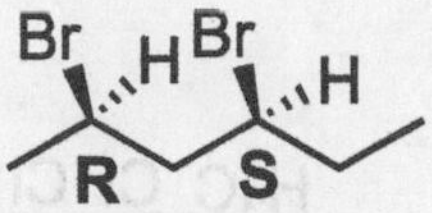

The molecule has two chiral centers, so take them one at a time. Starting with the left chiral center, prioritize the substituents. Because the fourth-priority substituent is in the back, you can draw the curve to obtain the configuration. In this case, the curve is clockwise, so the left chiral center is R (see *a*). Now do the right chiral center. You can draw the curve after assigning the substituent priorities because the fourth-priority substituent is already in the back. The curve is counterclockwise, so this chiral center is S (see *b*).

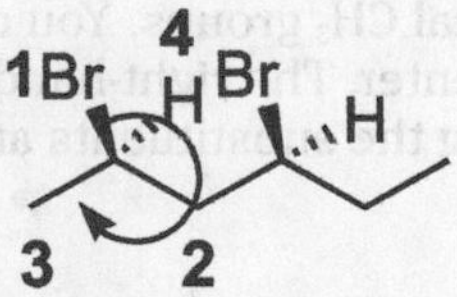

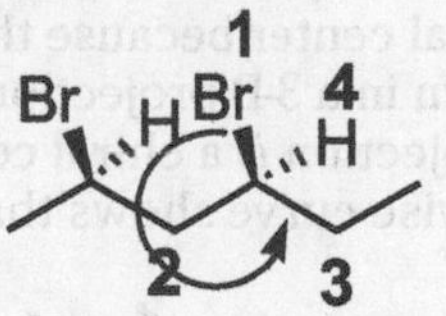

9

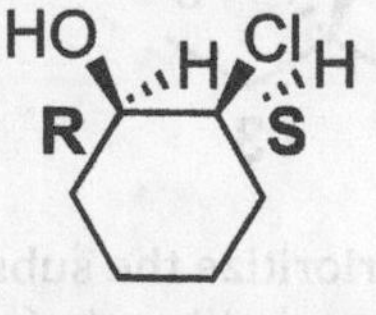

Take each chiral center one at a time. Starting with the left chiral center, the OH gets the highest priority and the H gets the fourth priority based on atomic number. The right side of the ring has a higher priority over the left side of the ring because the first carbon on the right side is attached to chlorine, which breaks the tie. The fourth-priority substituent is in the back, so drawing the clockwise curve yields an R configuration.

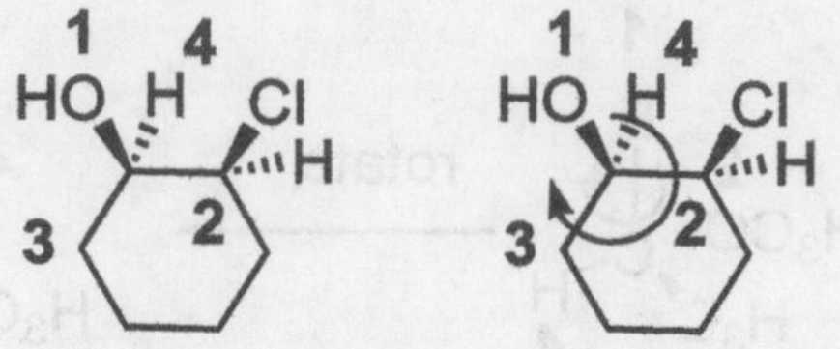

Determine the configuration of the right chiral center in the same manner. The chlorine gets the highest priority and the hydrogen, the fourth; the left-hand side of the ring gets higher priority over the right-hand side of the ring because the first carbon on the left side of the ring is attached to an OH, which breaks the tie. Drawing the counterclockwise curve yields an S configuration.

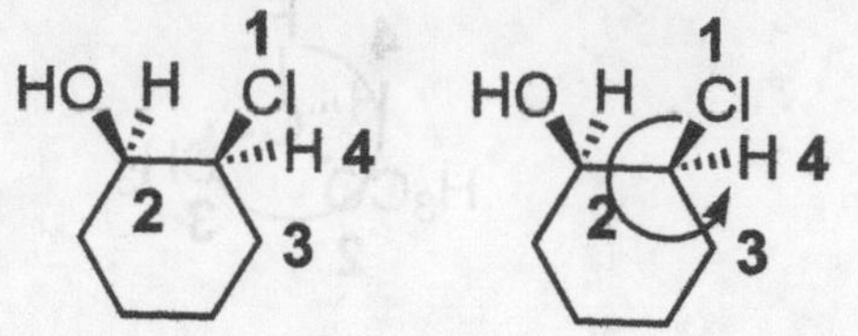

10

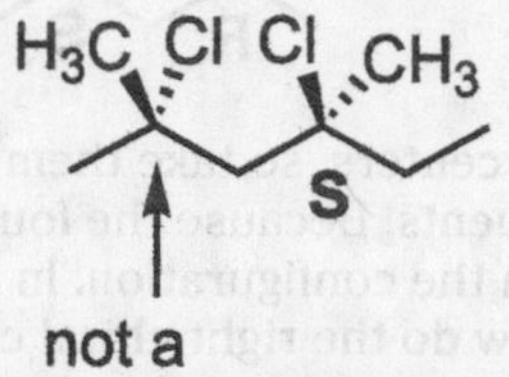

not a chiral center!

The first trick with this problem is to notice that the left-hand carbon that's drawn in a 3-D projection isn't a chiral center because this carbon has two identical CH_3 groups. You can't assume that an atom drawn in a 3-D projection is necessarily a chiral center. The right-hand carbon drawn in a 3-D projection *is* a chiral center, however. Prioritizing the substituents and drawing the counterclockwise curve shows that this chiral center is S.

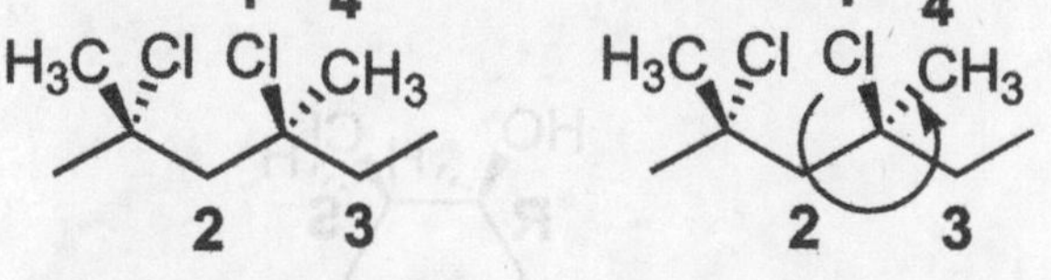

11 **The configuration is R.** First, prioritize the substituents (as in *a*). Then draw the curve from first- to second- to third-priority substituents (as in *b*). Because the fourth-priority substituent is on a vertical bar and the curve is clockwise, the configuration is R.

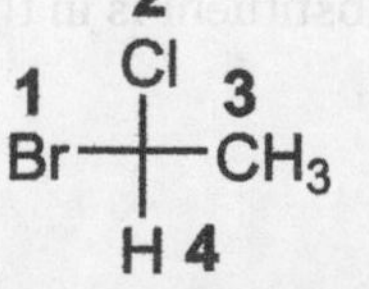

a)

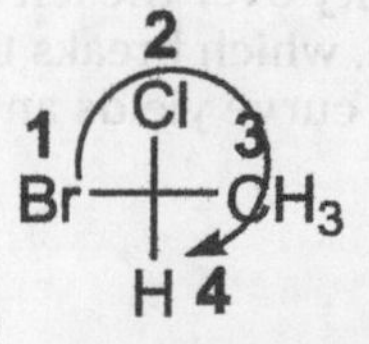

b)

12 **The configuration is S.** Begin by prioritizing the substituents (as in *a*). Now draw the curve from first to second to third (as in *b*). Because the fourth-priority substituent is on a horizontal bar, the configuration is exactly the opposite of what you may expect. The curve is clockwise, so the configuration is S.

a) b)

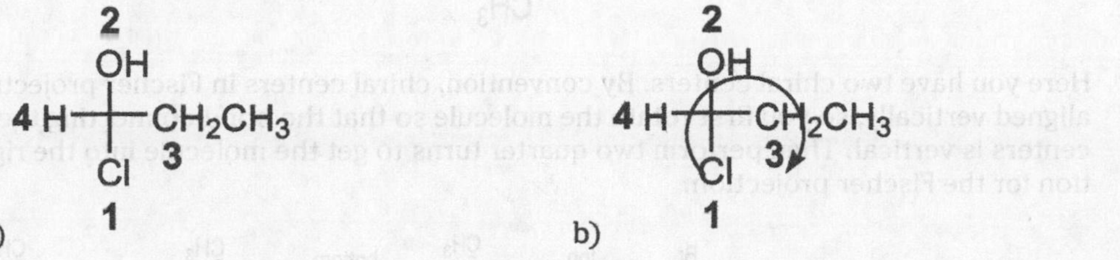

13

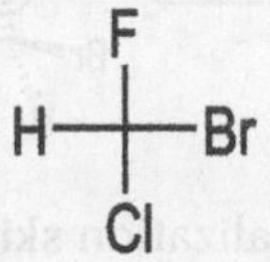

To convert the 3-D projection to the Fischer structure, perform the quarter turn so that two substituents come out of the plane on the horizontal axis and two substituents go back into the plane on the vertical axis.

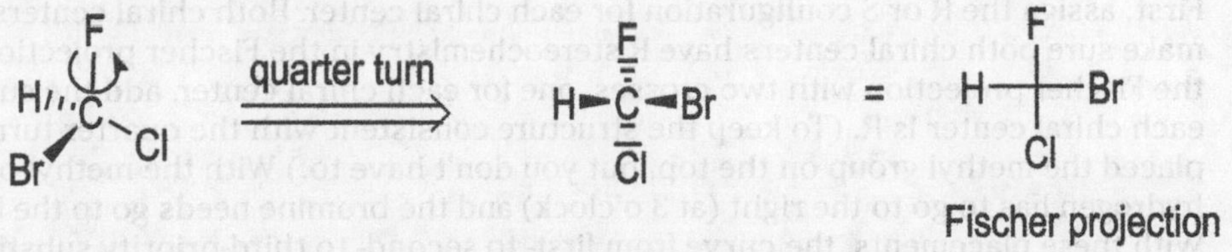

quarter turn

Fischer projection

Double-check your work by making sure the configuration didn't change. First, find the configuration of the chiral center in the 3-D projection by assigning priorities to the substituents. The fourth-priority substituent is in the back, so draw the curve from the first- to second- to third-priority substituents; the curve is counterclockwise, so the configuration is S. Next, assign the configuration to the Fischer projection. Because the fourth-priority substituent is on a horizontal bar and the curve from the first- to second- to third-priority substituents is clockwise, the configuration is S.

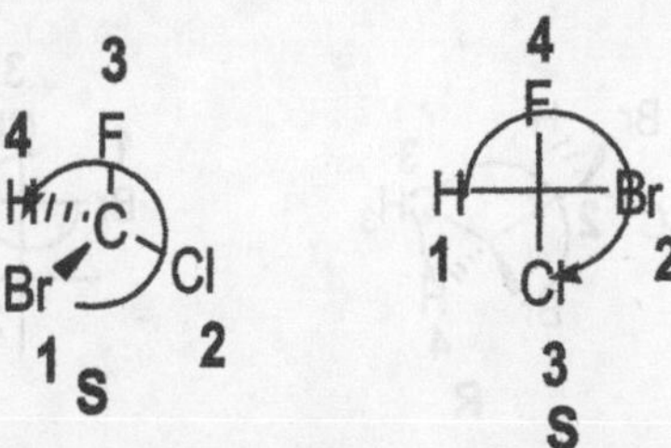

Note that you may have a Fischer projection that looks different from the one shown here. As long as the configuration is correct (S), placing substituents in alternative locations is perfectly fine.

14

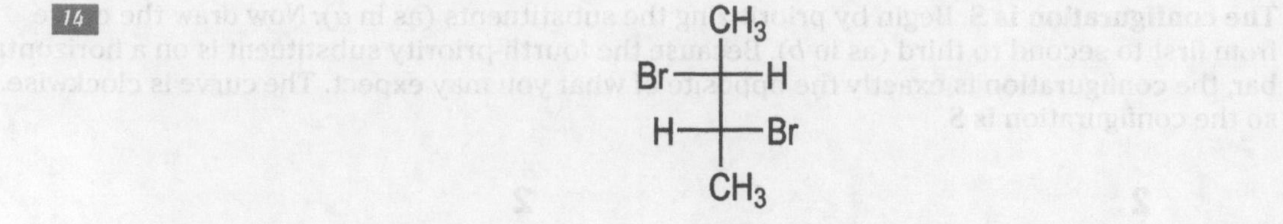

Here you have two chiral centers. By convention, chiral centers in Fischer projections are aligned vertically, so you first rotate the molecule so that the bond connecting the two chiral centers is vertical. Then perform two quarter turns to get the molecule into the right orientation for the Fischer projection.

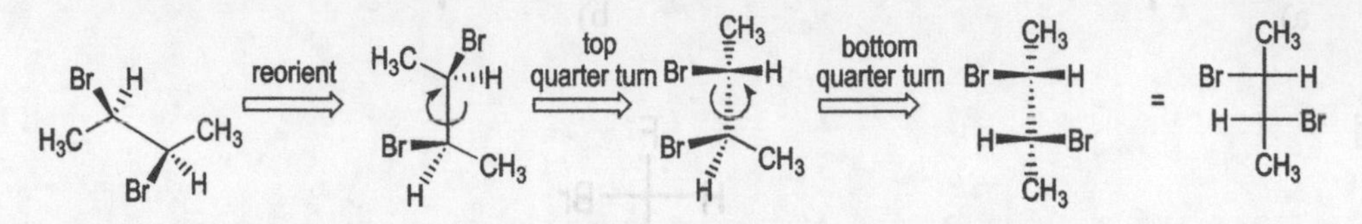

Even for people with excellent visualization skills, this is a difficult process. Molecules with more than one chiral center are often easier to convert to the Fischer projection by bypassing the quarter turn and 3-D visualization business: Assign the R or S configurations to the chiral centers in the 3-D projection and then add the substituents to a Fischer projection to make sure that the configurations remain unchanged.

First, assign the R or S configuration for each chiral center. Both chiral centers are R. Next, make sure both chiral centers have R stereochemistry in the Fischer projection. After drawing the Fischer projection with two crosses, one for each chiral center, add the substituents so that each chiral center is R. (To keep the structure consistent with the quarter-turn method, I've placed the methyl group on the top, but you don't have to.) With the methyl on the top, the hydrogen has to go to the right (at 3 o'clock) and the bromine needs go to the left (at 9 o'clock). With these placements, the curve from first- to second- to third-priority substituents is counterclockwise, which indicates the chiral center is R because the fourth-priority substituent is on a horizontal bar.

Next, add substituents to the second chiral center on the Fischer projection. For consistency, I've placed the methyl (CH_3) on the bottom at 6 o'clock. For an R configuration, the bromine needs to go to the right at 3 o'clock and the hydrogen has to go to the left at 9 o'clock. With this placement, the priority curve is counterclockwise, indicating an R stereochemistry because the fourth-priority substituent is on a horizontal bar. Thus, both methods yield the same solution. You can use whichever one works best for you.

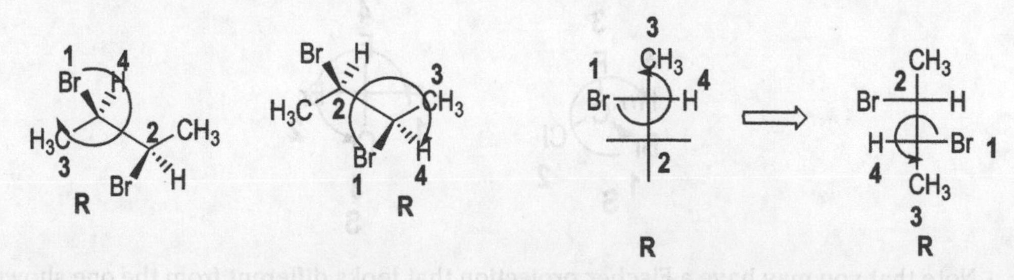

You may have a Fischer projection that looks different from the one shown here, but as long as the two chiral center configurations are correct (both R), placing substituents in alternative locations is perfectly fine.

15 **Enantiomers.** These Fischer projections are mirror images of each other (you can draw a mirror plane between the two Fischer projections). Moreover, switching the position of any two substituents in a Fischer projection inverts the stereochemistry (takes it from R to S or S to R).

16

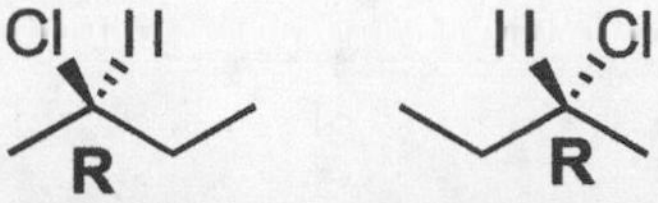

Cl H R H Cl R

The two structures are identical. The two molecules look different, but don't be fooled by this. Assigning the stereochemistry to both structures shows that the chiral center is R in both cases. Therefore, the two molecules have to be identical but merely drawn in a different way.

17 **The two molecules are diastereomers.** Two chiral centers are in the molecule, but only the top one switches upon going from the left-hand structure to the right-hand structure. Therefore, the two molecules aren't mirror images of each other. Both chiral centers need to be switched to make the molecules enantiomers (mirror images). Thus, these molecules are diastereomers.

18 **The molecule is meso and achiral.** Even though two chiral centers exist, a plane of symmetry bisects the ring so that the top half is a reflection of the bottom half.

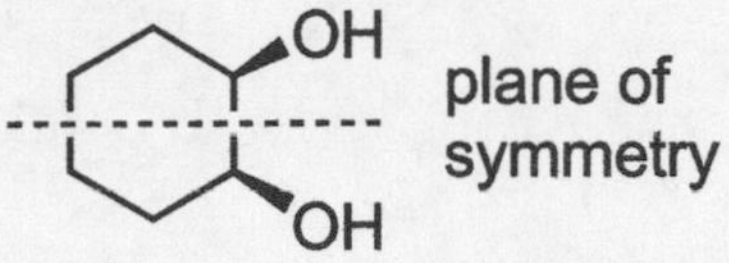

OH plane of symmetry OH

19 **This molecule is chiral.** This molecule has two chiral centers, but it doesn't have a plane of symmetry that bisects the molecule. Therefore, the molecule is chiral.

20 **The molecule is meso and achiral.** This problem is very tricky. Although the molecule as initially drawn has no plane of symmetry, you can rotate around the central bond to give a structure that *does* have a plane of symmetry, making the molecule meso and achiral.

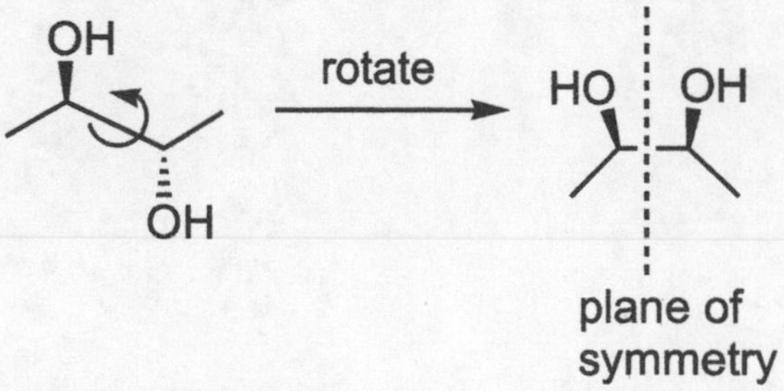

OH OH rotate HO OH plane of symmetry

Chapter 6

The Skeletons of Organic Molecules: The Alkanes

In This Chapter

- Naming alkanes using the IUPAC nomenclature
- Drawing a structure from a name

Ever wondered what octane is, the fuel you put in your car, or methane, the fuel in natural gas used to heat many homes? Those are the names of the organic molecules in those products. And for chemists, knowing what to call different molecules is an important part of keeping everyone on the same page. It's important enough, in fact, that the International Union of Pure and Applied Chemistry (IUPAC) formed a committee to create a systematic set of rules for naming organic molecules so that every organic molecule would have the same name all over the world.

In order for you to get in on the conversation, though, you need to know how to name organic molecules yourself. The best place to start honing your skills is with the simplest of organic molecules, the alkanes. *Alkanes,* the bones of organic molecules, are organic compounds containing only carbon-carbon and carbon-hydrogen single bonds.

This chapter gives you practice with the naming, or *nomenclature,* of alkanes, and it lays the foundation for the nomenclature of all organic molecules. In addition to giving proper names to molecules, you see how to draw a structure from a given name. Such naming can appear difficult at first, given that organic structures often have long, complex names — and frankly, somewhat difficult-to-pronounce names — but with a bit of practice, this process is actually quite straightforward (and kind of fun, too).

Understanding How to Name Alkanes

When you name straight-chain alkanes, use the four following steps:

1. **Find the parent chain.**

   Identify the *parent chain,* which is the longest continuous chain of carbon atoms. Table 6-1 shows the parent names for compounds, based on the number of carbons in the parent chain.

Table 6-1 Parent and Substituent Names

Number of Carbons	*Parent Name (-ane)*	*Substituent Name (-yl)*
1	Methane	Methyl
2	Ethane	Ethyl
3	Propane	Propyl
4	Butane	Butyl
5	Pentane	Pentyl
6	Hexane	Hexyl
7	Heptane	Heptyl
8	Octane	Octyl
9	Nonane	Nonyl
10	Decane	Decyl

2. **Number the parent chain.**

   Number the parent chain starting from the side that reaches the first substituent sooner. (A *substituent* is an atom or group that attaches to the parent chain.)

3. **Name each of the substituents.**

   You can name straight-chain substituents using Table 6-1, but you should also know the names of the three common complex substituents shown in Figure 6-1. The dashed line indicates where that substituent joins the parent chain.

Figure 6-1: Common names for complex substituents.

4. **Order the substituents alphabetically, using a number to indicate the position of each substituent on the parent chain.**

   Additionally, when you have two substituents of the same kind — for example, two methyl substituents on a molecule — they're combined in the name as *dimethyl* to save space. For example, instead of writing *2-methyl-3-methylbutane,* you'd write *2,3-dimethylbutane.* The *di-* indicates two substituents of the same kind; for three like substituents, you use the prefix *tri-;* for four, *tetra-;* for five, *penta-;* and so on.

Remember the following quirks concerning alphabetization and prefixes:

- **Ignore *tert-* (t) and *sec-* (s).** Tert-butyl (or t-butyl) and sec-butyl (or s-butyl) go under the letter *b*.
- **Ignore prefixes that specify the number of identical substituents, such as *di-*, *tri-*, or *tetra-*.** For example, dimethyl falls under *m*.
- ***Cyclo-* or *iso-* prefixes are alphabetized under *c* and *i*, respectively.** Cyclopropyl goes under *c*, and isopropyl goes under *i*.

Don't put a space between the final substituent name and the parent name.

Cycloalkanes, or alkanes in rings, are named in a similar way. The principal difference comes in selecting the parent chain. For a cycloalkane, the parent chain is either the ring or a substituent attached to the ring. To determine which of the two is the parent chain, you count the number of carbons in the ring and then count the longest continuous chain of carbons in the substituent. Whichever has more carbons is labeled the parent chain, and the other is considered a substituent. Here's how you indicate the ring:

- To indicate a ring in a name, add the prefix *cyclo-*. A three-carbon ring, for example, is called *cyclopropane*.
- When naming rings as substituents, add the suffix *-yl*. A three-membered ring as a substituent would be called a *cyclopropyl* substituent.

Q. Name the following alkane:

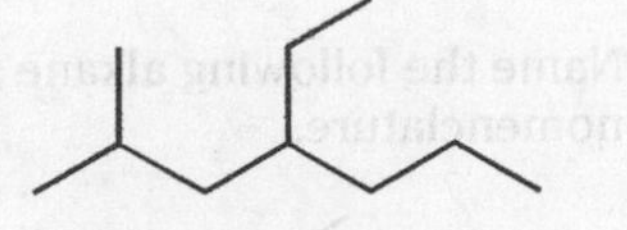

A. **4-ethyl-2-methylheptane.** The parent chain is the longer chain that goes from left to right that consists of seven carbons. Because the parent chain consists of seven carbons, the parent name is *heptane* (see Table 6-1). After you find the parent chain, you can number the parent chain in two possible ways. By convention, you start numbering from the end that reaches the first substituent sooner. In this case, numbering from left to right reaches the first substituent (the one-carbon substituent) at the second carbon, whereas numbering from right to left reaches the first substituent at the fourth carbon. Thus, the correct numbering in this case is from left to right.

1 2 3 4 5 6 7

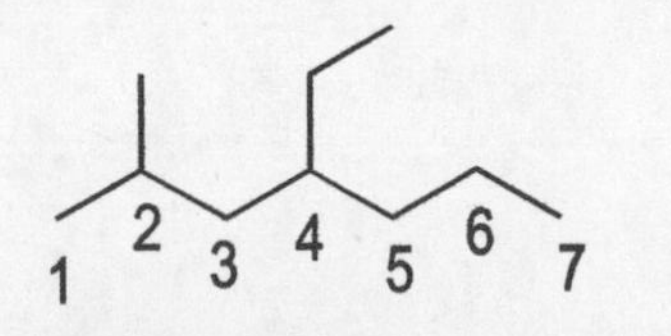

Now name the substituents. You name hydrocarbon substituents by taking the parent name and replacing the suffix *-ane* with *-yl*. Therefore, this one-carbon substituent is called a meth*yl* substituent (not a methane substituent), and the two-carbon substituent is called an ethyl substituent.

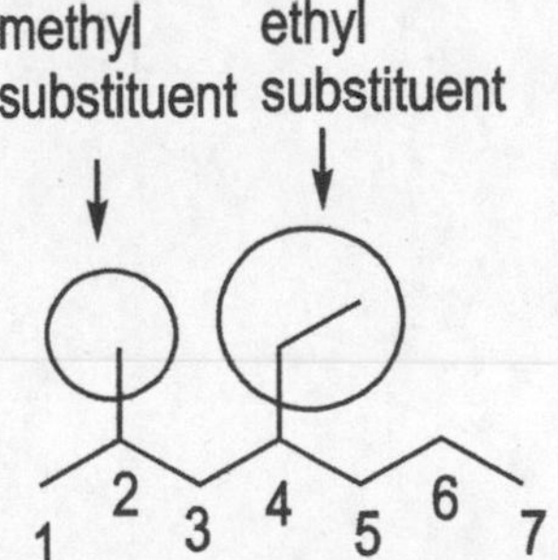

Finally, alphabetize the substituents in front of the parent name (heptane), with a number indicating the position. Because *ethyl* comes before *methyl* alphabetically, the name is 4-ethyl-2-methylheptane. Use dashes to separate the numbers from the substituents, and don't put a space between the final substituent name (methyl) and the parent name (heptane).

Q. Name the following alkane.

A. **1-butylcyclopentane.** The first step is to find the parent chain. Because there's both a ring and a straight chain portion, you count the number of carbons in each. The ring has five carbons, and the chain portion has four; thus, the parent chain is the ring because it has more carbons. Because the ring has five carbons, the parent name is *cyclopentane*.

Number the ring so as to give the lowest number to the substituents. In this case, you number so that the first carbon is the carbon with the substituent.

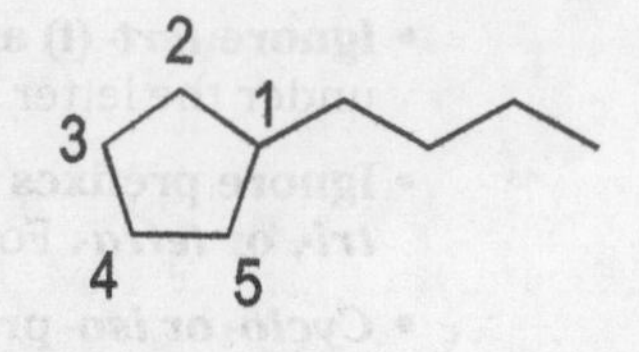

Then name the substituent. A four-carbon chain as a substituent is a *butyl* substituent. Lastly, place the substituent in front of the parent name with the number 1 indicating the position of the substituent to give you 1-butylcyclopentane.

butyl substituent

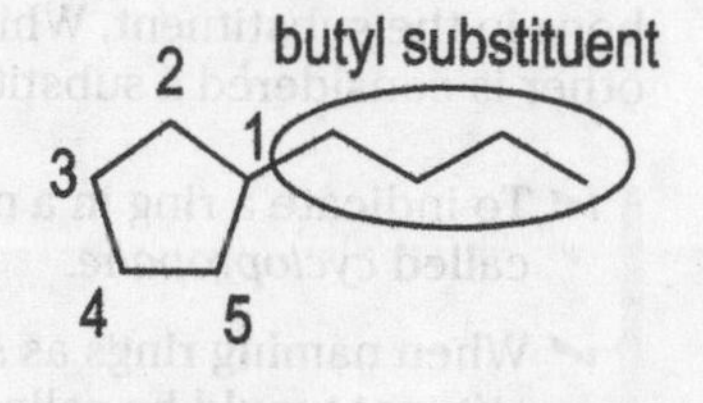

1. Name the following alkane using the IUPAC nomenclature.

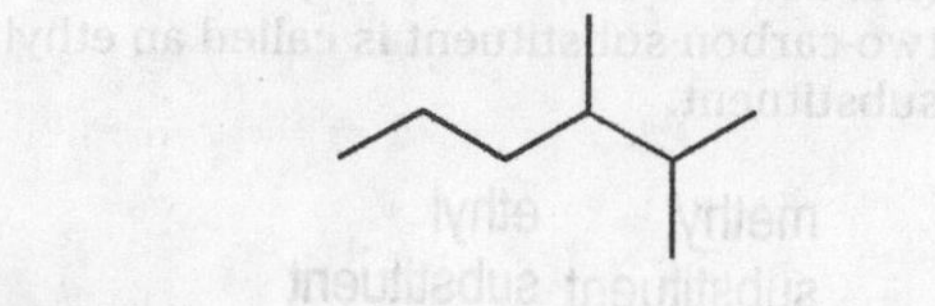

Solve It

2. Name the following alkane using the IUPAC nomenclature.

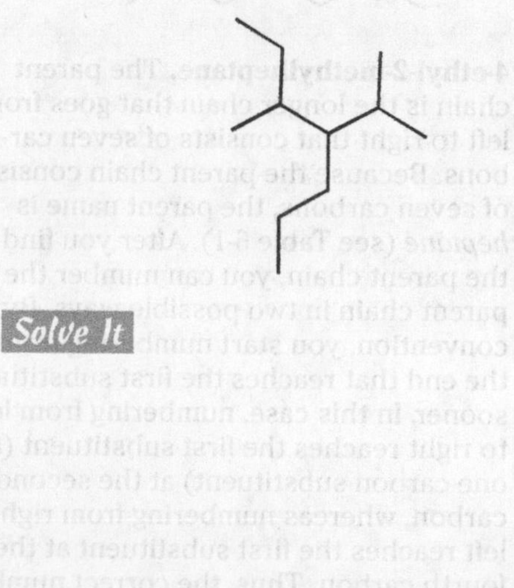

Solve It

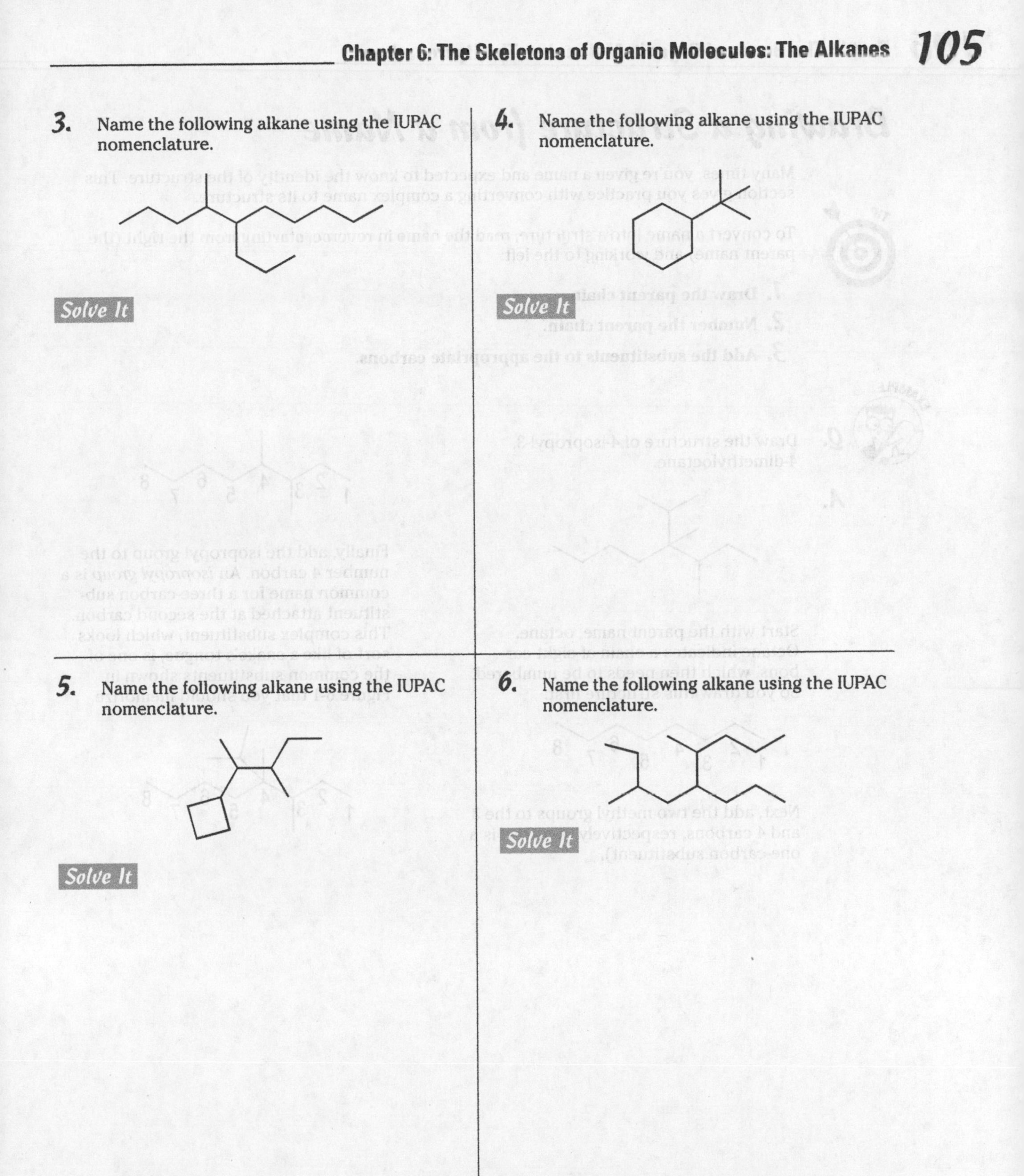

3. Name the following alkane using the IUPAC nomenclature.

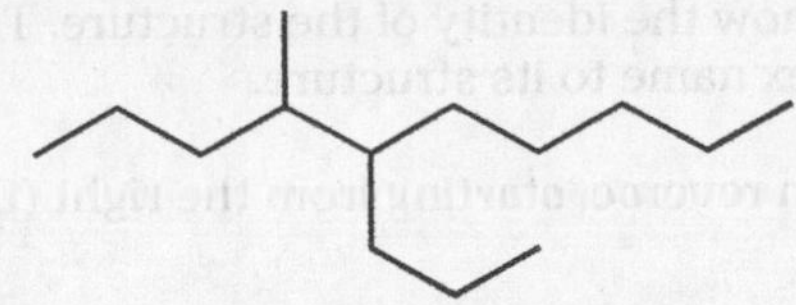

Solve It

4. Name the following alkane using the IUPAC nomenclature.

Solve It

5. Name the following alkane using the IUPAC nomenclature.

Solve It

6. Name the following alkane using the IUPAC nomenclature.

Solve It

Drawing a Structure from a Name

Many times, you're given a name and expected to know the identity of the structure. This section gives you practice with converting a complex name to its structure.

To convert a name into a structure, read the name in reverse, starting from the right (the parent name) and working to the left:

1. **Draw the parent chain.**
2. **Number the parent chain.**
3. **Add the substituents to the appropriate carbons.**

Q. Draw the structure of 4-isopropyl-3, 4-dimethyloctane.

A.

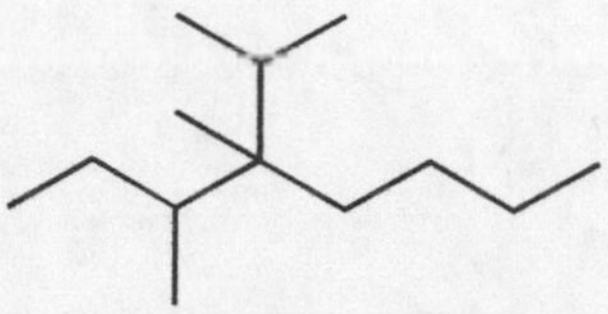

Start with the parent name, octane. Octane indicates a chain of eight carbons, which then needs to be numbered. So you draw this structure first:

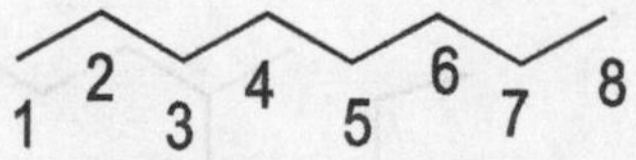

Next, add the two methyl groups to the 3 and 4 carbons, respectively (a *methyl* is a one-carbon substituent).

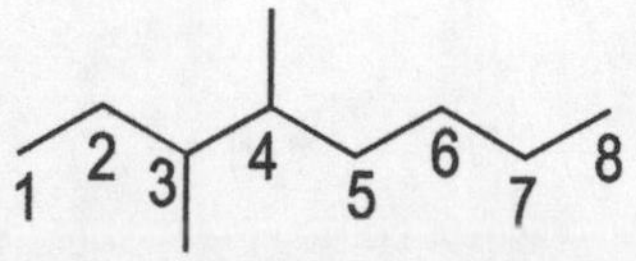

Finally, add the isopropyl group to the number 4 carbon. An *isopropyl group* is a common name for a three-carbon substituent attached at the second carbon. This complex substituent, which looks sort of like a snake's tongue, is one of the common substituents shown in Figure 6-1 that you should memorize.

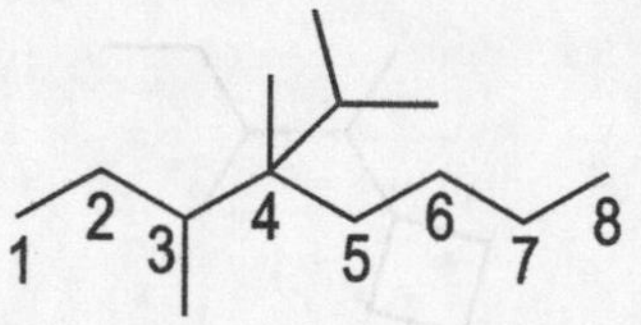

Q. Draw the structure of 2-t-butyl-1,1-dimethylcyclohexane.

A.

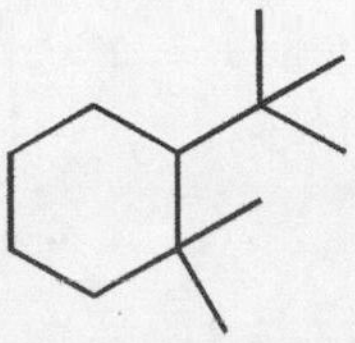

Reading from right to left on the name, the parent name is cyclohexane, which indicates a ring of six carbons. Drawing this ring and arbitrarily numbering it gives the following:

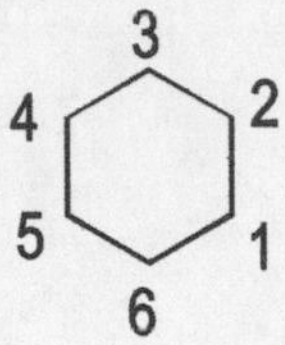

Next, add two methyl (one-carbon) groups to the number 1 carbon:

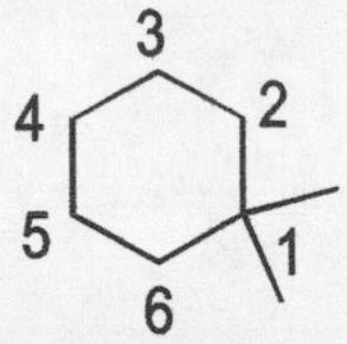

Last, add the t-butyl group to the number 2 carbon. A t-butyl group is the common name for a four-carbon substituent that sort of looks like a trident (and is one of the common substituent names you should memorize from Figure 6-1).

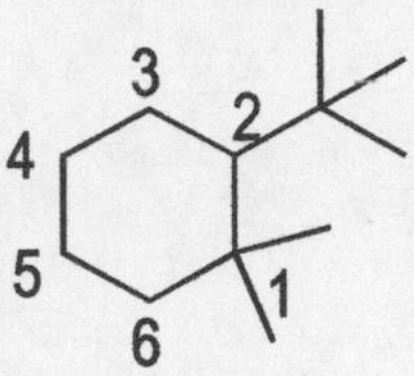

7. Draw the structure of 3-isopropyl-2,2,4-trimethylhexane.

Solve It

8. Draw the structure of 1-cyclopropyl-1,3-diethylcyclooctane.

Solve It

9. Draw the structure of 4-sec-butyl-5-cyclopentylnonane.

Solve It

10. Draw the structure of 1,1,2,2,3,3,4,4-octamethylcyclobutane.

Solve It

Answer Key

The following are the answers to the practice questions presented in this chapter.

1 **2,3-dimethylhexane.** The longest chain is the six-carbon chain that goes from right to left. Because the parent chain is six carbons long, the parent name for this alkane is hexane. Next, number the carbons from the side that reaches the first substituent sooner. In this case, you number from right to left so you reach the first substituent at the number 2 carbon (rather than the number 4, if you were to number left to right) (see *a*). Next, name each of the substituents (as in *b*). All four substituents are one-carbon groups, which are methyl groups.

a) 6 5 4 3 2 1

b) methyl substituent, 6 5 4 3 2 1, methyl substituent

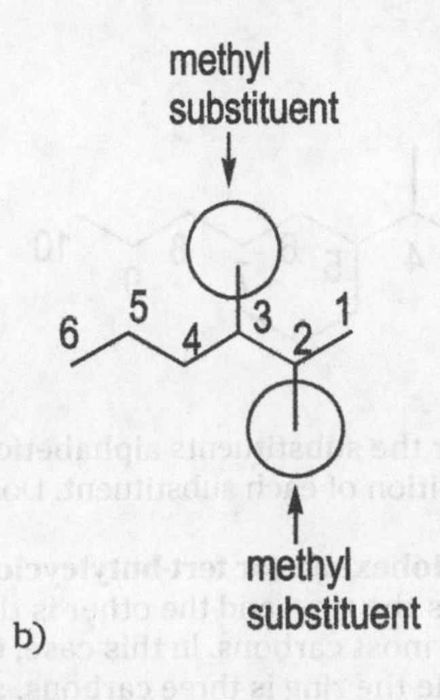

Finally, order the substituents in front of the parent name. Wherever you have two or more identical substituents, cluster them together and give them a prefix to show how many you have. Here, there are two methyl groups, so add the prefix *di-* to *methyl,* with two numbers (separated by commas) to give the location of each methyl on the parent chain. Thus, the proper name is 2,3-dimethylhexane rather than 2-methyl-3-methylhexane.

2 **4-isopropyl-3-methylheptane.** As is often the case, the trickiest part of this problem is simply finding the longest consecutive chain of carbon atoms. In this molecule, the longest chain is seven carbons long, giving you a parent name of heptane. Next, number the parent chain, which looks like figure *a*. Next, name the substituents (like in *b*). The three-carbon substituent that looks like a snake's tongue is an isopropyl substituent (see Figure 6-1), and the one-carbon substituent is a methyl substituent.

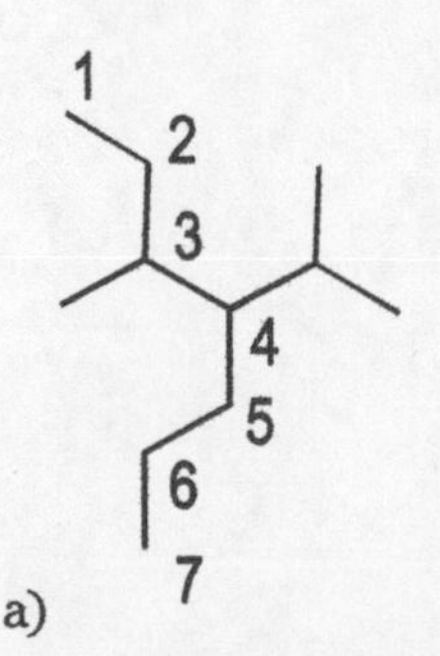

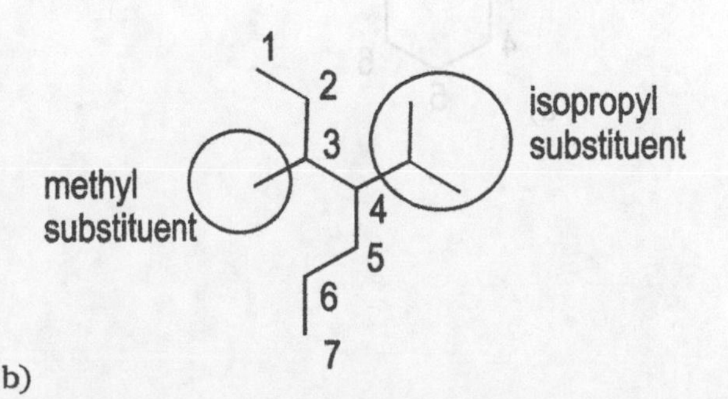

a) 1 2 3 4 5 6 7

b) 1 2 3 4 5 6 7, methyl substituent, isopropyl substituent

Then alphabetize these substituents in front of the parent name, with a number to indicate the position of each substituent on the chain. That gives you 4-isopropyl-3-methylheptane.

3 **4-methyl-5-propyldecane.** The longest chain of carbons is the chain that follows from left to right and is ten carbons long. The parent name of a ten-carbon chain is decane. Next, number the parent chain from left to right so you give the lower number to the first substituent (see *a*). Then name the substituents (see *b*). The one-carbon substituent at carbon 4 is a methyl substituent, and the three-carbon substituent at carbon 5 is a propyl substituent.

a) 1 2 3 4 5 6 7 8 9 10

b) methyl substituent; 1 2 3 4 5 6 7 8 9 10; propyl substituent

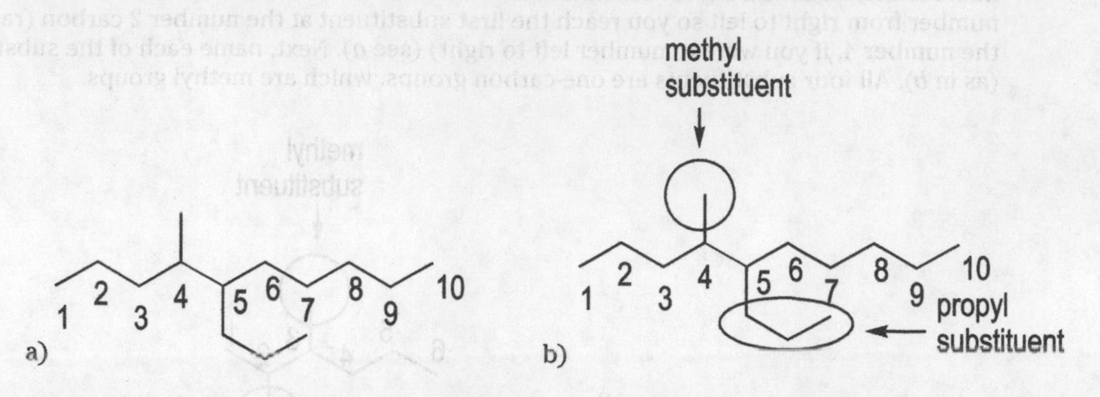

Lastly, order the substituents alphabetically in front of the parent name, using a number to indicate the position of each substituent. Doing so gives you the name 4-methyl-5-propyldecane.

4 **1-t-butylcyclohexane (or tert-butylcyclohexane).** Here you have two options for the parent chain: One is the ring, and the other is the straight carbon chain. The parent chain is the one that has the most carbons. In this case, the ring has six carbons, and the longest consecutive chain outside the ring is three carbons, so the ring is the parent chain. Number the ring to give the substituent the lowest number (number 1). ***Note:*** You can't number so the parent chain includes both the ring and the chain — the parent chain is *either* one or the other. Because the parent chain is a six-membered ring (see *a*), the parent name is cyclohexane. Finally, name the substituent (see *b*). In this case, the substituent is a tertiary butyl (t-butyl) group — one of the common substituents from Figure 6-1 (it looks like a trident or a chicken's foot). Thus, the name is 1-t-butylcyclohexane (t-butylcycohexane, without the number, is also acceptable because it's assumed that if only one substituent is present, the substituent is at the number 1 carbon).

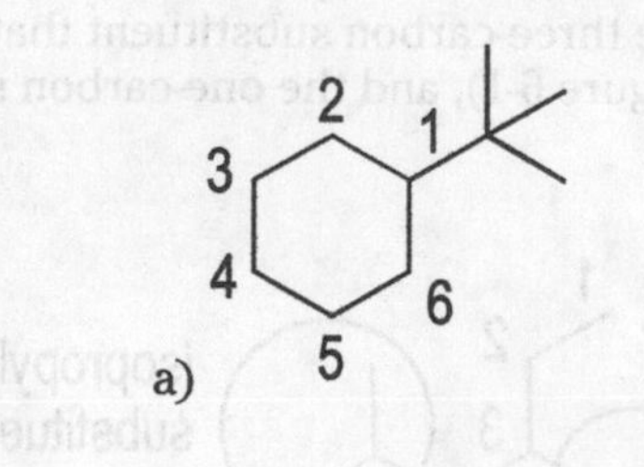

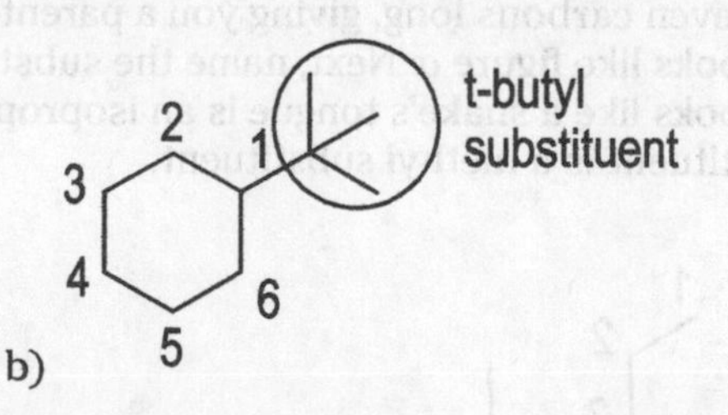

a) 1 2 3 4 5 6

b) t-butyl substituent; 1 2 3 4 5 6

5 **2-cyclobutyl-3-methylpentane.** In this molecule, you have two possibilities — the ring or the chain of carbons — for the parent chain. The ring has four carbons, and the longest chain has five carbons. Because you choose the parent chain with the most carbons, the parent chain is the straight chain of five carbons, so the parent name is pentane. Next, number the chain so you reach the first substituent sooner, which you accomplish by numbering from left to right.

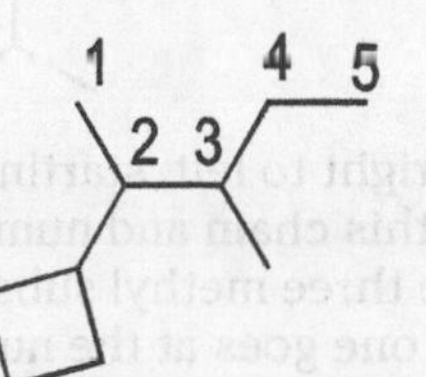

Next, name each of the substituents. A one-carbon substituent is a methyl substituent, and a four-membered ring as a substituent is called a cyclobutyl group (don't forget the *cyclo-* to indicate that this substituent is a ring).

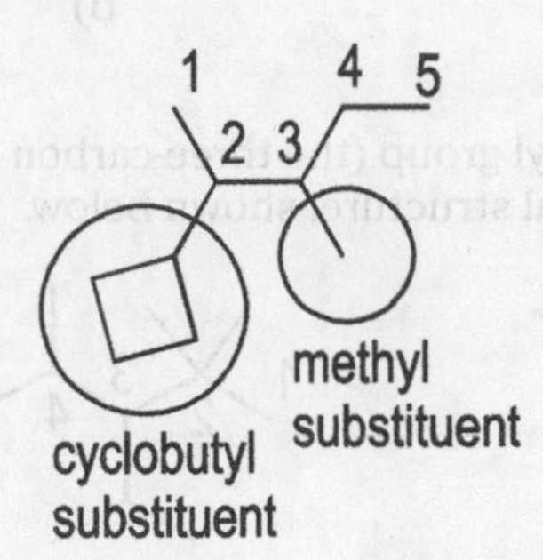

Finally, place the substituents alphabetically in front of the parent name to give you 2-cyclobutyl-3-methylpentane.

6 **3,6-dimethyl-5-propylnonane.** The longest chain of carbons is nine carbons long. Numbering the chain so the first substituent gets the lowest possible number gives you the following:

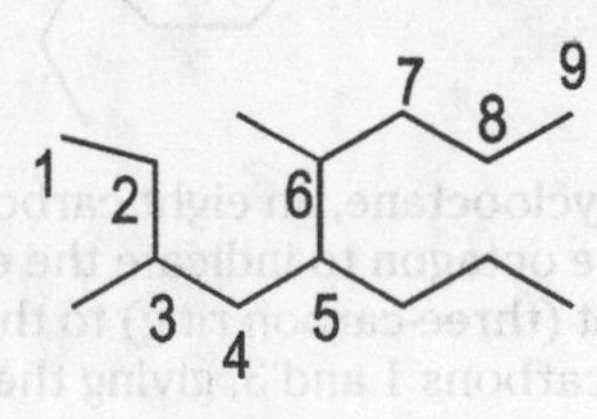

Then name the substituents. The two one-carbon substituents at carbons 3 and 6 are methyl substituents, and the three-carbon substituent at carbon 5 is a propyl substituent.

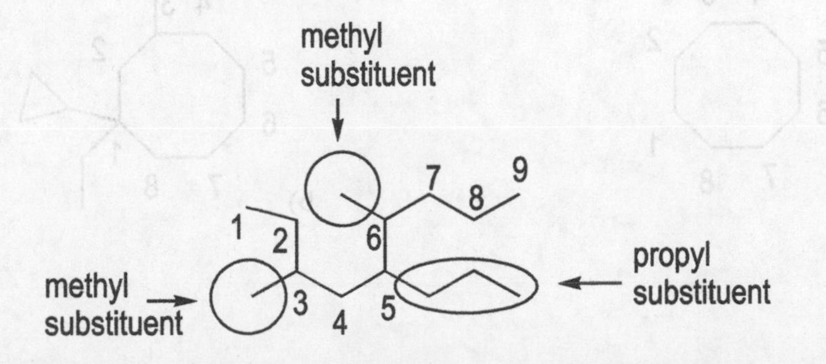

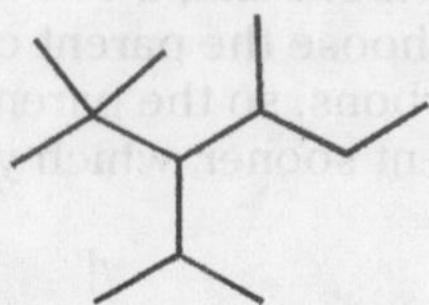

First, read the name from right to left, starting with the parent name, hexane. Hexane is a six-carbon chain, so drawing this chain and numbering it looks like figure *a*. Next, add the substituents, starting with the three methyl substituents that need to be placed (see *b*). Two go at the number 2 carbon, and one goes at the number 4 carbon.

a) 1 2 3 4 5 6 b) 1 2 3 4 5 6

Finally, add the isopropyl group (the three-carbon group that looks like a snake's tongue) at carbon 3 to give the final structure, shown below.

1 2 3 4 5 6

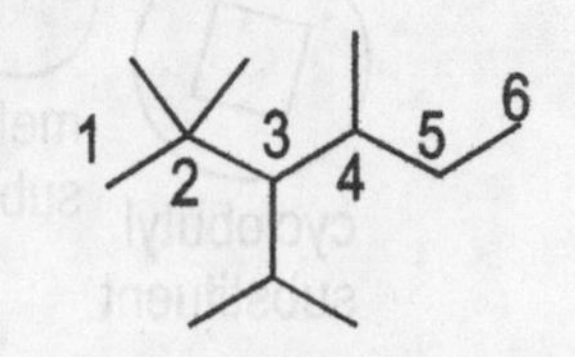

8

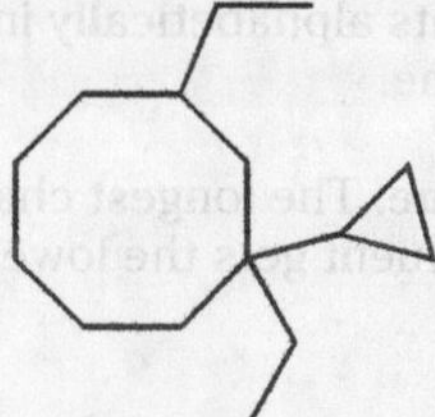

Draw the parent chain, cyclooctane, an eight-carbon ring (which admittedly can be tough to draw). After you draw the octagon to indicate the eight-membered ring, number it (see *a*). Add a cyclopropyl substituent (three-carbon ring) to the number 1 carbon and add two ethyl (two-carbon) substituents to carbons 1 and 3, giving the final structure (see *b*).

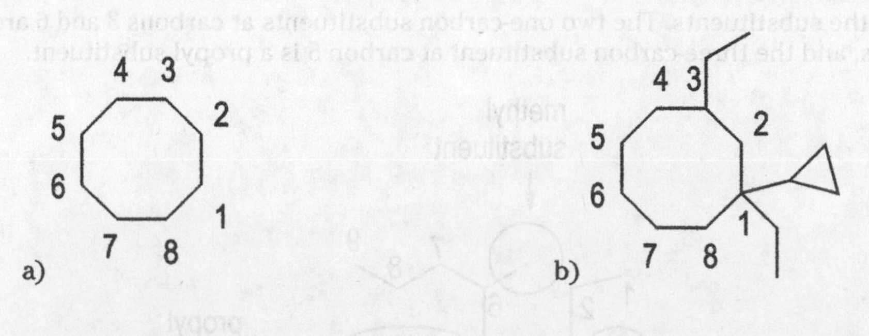

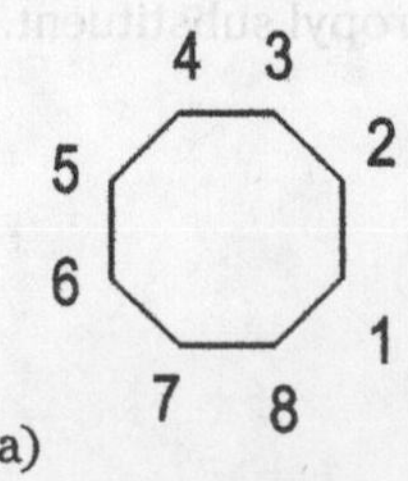

4 3 2 5 6 1 7 8 b)

9

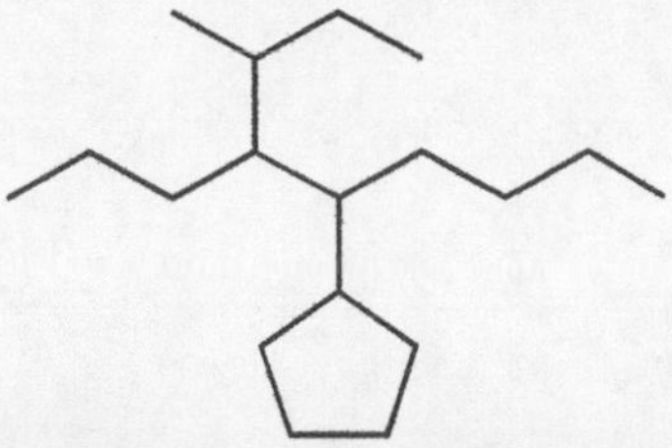

Draw the parent chain, nonane, which is a nine-carbon chain. Then number it (see *a*). Add the substituents. Start by adding a sec-butyl group (a butyl group attached to the secondary, or *sec,* carbon) to the number 4 carbon (see *b*).

a) 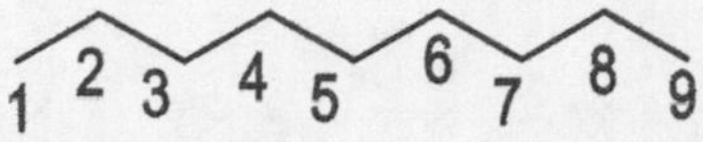 1 2 3 4 5 6 7 8 9 b) 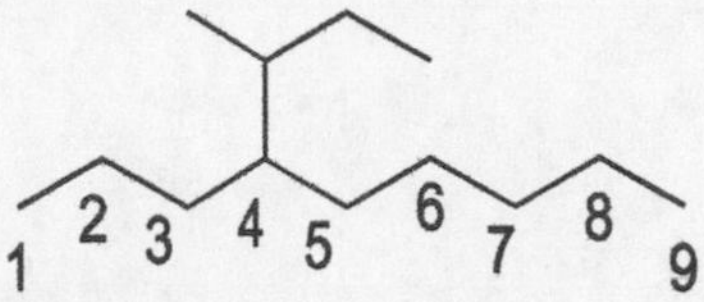 1 2 3 4 5 6 7 8 9

Add a cyclopentyl substituent (a five-carbon ring) to the number 5 carbon to give the final structure.

1 2 3 4 5 6 7 8 9

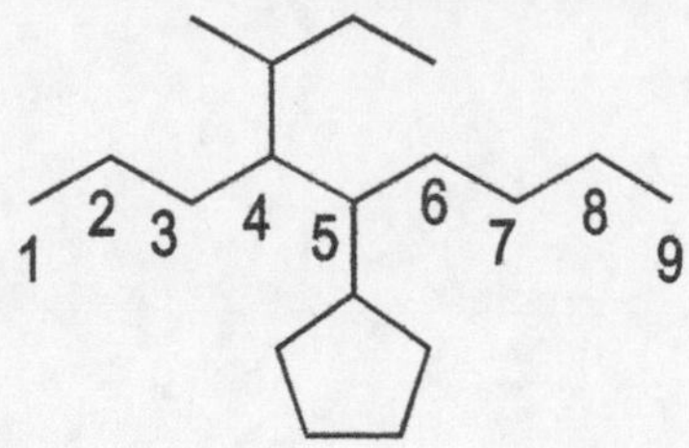

10

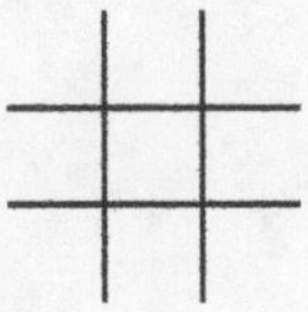

First, draw the parent chain, cyclobutane, which is a four-membered ring. Then number it (see *a*). Add eight methyl substituents (*octamethyl* indicates eight methyl substituents), two to each of the carbons, to give the final structure (see *b*).

a) 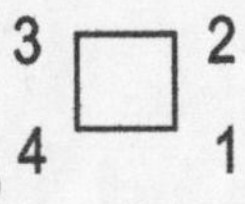 3 2 4 1 b) 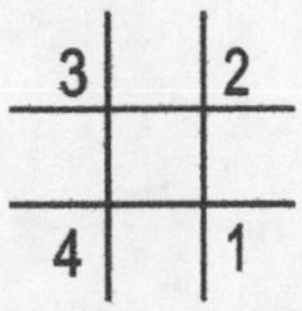 3 2 4 1

Hmm. . . . Anyone up for a game of tic-tac-toe?

Chapter 7

Shaping Up with Bond Calisthenics and Conformation

In This Chapter

- Working with Newman projections
- Drawing cyclohexane chair structures
- Seeing ring flips
- Making conformation energy predictions

The structure of a molecule on paper is a little misleading, kind of like a photograph of a child who's just gorged on candy — the picture makes the child look still and well-behaved, but in real life, the kid's jazzed up and going bananas, tearing things up and buffeting around the house like a whirling dervish. Molecules are very much the same because the little things appear motionless in drawn Lewis structures, but in reality, the atoms are constantly moving and kicking about. This movement results because atoms are often attached to each other with single bonds, and single bonds let atoms rotate freely — like wheels spinning around a car axle. This ability to rotate allows a molecule to fold into a number of different shapes, called *conformations.*

In this chapter, you see how chemists represent these three-dimensional folded shapes on a two-dimensional piece of paper, and you find out how to predict the relative energies of these different conformations.

Setting Your Sights on Newman Projections

One of the best ways to view the different conformations of molecules is through a Newman projection. In a Newman projection, you focus in on the conformation around a particular C-C bond by sighting down the bond to see the arrangement of the atoms on the carbons relative to each other. Sighting down the bond is similar to sighting down the barrel of a rifle (where the barrel is the C-C bond).

Before getting to Newman projections, though, here's the key for decoding depictions of bonds in three-dimensional space:

- **Solid wedge:** A bond coming out of the plane of the paper
- **Dashed line:** A bond that's going back into the paper
- **Normal line:** A bond that's in the plane of the paper

Figure 7-1 shows a Newman projection for the molecule ethane (CH_3CH_3).

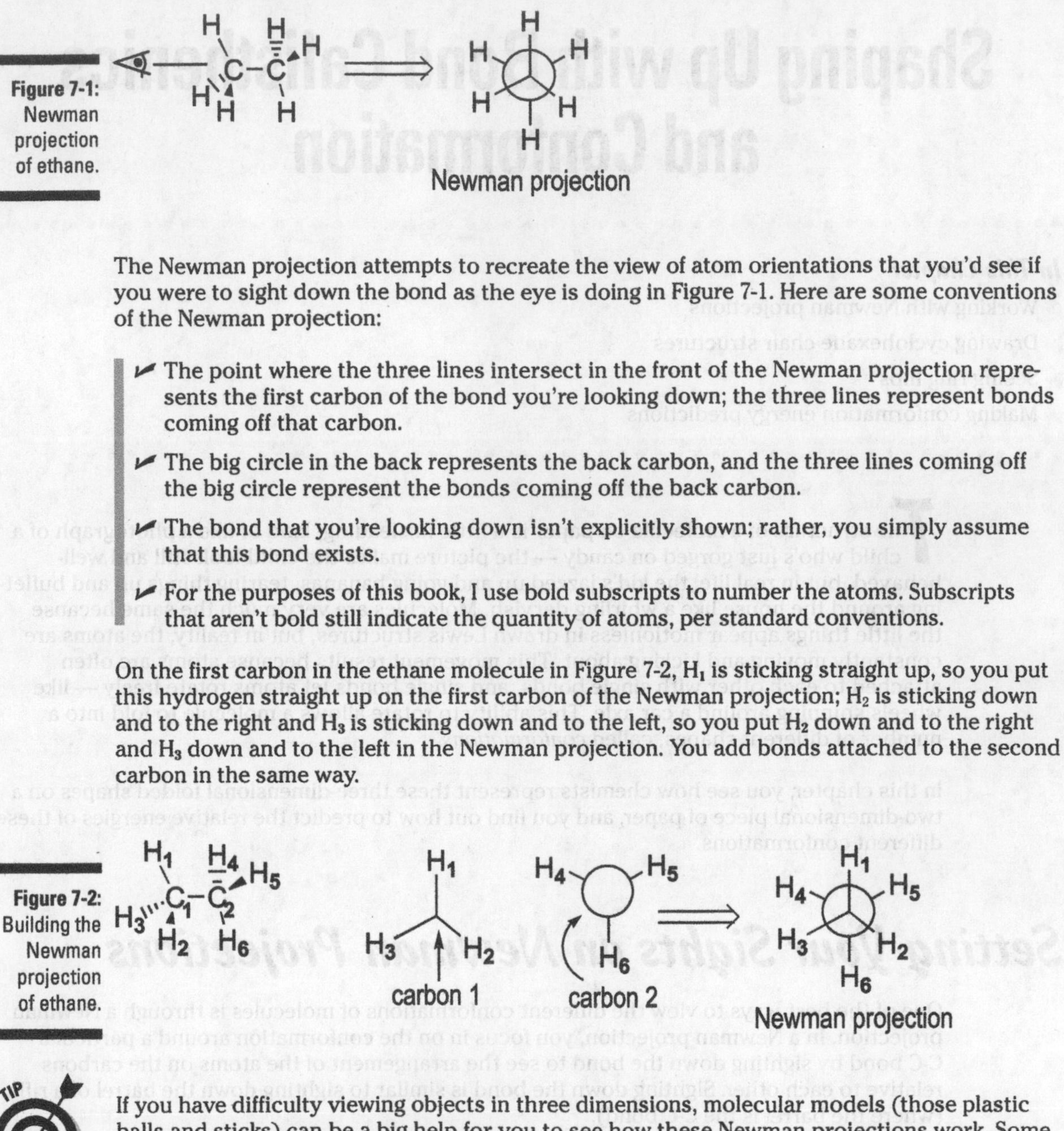

Figure 7-1: Newman projection of ethane.

The Newman projection attempts to recreate the view of atom orientations that you'd see if you were to sight down the bond as the eye is doing in Figure 7-1. Here are some conventions of the Newman projection:

- The point where the three lines intersect in the front of the Newman projection represents the first carbon of the bond you're looking down; the three lines represent bonds coming off that carbon.
- The big circle in the back represents the back carbon, and the three lines coming off the big circle represent the bonds coming off the back carbon.
- The bond that you're looking down isn't explicitly shown; rather, you simply assume that this bond exists.
- For the purposes of this book, I use bold subscripts to number the atoms. Subscripts that aren't bold still indicate the quantity of atoms, per standard conventions.

On the first carbon in the ethane molecule in Figure 7-2, H_1 is sticking straight up, so you put this hydrogen straight up on the first carbon of the Newman projection; H_2 is sticking down and to the right and H_3 is sticking down and to the left, so you put H_2 down and to the right and H_3 down and to the left in the Newman projection. You add bonds attached to the second carbon in the same way.

Figure 7-2: Building the Newman projection of ethane.

If you have difficulty viewing objects in three dimensions, molecular models (those plastic balls and sticks) can be a big help for you to see how these Newman projections work. Some professors may even let you bring them to your exam.

You need to worry about two primary types of conformations when performing conformational analysis of alkanes (see Figure 7-3):

- **The staggered conformation:** In the staggered conformation, the bonds are offset from each other by 60°.
- **The eclipsed conformation:** In the eclipsed conformation, the atoms on the adjacent carbons are in the same plane when viewed down the bond, although on paper, the back bonds are tilted slightly to allow the back atoms to be seen.

staggered conformation

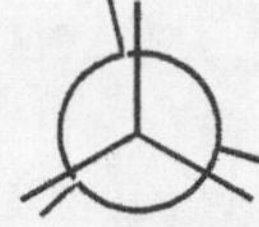

eclipsed conformation

Figure 7-3: Eclipsed and staggered conformations.

Q. Draw the Newman projection of the shown conformation of propane, sighting down the C_1-C_2 bond.

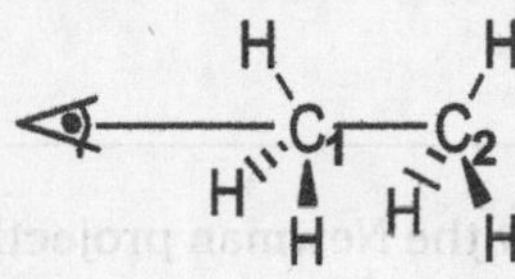

A. In your mind's eye, sight down the C_1-C_2 bond as the eye is doing. You can start the Newman projection by adding the hydrogens onto the first carbon: One hydrogen atom sticks straight up and toward the viewer, one sticks down and to the right, and one sticks down and to the left. Then add the substituents to the second carbon. All the hydrogens on the back carbon line up with the hydrogens on the front carbon, so this is an eclipsed conformation. As in all eclipsed conformations, you tilt the atoms on the back carbon slightly so you can see the atoms.

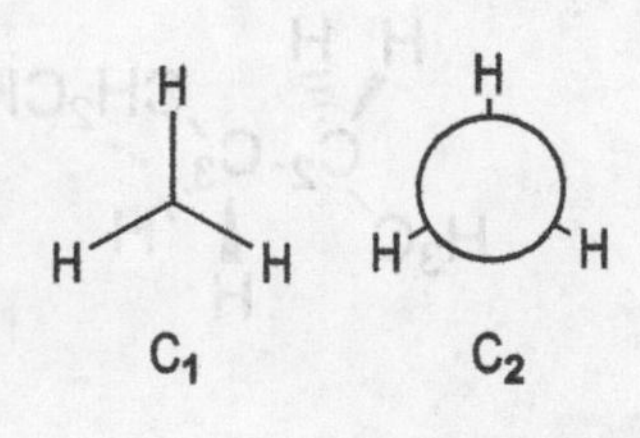

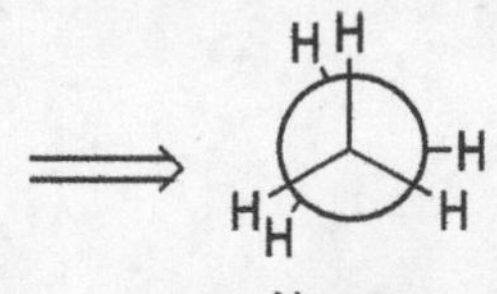

C_1 C_2

Newman Projection

1. Draw the Newman projection of the shown conformation of propane, sighting down the C_1-C_2 bond.

H, H, H, H, H, C_1-C_2, CH_3

Solve It

2. Draw the Newman projection of the shown conformation of butane, sighting down the C_1-C_2 bond.

H, CH_2CH_3, C_1-C_2, H, H, H, H

Solve It

3. Draw the Newman projection of the shown conformation of butane, sighting down the C_2-C_3 bond.

H_3C, CH_3, C_2-C_3, H, H, H, H

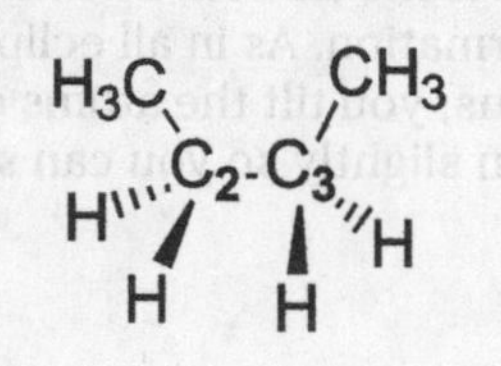

Solve It

4. Draw the Newman projection of the shown conformation of pentane, sighting down the C_2-C_3 bond.

H, H, CH_2CH_3, C_2-C_3, H_3C, H, H

Solve It

Comparing Conformational Stability

Some conformations are more stable than others. The conformation that puts the atoms or groups as far as possible from each other is the *staggered* conformation; it's generally lower-energy and more stable. The *eclipsed* conformation places groups on the adjacent carbons in direct contact, and it's the highest-energy conformation. (***Note:*** The Newman projection for the eclipsed conformation rotates the bonds slightly away from each other so you can see all the atoms on paper, but it's assumed the atoms are aligned.)

When you have two big groups or atoms attached to the two carbons you're sighting down in a Newman projection, not all eclipsed or staggered conformations are equal in energy. In general, if you have two big groups (such as two CH_3 groups) or big atoms on each carbon, the lowest-energy conformation puts these two groups as far from each other as possible. Here are the conformations you need to worry about, from most to least stable (see Figure 7-4):

- **Anti conformation:** The staggered conformation with the big groups opposite each other; it's the lowest-energy conformation
- **Gauche conformation:** The staggered conformation with the big groups positioned next to each other; it's slightly higher in energy than the anti conformation
- **Eclipsed conformation:** This basic eclipsed conformation doesn't put the big groups on top of each other; it's the most stable eclipsed conformation but is less stable than the gauche or anti conformation
- **Totally eclipsed conformation:** The eclipsed conformation with the big groups on top of each other; it's the most unstable conformation

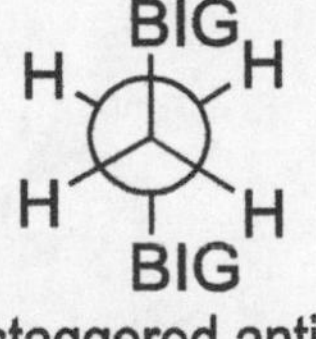

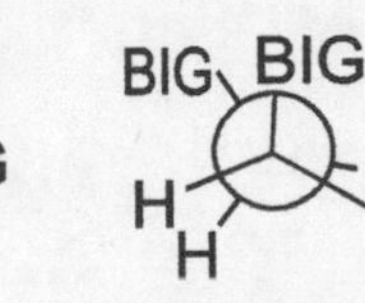

Figure 7-4: Anti, gauche, eclipsed, and totally eclipsed conformations.

Q. Using a Newman projection, draw the lowest-energy conformation for butane ($CH_3CH_2CH_2CH_3$), sighting down the C_2-C_3 bond.

A.

$H_3C-C_2-C_3-CH_3$ (each of C_2 and C_3 bearing two H)

Newman projection: front carbon with CH_3, H, H; back carbon with CH_3, H, H; the two CH_3 groups opposite each other.

staggered anti

The first thing I do with these problems is draw out all the substituents of the two carbons in the bond I'm looking down — in this case, C_2 and C_3. Next, because you want the most stable Newman conformation, you draw a staggered conformation, because staggered conformations are more stable (lower energy) than eclipsed conformations. On C_2, you have three substituents: two hydrogens and a CH_3 group. Likewise, on C_3, which is the back carbon in the Newman projection, you have two hydrogens and a CH_3 group. The CH_3 groups are big, so the most stable conformation puts these groups as far away from each other as possible, in a staggered anti conformation.

5. Using a Newman projection, draw the lowest-energy conformation for propane ($CH_3CH_2CH_3$), sighting down the C_1-C_2 bond.

Solve It

6. Using a Newman projection, draw the highest-energy conformation for butane ($CH_3CH_2CH_2CH_3$), sighting down the C_2-C_3 bond.

Solve It

7. Using a Newman projection, draw the least-stable *staggered* conformation for pentane ($CH_3CH_2CH_2CH_2CH_3$), sighting down the C_2-C_3 bond.

Solve It

8. Using a Newman projection, draw the lowest-energy conformation of the following molecule, sighting down the C-C bond.

Br, Br, Cl

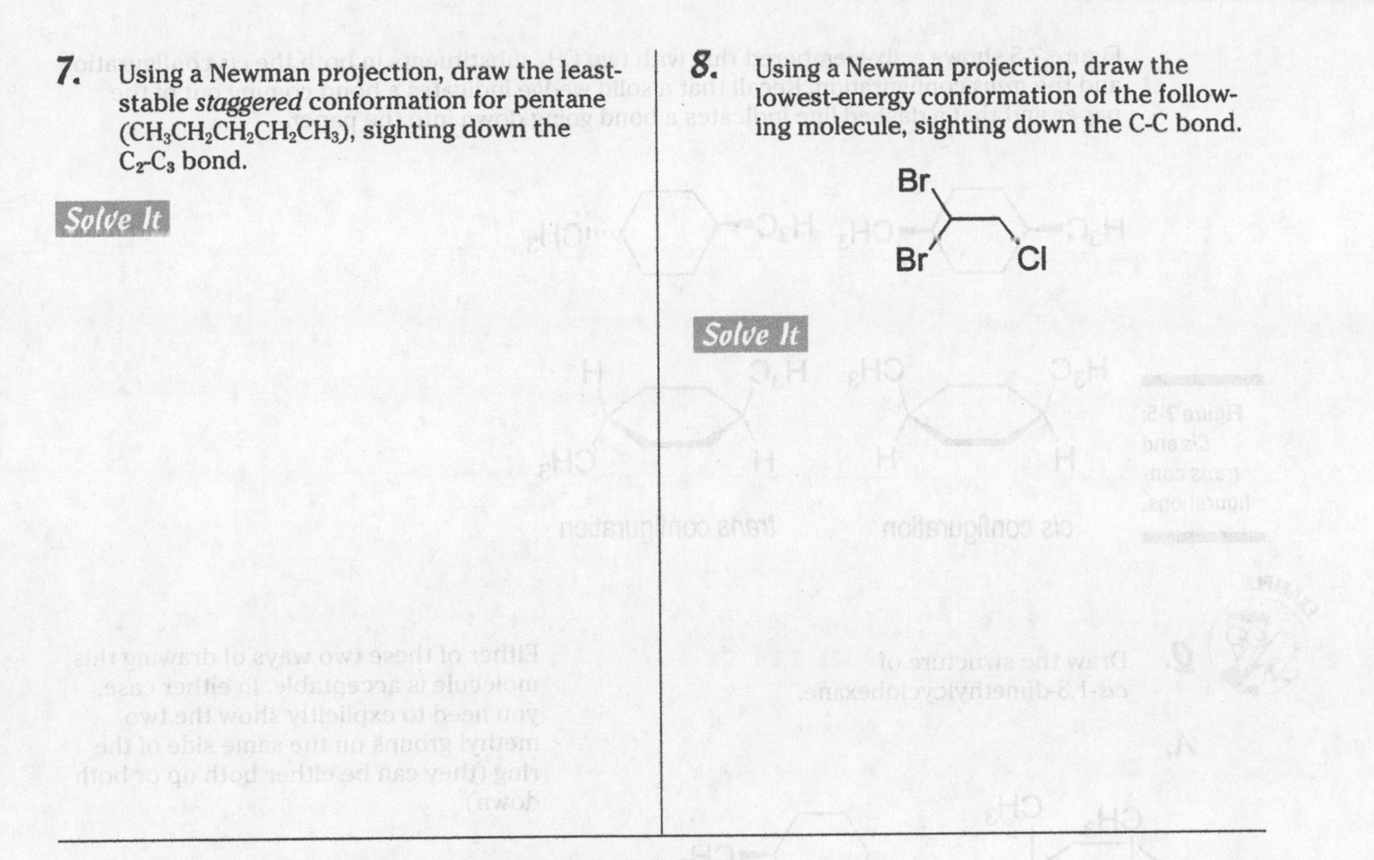

Solve It

Choosing Sides: The Cis-Trans Stereochemistry of Cycloalkanes

Like carbon chains, rings can exist in different conformations. But rings have a separate important feature: A ring has two distinct faces, a top face and a bottom face. Therefore, if you have two substituents on a ring, two configurations for the substituents are possible. You can have both substituents point off the same face of the ring (called the *cis* configuration), or you can have the substituents point off opposite faces of the ring (called the *trans* configuration).

Unlike straight-chain molecules, you can't rotate around a carbon-carbon bond constrained in a ring without breaking any bonds, so you can't convert a *cis* configuration to a *trans* configuration simply by rotation. You can try this rotation on a molecular model of a ring to prove this to yourself. A chemical reaction needs to take place to convert a ring in a *cis* configuration to one in a *trans* configuration.

Figure 7-5 shows a six-membered ring with two CH_3 substituents in both the *cis* configuration and the *trans* configuration. Recall that a solid wedge indicates a bond coming out of the paper and that a dashed line indicates a bond going down into the paper.

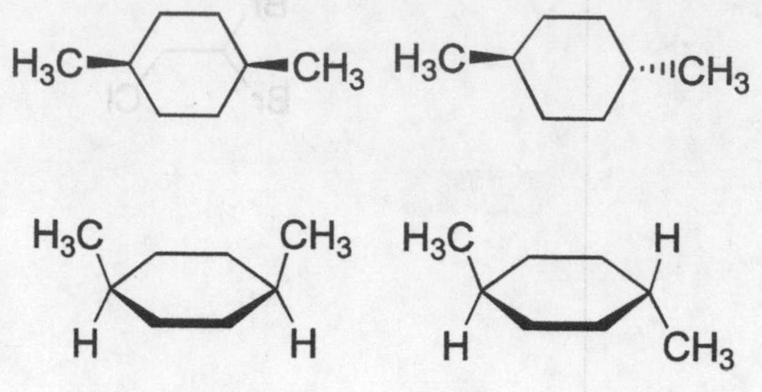

cis configuration *trans* configuration

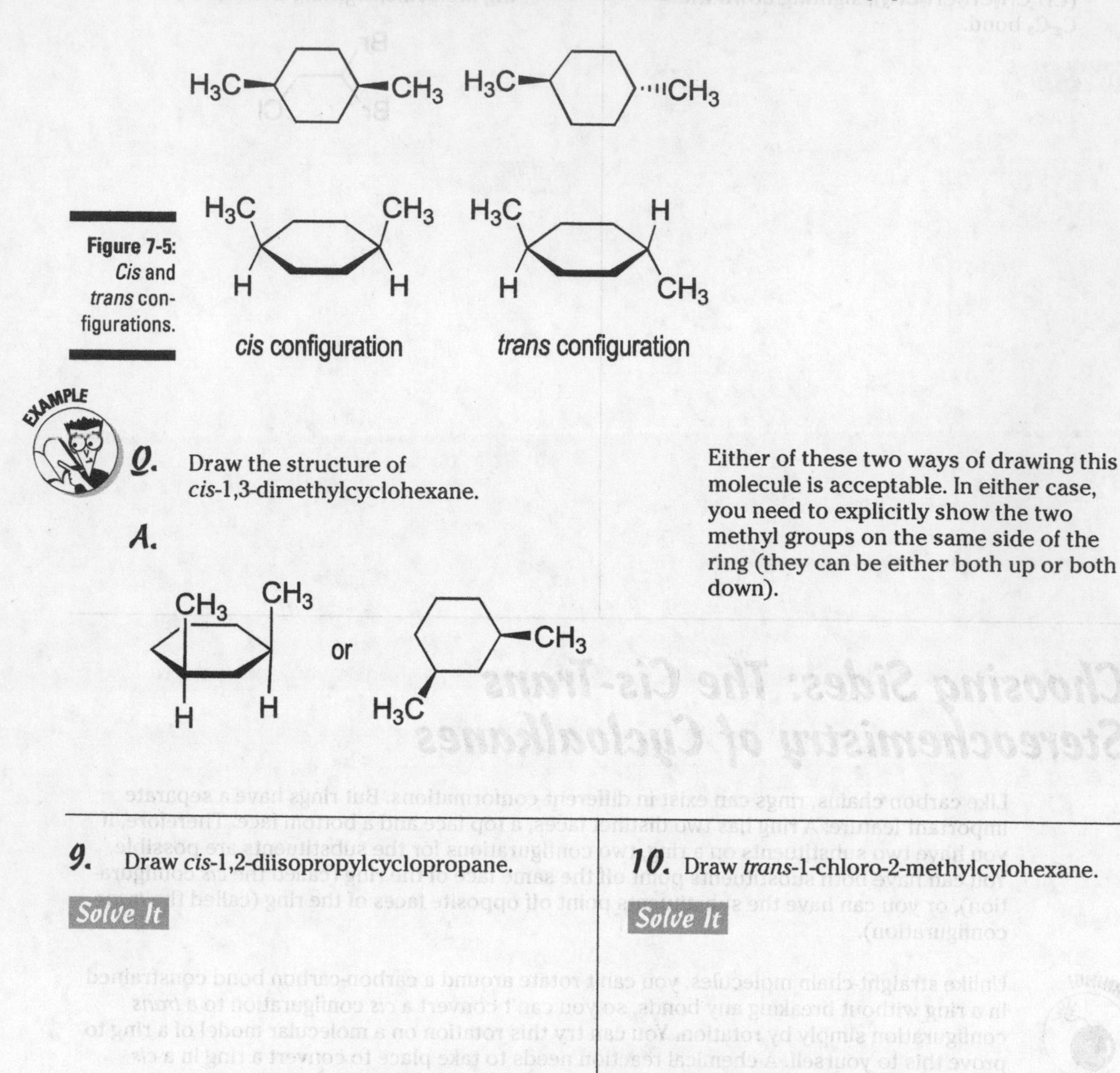

Figure 7-5: *Cis* and *trans* configurations.

Q. Draw the structure of *cis*-1,3-dimethylcyclohexane.

A. Either of these two ways of drawing this molecule is acceptable. In either case, you need to explicitly show the two methyl groups on the same side of the ring (they can be either both up or both down).

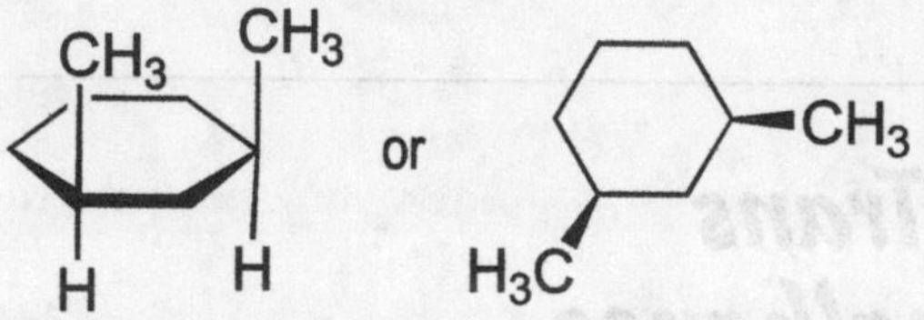

9. Draw *cis*-1,2-diisopropylcyclopropane.

Solve It

10. Draw *trans*-1-chloro-2-methylcylohexane.

Solve It

Getting a Ringside Seat with Cyclohexane Chair Conformations

The most common ring that you see in organic chemistry is *cyclohexane,* a ring of six carbons (C_6H_{12}). Cyclohexane is often drawn as a flat hexagon on paper, but the lowest energy conformation of a cyclohexane ring isn't flat but rather is folded into a conformation called a *chair* (because this conformation looks like a lounge chair — sort of; see Figure 7-6). When a cyclohexane molecule is in the chair conformation, you have two distinct types of hydrogens:

- **Axial hydrogens:** Hydrogens that stick up and down roughly perpendicular to the plane of the ring
- **Equatorial hydrogens:** Hydrogens that point out roughly in the plane of the ring (or around the belt or equator of the ring)

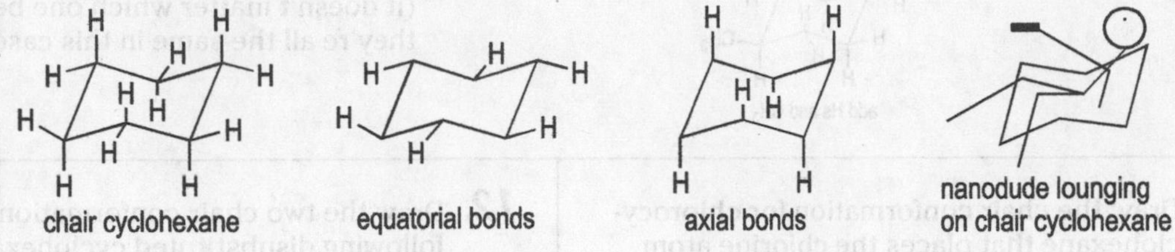

Figure 7-6: A chair cyclohexane with axial and equatorial hydrogens.

Because bonds in a ring can't rotate all the way around like they can in a straight-chain molecule, the way a chair can change conformation is by undergoing what's called a *ring flip.* I show this in Figure 7-7, with all but one of the hydrogens omitted for clarity. To perform a ring flip, you push the nose of the chair down and lift the tail of the chair up, as shown by the arrows, giving a new cyclohexane conformation. (To really appreciate this ring flip, try performing it with molecular models.) Upon undergoing a ring flip, all axial bonds become equatorial, and all equatorial bonds become axial.

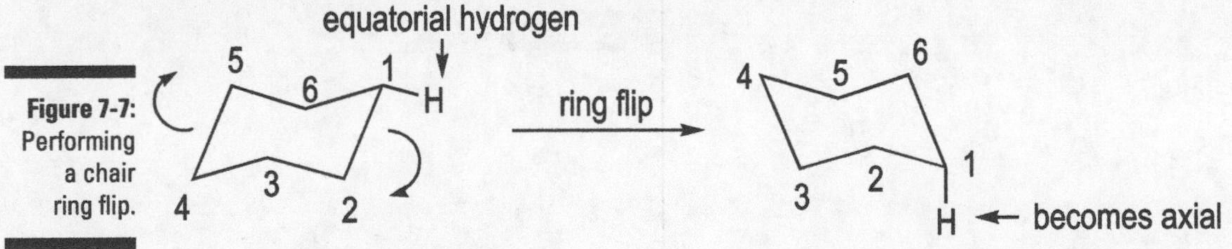

Figure 7-7: Performing a chair ring flip.

Q. Draw the chair conformation of methylcyclohexane. Draw a conformation that places the CH_3 group into an equatorial position.

A. Drawing a chair is hard at first, but with a bit of practice, it'll soon become second nature to you. First draw two slightly separated parallel lines that are slightly tilted from the vertical, and then complete the chair by adding the nose and the tail. You then add the bonds attached to the chair. I like to draw the axial bonds first; if the carbon on the chair points up, draw the axial bond up on that carbon, and if the carbon on the chair points down, draw the axial bond sticking straight down on that carbon. After you've drawn the axial bonds, adding the equatorial bonds around the equator of the chair is a fairly straightforward task. Then add the atoms, making sure to put the CH_3 in one of the equatorial bonds (it doesn't matter which one because they're all the same in this case).

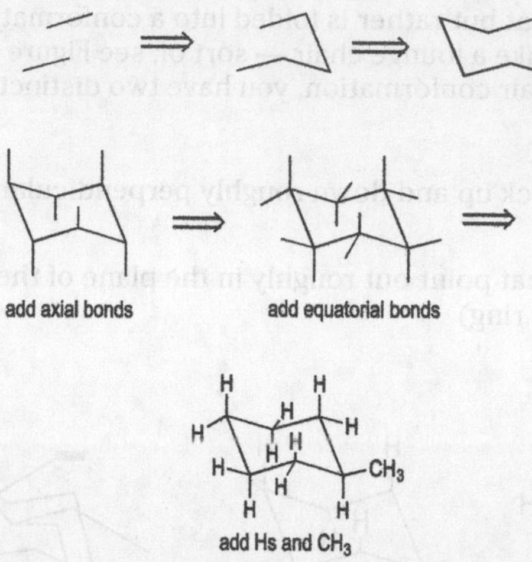

add axial bonds

add equatorial bonds

add Hs and CH_3

11. Draw the chair conformation for chlorocyclohexane that places the chlorine atom in the equatorial position. Then perform the ring flip to generate the other chair conformation.

Solve It

12. Draw the two chair conformations for the following disubstituted cyclohexane. ***Hint:*** Make sure your chair drawings show the two substituents with the correct *trans* stereochemistry. (I discuss *trans* and *cis* configurations in the preceding section.)

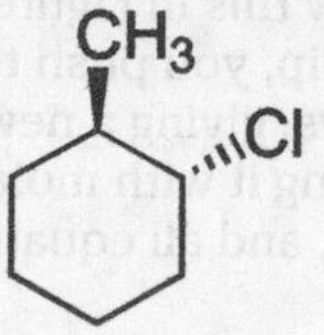

CH_3

Cl

Solve It

Predicting Cyclohexane Chair Stabilities

When a cyclohexane has substituents, the two possible chair conformations often have different stabilities. To minimize strain, cyclohexanes prefer to exist in a chair conformation that places the large substituent groups in the equatorial position rather than the axial position.

To predict which of the two chair conformations is lower-energy, do the following:

1. **Draw the cyclohexane ring in the chair with the substituents in the proper orientation to each other.**
2. **Find the other conformation by performing a ring flip.**
3. **Determine which of the two conformations is lower-energy by seeing which conformation places the large substituents in the equatorial position.**

Q. Draw both chair conformations of *trans*-1,2-dibromocyclohexane. Then determine which conformation is more stable.

A.

up relative to hydrogen

down relative to hydrogen

ring flip

diequatorial
more stable

diaxial
less stable

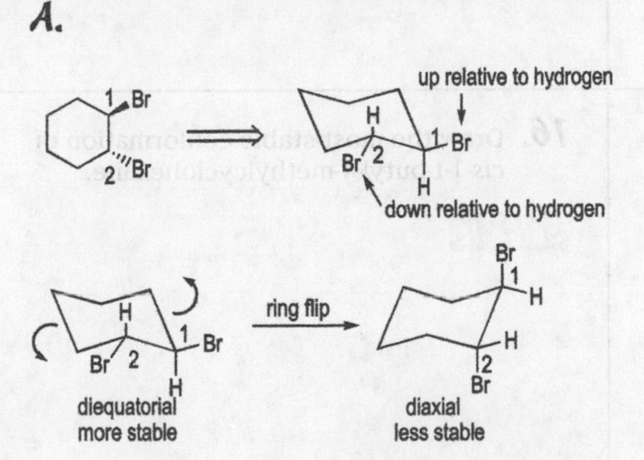

Before you worry about which conformation is lower in energy, draw a chair and then place the substituents on the chair with the correct configuration (that is, place the substituents on the ring *trans* to each other, not *cis*). I've arbitrarily numbered the cyclohexane ring carbons, but of course, you can start numbering anywhere. The bromine on the first carbon can go either axial or equatorial on the first carbon (I've arbitrarily placed it equatorial), but the next bromine has to be *trans* relative to the first bromine. The first bromine is up relative to the hydrogen on the same carbon, so to make a *trans* configuration, the next bromine needs to go down relative to the hydrogen on carbon 2, which is also an equatorial position. This gives a diequatorial conformation. Performing the ring flip to get the other conformation leads to a diaxial conformation because all equatorial bonds become axial (and all axial bonds become equatorial). Because the lower-energy conformation is the one in which you place the substituents in the equatorial position, the diequatorial conformation is the more stable.

13. Draw both chair conformations of *trans*-1,4-diethylcyclohexane and predict which one is more stable.

Solve It

14. Draw the least-stable conformation of *cis*-1,3-dimethylcyclohexane.

Solve It

15. Draw the most-stable conformation of *trans*-1,2-dimethylcyclohexane.

Solve It

16. Draw the most-stable conformation of *cis*-1-t-butyl-4-methylcyclohexane.

Solve It

Answer Key

The following are the answers to the practice questions presented in this chapter.

1

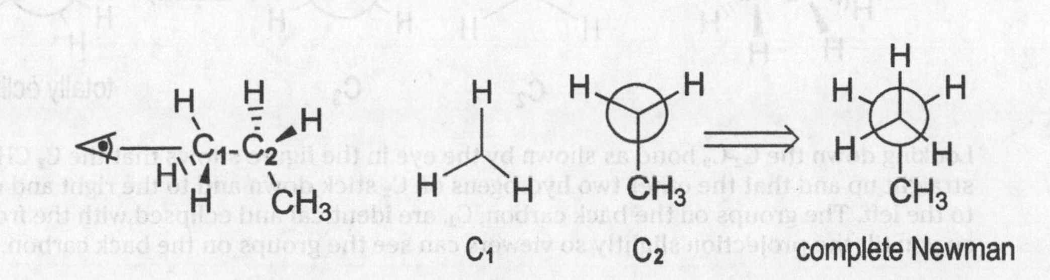

In your mind's eye, sight down the C_1-C_2 bond as the eye is doing. You can start the Newman projection by adding the hydrogens onto the C_1: One hydrogen atom sticks straight up and toward the viewer, one sticks down and to the right, and one sticks down and to the left. Then add the substituents to C_2. The CH_3 group sticks straight down, and the other two hydrogens stick up and to the right and up and to the left.

2

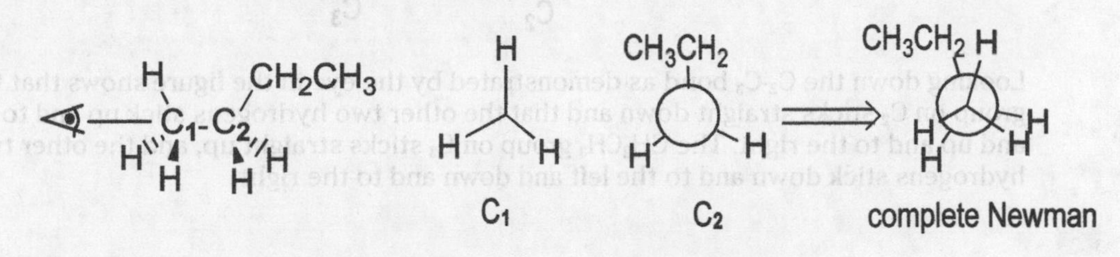

The first thing to do is to visualize your eye looking down the C_1-C_2 bond as shown. Then draw the Newman projection. On C_1, hydrogen is sticking straight up, and the other two hydrogens are sticking down and to the right and down and to the left. C_2 has a CH_2CH_3 group sticking straight up, so you draw this group straight up, eclipsed with the front H. Note that the bonds in the Newman projection are tilted slightly to the side to allow the atoms to be shown, but you assume that the bonds are eclipsed.

3

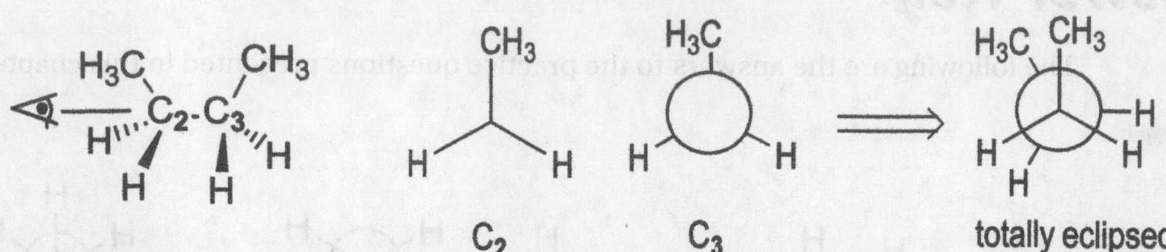

Looking down the C_2-C_3 bond as shown by the eye in the figure shows that the C_2 CH_3 is sticking straight up and that the other two hydrogens on C_2 stick down and to the right and down and to the left. The groups on the back carbon, C_3, are identical and eclipsed with the front groups, so you tilt the projection slightly so viewers can see the groups on the back carbon.

4

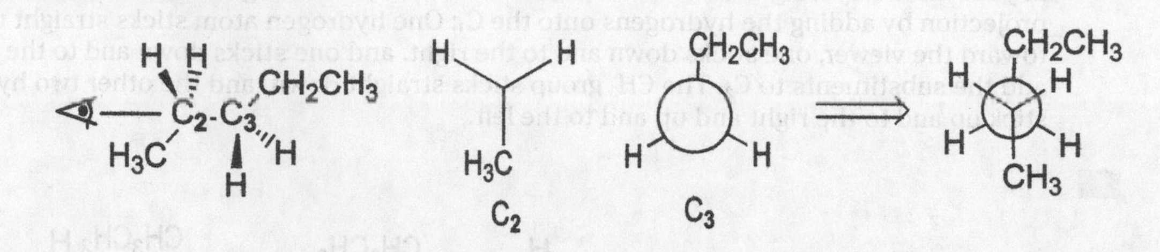

Looking down the C_2-C_3 bond as demonstrated by the eye in the figure shows that the CH_3 group on C_2 sticks straight down and that the other two hydrogens stick up and to the left and up and to the right. The CH_2CH_3 group on C_3 sticks straight up, and the other two back hydrogens stick down and to the left and down and to the right.

5

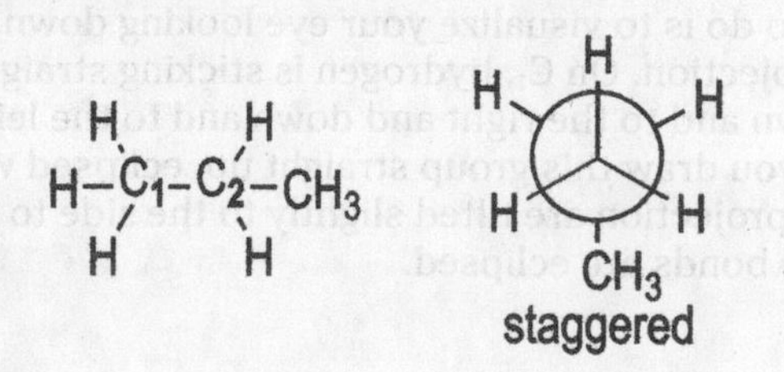

First draw out all the substituents on each of the two carbons you're sighting down — in this case, the C_1-C_2 carbons. Next, draw the staggered conformation Newman projection, because staggered conformations are lower in energy than eclipsed conformations. Add the C_1 substituents and then the C_2 substituents. In this example, you don't need to worry about where you place the substituents in the Newman projection because there's only one non-hydrogen group (the CH_3), so you can't have gauche or anti conformations.

6

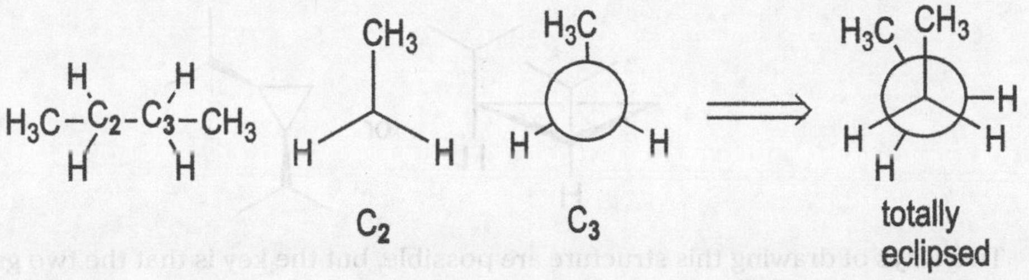

$H_3C-C_2-C_3-CH_3$ (H, H above; H, H below)

C_2 C_3 totally eclipsed

The first thing I recommend is to draw all the substituents on the two carbons that you're sighting down, as shown. That drawing shows you that you have two H's and a CH_3 that need to be attached to both C_2 and C_3. Because you want the highest-energy conformation, draw an eclipsed conformation. To give the highest-energy eclipsed conformation, you eclipse the two CH_3 groups (shown here both sticking up).

7

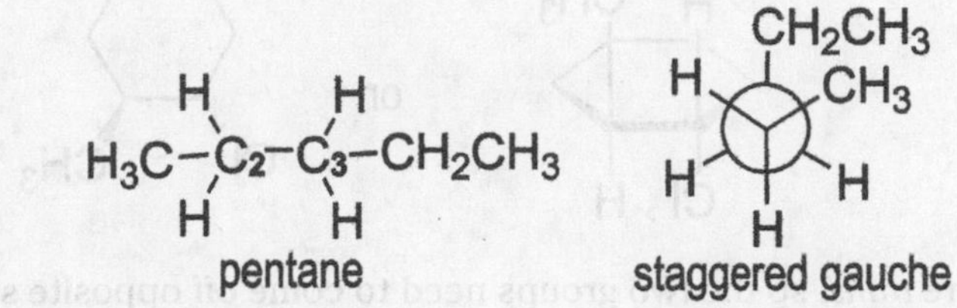

$H_3C-C_2-C_3-CH_2CH_3$

pentane staggered gauche

The first thing to do is draw out the structure of pentane to show what groups are attached to the two atoms you're sighting down in the Newman projection (C_2 and C_3). You want the least-stable staggered conformation, so you want to put the two big groups (the CH_3 and the CH_2CH_3) as close as possible to give a gauche conformation.

8

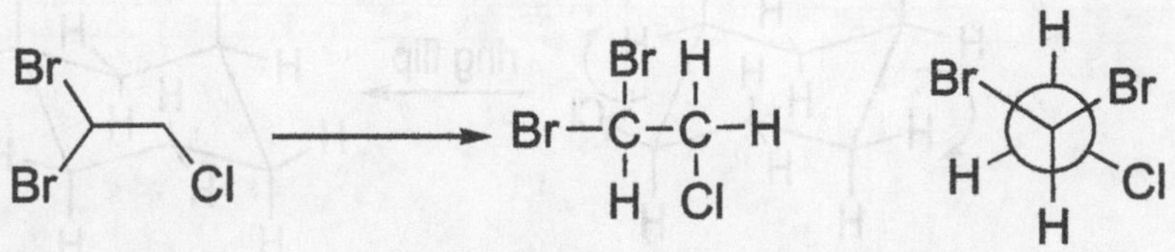

Drawing out all the atoms attached to both carbons tells you which groups you need to attach to the Newman projection. Because you want the most stable conformation, you need to draw a staggered conformation. After you've added the front atoms (the two Br's and the H), you need to place the back Cl to make the most stable structure. The lowest-energy arrangement places the Cl so that it neighbors only one Br and is not stuck between the two Br's.

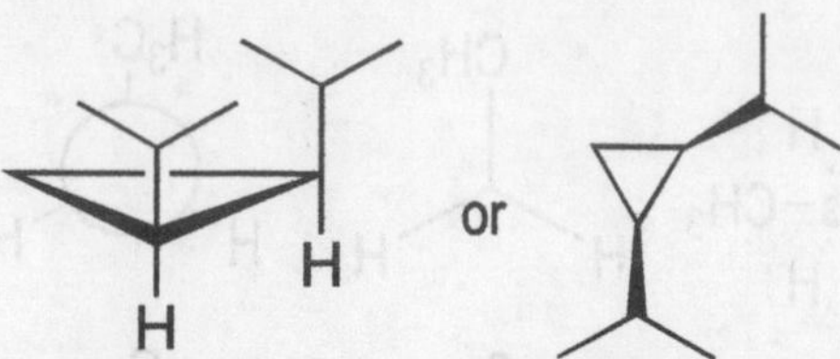

Two ways of drawing this structure are possible, but the key is that the two groups are *cis*, so you need to attach them to the same side of the ring. You can draw the canted ring that shows both isopropyl groups sticking up (or both sticking down). An *isopropyl group* is a three-carbon substituent that looks like a snake's tongue. Alternatively, you can draw a flat three-membered ring and use two wedges (or dashes) to show that the two isopropyl groups are on the same side of the ring.

10

H CH_3 or Cl H Cl CH_3

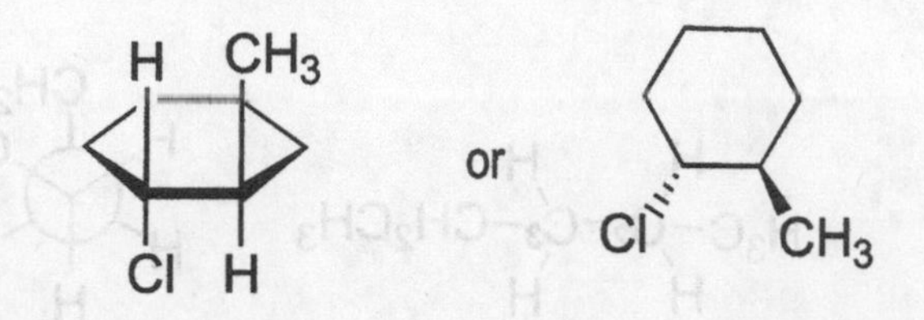

The groups are *trans*, so the two groups need to come off opposite sides of the ring. You can draw the canted six-membered ring that shows one group going up and one going down, or you can draw the flat ring and use a dash and a wedge to show that the groups are on opposite sides of the ring.

11

ring flip

Cl equatorial
most stable

Cl axial
least stable

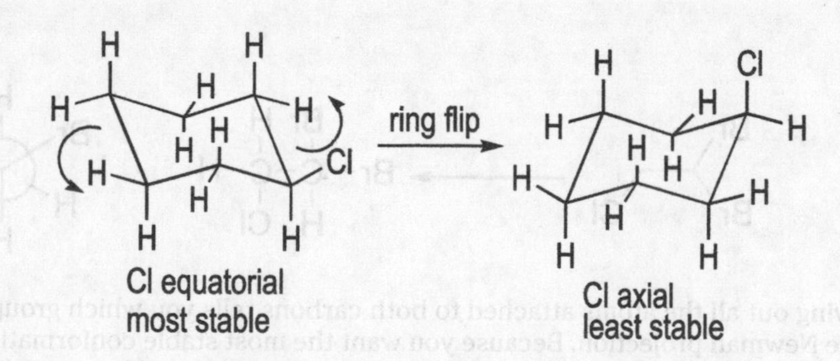

The first order of business is to draw a cyclohexane chair. Next, add the chlorine to one of the equatorial positions. Performing the ring flip moves the chlorine atom from an equatorial position to an axial position because on ring flips, all axial bonds become equatorial and all equatorial bonds become axial.

up relative to hydrogen

down relative to hydrogen

ring flip

diequatorial
most stable

diaxial
least stable

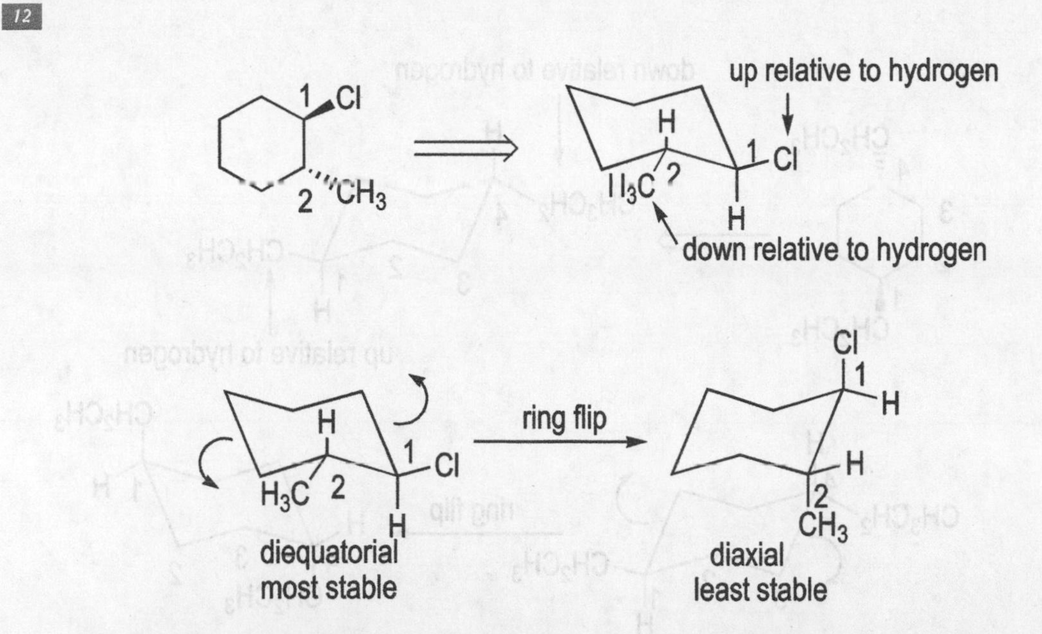

After you draw a chair cyclohexane, you have to add the substituents. I've arbitrarily chosen bottom right carbons on the chair, but you can add the substituents anywhere on the chair you want. It doesn't matter where you place the first substituent, but I've placed the first chlorine equatorial on the first carbon. For clarity, I've omitted the other hydrogens, but you can draw out all the atoms if you want. The key is that after you place the first substituent, you have to place the second one so that you get the correct stereochemistry (*trans* in this case). The chlorine in the equatorial position is up relative to the hydrogen on that carbon, so to make the stereochemistry *trans*, you have to place the methyl (CH_3) group down relative to the hydrogen on the second carbon. This down position turns out to be the equatorial position. Performing the ring flip switches both groups from equatorial to axial. The most stable conformation is the conformation that places the substituents equatorial.

13

down relative to hydrogen

up relative to hydrogen

ring flip

diequatorial
more stable

diaxial
less stable

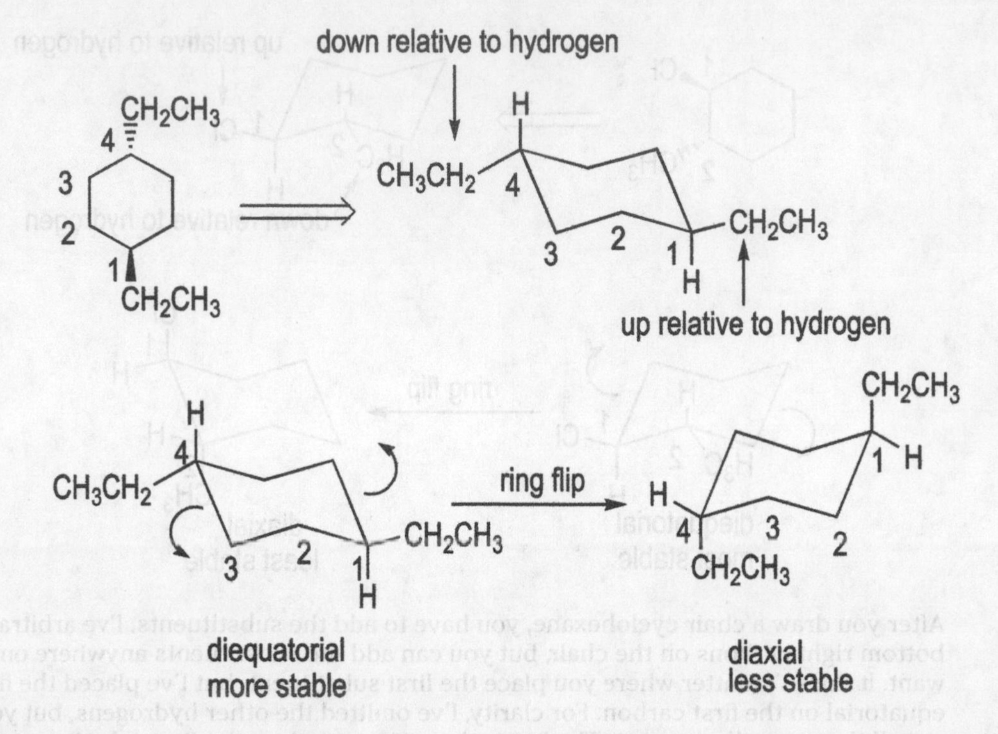

After you draw the chair, the next step is to add the substituents so they have the right stereochemistry (*trans* in this case). I've arbitrarily numbered the carbons starting from the bottom right carbon, but you can start anywhere on the ring you'd like. It doesn't matter which position (axial or equatorial) you choose for the ethyl (CH_3CH_2) substituent on the first carbon, so I show it in the equatorial position. The key is that the next ethyl group you add has to be *trans* to the first substituent. In the equatorial position, the ethyl group on the first carbon is up relative to the hydrogen, so on the fourth carbon, the other ethyl group needs to go down relative to the hydrogen, which is also an equatorial position. Doing the ring flip changes all equatorial bonds to axial (and all axial bonds to equatorial), so the ring flip structure has both ethyl groups axial. Because the lower-energy conformation places the groups in the equatorial position, the first conformation I drew is the lower-energy conformation.

14

up relative to hydrogen

up relative to hydrogen

CH_3 1 2 3 CH_3 ⟹ H_3C 3 2 1 CH_3 H H

H_3C 3 2 1 CH_3 H H ring flip CH_3 1 H CH_3 H 3 2

diequatorial more stable

diaxial less stable

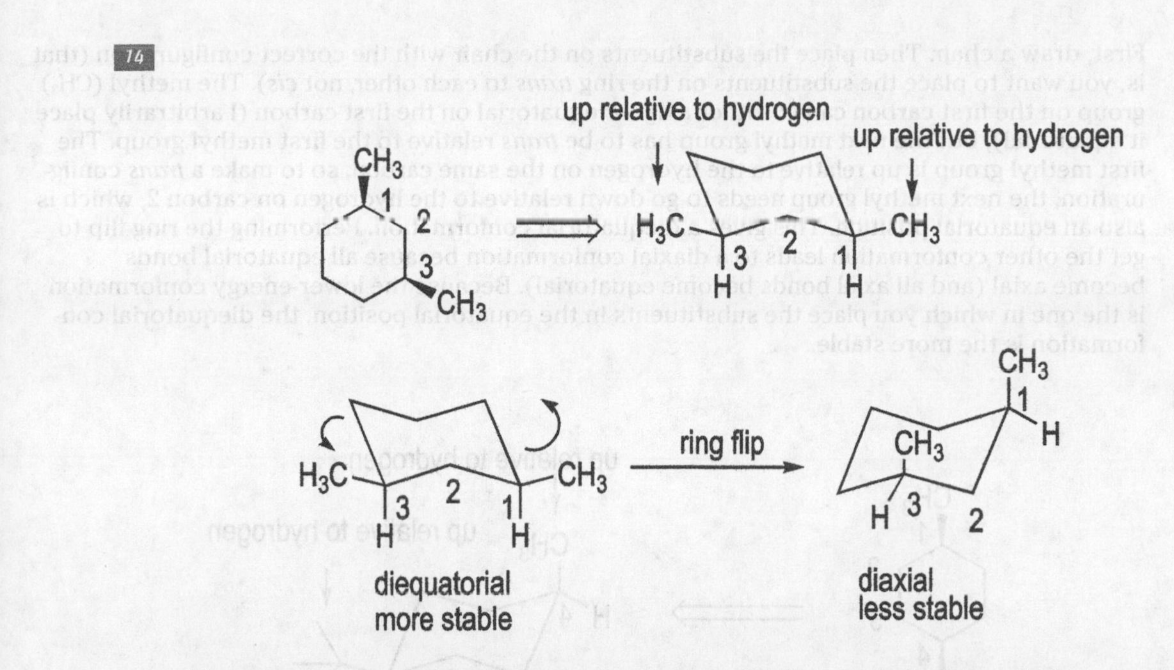

After drawing the chair and numbering it, place the first methyl arbitrarily (I show it in the equatorial position) and make sure that the next methyl group you add to C_3 is on the same side as the first methyl. Because the first methyl in the equatorial position is up relative to the hydrogen, the next methyl group needs to be up relative to the hydrogen, which in this case is the equatorial position. Undergoing the ring flip converts both equatorial methyl groups to axial methyl groups. Axial placement of big groups raises the energy, so the diaxial conformation is the conformation that's the higher in energy and the less stable.

15

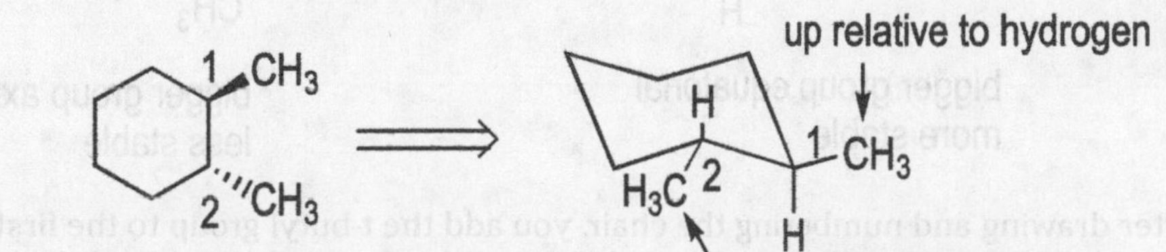

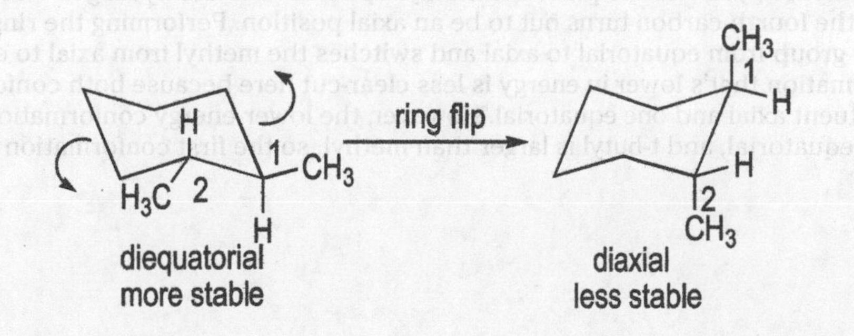

1 CH_3 2 CH_3 ⟹ up relative to hydrogen H 1 CH_3 H_3C 2 H down relative to hydrogen

H H_3C 2 1 CH_3 H ring flip CH_3 1 H H 2 CH_3

diequatorial more stable

diaxial less stable

First, draw a chair. Then place the substituents on the chair with the correct configuration (that is, you want to place the substituents on the ring *trans* to each other, not *cis*). The methyl (CH_3) group on the first carbon can go either axial or equatorial on the first carbon (I arbitrarily place it equatorial), but the next methyl group has to be *trans* relative to the first methyl group. The first methyl group is up relative to the hydrogen on the same carbon, so to make a *trans* configuration, the next methyl group needs to go down relative to the hydrogen on carbon 2, which is also an equatorial position. This gives a diequatorial conformation. Performing the ring flip to get the other conformation leads to a diaxial conformation because all equatorial bonds become axial (and all axial bonds become equatorial). Because the lower-energy conformation is the one in which you place the substituents in the equatorial position, the diequatorial conformation is the more stable.

16

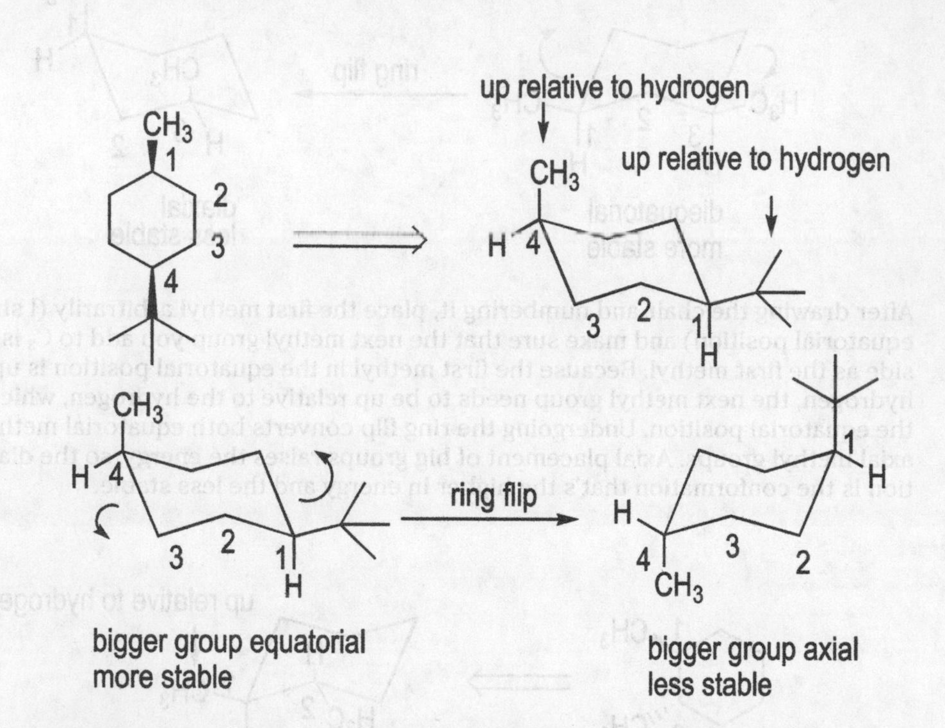

After drawing and numbering the chair, you add the t-butyl group to the first carbon (the t-butyl group is a four-carbon substituent that looks like a trident). I've arbitrarily added the t-butyl group to the equatorial position, which is up relative to the hydrogen, so to make the *cis* isomer, you need to place the methyl up relative to the hydrogen on the fourth carbon. *Up* on the fourth carbon turns out to be an axial position. Performing the ring flip switches the t-butyl group from equatorial to axial and switches the methyl from axial to equatorial. The conformation that's lower in energy is less clear-cut here because both conformations have one substituent axial and one equatorial. However, the lower-energy conformation places the larger group equatorial, and t-butyl is larger than methyl, so the first conformation is lower in energy.

Chapter 8

Doubling Down: The Alkenes

In This Chapter

- Naming alkenes
- Working with alkene reactions
- Proposing alkene reaction mechanisms
- Seeing carbocation rearrangements

Carbon-carbon double bonds, or *alkenes,* are one of the most important functional groups of organic molecules. The reactions of this functional group provide the first introduction to the reactions of organic compounds for most students. And for the majority of the reactions you encounter in organic chemistry, you're expected to know how to do at least two things:

- Predict the product of the reaction given the starting materials and the reagents
- Show the mechanism of the reaction

The *mechanism* of a reaction is the step-wise process of how the starting material changes into the product; it shows which bonds are formed and broken and in what order. You also include any intermediate molecules formed along the way in the reaction in a mechanism. On paper, you show how bonds are formed and broken by using arrows to show the movement of electrons in a reaction. Getting the hang of drawing mechanisms, or *arrow-pushing,* takes practice, and in this chapter, you begin to master the drawing of mechanisms and predicting reaction products. But before getting to the alkene reactions, I first take you on a quick detour to show how to name alkenes.

Giving Alkenes a Good Name

For the most part, the nomenclature of alkenes is a fairly straightforward extension from the naming system of alkanes (see Chapter 6). Instead of ending with the suffix *-ane,* however, the names of alkenes end with the suffix *-ene.* The most challenging aspect of naming alkenes comes from the possibility that the double bond will have stereochemistry.

Here's how the alkene-naming process works:

1. **Identify the parent chain.**

   With alkenes, the parent chain is the longest chain of carbons that contains the carbon-carbon double bond. Table 8-1 lists the parent names for alkenes.

Table 8-1 Naming the Parent Chains

Number of Carbons	Parent Name	Number of Carbons	Parent Name
2	Ethene	7	Heptene
3	Propene	8	Octene
4	Butene	9	Nonene
5	Pentene	10	Decene
6	Hexene		

2. **Number the parent chain, starting from the side that reaches the double bond first.**

   Add a number to the parent name to indicate the first carbon of the carbon-carbon double bond. In a ring, the double bond has to contain the number 1 and number 2 carbons, and after including the double bond as the first two carbons, you number to give the substituents the lowest possible number.

3. **Add the substituents.**

   Alphabetize the names of the substituents and place them before the parent name. Use numbers to indicate the location of each substituent.

4. **Use a prefix to indicate any stereochemistry of the alkene.**

   Double bonds may have stereochemistry. If identical substituents stick off the double bond, use the *cis-trans* system (see Chapter 7 for details):

   - ***Cis*** **stereochemistry:** The identical substituents stick off the same side of the double bond.
   - ***Trans*** **stereochemistry:** The identical substituents stick off opposite sides of the double bond.

   If the two double-bonded substituents aren't identical, use the E/Z system. Divide the double bond in half with a line. Taking one half at a time, determine which substituent (the top or the bottom) has higher priority using the Cahn-Ingold-Prelog priority rules (which also determine priority in R/S nomenclature; see Chapter 5). Basically, under these rules, the substituent attached to the alkene by the atom with the higher atomic number has higher priority. Here's how the naming works:

   - **Z stereochemistry:** The high-priority substituents are on the same side ("ze zame zide" — top or bottom) of the double bond.
   - **E stereochemistry:** The high-priority substituents are on opposite sides of the double bond.

Draw out any hydrogens that aren't shown when assigning stereochemistry to double bonds. You don't need to name the stereochemistry of double bonds in small rings because *trans* and E double bonds in rings are unstable, but you can if you want to.

Q. Name the following alkene:

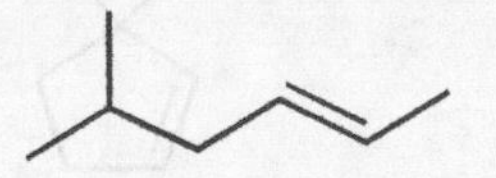

A. ***trans*-5-methyl-2-hexene.** Begin by identifying the parent chain, the longest chain of carbons that contains the carbon-carbon double bond. In this case, you number the alkene from right to left to give the first carbon of the alkene group the lowest number possible (as in *a*). Because the parent chain is six carbons long, the parent name for the molecule is 2-hexene, with the 2 indicating the position of the carbon-carbon double bond on the parent chain and the *-ene* suffix identifying the molecule as an alkene.

a)

6 5 4 3 2 1

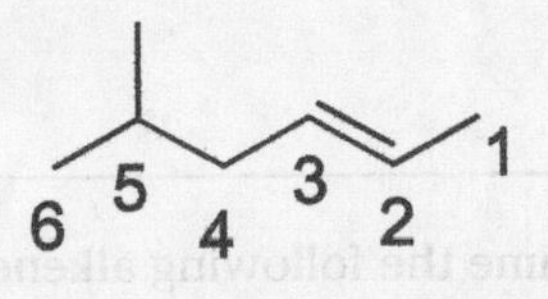

Next, add the substituents. As in *b,* the only substituent is a one-carbon group (a methyl) at the 5 carbon. Adding the substituent to the name gives you 5-methyl-2-hexene. Finally, indicate the stereochemistry. In this case, the double bond has two identical groups (the hydrogens) sticking off opposite sides of the double bond, giving the double bond *trans* stereochemistry.

b)

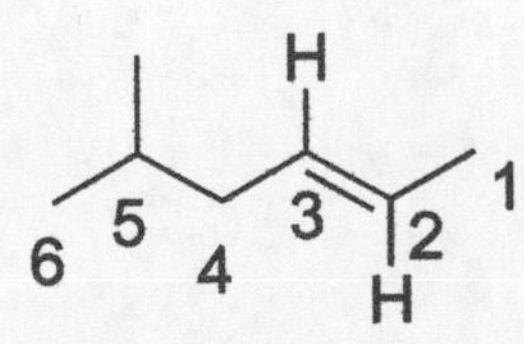

H

6 5 4 3 2 1

H

Q. Name the following alkene.

F

Cl

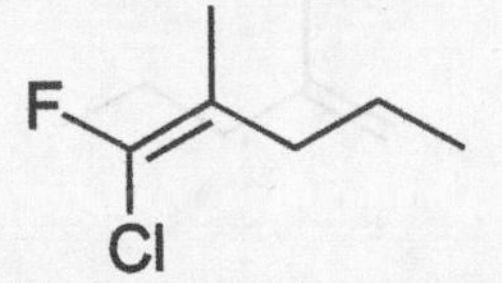

A. **Z-1-chloro-1-fluoro-1-pentene.** Find and number the parent chain, giving the lowest number possible to the double bond (as in *a*). The parent chain is a five-carbon alkene with the double bond at the first carbon, so the parent name is 1-pentene.

a)

F 1 2 3 4 5

Cl

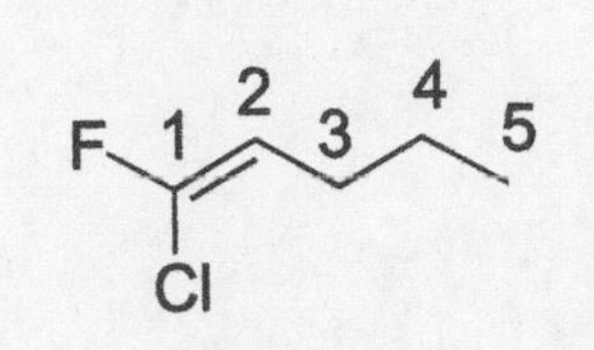

For substituent halogens, you replace *-ine* with *-o*, so fluorine as a substituent is called *fluoro* and chlorine as a substituent is called *chloro;* both attach to the number 1 carbon. After placing the substituents alphabetically in front of the parent name, you get 1-chloro-1-fluoro-1-pentene. Now the tricky part — the double bond stereochemistry. You can't use the *cis-trans* nomenclature because you don't have two identical substituents. Thus, you have to use the E/Z system. Divide the double bond in half with a dashed line and then determine priority on each side of the double bond. On the left side, chlorine has higher priority than fluorine because chlorine has a higher atomic number; likewise, the carbon chain gets higher priority on the right side of the double bond because carbon has a higher atomic number than hydrogen (see *b*). In this case, the highs are on the same side, so the alkene has Z stereochemistry.

b)

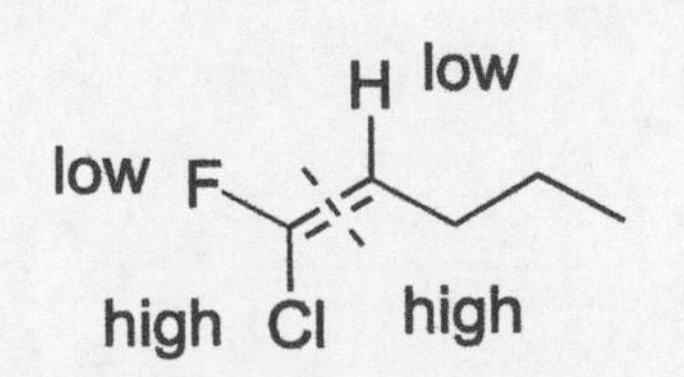

H low

low F

high Cl high

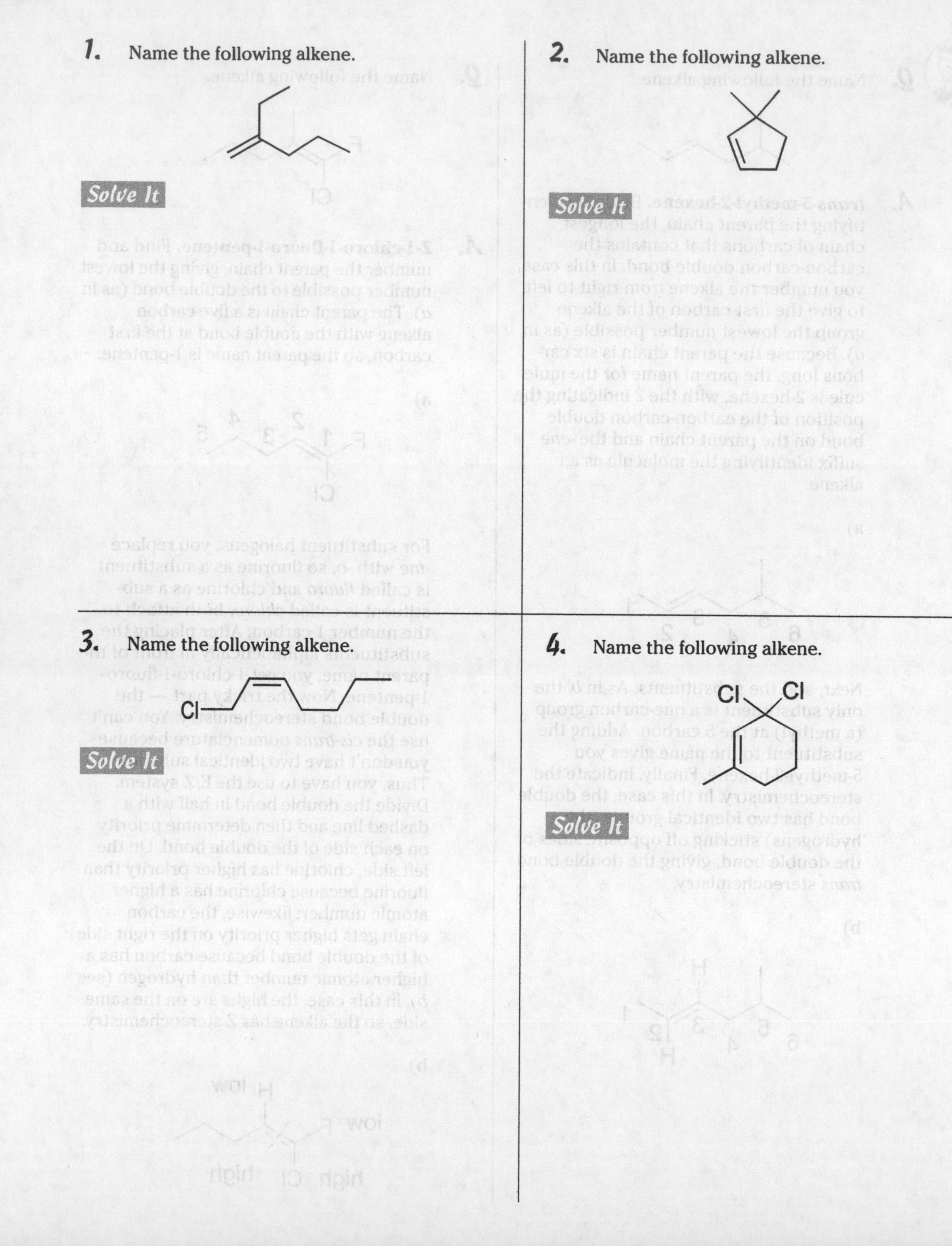

1. Name the following alkene.

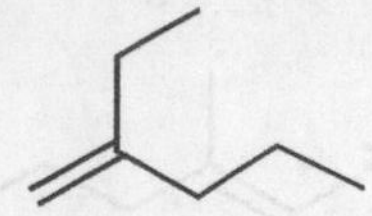

Solve It

2. Name the following alkene.

Solve It

3. Name the following alkene.

Cl

Solve It

4. Name the following alkene.

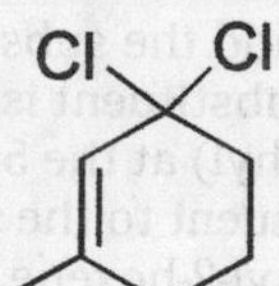

Solve It

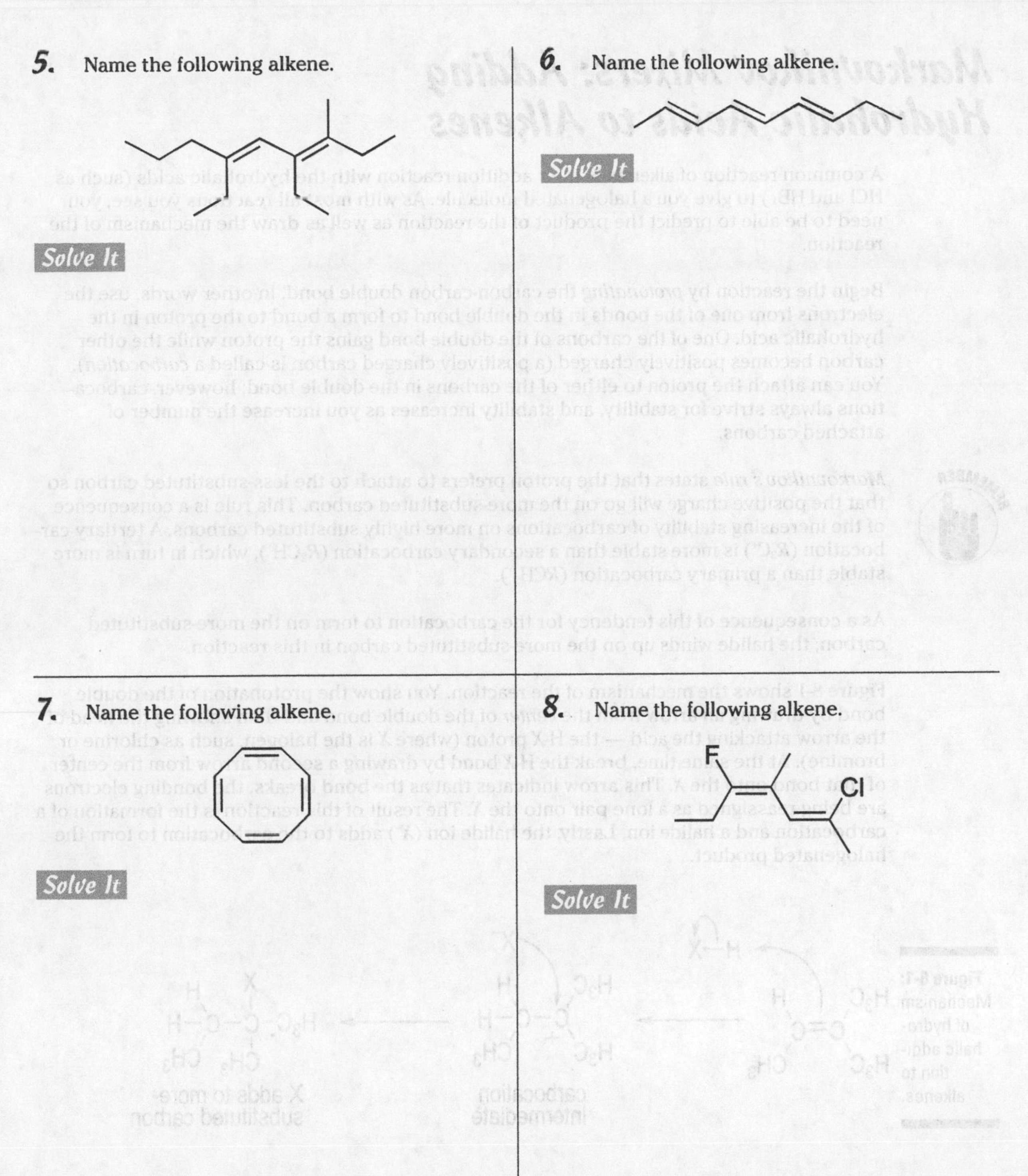

5. Name the following alkene.

Solve It

6. Name the following alkene.

Solve It

7. Name the following alkene.

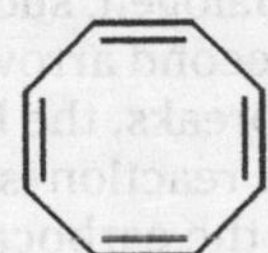

Solve It

8. Name the following alkene.

Solve It

Markovnikov Mixers: Adding Hydrohalic Acids to Alkenes

A common reaction of alkenes is their addition reaction with the hydrohalic acids (such as HCl and HBr) to give you a halogenated molecule. As with most all reactions you see, you need to be able to predict the product of the reaction as well as draw the mechanism of the reaction.

Begin the reaction by *protonating* the carbon-carbon double bond. In other words, use the electrons from one of the bonds in the double bond to form a bond to the proton in the hydrohalic acid. One of the carbons of the double bond gains the proton while the other carbon becomes positively charged (a positively charged carbon is called a *carbocation*). You can attach the proton to either of the carbons in the double bond; however, carbocations always strive for stability, and stability increases as you increase the number of attached carbons.

Markovnikov's rule states that the proton prefers to attach to the less-substituted carbon so that the positive charge will go on the more-substituted carbon. This rule is a consequence of the increasing stability of carbocations on more highly substituted carbons. A tertiary carbocation (R_3C^+) is more stable than a secondary carbocation (R_2CH^+), which in turn is more stable than a primary carbocation (RCH_2^+).

As a consequence of this tendency for the carbocation to form on the more-substituted carbon, the halide winds up on the more-substituted carbon in this reaction.

Figure 8-1 shows the mechanism of the reaction. You show the protonation of the double bond by drawing an arrow from the *center* of the double bond and then showing the head of the arrow attacking the acid — the H-*X* proton (where *X* is the halogen, such as chlorine or bromine). At the same time, break the H-*X* bond by drawing a second arrow from the center of that bond onto the *X*. This arrow indicates that as the bond breaks, the bonding electrons are being reassigned as a lone pair onto the *X*. The result of this reaction is the formation of a carbocation and a halide ion. Lastly, the halide ion (X^-) adds to the carbocation to form the halogenated product.

carbocation intermediate

X adds to more-substituted carbon

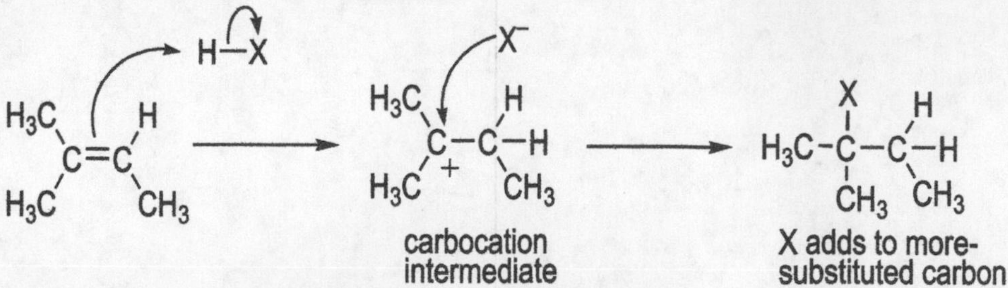

Figure 8-1: Mechanism of hydrohalic addition to alkenes.

The overall effect is that you add the H and *X* atoms to your former alkene. Remember that adding a hydrohalic acid to an alkene involves *Markovnikov addition.*

Q. Rank the following carbocations from most stable to least stable:

$H_3C-C^+(CH_3)(CH_3)$ $\quad$ $H_3C-C^+(H)(CH_3)$ $\quad$ $H_3C-C^+(H)(H)$

A.

$H_3C-C^+(CH_3)(CH_3)$ $\quad$ $H_3C-C^+(H)(CH_3)$ $\quad$ $H_3C-C^+(H)(H)$

teriary (3°) cation most stable $\quad$ secondary (2°) cation $\quad$ primary (1°) cation least stable

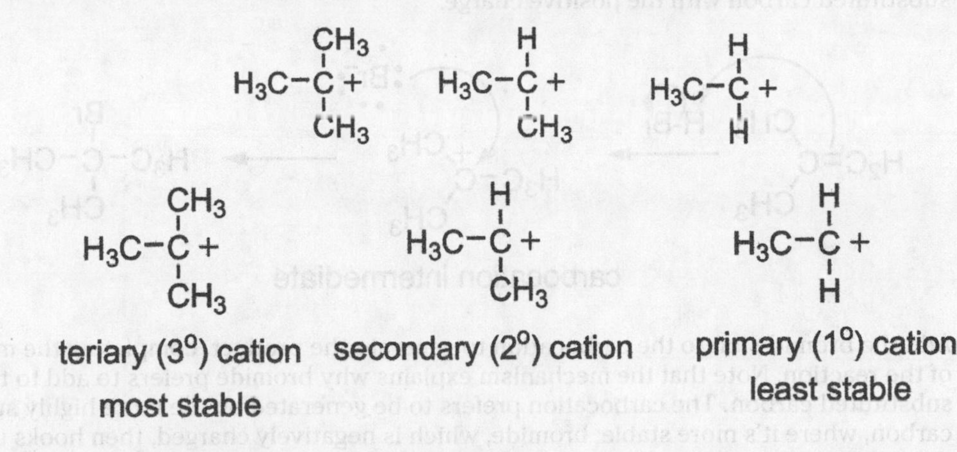

For carbocations, the stability increases as you increase the number of attached carbons. Therefore, the left-hand structure, which is tertiary (3°), is the most stable carbocation, and the right-hand structure, which is primary (1°), is the least stable carbocation of the three.

Q. Draw the product of the following reaction. Then draw the mechanism of the reaction, using arrows to show the movement of electrons that changes the starting material into the product.

$$H_2C=C(CH_3)_2 \xrightarrow{\text{H-Br}}$$

A. HBr addition to an alkene involves a Markovnikov addition of hydrogen bromide. *Markovnikov addition* indicates that the larger atom (bromine) adds to the more-substituted carbon and that the hydrogen adds to the less-substituted carbon (the left-most carbon).

$$H_2C=C(CH_3)_2 \xrightarrow{\text{H-Br}} H_3C-C(Br)(CH_3)-CH_3$$

Begin the mechanism of the reaction by protonating the double bond. Draw an arrow from the center of the double bond, with the head of the arrow attacking the H-Br proton. At the same time, break the H-Br bond by drawing a second arrow from the center of the H-Br bond onto the bromine. This arrow indicates that the two H-Br bonding electrons are being reassigned as a lone pair onto bromine and that the H-Br bond is being broken.

You have two possibilities of where to put the proton, but Markovnikov's rule states that the proton prefers to attach to the less-substituted carbon because doing so leaves the more-substituted carbon with the positive charge.

$$H_2C{=}C(CH_3)_2 \xrightarrow{H\text{-}Br} H_3C{-}\overset{+}{C}(CH_3)_2 \;\; :Br^- \longrightarrow H_3C{-}CBr(CH_3){-}CH_3$$

carbocation intermediate

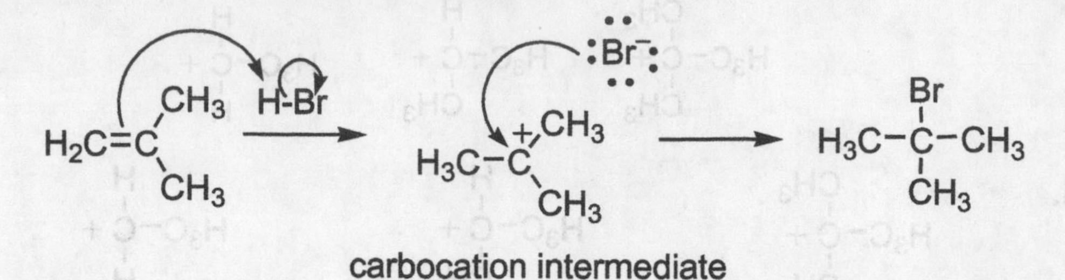

Add the bromide ion to the carbocation to generate the product, completing the mechanism of the reaction. Note that the mechanism explains why bromide prefers to add to the more-substituted carbon. The carbocation prefers to be generated on the more highly substituted carbon, where it's more stable; bromide, which is negatively charged, then hooks up with that positively charged carbon.

9. Show the major product of the following reaction. Then draw the mechanism of the reaction (using arrows) to show how the starting material is converted into the product.

H-Cl

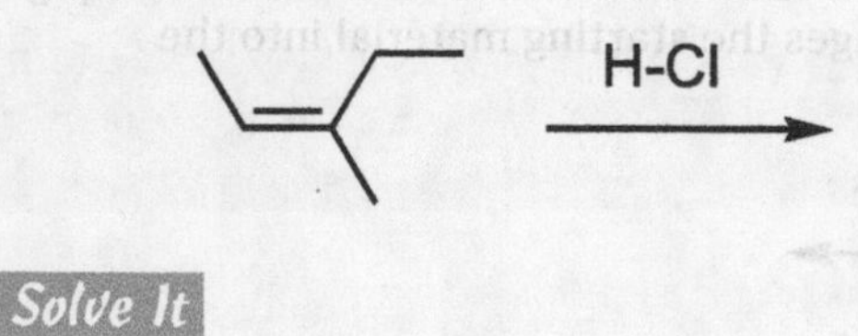

Solve It

10. Show the major product of the following reaction. Then draw the mechanism of the reaction (using arrows) to show how the starting material is converted into the product.

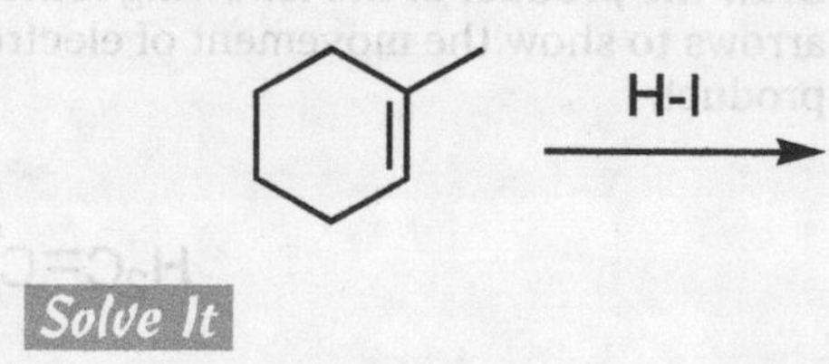

H-I

Solve It

11. Predict which of the following alkenes will react the fastest with HCl.

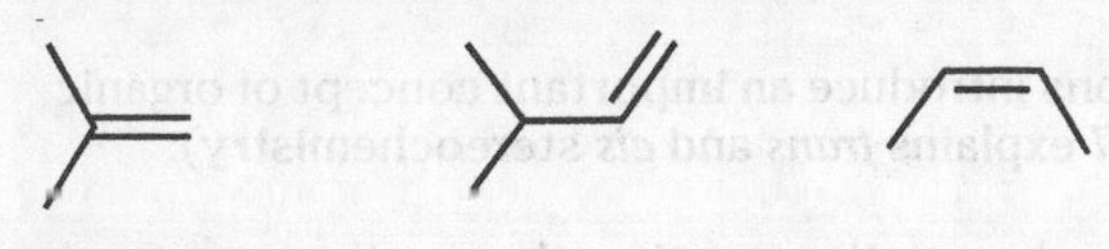

Solve It

12. Draw the mechanism for the following reaction. (***Hint:*** Begin by protonating the double bond outside of the ring.)

H-Cl

Cl

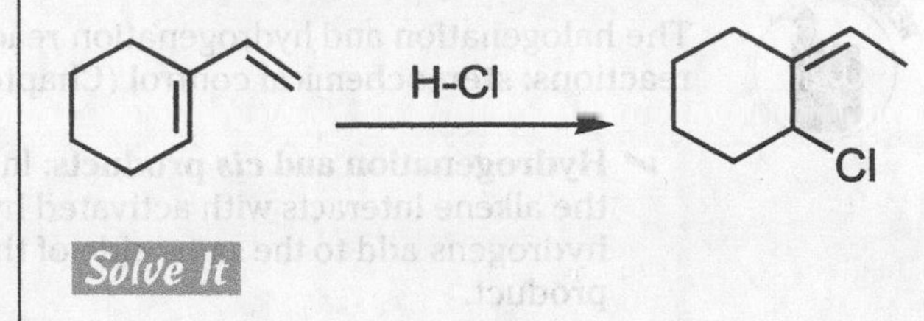

Solve It

Adding Halogens and Hydrogen to Alkenes

The halogenation (addition of Br_2 or Cl_2) and hydrogenation (addition of H_2) reactions of alkenes are two significant addition reactions that you see for alkenes. Both of these reaction types add two identical atoms across the double bond, with one atom attaching to each of the double-bond carbons (see Figure 8-2, in which *X* represents a halogen).

H_2 / Pd, C

syn addition
H's add to same side of double bond

halonium ion

X^- attacks backside of halonium ion

anti addition
X's add to opposite sides of double bond

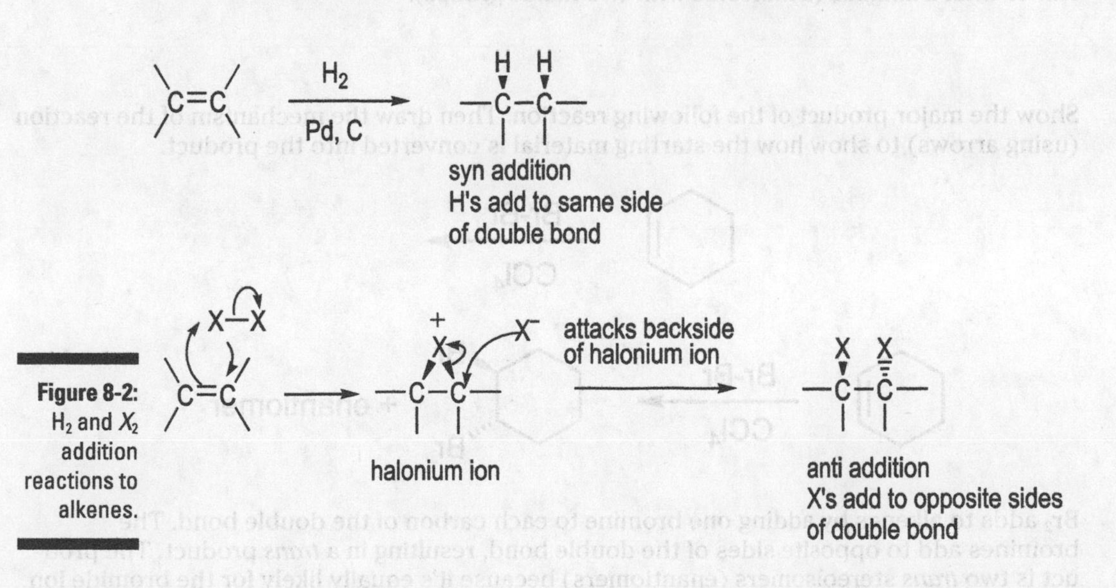

Figure 8-2: H_2 and X_2 addition reactions to alkenes.

Hydrogenation reactions occur by adding two hydrogen atoms across the double bond in the presence of a metal catalyst. Similarly, *halogenation reactions* occur by addition of a halogen (such as Cl_2 or Br_2) to an alkene, which adds two halogen atoms across a double bond.

The halogenation and hydrogenation reactions introduce an important concept of organic reactions: stereochemical control (Chapter 7 explains *trans* and *cis* stereochemistry):

- **Hydrogenation and *cis* products:** In a hydrogenation reaction, the reaction occurs when the alkene interacts with activated hydrogen atoms on the surface of the catalyst, so the hydrogens add to the same side of the double bond (called *syn addition*), resulting in a *cis* product.
- **Halogenation and *trans* products:** In a halogenation reaction, as a result of the reaction mechanism, the two halides add to opposite sides of the double bond — called *anti addition* — to give the product a defined *trans* stereochemistry.

  The mechanism of this reaction involves a strange intermediate, a three-membered ring called a *halonium ion* (specifically, a *bromonium ion* if it's bromine, *chloronium ion* if it's chlorine). You reach this intermediate by simultaneous attack of the double bond on one of the halogen atoms and attack of the same halogen atom on one of the double-bond carbons. At the same time, the *X-X* bond breaks and X^- is expelled. The next step is attack of the newly formed halide ion (X^-) on one of the halonium ion carbons, breaking the C-X bond. Because the positively charged halogen atom blocks attack of the halide ion from the front, the halide ion must attack the carbon on the other side of the halonium ion (called *backside attack*), which results in the halides' adding on opposite sides of the double bond.

Thus, for these reactions, in addition to giving the structure of the product, you often need to show the stereochemistry of the product.

This reaction has a twist when run in H_2O solvent rather than in CCl_4 solvent. In H_2O solvent, the second step is water addition to the halonium ion rather than halide (X^-) addition. Thus, the product in water is a *halohydrin* (a molecule with one halide and one alcohol [OH] group) rather than a *dihalide* (a molecule with two halide groups).

Q. Show the major product of the following reaction. Then draw the mechanism of the reaction (using arrows) to show how the starting material is converted into the product.

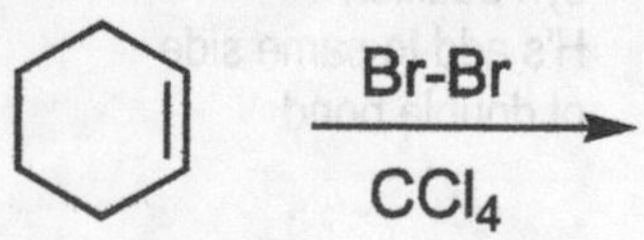

Br-Br
CCl$_4$

A.

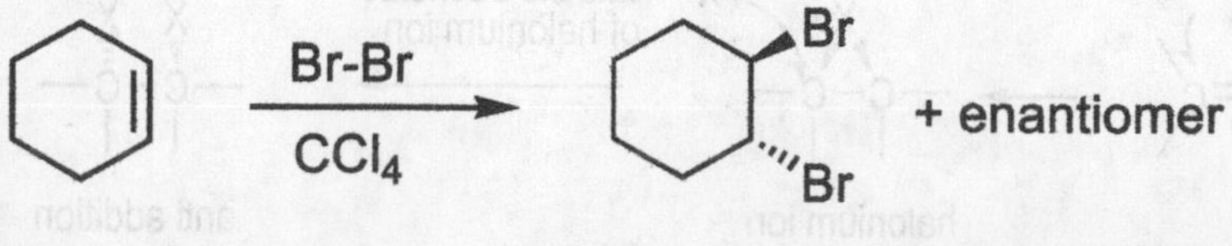

Br-Br
CCl$_4$
Br
Br
+ enantiomer

Br_2 adds to alkenes by adding one bromine to each carbon of the double bond. The bromines add to opposite sides of the double bond, resulting in a *trans* product. The product is two *trans* stereoisomers (enantiomers) because it's equally likely for the bromide ion to attack the bottom carbon of the bromonium ion (shown here) as it is to attack the top carbon (not shown), which would lead to the other enantiomer (in which the top bromine sticks down and the bottom bromine sticks up).

:Br-Br: → bromonium ion + :Br⁻: → Br / Br + enantiomer

bromonium ion

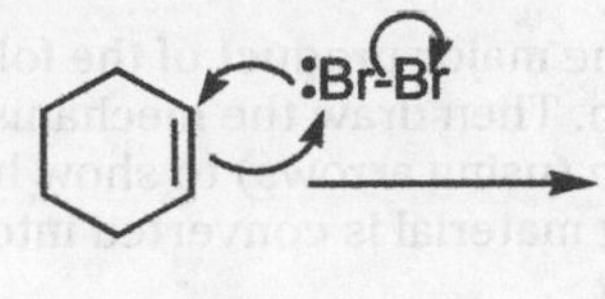

Q. Show the product of the following reaction, making sure to indicate the stereochemistry of the product.

$\xrightarrow[\text{Pd, C}]{H_2}$

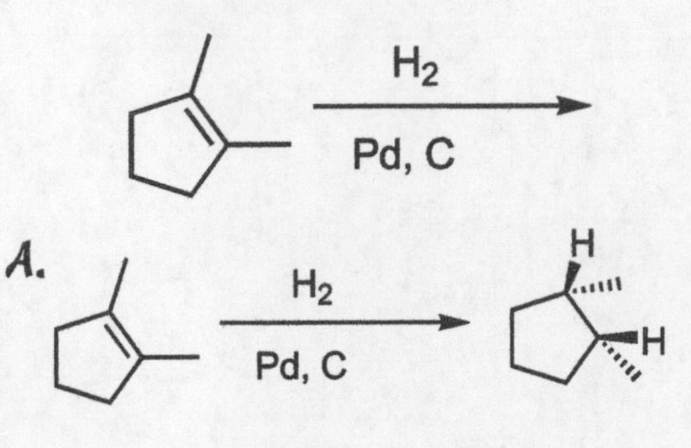

A. This reaction occurs when hydrogen gas is bubbled through a solution containing the alkene and a metal catalyst (often Pd, C or Pt). Because the reaction occurs when the alkene interacts with activated hydrogen atoms on the surface of the catalyst, the two hydrogens add to the same side of the double bond, which is called syn addition, leading to the *cis* product.

$\xrightarrow[\text{Pd, C}]{H_2}$ H, H

13. Show the product of the following reaction. Then draw the mechanism of the reaction (using arrows) to show how the starting material is converted into the product.

$\xrightarrow[CCl_4]{\text{Cl-Cl}}$

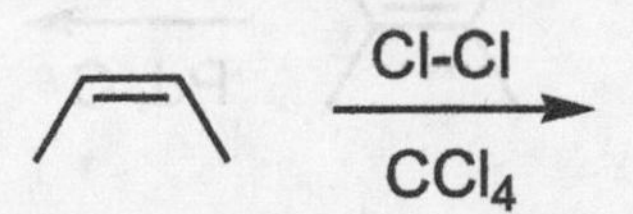

Solve It

14. Show the major product of the following reaction. Then draw the mechanism of the reaction (using arrows) to show how the starting material is converted into the product.

$\xrightarrow[CCl_4]{\text{Cl-Cl}}$

Solve It

15. Show the major product of the following reaction. Then draw the mechanism of the reaction (using arrows) to show how the starting material is converted into the product.

Br_2 / CCl_4

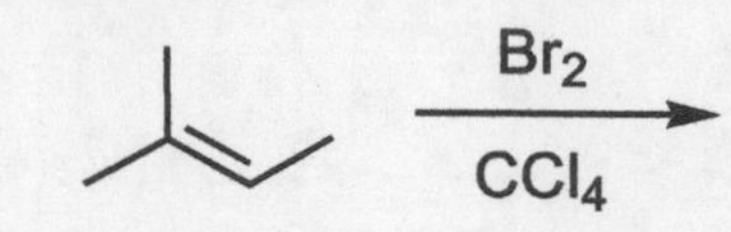

Solve It

16. Show the major product of the following reaction. Then draw the mechanism of the reaction (using arrows) to show how the starting material is converted into the product.

Br_2 / H_2O

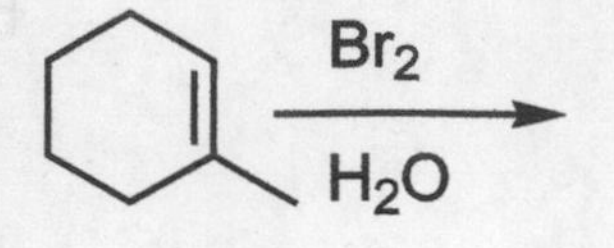

Solve It

17. Show the major product of the following reaction. Then draw the mechanism of the reaction (using arrows) to show how the starting material is converted into the product.

Cl-Cl / H_2O

Solve It

18. Show the major product of the following reaction.

H_2 / Pd, C

Solve It

Just Add Water: Adding H_2O to Alkenes

Addition of water to alkenes comes in two main flavors — *Markovnikov addition,* in which the OH adds to the more-substituted double-bond carbon, and *anti-Markovnikov* addition, in which the OH adds to the less-substituted carbon.

Markovnikov addition of water to alkenes is accomplished by the *oxymercuration/demurcuration reaction* like the one shown in Figure 8-3. In this reaction, mercuric acetate, or $Hg(OAc)_2$, is added to the alkene in water, followed by addition of sodium borohydride ($NaBH_4$), to give the Markovnikov alcohol product. Here's what happens:

1. **The double bond attacks the mercury acetate, and the mercury likewise attacks the double bond, kicking off acetate ion (AcO^-).**

   This forms a three-membered mercury intermediate called a *mercurinium ion* (reminiscent of the halonium ions formed from the addition of halogens to alkenes in the preceding section).

2. **Water attacks the *most highly substituted carbon* of the mercurinium ion, and the three-membered ring is opened (just like opening a halonium ion).**

   This action results in anti addition, with the water and the mercury on opposite sides of each other.

3. **Sodium borohydride is added, which replaces the mercury acetate (HgOAc) with hydrogen by a mechanism you probably don't need to know.**

   The end result is that the hydroxyl (OH) group is added to the most-substituted carbon of the double bond.

R₂C=CHR —(1. $Hg(OAc)_2$, H_2O; 2. $NaBH_4$)→ R–C(OH)(R)–C(H)(R)–H

Markovnikov addition (OH adds to more-substituted C)

AcO–Hg–OAc + R₂C=CHR → mercurinium ion —(H_2O, backside attack)→ R–C(R)(+OH₂)–C(R)(HgOAc)–H —(AcO^-)→ R–C(R)(OH)–C(R)(HgOAc)–H —($NaBH_4$)→ R–C(R)(OH)–C(R)(H)–H

anti addition (H and OH on opposite sides)

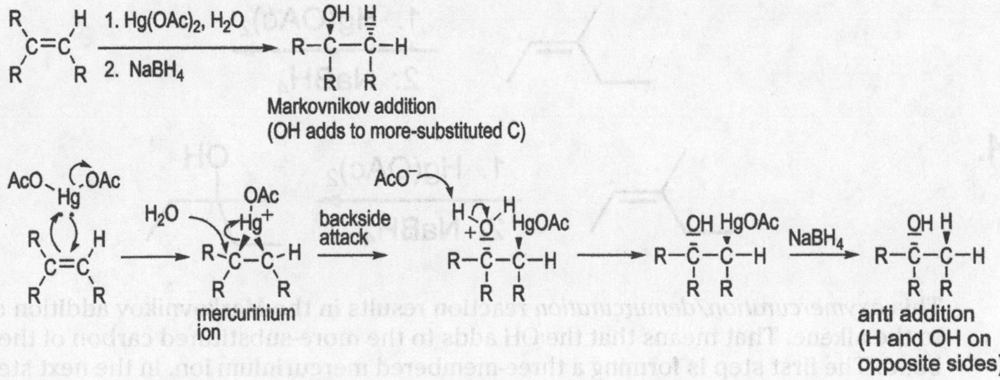

Figure 8-3: Markovnikov addition of water to an alkene via the oxymercuration/demurcuration reaction.

Conversely, anti-Markovnikov addition of water to alkenes is accomplished by the *hydroboration reaction* (Figure 8-4). In this reaction, borane (BH_3) in THF solvent is added to the alkene, followed by addition of hydrogen peroxide (H_2O_2), to give the anti-Markovnikov alcohol product. The mechanism of borane addition to the alkene is a *concerted* mechanism, meaning that all the steps occur at one time. The alkene attacks the borane, and a hydride is transferred to the more highly substituted carbon, adding the borane to the less-substituted carbon of the double bond. Because this occurs in a concerted fashion, the H and BH_2 add to the same side of the double bond (called *syn addition*).

Then hydrogen peroxide is added, replacing the borane (BH_2) unit with a hydroxyl group by a mechanism you probably don't need to know. The end result is that the hydroxyl (OH) is added to the less-substituted carbon of the double bond.

$$R_2C{=}CHR \xrightarrow[\text{2. } H_2O_2,\ OH^-]{\text{1. } BH_3,\ THF} R{-}\underset{R}{\overset{H}{C}}{-}\underset{R}{\overset{OH}{C}}{-}H$$

anti-Markovnikov addition
(OH adds to least-substituted C)

$$R_2C{=}CHR \ (H{-}BH_2) \longrightarrow R{-}\underset{R}{\overset{H}{C}}{-}\underset{R}{\overset{BH_2}{C}}{-}H \xrightarrow{H_2O_2,\ OH^-} R{-}\underset{R}{\overset{H}{C}}{-}\underset{R}{\overset{OH}{C}}{-}H$$

syn addition
(H and OH on same side)

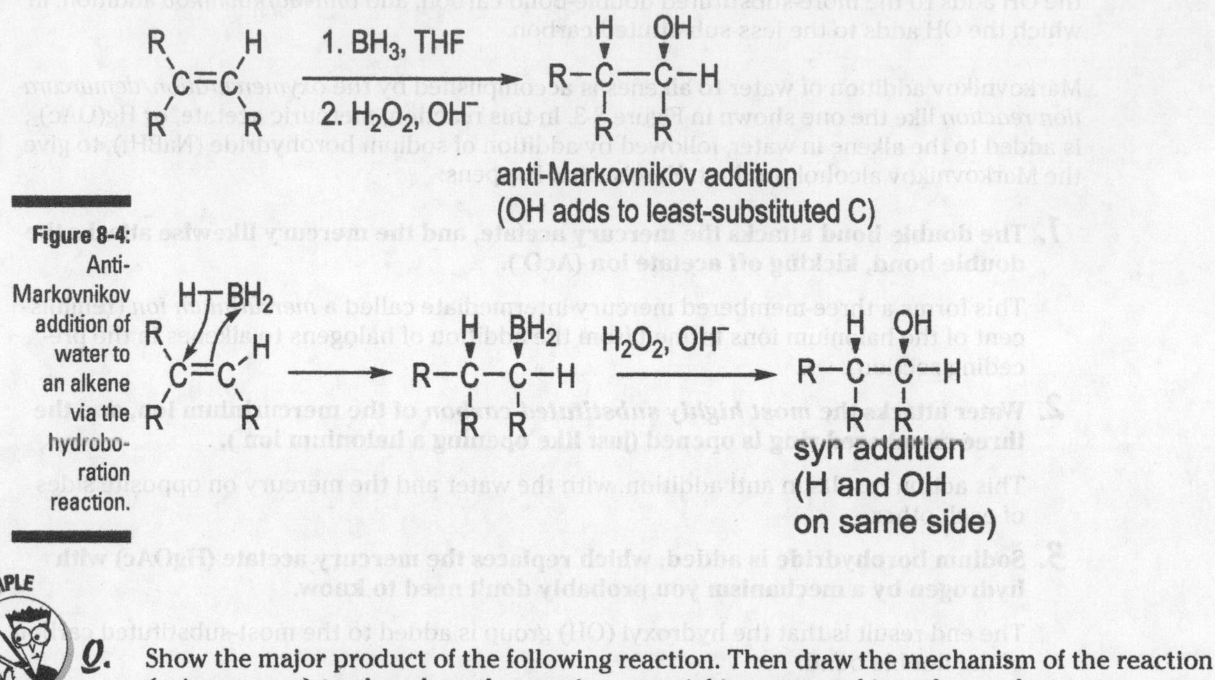

Figure 8-4: Anti-Markovnikov addition of water to an alkene via the hydroboration reaction.

Q. **Show the major product of the following reaction. Then draw the mechanism of the reaction (using arrows) to show how the starting material is converted into the product.**

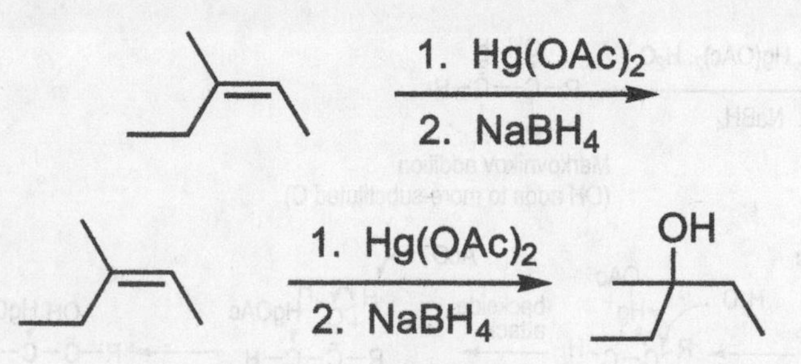

$$\xrightarrow[\text{2. } NaBH_4]{\text{1. } Hg(OAc)_2}$$

A.

$$\xrightarrow[\text{2. } NaBH_4]{\text{1. } Hg(OAc)_2} \text{OH}$$

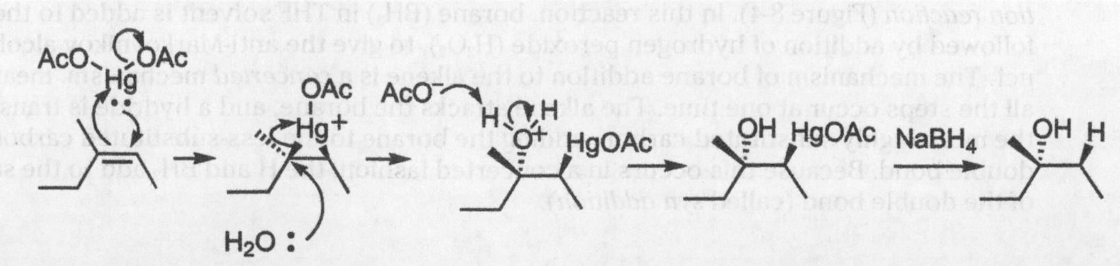

This *oxymercuration/demurcuration* reaction results in the Markovnikov addition of water to the alkene. That means that the OH adds to the more-substituted carbon of the double bond. The first step is forming a three-membered mercurinium ion. In the next step, water attacks the more highly substituted carbon. Loss of a proton from the oxygen then leads to a mercury-substituted alcohol. Addition of $NaBH_4$ in the final step removes the mercury and replaces it with an H.

AcO–Hg–OAc → OAc, Hg^+, H_2O: , AcO^- → H–O^+–H, HgOAc → OH HgOAc → ($NaBH_4$) → OH H

Q. Show the major product of the following reaction. Then draw the mechanism of the reaction (using arrows) to show how the starting material is converted into the product.

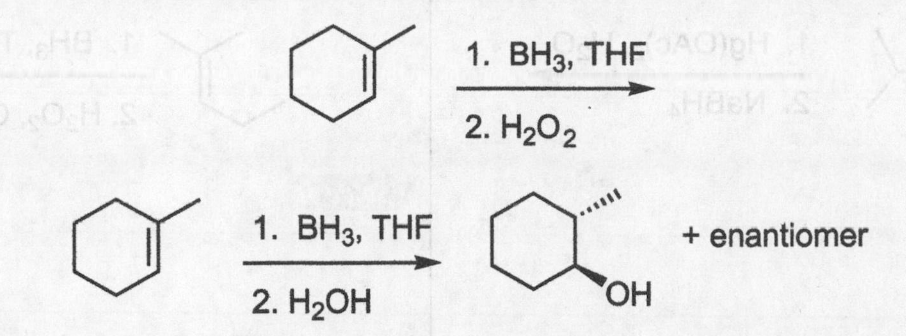

1. BH_3, THF
2. H_2O_2

A.

1. BH_3, THF
2. H_2OH

+ enantiomer

OH

This *hydroboration* reaction results in the anti-Markovnikov addition of water to the alkene (the OH adds to the less-substituted carbon). The double bond attacks the boron, and the hydride transfers from the boron to the more-substituted carbon in the same step. This leads to a BH_2 unit attached to the less-substituted carbon and an H attached to the more-substituted carbon. Because this all happens at once, the BH_2 and the H add to the same face of the double bond (syn addition). Addition of peroxide (H_2O_2) replaces the BH_2 with OH. Because it's equally likely for the borane to add to the top or bottom face of the double bond, the enantiomer is also formed by attack of the borane on the bottom face of the double bond (not shown).

H, BH_2 → H, BH_2 → (H_2O_2, OH^-) → H, OH + enantiomer

19. Show the major product of the following reaction. Then draw the mechanism of the reaction.

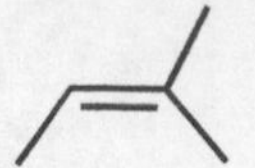

1. $Hg(OAc)_2$, H_2O
2. $NaBH_4$

Solve It

20. Show the major product of the following reaction. Then draw the mechanism of the reaction.

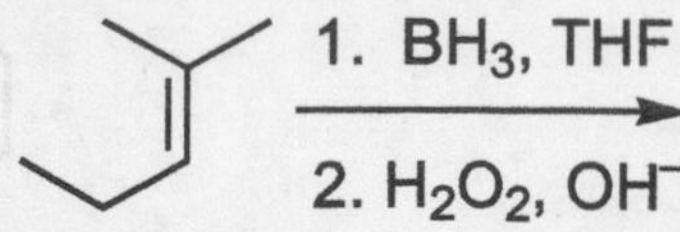

Solve It

21. Show the major product of the following reaction. Then draw the mechanism of the reaction.

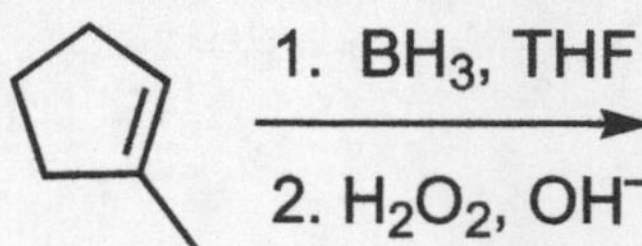

Solve It

22. An alternative reaction to the oxymercuration/demurcuration reaction for Markovnikov addition of water is the acid-catalyzed addition of water to alkynes. Show the major product of the following reaction. Then draw the mechanism of the reaction.

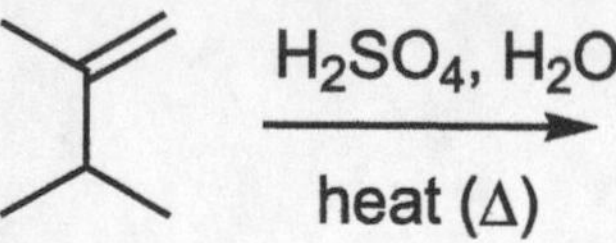

Solve It

Seeing Carbocation Rearrangements

Carbocations (positively charged carbons) are mischievous reaction intermediates. These intermediates rearrange when doing so leads to a more stable carbocation. The stability of carbocations increases as you increase the number of carbons attached to the charged carbon. Here are the classifications of carbocations, from least to most stable:

- **Primary (1°):** Carbocations with just a single attached carbon atom; they're quite unstable and rarely formed
- **Secondary (2°):** Carbocations attached to two other carbons; these cations are more stable than primary but less stable than tertiary carbocations
- **Tertiary (3°):** Carbocations attached to three other carbons; these are the most stable of the three

Carbocations rearrange principally for two reasons:

- **To obtain a more stable carbocation:** The most frequent cause of carbocation rearrangements is to form a more-substituted (stable) carbocation — for example, if possible, a secondary carbocation rearranges to a more stable tertiary one. This can be accomplished by either a hydride shift or an alkyl shift (see Figure 8-5):
  - In a *hydride shift,* a hydrogen on an adjacent, more highly substituted carbon moves to the cationic carbon, taking the hydrogen and the two electrons in the C-H bond with it. A hydride shift is preferred over an alkyl shift.
  - If no hydrogens are on the adjacent carbon, an *alkyl shift* can occur, in which an *R* group (such as a methyl) moves over to the positively charged carbon, taking the two electrons in the C-C bond with it.
- **To relieve ring strain:** When next to small rings (principally three-, four-, and five-carbon rings), carbocations can also undergo ring expansions to relieve ring strain (see Figure 8-6). These ring expansions are often challenging to work through, but you can manage them if you take them one step at a time.

R–C(H)(R)–C⁺(R)–H —hydride shift→ R–C⁺(R)–C(H)(R)–H

secondary carbocation → tertiary carbocation

R–C(R)(R)–C⁺(R)–H —alkyl shift→ R–C⁺(R)–C(R)(R)–H

secondary carbocation → tertiary carbocation

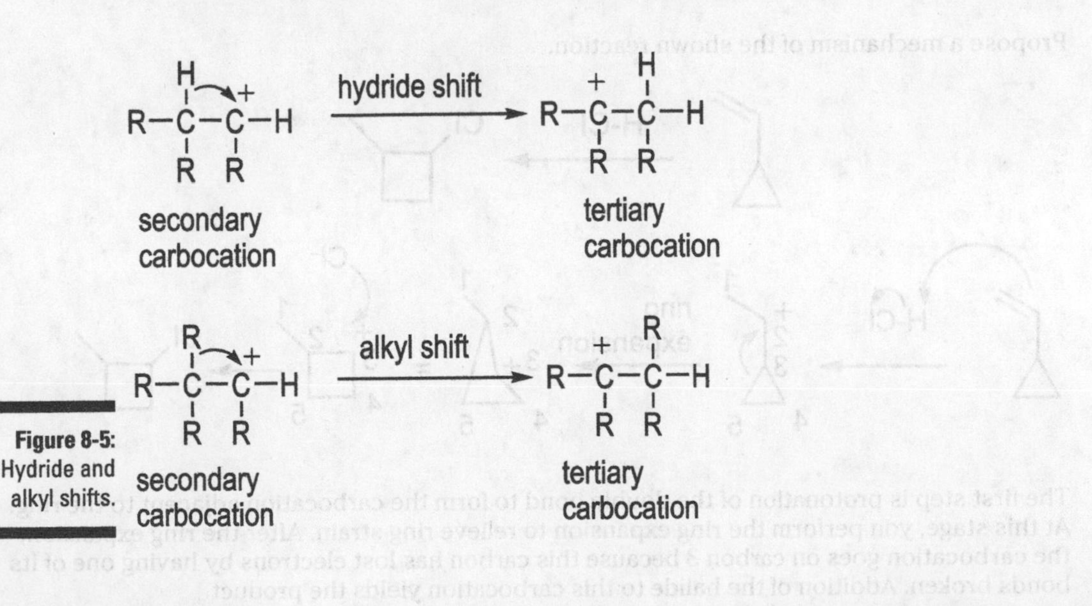

Figure 8-5: Hydride and alkyl shifts.

ring expansion

redrawn

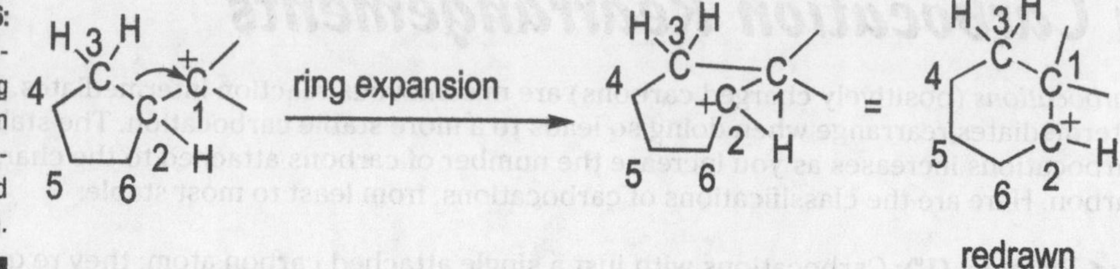

Figure 8-6: A carbocation ring expansion of a five-membered ring.

To perform a ring expansion, move the electrons from the adjacent C-C ring bond to the carbocation. This increases the ring size by one carbon and moves the carbocation to a ring carbon. I recommend that you number all the carbons during these ring expansion mechanisms so you can keep track of where atoms and bonds end up.

Q. **Is the following carbocation stable to rearrangement? If not, show the rearranged cation.**

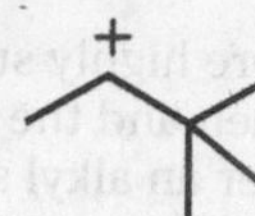

A. **This carbocation is secondary, but it's not stable to rearrangement because the molecule can undergo a methyl shift from the adjacent carbon to lead to a more stable tertiary carbocation.**

methyl shift

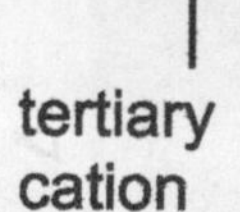

secondary cation

tertiary cation

Q. **Propose a mechanism of the shown reaction.**

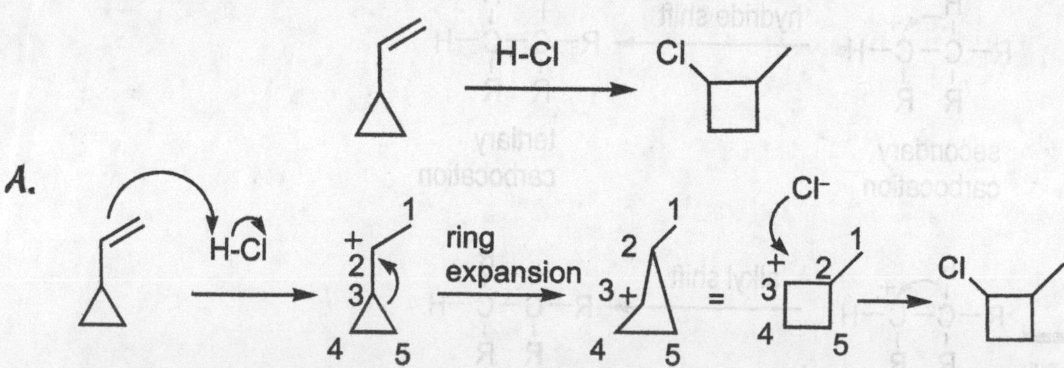

H-Cl

A.

H-Cl

ring expansion

The first step is protonation of the double bond to form the carbocation adjacent to the ring. At this stage, you perform the ring expansion to relieve ring strain. After the ring expansion, the carbocation goes on carbon 3 because this carbon has lost electrons by having one of its bonds broken. Addition of the halide to this carbocation yields the product.

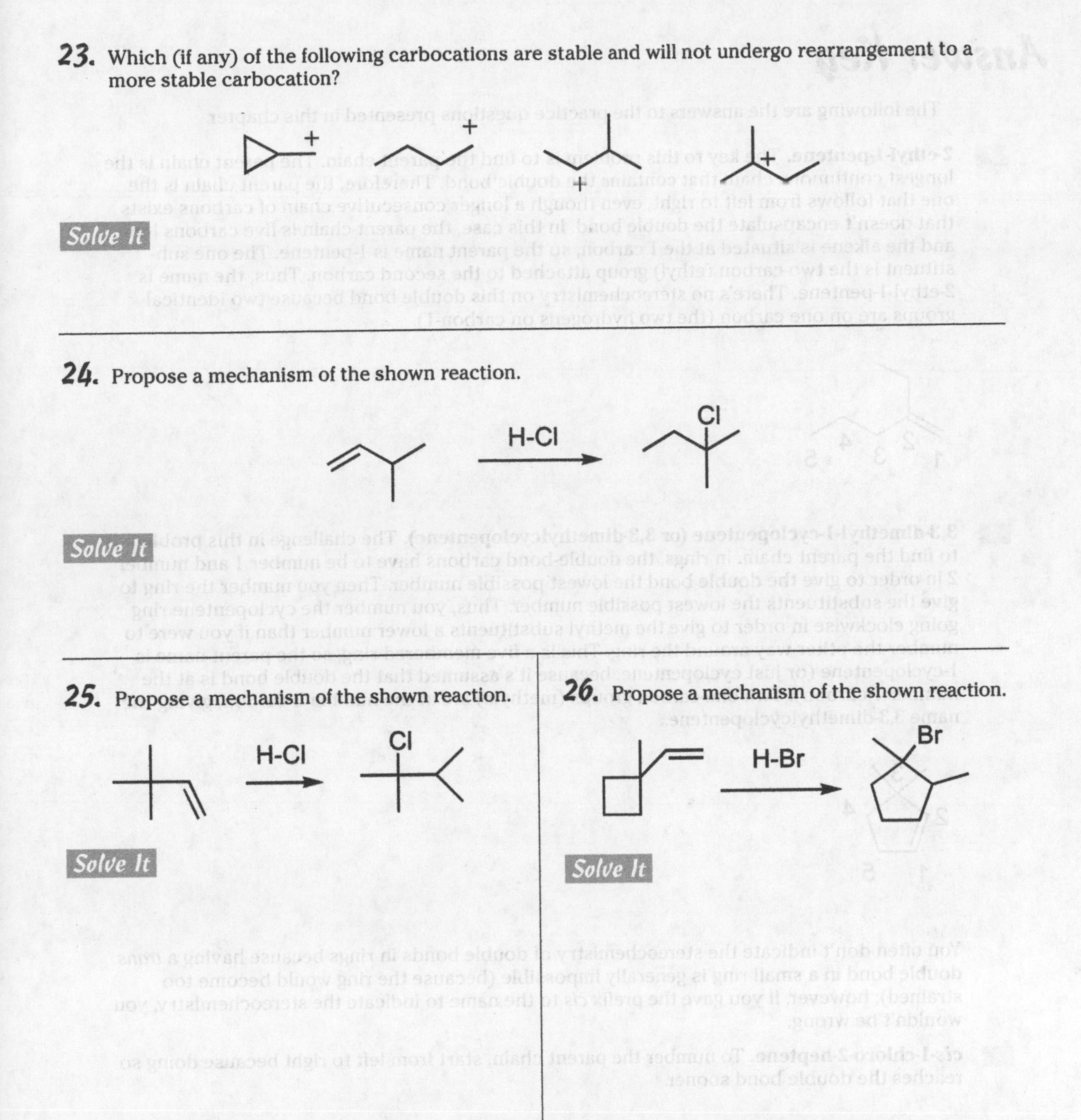

23. **Which (if any) of the following carbocations are stable and will not undergo rearrangement to a more stable carbocation?**

+ + + +

Solve It

24. **Propose a mechanism of the shown reaction.**

H-Cl → Cl

Solve It

25. **Propose a mechanism of the shown reaction.**

H-Cl → Cl

Solve It

26. **Propose a mechanism of the shown reaction.**

H-Br → Br

Solve It

Answer Key

The following are the answers to the practice questions presented in this chapter.

1 **2-ethyl-1-pentene.** The key to this problem is to find the parent chain. The parent chain is the longest continuous chain that contains the double bond. Therefore, the parent chain is the one that follows from left to right, even though a longer consecutive chain of carbons exists that doesn't encapsulate the double bond. In this case, the parent chain is five carbons long, and the alkene is situated at the 1 carbon, so the parent name is 1-pentene. The one substituent is the two-carbon (ethyl) group attached to the second carbon. Thus, the name is 2-ethyl-1-pentene. There's no stereochemistry on this double bond because two identical groups are on one carbon (the two hydrogens on carbon-1).

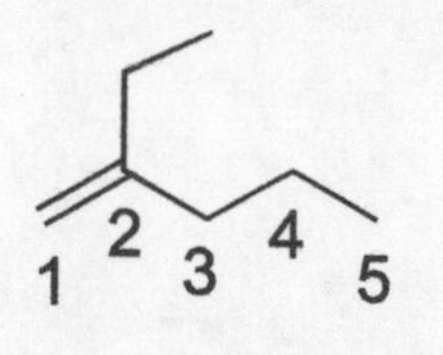

2 **3,3-dimethyl-1-cyclopentene (or 3,3-dimethylcyclopentene).** The challenge in this problem is to find the parent chain. In rings, the double-bond carbons have to be number 1 and number 2 in order to give the double bond the lowest possible number. Then you number the ring to give the substituents the lowest possible number. Thus, you number the cyclopentene ring going clockwise in order to give the methyl substituents a lower number than if you were to number the other way around the ring. This is a five-membered ring, so the parent name is 1-cyclopentene (or just cyclopentene, because it's assumed that the double bond is at the number 1 carbon). Two one-carbon groups (methyls) are at the number 3 carbon, giving the name 3,3-dimethylcyclopentene.

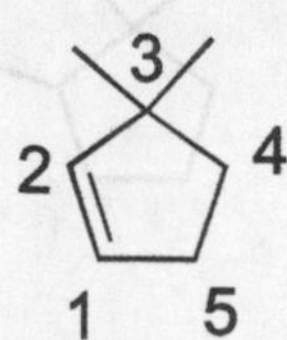

You often don't indicate the stereochemistry of double bonds in rings because having a *trans* double bond in a small ring is generally impossible (because the ring would become too strained); however, if you gave the prefix *cis* to the name to indicate the stereochemistry, you wouldn't be wrong.

3 ***cis*-1-chloro-2-heptene.** To number the parent chain, start from left to right because doing so reaches the double bond sooner.

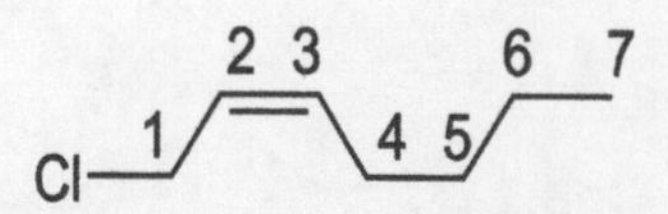

This is a seven-carbon alkene with the double bond at the second carbon, giving a parent name of 2-heptene. The only substituent is a chlorine at the first carbon (a chloro), giving you 1-chloro-2-pentene. Finally, identify the stereochemistry. The two identical substituents (the hydrogens) are pointing off the same side of the double bond, so the stereochemistry is *cis*. The full name is *cis*-1-chloro-1-pentene.

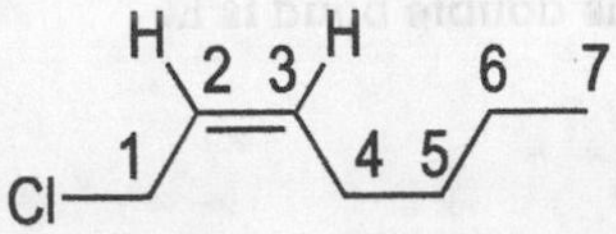

4 **3,3-dichloro-1-methyl-1-cyclohexene.** Start with numbering the alkene. Remember that in a ring, the double bond has to contain the number 1 and number 2 carbons. After including the double bond as the first two carbons, you number clockwise, giving the substituents the lowest possible number.

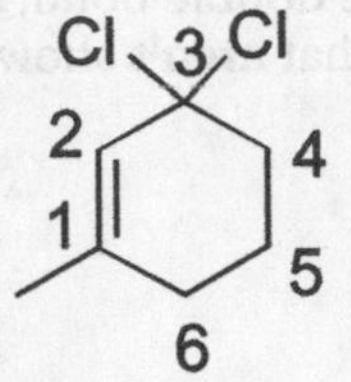

Next, name the substituents. Chlorine as a substituent is chloro, and the one-carbon group is a methyl substituent. Put the two chlorines together to give you dichloro, and then put the substituents alphabetically in front of the parent name (ignore numerical prefixes such as *di-*, *tri-*, *tetra-*, and so on, so alphabetize *dichloro* under the letter *c*). This gives you 3, 3-dichloro-1-methyl-1-cyclohexene. Note that you don't need to specify the stereochemistry of double bonds in small rings (because *trans* or E double bonds in rings are usually unstable), but you can if you want to.

5 **(3E,5E)-4,6-diethyl-3-methyl-3,5-nonadiene.** This is a tricky one because this molecule has two alkenes (making the molecule a *di*ene). The parent chain is the longest chain that contains both alkenes. You then number the parent chain, giving the two double bonds the lower numbers (right to left in this case). Nine carbons are in the parent chain, and the double bonds come at carbons 3 and 5, so the parent name is 3,5-nonadiene. Note that instead of using the suffix *-ene,* you use the suffix *diene* to indicate the number of double bonds in the molecule.

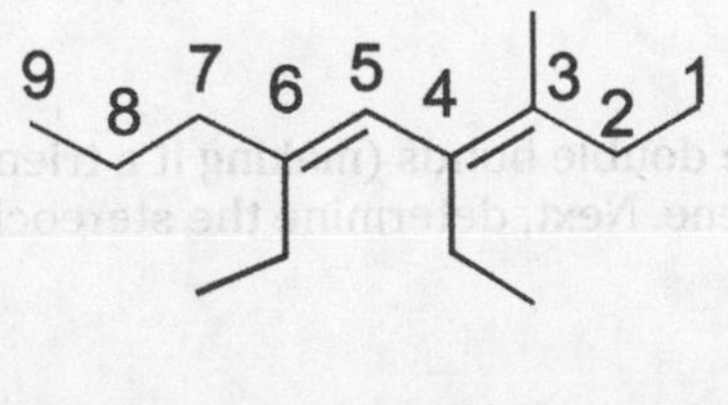

Next, name the substituents. The two two-carbon substituents at carbons 4 and 6 are ethyl groups, and these are clustered together to make a diethyl (still alphabetize under the letter *e*). The one-carbon group at carbon 3 is a methyl group. Alphabetizing the substituents in front of the parent name gives you 4,6-diethyl-3-methyl-3,5-nonadiene. Lastly, name the stereochemistry. You can start with the right-most double bond and divide the double bond in half (shown with the dashed line). Then prioritize the substituents on the right side using the Cahn-Ingold-Prelog priority rules. Here, the bottom gets higher priority over the top because a two-carbon chain gets priority over a one-carbon chain. On the left side, the double bond gets higher priority over the two-carbon chain. Because the two high priorities are on opposite sides of the double bond, the stereochemistry of this double bond is E.

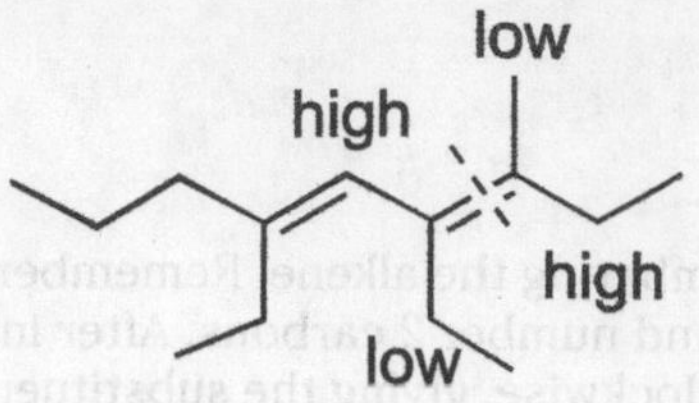

Next, consider the second double bond. Here on the right side of the double bond, the carbon chain has priority over the hydrogen. (***Tip:*** Draw in any hydrogens that aren't shown.)

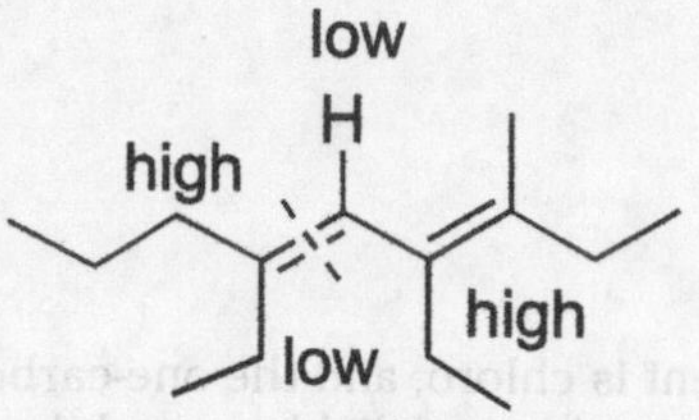

On the left side, the three-carbon chain has higher priority than the two-carbon chain. Because the highs are on opposite sides of the double bond, this double bond has E stereochemistry. Finally, add the stereochemistry to the name by placing the two double-bond configurations in parentheses at the beginning of the name and using a number to indicate which double bond each configuration refers to. This yields the name (3E,5E)-4,6-diethyl-3-methyl-3,5-nonadiene.

6 **all-*trans*-2,4,6-nonatriene (or 2E,4E,6E-2,4,6-nonatriene).** First number the parent chain, giving the double bonds the lowest numbers possible.

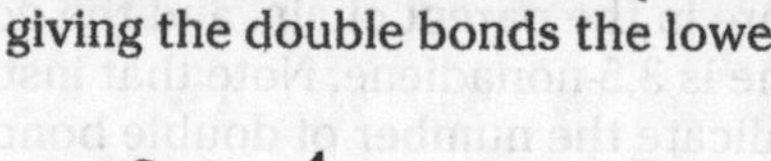

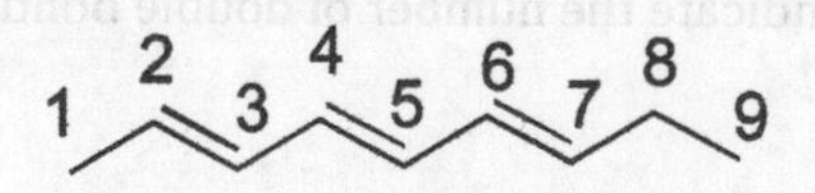

The parent chain is nine carbons long and has three double bonds (making it a triene) at the 2, 4, and 6 positions. The parent name is 2,4,6-nonatriene. Next, determine the stereochemistry.

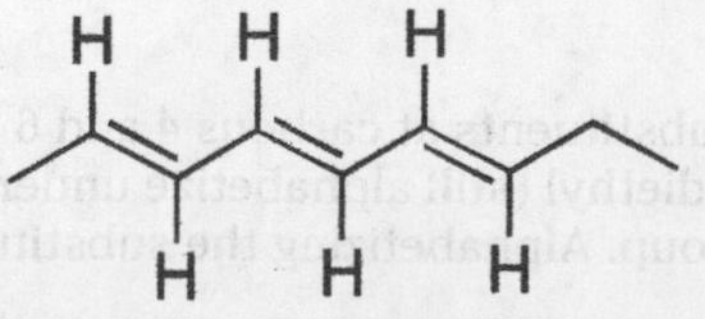

In each case, the two identical substituents (hydrogens) are on opposite sides of the double bond, making each double bond have *trans* stereochemistry (alternatively, using the E/Z nomenclature, each double bond has E stereochemistry). The final name is all-*trans*-2,4,6-nonatriene (alternatively, you can specify each double bond individually as *trans*).

7 **1,3,5,7-cyclooctatetraene.** This is a cyclic alkene that contains four double bonds, making it a tetraene. Because all the carbons are identical, where you start numbering it doesn't matter, so long as the first two carbons encompass a double bond. The parent ring has eight carbons, with alkenes at the 1, 3, 5, and 7 carbons, giving the name 1,3,5,7-cyclooctatetraene. As with most double bond–containing rings, you don't have to specify the stereochemistry because all double bonds are assumed to be *cis*.

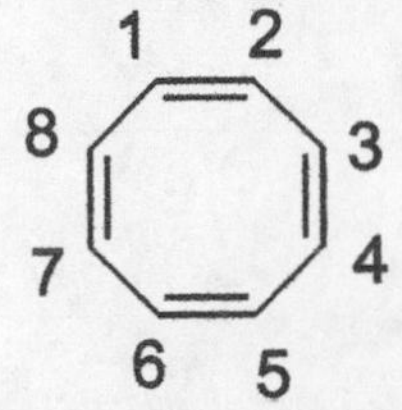

8 **(2Z,4E)-2-chloro-5-fluoro-4-methyl-2,4-heptadiene.** This is a tough one, though you can work it out if you tackle it systematically. First, find the parent chain and number it. In this case, it's the seven-carbon chain containing the two double bonds. Numbering the chain from right to left gives the lowest possible numbers to the double bonds. The parent chain is seven carbons long, and it has two double bonds at the 2 and 4 carbons, and so the parent name is 2,4-heptadiene.

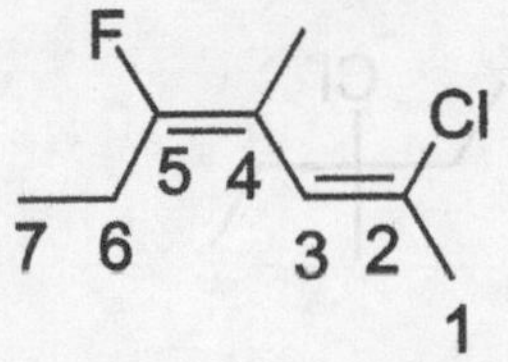

Next, add the substituents. The chlorine and fluorine are called chloro and fluoro, respectively, and the one-carbon group is a methyl. Alphabetizing the substituents in front of the parent name gives you 2-chloro-5-fluoro-4-methyl-2,4-heptadiene.

Last, do the stereochemistry. Four different groups are coming off double bond, so assign the double bonds using the E/Z system. Starting with the right double bond, cut the double bond in half and then assign the priorities (making sure to add the not-shown hydrogen), which gives you the following:

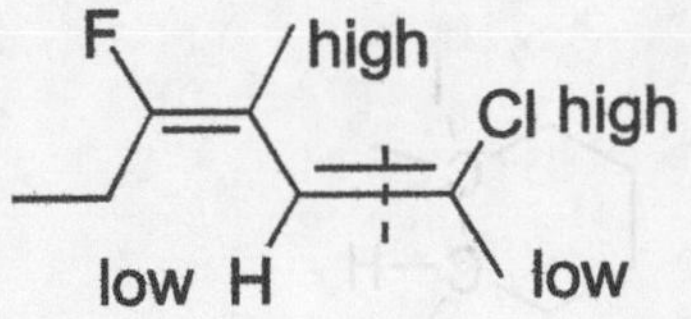

The two highs are on the same side, so the stereochemistry is Z. Prioritizing the substituents to the left side gives you the following:

high F low Cl low high H

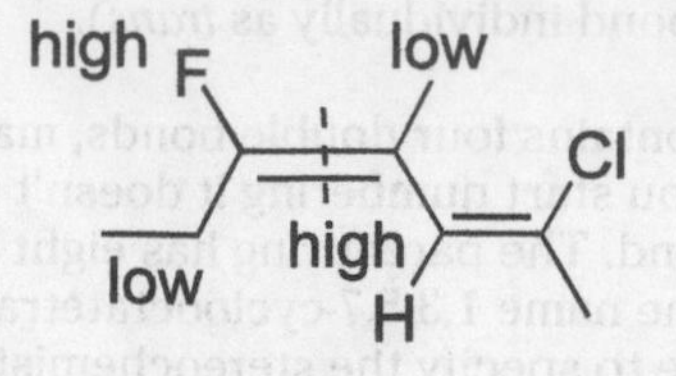

The two highs are on opposite sides of the double bond, indicating E stereochemistry.

9 H-Cl Cl

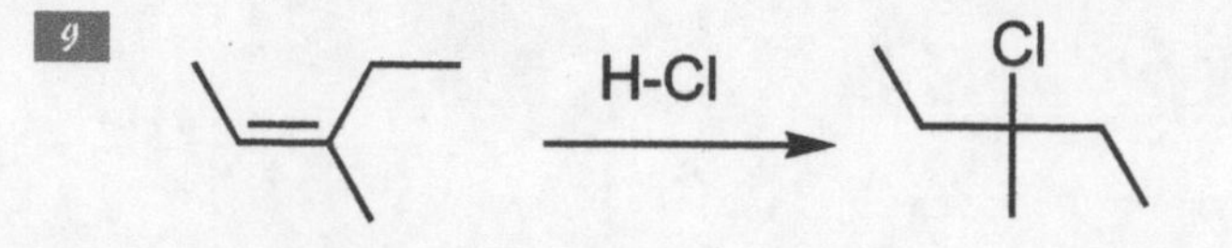

The reaction of the alkene with HCl yields addition of the H to the less-substituted carbon and addition of the chlorine to the more-substituted carbon of the double bond (Markovnikov addition). The mechanism of the reaction occurs in two steps. In the first step, show the protonation of the double bond by drawing an arrow from the double bond to the proton on H-Cl. The hydrogen adds to the less-substituted carbon, which places the positive charge on the more highly substituted carbon (where the charge is more stable). Finally, the chloride adds to yield the product.

H-Cl :Cl:⁻ + Cl

10 H-I I

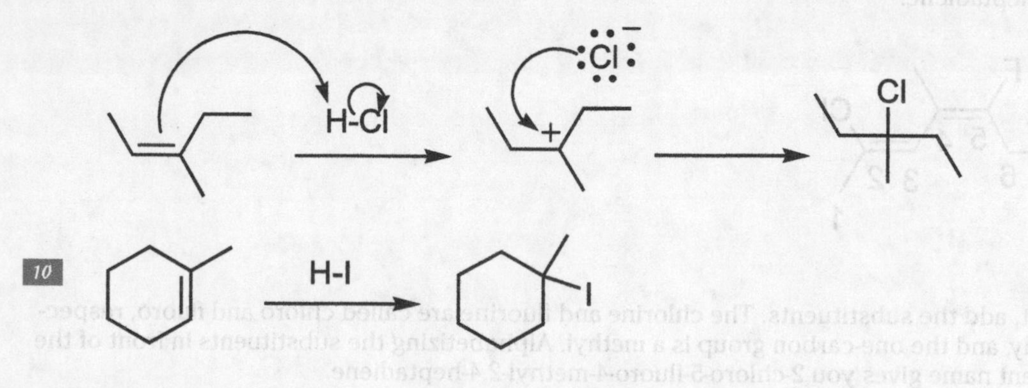

HI adds in Markovnikov fashion to the double bond, which means that the H adds to the less-substituted carbon and the I adds to the more-substituted carbon. The mechanism involves the protonation of the double bond to give the carbocation on the more highly substituted carbon, followed by attack of the iodide ion onto the cationic carbon to give the product.

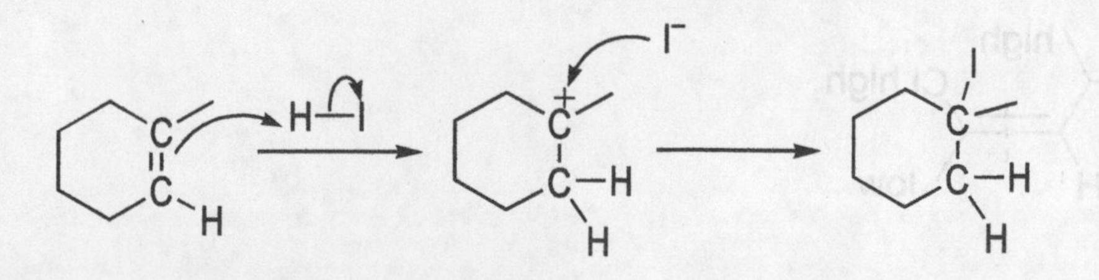

C C H H—I I⁻ C+ C—H H I C C—H H

11

HCl HCl HCl

tertiary secondary secondary

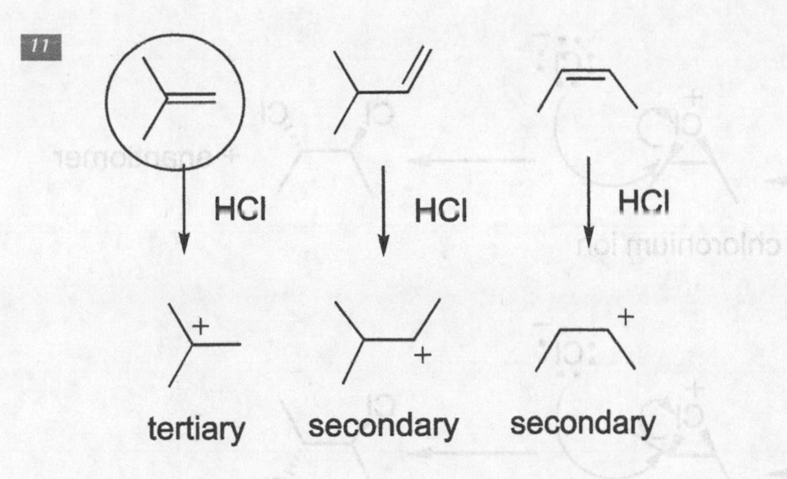

The circled alkene reacts the fastest with HCl because upon protonation, you obtain the most stable carbocation (tertiary carbocations are more stable than secondary).

12

H-Cl

Cl

resonance structures

:Cl:

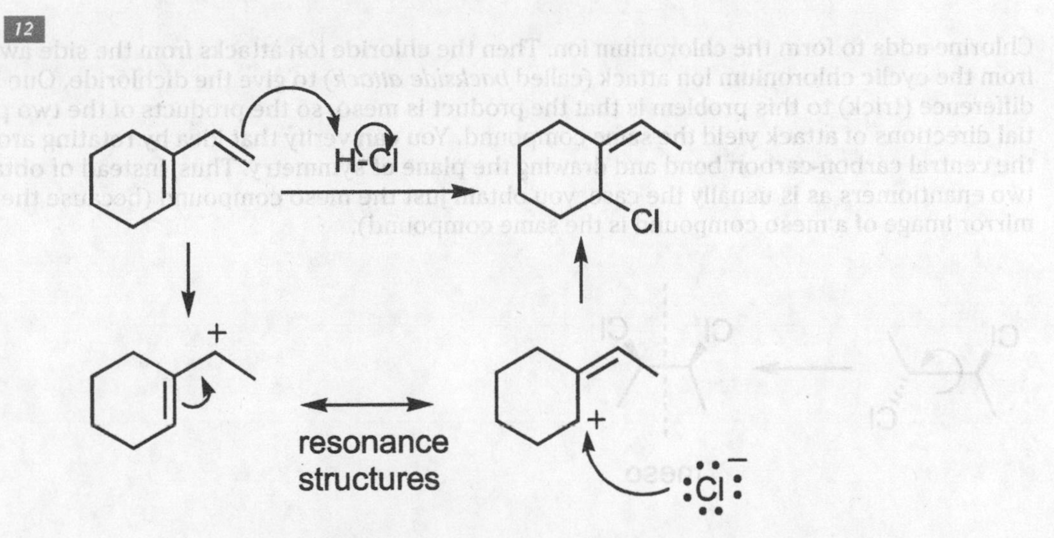

This is a challenging problem. Protonation of the double bond, as usual, leads to the carbocation; however, addition of the chloride ion at this point doesn't lead to the correct structure. Nonetheless, this cation has a resonance structure in which you place the cation on a ring carbon and move up the double bond. Addition of the chloride ion to this carbocation yields the product.

13

Cl-Cl / CCl_4 → Cl Cl + enantiomer

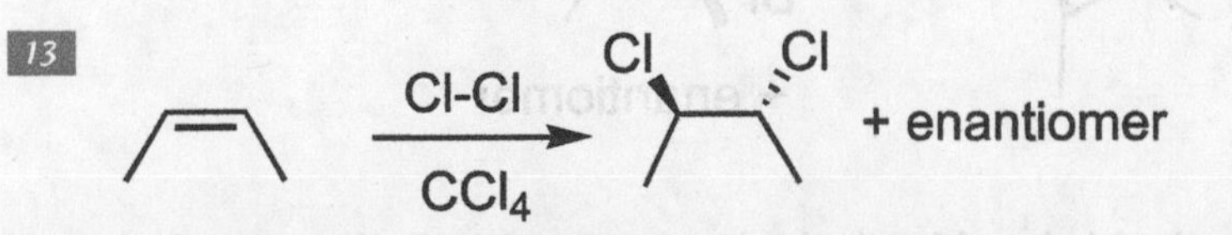

The reaction of the alkene with chlorine leads to the *trans*-dichloride. The mechanism proceeds through a three-membered chloronium ion intermediate. Because the positively charged chlorine blocks frontside attack, the chloride ion undergoes backside attack to give the *trans* dichloride. You also obtain the other *trans* enantiomer, which results from the chlorine's attacking the bottom side of the double bond to form the chloronium ion sticking down rather than sticking up as shown here.

14

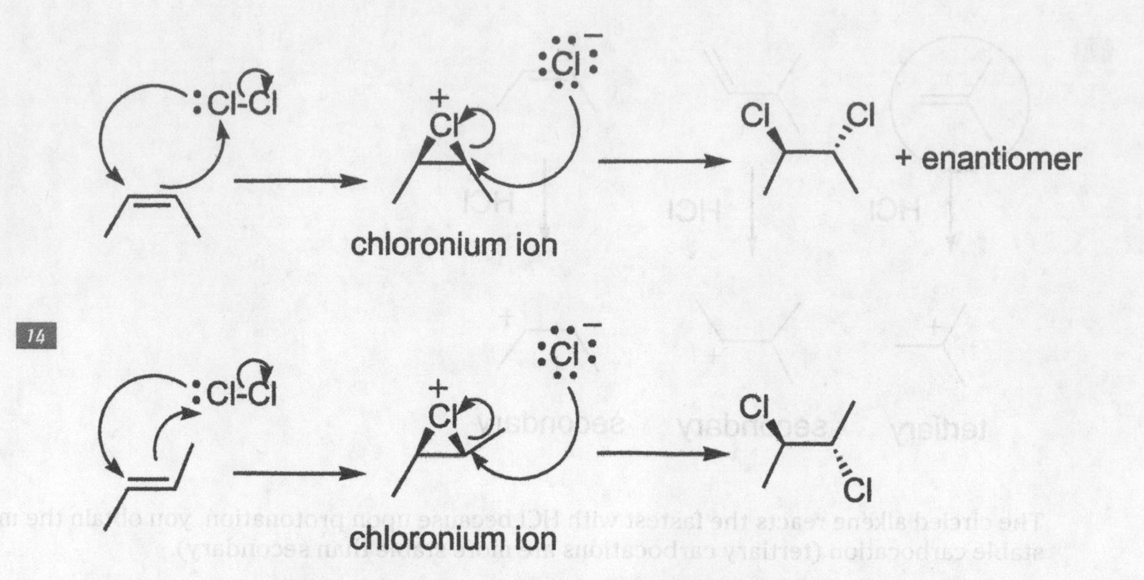

Chlorine adds to form the chloronium ion. Then the chloride ion attacks from the side away from the cyclic chloronium ion attack (called *backside attack*) to give the dichloride. One little difference (trick) to this problem is that the product is meso, so the products of the two potential directions of attack yield the same compound. You can verify that idea by rotating around the central carbon-carbon bond and drawing the plane of symmetry. Thus, instead of obtaining two enantiomers as is usually the case, you obtain just the meso compound (because the mirror image of a meso compound is the same compound).

15

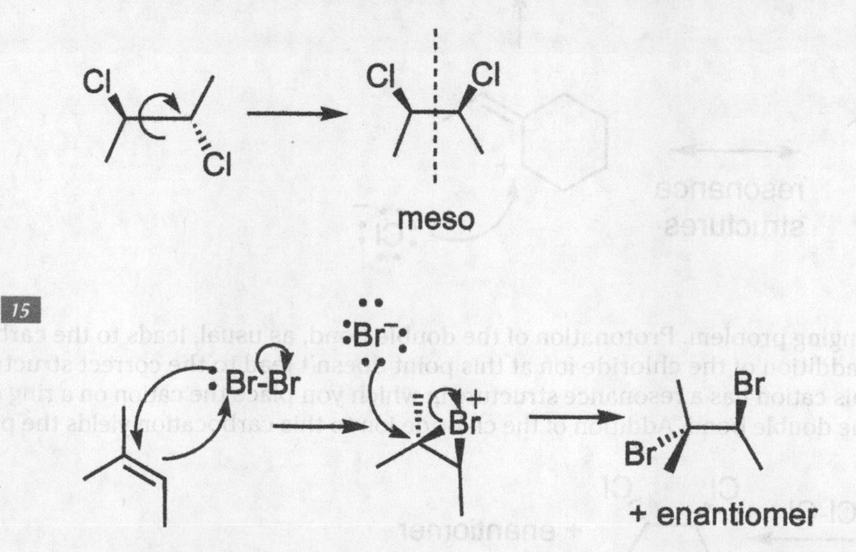

Addition of bromine to the double bond forms the bromonium ion. In the second step, the bromide ion attacks the bromonium ion to yield the *trans* dibromide. As usual with halogen (or hydrogen) addition to alkenes, you obtain two enantiomers because you can add the bromine to either side of the flat double bond, giving the bromonium ion sticking up or down.

16

Br_2, H_2O → + enantiomer + HBr

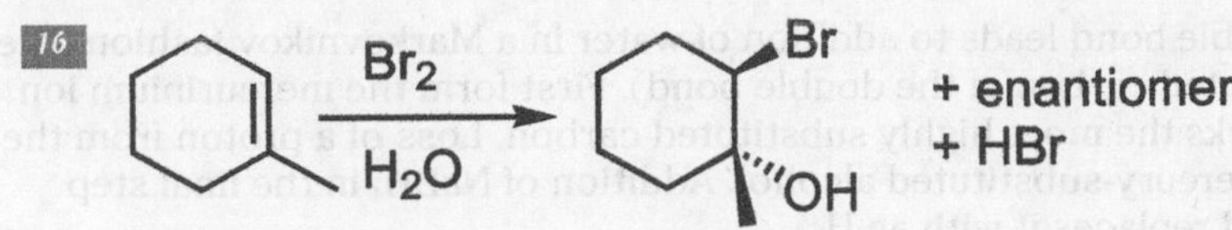

Here, you add bromine to the double bond in the presence of water to obtain a *bromohydrin* (a molecule with a bromine and an OH group). The mechanism goes through the bromonium ion intermediate, followed by the water's attack on the more highly substituted carbon. Loss of a proton on the oxygen gives the product.

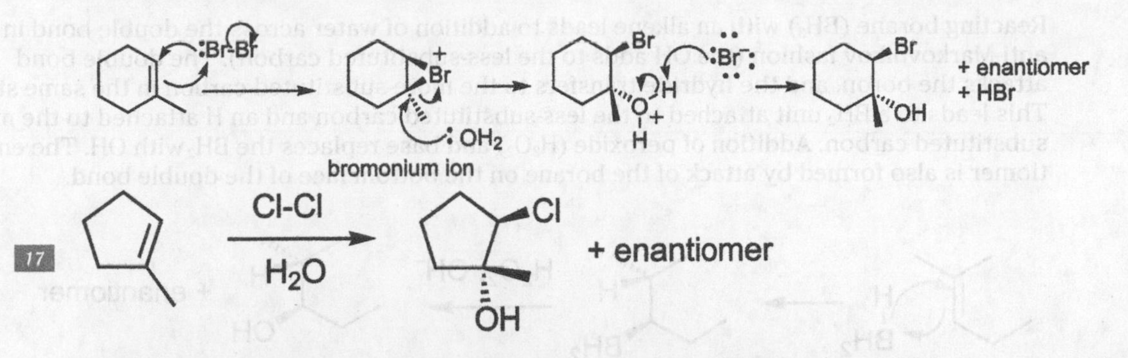

bromonium ion

+ enantiomer + HBr

17

Cl-Cl, H_2O → + enantiomer

When you add chlorine in the solvent water (instead of in the solvent CCl_4), you get a *chlorohydrin* (a molecule with a chlorine and an OH group) instead of obtaining the typical dichloride. You can see why from the mechanism. Chlorine adds, as usual, to give you the chloronium ion. But then, instead of the chloride ion's attack, water attacks the more highly substituted carbon. Loss of a proton on the oxygen gives the product. A key point to remember is that the water adds to the more highly substituted carbon of the chloronium ion.

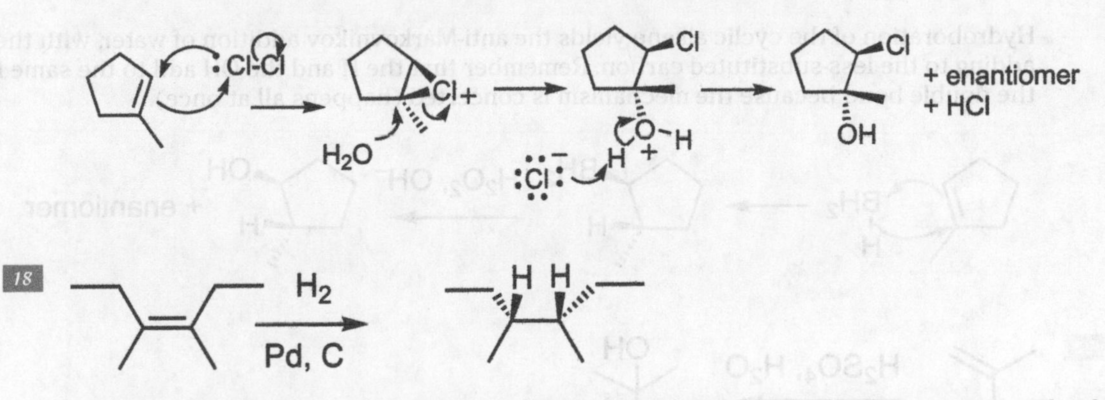

+ enantiomer + HCl

18

H_2, Pd, C →

Addition of hydrogen in the presence of a Pd, C catalyst adds the hydrogens across the double bond in a syn fashion (meaning that both hydrogens add to the same side of the double bond). You don't get an enantiomer in this case because the product is meso (which is superimposable on its mirror image and therefore doesn't have an enantiomer).

19

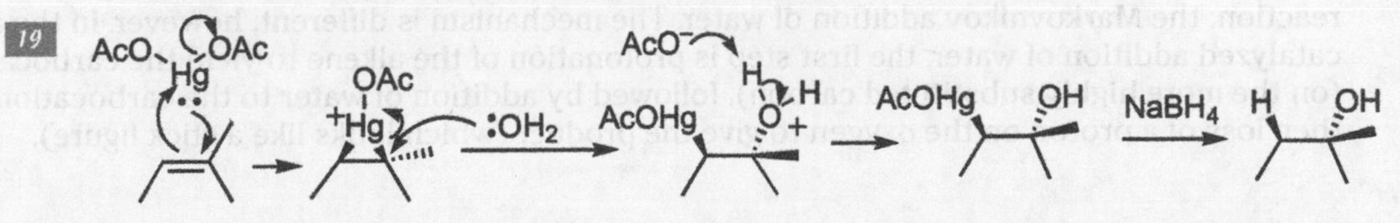

AcO–Hg–OAc → :OH_2 → AcOHg → AcOHg → $NaBH_4$ →

Oxymercuration of a double bond leads to addition of water in a Markovnikov fashion (the OH adds to the more-substituted carbon of the double bond). First form the mercurinium ion. In the next step, water attacks the more highly substituted carbon. Loss of a proton from the oxygen then leads to a mercury-substituted alcohol. Addition of $NaBH_4$ in the final step removes the mercury and replaces it with an H.

20 1. BH_3, THF
2. H_2O_2, OH^-
H
OH
+ enantiomer

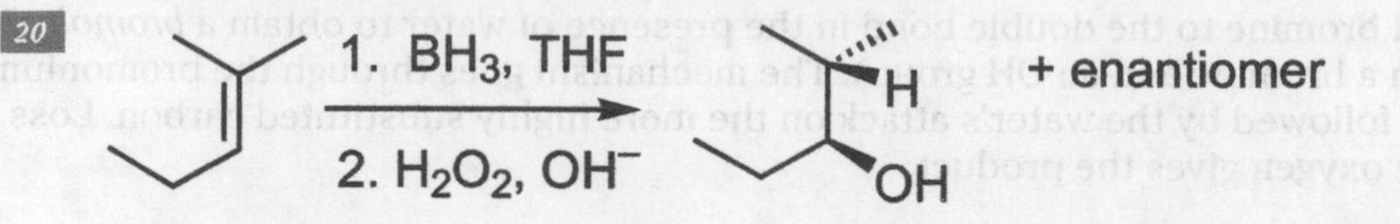

Reacting borane (BH_3) with an alkene leads to addition of water across the double bond in an anti-Markovnikov fashion (the OH adds to the less-substituted carbon). The double bond attacks the boron, and the hydride transfers to the more-substituted carbon in the same step. This leads to a BH_2 unit attached to the less-substituted carbon and an H attached to the more-substituted carbon. Addition of peroxide (H_2O_2) and base replaces the BH_2 with OH. The enantiomer is also formed by attack of the borane on the bottom face of the double bond.

H
BH_2
H
BH_2
H_2O_2, OH^-
H
OH
+ enantiomer

21 1. BH_3, THF
2. H_2O_2, OH^-
OH
H
+ enantiomer

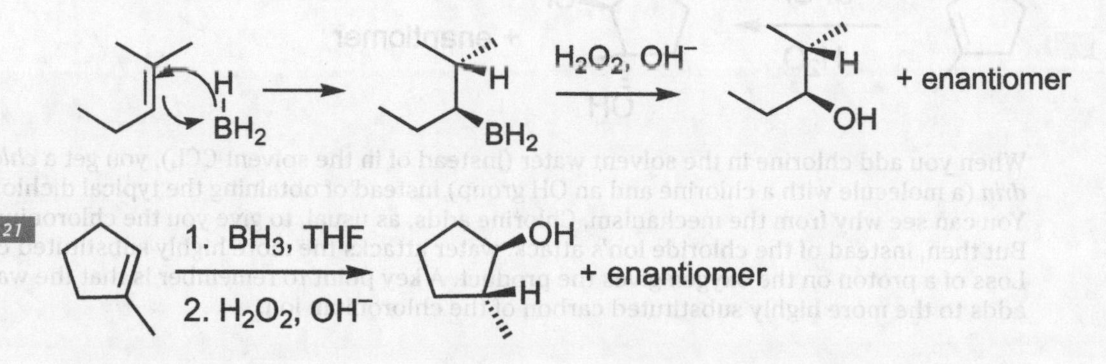

Hydroboration of the cyclic alkene yields the anti-Markovnikov addition of water, with the OH adding to the less-substituted carbon. Remember that the H and the OH add to the same face of the double bond because the mechanism is concerted (happens all at once).

BH_2
H
BH_2
H
H_2O_2, OH^-
OH
H
+ enantiomer

22 H_2SO_4, H_2O
heat (Δ)
OH

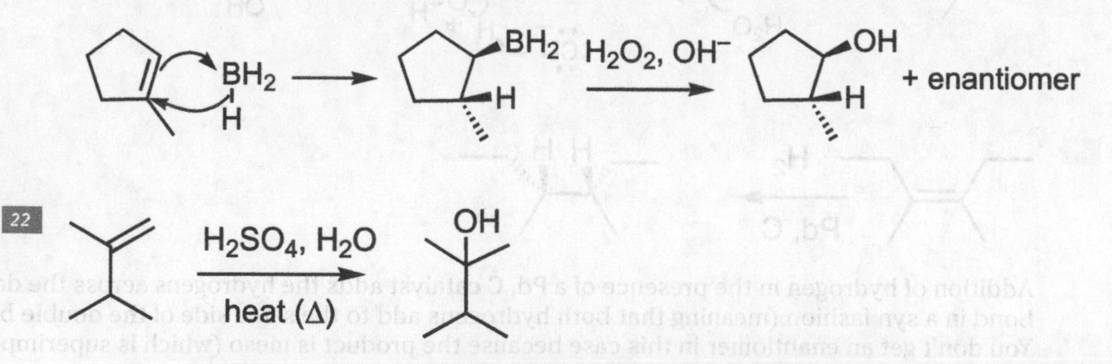

Addition of acid and water across a double bond yields the same product as the oxymercuration reaction: the Markovnikov addition of water. The mechanism is different, however. In the acid-catalyzed addition of water, the first step is protonation of the alkene to yield the carbocation (on the more highly substituted carbon), followed by addition of water to the carbocation and then loss of a proton on the oxygen to give the product (which looks like a stick figure).

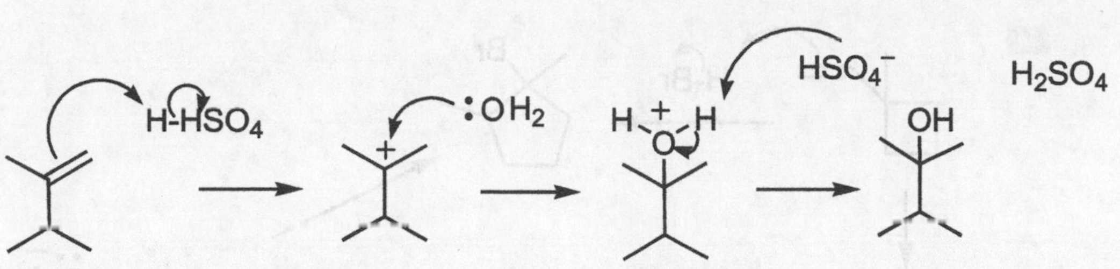

H-HSO4 :OH2 H-O+-H HSO4− H2SO4 OH

23

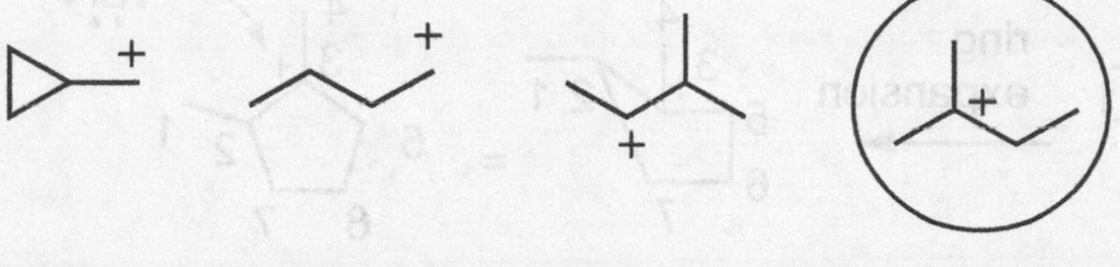

The only stable carbocation is the one circled. In the left-most structure, the carbocation is adjacent to a small ring, which can undergo a ring expansion to relieve strain. The two middle structures can both undergo hydride shifts to the adjacent carbons to yield a more-substituted (and thus more stable) carbocation.

24

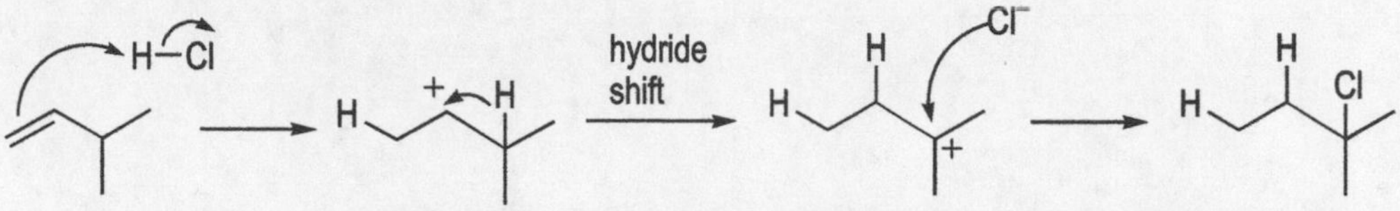

H–Cl hydride shift Cl−

Addition of HCl leads first to the carbocation. Attack of the carbocation at this point, however, doesn't lead to the correct product. Instead, the molecule undergoes a hydride shift, moving the carbocation from the secondary carbon to the more stable tertiary site. Addition of chloride ion then yields the product.

25

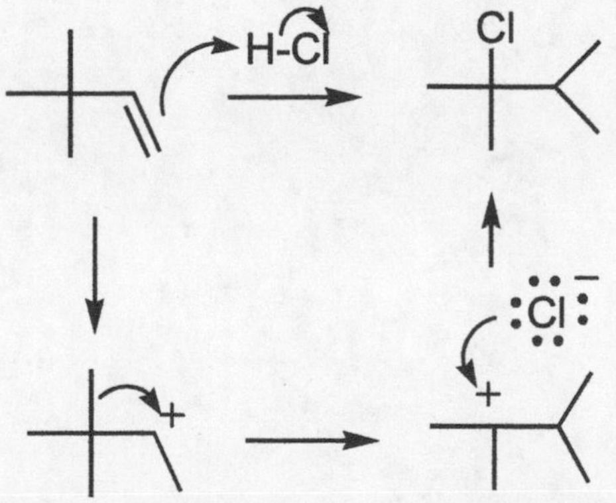

H-Cl Cl :Cl:−

First, add HCl to give the carbocation. The next step is a methyl shift from the adjacent carbon to yield the more stable tertiary carbocation. In the last step, the chloride adds to the rearranged carbocation to give the product.

26

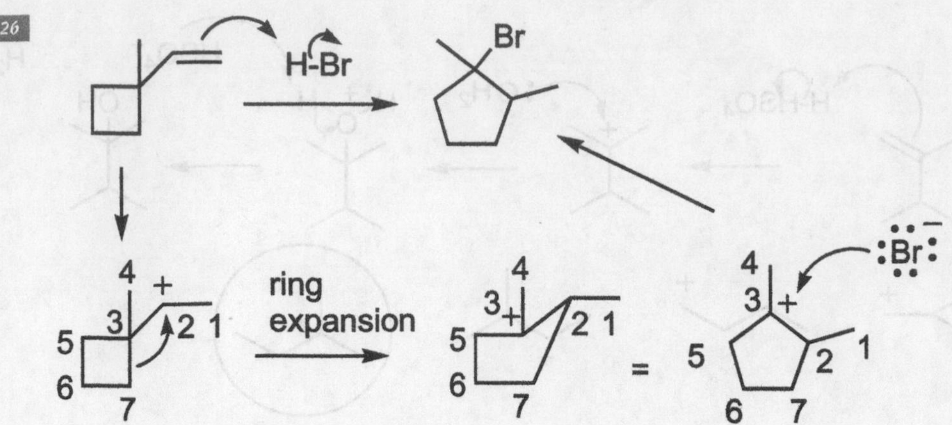

Addition of HBr first gives the carbocation adjacent to the ring. Then the molecule undergoes a ring expansion. Number the carbons to help you keep track of all the atoms, bonds, and charges. Finally, bromide adds to the rearranged carbocation to give you the product.

Chapter 9

Tripling the Fun: Alkyne Reactions and Nomenclature

In This Chapter

- Naming alkynes
- Making reactions with alkynes
- Practicing with some multistep synthesis problems

The reactions and nomenclature of alkynes are very similar to the reactions and nomenclature of the alkenes. As a result, you're sitting pretty if you understand alkenes because many parts of this chapter may feel like déjà vu. (If you're not confident with alkene nomenclature and reactions, you may want to check out Chapter 8.)

Alkynes are molecules that contain carbon-carbon triple bonds. The big concept introduced with alkynes is the addition of simple multistep synthesis problems. A *multistep synthesis* is a sequence of two or more individual reactions that converts a starting material into a desired product with a more complicated structure. Of course, you can't practice multistep synthesis until you understand the reactions, so this chapter starts with individual reactions of alkynes and then progresses to ordering reactions you already know into sequences in multistep synthesis problems.

Playing the Name Game with Alkynes

Naming alkynes is a lot like naming alkenes, only simpler because you don't have to worry about the stereochemistry around a carbon-carbon triple bond as you do around a carbon-carbon double bond.

To name alkynes, follow these steps:

1. **Locate the parent chain.**

   The parent chain is the longest chain of carbons that contains the carbon-carbon triple bond. The parent name of an alkyne ends with the suffix *-yne.* (See Chapter 8 for more on parent names.)

   If you have both an alkene and an alkyne in the molecule, the parent name is *-enyne*, with a number in front of the parent name to indicate the location of the alkene and a number in front of *-yne* to indicate the location of the alkyne. For example, 1-penten-3-yne indicates a double bond at carbon 1 and a triple bond at carbon 3.

2. **Number the chain to give the alkyne the lowest possible number.**

3. **Name the substituents.**

   Hydrocarbon substituents end with the suffix *-yl*, and halogen substituents end with the suffix *-o* (as in *chloro, fluoro, iodo,* and so forth).

4. **Place the substituents in alphabetical order in front of the parent name.**

   Use numbers to indicate the positions of the substituents on the parent chain.

Q. Name the following alkyne.

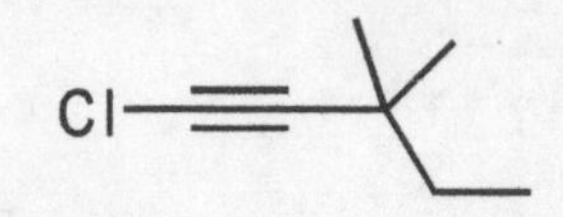

A. **1-chloro-3,3-dimethyl-1-pentyne.** After you first locate the parent chain, number the chain to give the alkyne the lowest possible number. In this case, the numbering starts from left to right, giving you a five-carbon parent chain. Because the alkyne comes at the first carbon, the name of the parent chain is 1-pentyne.

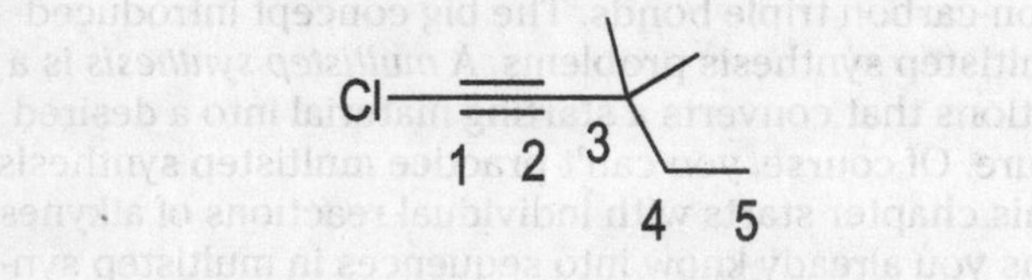

The chlorine is named *chloro* as a substituent, and the two methyl groups are clustered together to make dimethyl. Place the substituents alphabetically in front of the parent name, with numbers indicating their position on the parent chain.

Q. Name the following alkyne.

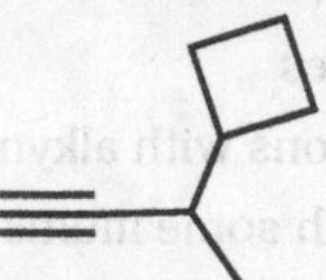

A. **4-cyclobutyl-2-pentyne.** The parent chain is the five-carbon unit containing the triple bond (not the four-carbon ring, because the parent chain needs to contain the triple bond). Number the parent chain to give the alkyne the lower number. In this case, the parent chain is five carbons long and the alkyne comes at the second carbon, so the parent name is 2-butyne. The only substituent is the four-membered ring, which as a substituent is called a *cyclobutyl.* Placing this substituent in front of the parent name gives you 4-cyclobutyl-2-pentyne.

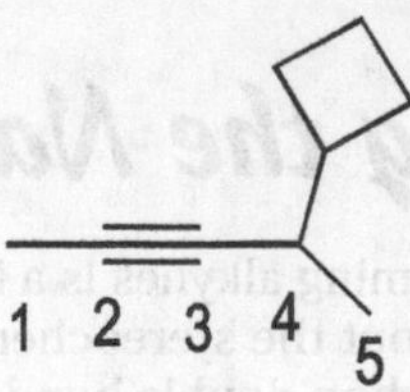

1. Name the following alkyne.

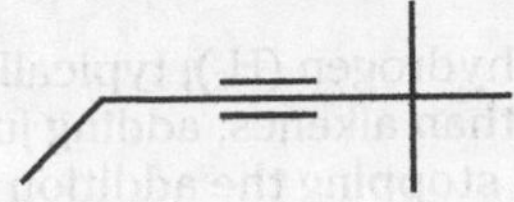

Solve It

2. Name the following alkyne.

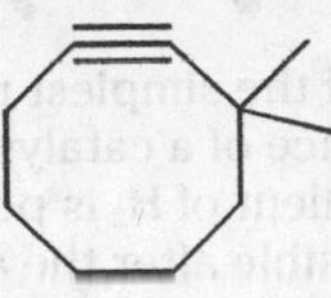

Solve It

3. Name the following alkyne.

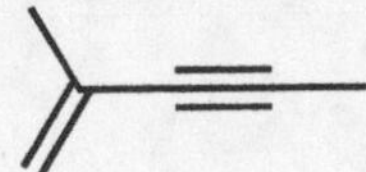

Solve It

4. Name the following alkyne.

Solve It

Adding Hydrogen and Reducing Alkynes

One of the simplest reactions of alkynes is the addition of hydrogen (H_2), typically in the presence of a catalyst. Because alkynes are more reactive than alkenes, adding just one equivalent of H_2 is possible with the right catalyst. That is, stopping the addition of hydrogen is possible after the alkyne has been reduced to an alkene without totally reducing the alkyne to an alkane. Of course, whether the alkene formed is of *cis* or *trans* stereochemistry depends on the reagents used.

Three common reagents and catalysts can reduce alkynes:

- **H_2/Lindlar catalyst:** This catalyst, which is a "poisoned" palladium catalyst, reduces alkynes to *cis*-alkenes. *Poisoned* here just means that the additives lead acetate and quinoline make the palladium catalyst less reactive in order to stop the reduction at the alkene stage.
- **Na/NH_3:** These reagents (which, unlike the other two, don't contain H_2 gas) reduce alkynes to *trans*-alkenes.
- **H_2/Pd (or H_2/Pd(C) or H_2/Pt):** These catalysts are highly active and reduce alkynes to alkanes, not stopping at the alkene stage.

Q. Predict the product of the following reaction.

A.

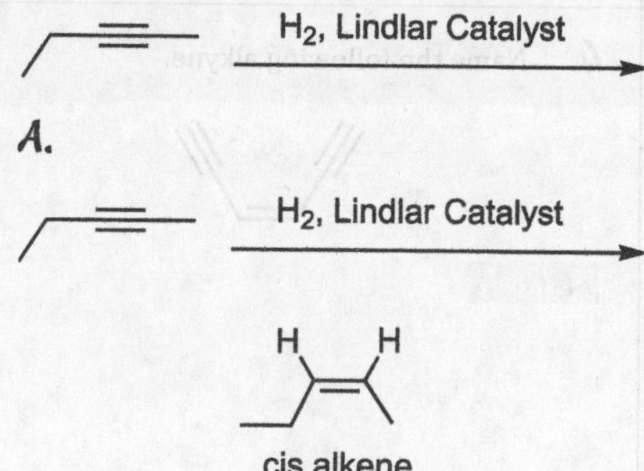

The addition of hydrogen in the presence of the Lindlar catalyst leads to reduction to the *cis* alkene. This catalyst selectively forms the *cis*-alkene over the *trans*-alkene.

Q. Predict the product of the following reaction.

A.

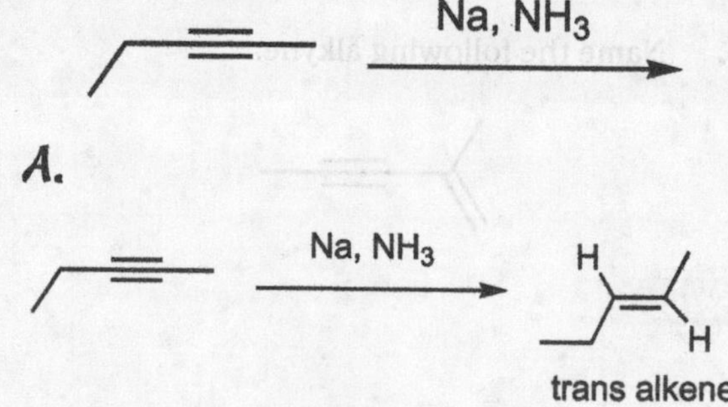

Addition of sodium (Na) in ammonia (NH_3) reduces the alkyne to the *trans* alkene.

5. Predict the product of the following reaction.

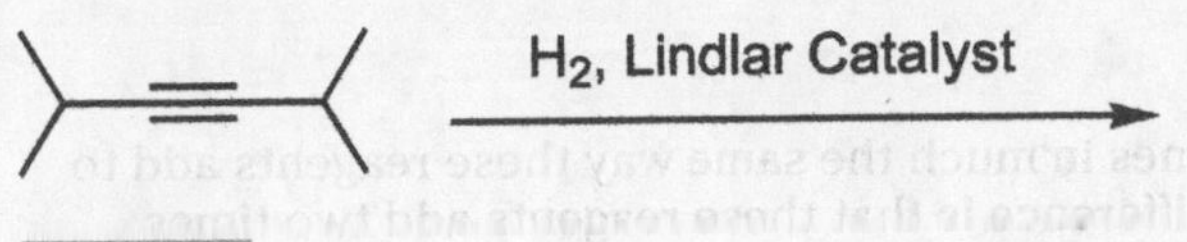

Solve It

6. Predict the product of the following reaction.

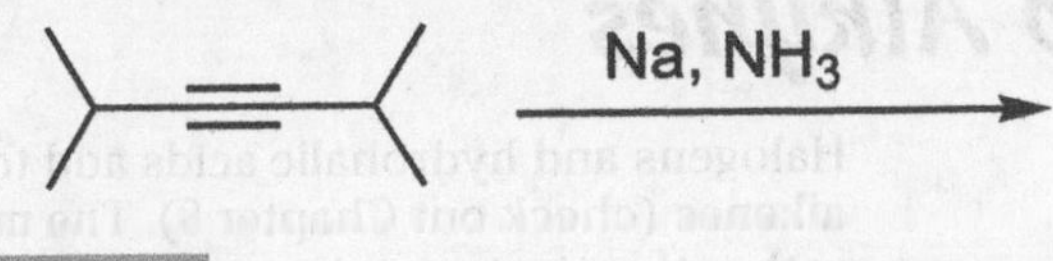

Solve It

7. Predict the reagent required to complete the following reaction.

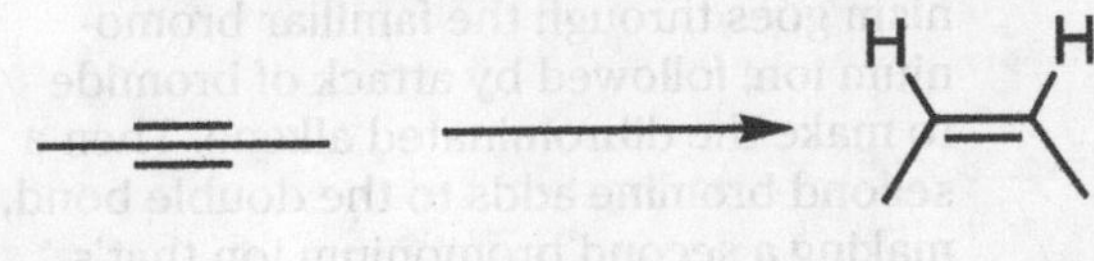

Solve It

8. Predict the product of the following reaction.

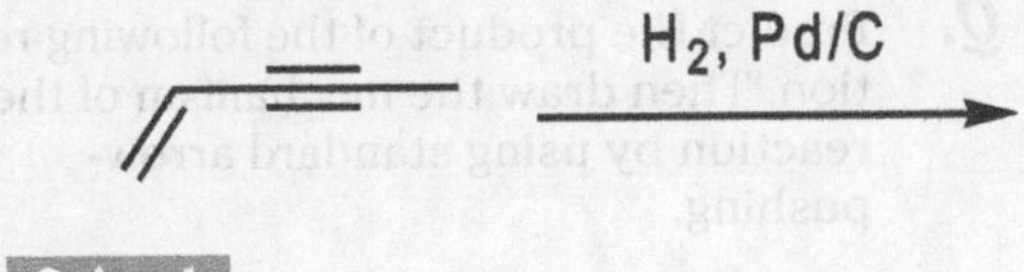

Solve It

Adding Halogens and Hydrohalic Acids to Alkynes

Halogens and hydrohalic acids add to alkynes in much the same way these reagents add to alkenes (check out Chapter 8). The main difference is that these reagents add two times rather than just once because each alkyne contains two reactive pi bonds, compared to just one pi bond in alkenes.

Figure 9-1 shows the addition of halogens and hydrohalic acids reactions. In the case of the halogenation reaction, the product is a *tetrahalide* (four halides). In the hydrohalic acid reaction, the product is a *dihalide,* with both halogen atoms attached to the same carbon (not one halogen to each carbon).

R−C≡C−R —(2 Br_2)→ $R(Br)_2C-C(Br)_2R$

R−C≡C−H —(2 HBr)→ $R(Br)_2C-CH_3$

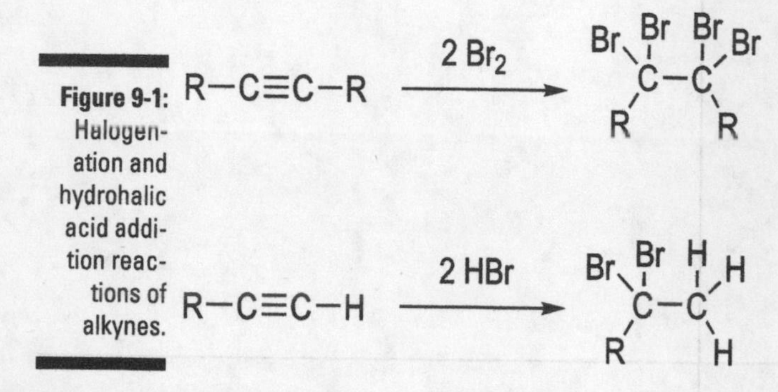

Figure 9-1: Halogenation and hydrohalic acid addition reactions of alkynes.

Q. Predict the product of the following reaction. Then draw the mechanism of the reaction by using standard arrow-pushing.

Br_2

A.

Br_2

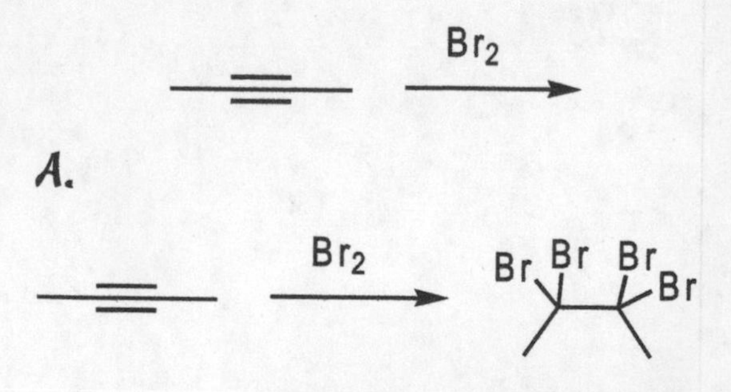

Alkynes react with halides (such as Br_2 or Cl_2) in the same fashion alkenes react with these reagents — except, of course, these reagents add twice to alkynes. The mechanism goes through the familiar bromonium ion, followed by attack of bromide to make the dibrominated alkene. Then a second bromine adds to the double bond, making a second bromonium ion that's attacked by bromide to make the tetrabrominated product.

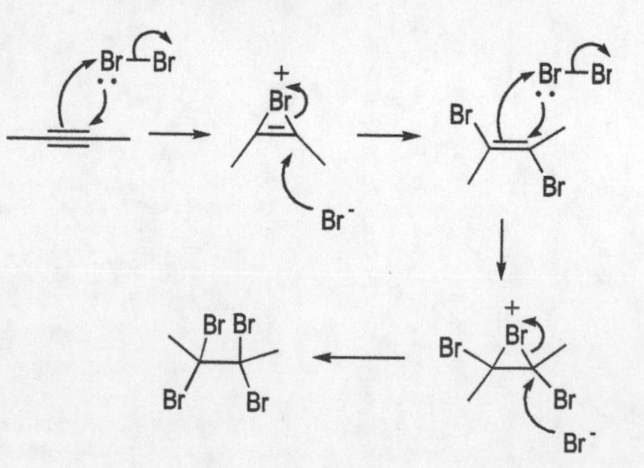

Q. Predict the product of the following reaction. Then draw the mechanism of the reaction by using standard arrow-pushing.

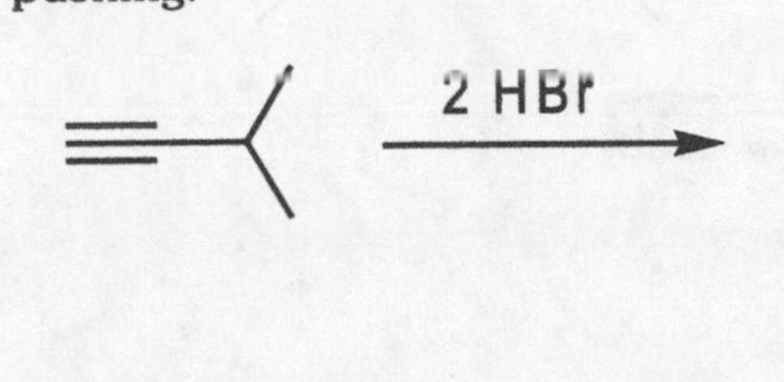

2 HBr

A.

2 HBr

Br Br

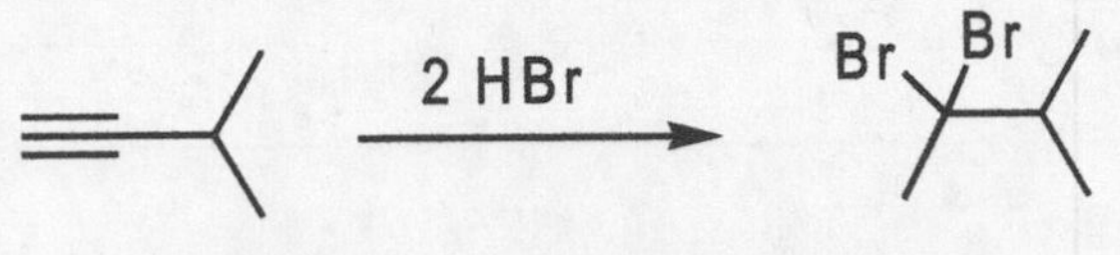

Alkynes react twice with HBr to form a dibromide, where both bromines add to the more-substituted carbon through Markovnikov addition (in other words, the large group — in this case, the halide — adds to the more highly substituted carbon). The mechanism of the reaction is quite similar to the addition of HBr to alkenes. The first step is protonation of the triple bond to form a carbocation (called a *vinyl* cation because the cation is on a double-bond carbon). The cation prefers to rest on the more-substituted carbon, where the charge is more stable. The bromide (Br^-) then attacks the cation to form a bromine substituted alkene. The next step is protonation of the double bond to form a new carbocation (again, on the more-substituted carbon). Lastly, the bromide attacks this carbocation to form the dibrominated product.

H—Br

Br⁻

vinyl cation

H—Br

Br

Br⁻

Br

Br Br

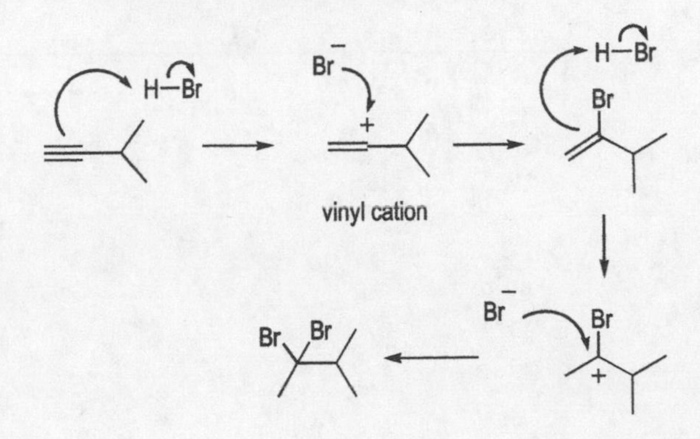

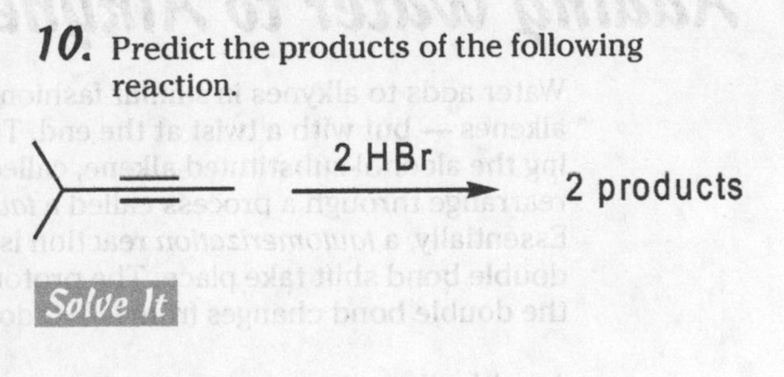

9. Predict the product of the following reaction. Then draw the mechanism of the reaction by using standard arrow-pushing.

2 HCl

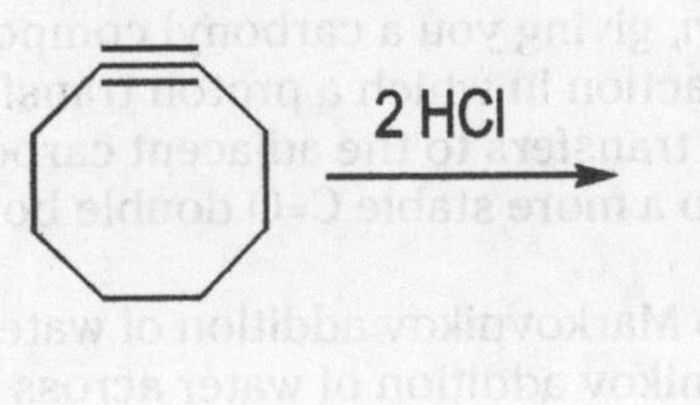

Solve It

10. Predict the products of the following reaction.

2 HBr

2 products

Solve It

11. Predict the product of the following reaction. Then draw the mechanism of the reaction by using standard arrow-pushing.

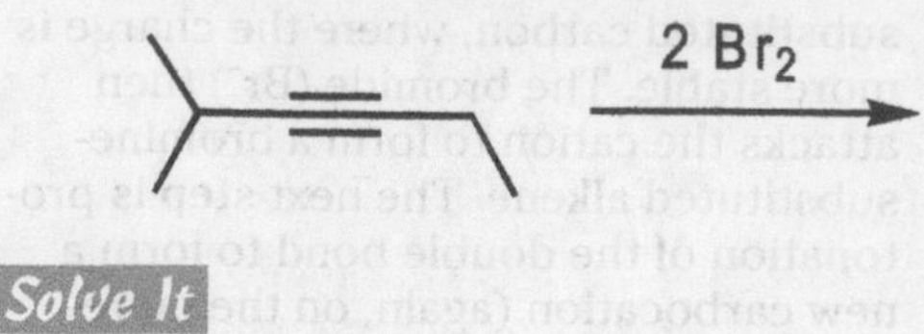

Solve It

12. Predict the product of the following reaction.

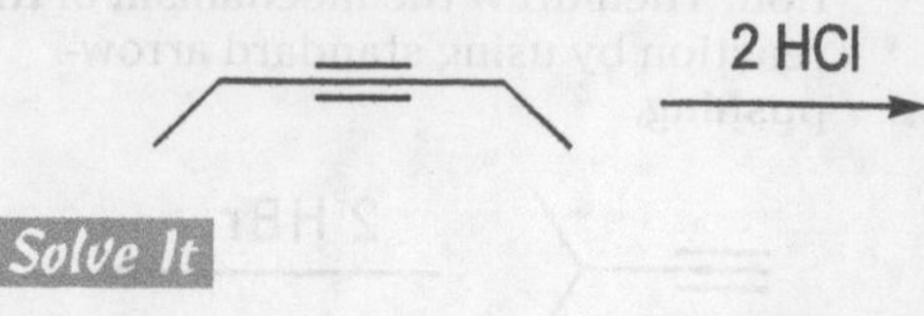

Solve It

Adding Water to Alkynes

Water adds to alkynes in similar fashion (and with similar reagents) to how it adds to alkenes — but with a twist at the end. The addition goes as you may expect at first, producing the alcohol-substituted alkene, called an *enol.* However, enols are unstable and rapidly rearrange through a process called a *tautomerization,* giving you a carbonyl compound. Essentially, a *tautomerization* reaction is simply a reaction in which a proton transfer and a double bond shift take place: The proton on oxygen transfers to the adjacent carbon while the double bond changes from a C=C double bond to a more stable C=O double bond.

As with alkenes, an *oxymercuration* reaction leads to Markovnikov addition of water across the alkyne, and a borane reaction gives anti-Markovnikov addition of water across the alkyne. The reagents are slightly different structurally from those used for alkenes, but they serve the same purpose. Figure 9-2 shows these reactions.

$$R{-}C{\equiv}C{-}H \xrightarrow[H_2O]{HgSO_4,\ H_2SO_4} R{-}C(OH){=}C(H){-}H \xrightarrow{\text{tautomerization}} R{-}C(=O){-}CH_2{-}H$$

unstable enol

Markovnikov addition (OH adds to more-substituted carbon)

$$R{-}C{\equiv}C{-}H \xrightarrow[2.\ H_2O_2,\ OH^-]{1.\ (Sia)_2BH} R{-}C(H){=}C(OH){-}H \xrightarrow{\text{tautomerization}} R{-}CH_2{-}C(=O){-}H$$

unstable enol

anti-Markovnikov addition (OH adds to less-substituted carbon)

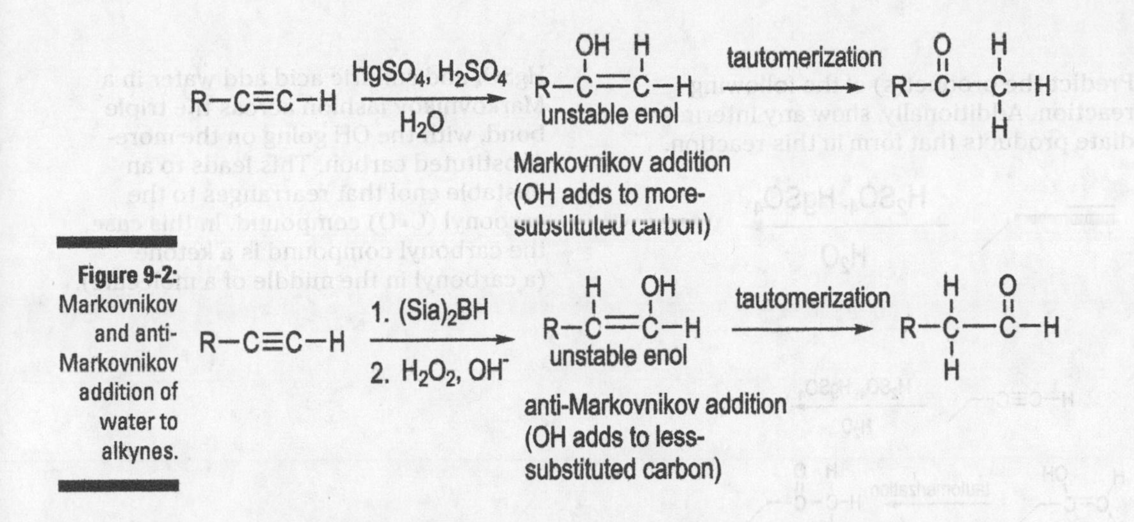

Figure 9-2: Markovnikov and anti-Markovnikov addition of water to alkynes.

Thus, these water-addition reactions to alkynes employ two steps:

1. **Add water across the triple bond to form the unstable enol.**

   Whether the hydroxyl (OH) group adds to the more-substituted carbon or to the less-substituted carbon depends on whether you added Markovnikov reagents (oxymercuration) or anti-Markovnikov reagents (borane).

2. **Perform the tautomerization that converts the enol into the carbonyl form.**

Q. Predict the product(s) of the following reaction. Additionally, show any intermediate products that form in this reaction.

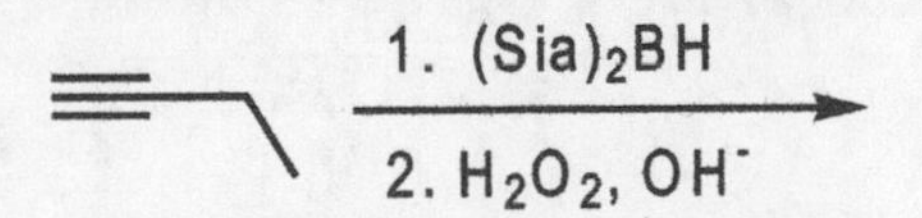

1. $(Sia)_2BH$
2. H_2O_2, OH^-

A. 1. $(Sia)_2BH$ 2. H_2O_2 → HO, H (unstable enol) → tautomerization → O (aldehyde)

Disiamyl borane $(Sia)_2BH$ adds to alkynes in much the same way that BH_3 adds to alkenes. You get anti-Markovnikov addition of water across the triple bond (which means that the OH goes on the less-substituted carbon and the H goes on the more-substituted carbon). This yields an enol, an OH group directly attached to an alkene. The unstable enol tautomerizes to a carbonyl product (a C=O group), with the carbonyl forming on the carbon that held the OH group. In this case, you get an aldehyde (a C=O group at the end of a chain) because the water adds in anti-Markovnikov fashion.

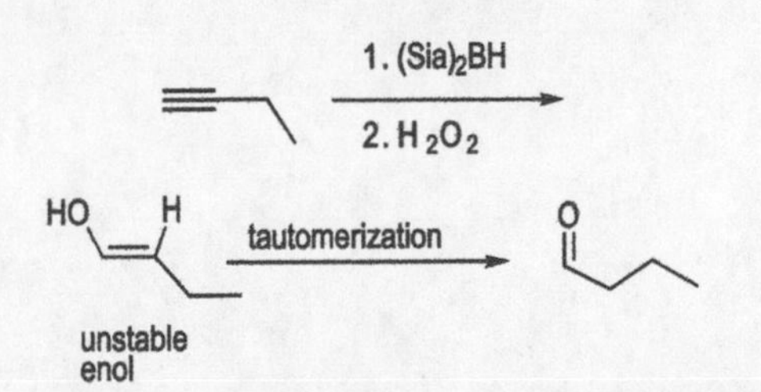

Q. Predict the product(s) of the following reaction. Additionally, show any intermediate products that form in this reaction.

$$\text{HC}\equiv\text{C–CH}_2\text{CH}_3 \xrightarrow[H_2O]{H_2SO_4,\ HgSO_4}$$

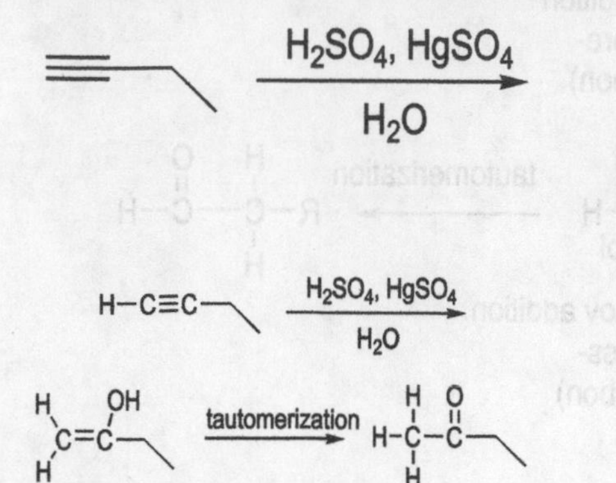

A. HgSO$_4$ and sulfuric acid add water in a Markovnikov fashion across the triple bond, with the OH going on the more-substituted carbon. This leads to an unstable enol that rearranges to the carbonyl (C=O) compound. In this case, the carbonyl compound is a ketone (a carbonyl in the middle of a molecule).

$$\text{H–C}\equiv\text{C–CH}_2\text{CH}_3 \xrightarrow[H_2O]{H_2SO_4,\ HgSO_4}$$

$$\text{H}_2\text{C=C(OH)–CH}_2\text{CH}_3 \xrightarrow{\text{tautomerization}} \text{H–CH}_2\text{–C(=O)–CH}_2\text{CH}_3$$

unstable enol

13. Predict the product(s) of the following reaction. Additionally, show any intermediate products that form in this reaction.

$$\text{HC}\equiv\text{C–C}_6\text{H}_{11} \xrightarrow[H_2SO_4]{HgSO_4}$$

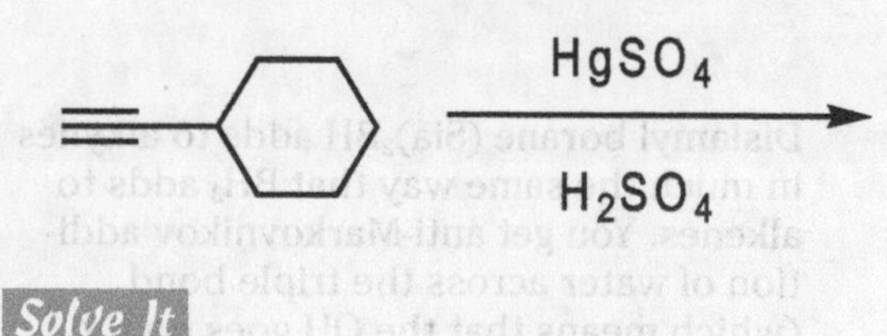

Solve It

14. Predict the product(s) of the following reaction. Additionally, show any intermediate products that form in this reaction.

$$\text{HC}\equiv\text{C–C}_6\text{H}_{11} \xrightarrow[\text{2. } H_2O_2,\ OH^-]{\text{1. (Sia)}_2\text{BH}}$$

Solve It

15. Predict the product(s) of the following reaction.

$$\text{alkyne} \xrightarrow[H_2SO_4]{HgSO_4} \text{2 products}$$

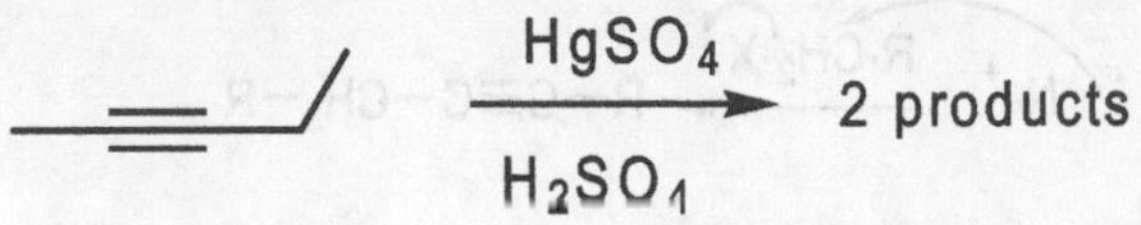

Solve It

16. Predict the product(s) of the following reaction. Additionally, draw the mechanism of the reaction by using standard arrow-pushing.

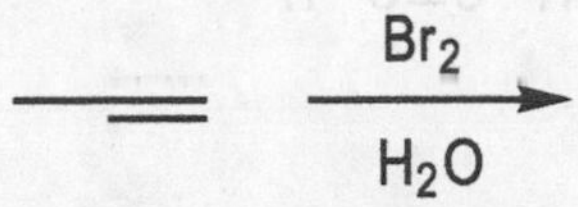

$$\text{alkyne} \xrightarrow[H_2O]{Br_2}$$

Solve It

Creating Alkynes

To make alkynes, you have two principal reactions. The first is a double-elimination reaction similar to the one you may have seen for making alkenes (more technically called a *double dehydrohalogenation reaction,* for those who enjoy tongue twisters). Figure 9-3 shows this reaction.

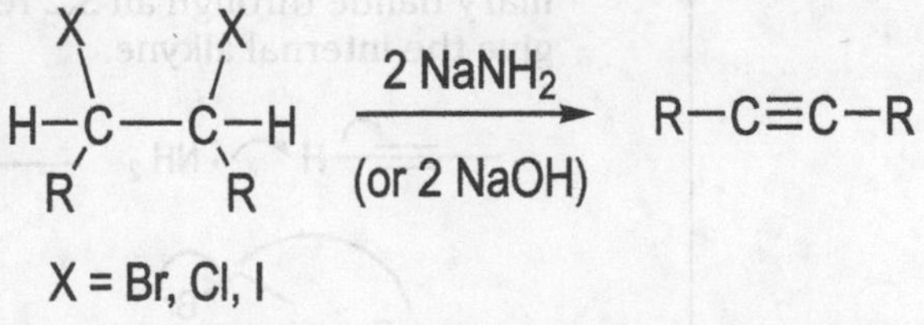

$$H{-}CXR{-}CXR{-}H \xrightarrow[\text{(or 2 NaOH)}]{2\ NaNH_2} R{-}C{\equiv}C{-}R$$

X = Br, Cl, I

Figure 9-3: Making alkynes by double dehydrohalogenation.

The other one is new chemistry, using acetylide ions (deprotonated alkynes) to form new alkyne-carbon bonds (see Figure 9-4). These reactions require *terminal alkynes,* which are alkynes at the end of a chain with a proton attached. Protons attached to alkynes are weakly acidic, so a very strong base, such as $NaNH_2$, can deprotonate the alkyne to generate an acetylide ion. This acetylide ion is a powerful nucleophile that adds to primary alkyl halides ($R\text{-}CH_2\text{-}X$) via an S_N2 reaction to give an *internal alkyne,* an alkyne in the middle of a molecule (for a briefing on S_N2 reactions, see Chapter 10). This reaction is especially important because it forms a carbon-carbon bond.

Figure 9-4: Making alkynes using acetylide chemistry.

$$R{-}C{\equiv}C{-}H \xrightarrow{:\overset{-}{N}H_2\ Na^+} R{-}C{\equiv}C:^-\ Na^+ \xrightarrow{R{-}CH_2{-}X} R{-}C{\equiv}C{-}CH_2{-}R$$

Don't mistake the deprotonating base $NaNH_2$ with the reagents Na/NH_3 for reducing an alkyne.

Q. Predict the product of the following reaction.

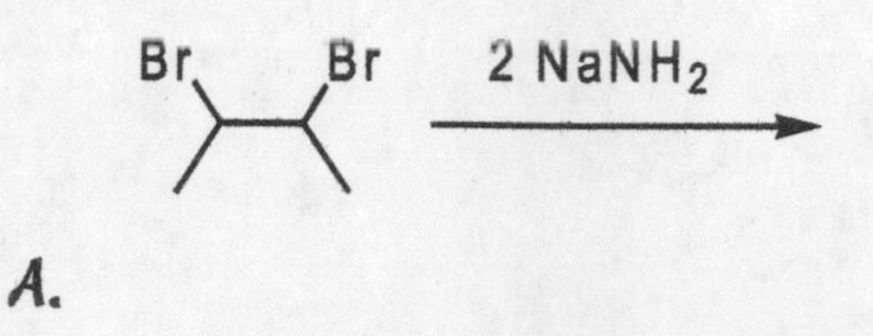

Br, Br (vicinal dibromide) $\xrightarrow{2\ NaNH_2}$

A.

Br, Br (vicinal dibromide) $\xrightarrow{2\ NaNH_2}$ alkyne

In this reaction, you eliminate two molecules of HBr to make the alkyne. This reaction occurs through the E2 mechanism (refer to Chapter 10) in similar fashion to the elimination of a single halide to give an alkene, except you have to eliminate twice to make an alkyne.

Q. Predict the product of the following reaction. Also draw the mechanism of the reaction.

terminal alkyne $\xrightarrow[2.\ CH_3CH_2Br]{1.\ NaNH_2}$

A.

terminal alkyne $\xrightarrow[2.\ CH_3CH_2Br]{1.\ NaNH_2}$ internal alkyne

Terminal alkynes are weakly acidic. In the presence of a really strong base (such as $NaNH_2$), the terminal proton is pulled off to generate an acetylide ion. This acetylide ion reacts with the primary halide through an S_N2 reaction to give the internal alkyne.

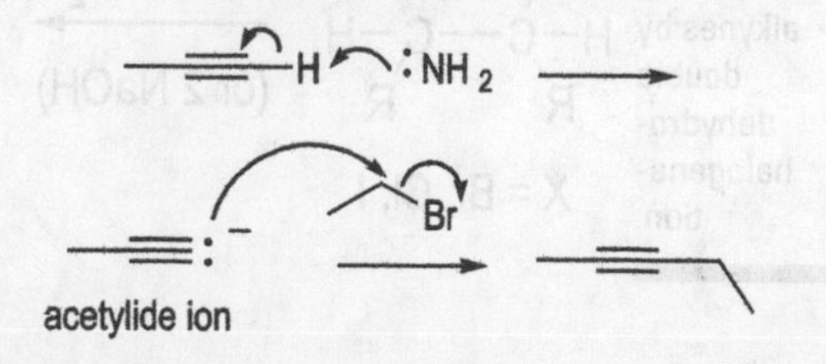

alkyne–H + $:NH_2^-$ →

acetylide ion + Br (ethyl bromide) → internal alkyne

acetylide ion

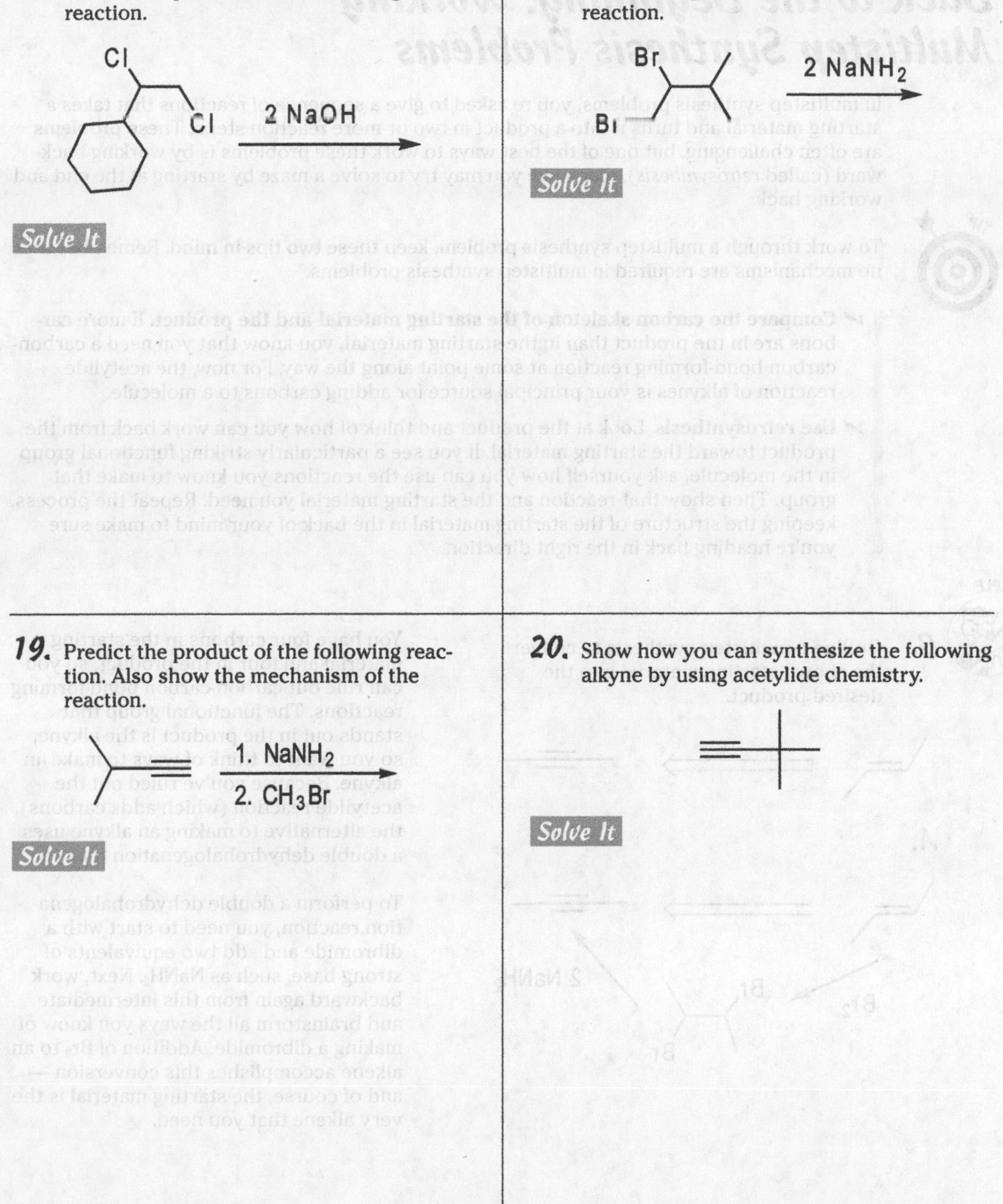

17. **Predict the product of the following reaction.**

Cl

Cl

$\xrightarrow{\text{2 NaOH}}$

Solve It

18. **Predict the product of the following reaction.**

Br

Br

$\xrightarrow{\text{2 NaNH}_2}$

Solve It

19. **Predict the product of the following reaction. Also show the mechanism of the reaction.**

$\xrightarrow[\text{2. CH}_3\text{Br}]{\text{1. NaNH}_2}$

Solve It

20. **Show how you can synthesize the following alkyne by using acetylide chemistry.**

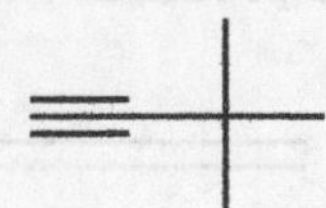

Solve It

Back to the Beginning: Working Multistep Synthesis Problems

In multistep synthesis problems, you're asked to give a sequence of reactions that takes a starting material and turns it into a product in two or more reaction steps. These problems are often challenging, but one of the best ways to work these problems is by working backward (called *retrosynthesis*), much like you may try to solve a maze by starting at the end and working back.

To work through a multistep synthesis problem, keep these two tips in mind. Remember that no mechanisms are required in multistep synthesis problems.

- **Compare the carbon skeleton of the starting material and the product.** If more carbons are in the product than in the starting material, you know that you need a carbon-carbon bond-forming reaction at some point along the way. For now, the acetylide reaction of alkynes is your principal source for adding carbons to a molecule.
- **Use retrosynthesis.** Look at the product and think of how you can work back from the product toward the starting material. If you see a particularly striking functional group in the molecule, ask yourself how you can use the reactions you know to make that group. Then show that reaction and the starting material you need. Repeat the process, keeping the structure of the starting material in the back of your mind to make sure you're heading back in the right direction.

EXAMPLE

Q. Outline a multistep synthesis to convert the shown starting material into the desired product.

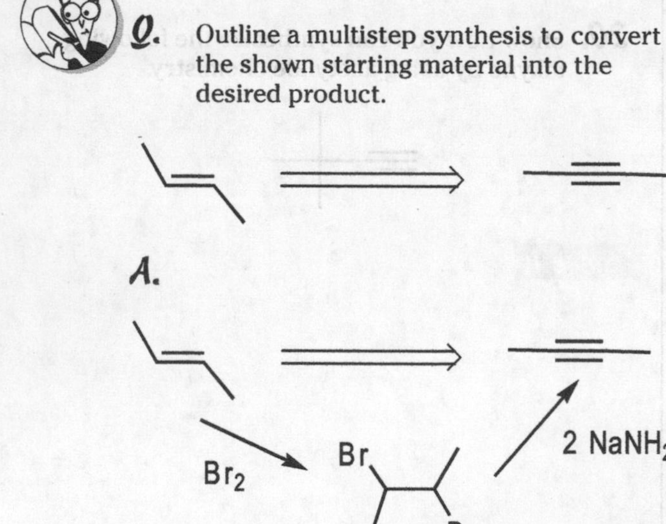

A. You have four carbons in the starting material and four in the product, so you can rule out carbon-carbon bond-forming reactions. The functional group that stands out in the product is the alkyne, so you want to think of ways to make an alkyne. Because you've ruled out the acetylide reaction (which adds carbons), the alternative to making an alkyne uses a double dehydrohalogenation reaction.

To perform a double dehydrohalogenation reaction, you need to start with a dibromide and add two equivalents of strong base, such as $NaNH_2$. Next, work backward again from this intermediate and brainstorm all the ways you know of making a dibromide. Addition of Br_2 to an alkene accomplishes this conversion — and of course, the starting material is the very alkene that you need.

Q. Outline a multistep synthesis to convert the shown starting material into the desired product.

A.

NaNH$_2$

1. $(Sia)_2BH$
2. H_2O_2, OH^-

Br

$\equiv\!:^{-}\ Na^{+}$

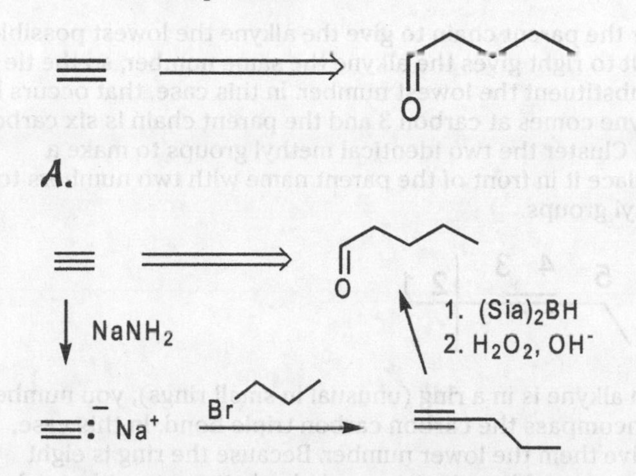

The first thing you may notice is that the product has three more carbons than the starting material, which means that one of the steps is to lengthen the carbon chain by using a C-C bond-forming reaction. Working backward, brainstorm all the ways you know of making an aldehyde: There's only one — anti-Markovnikov addition of water to an alkyne by using a borane reagent followed by peroxides.

Working back again, think how you could've made that alkyne. Double elimination or acetylide chemistry are both possible, but because the starting material has fewer carbons than the product, using acetylide chemistry seems the best route. The starting alkyne is acetylene (ethyne), and you want to add a three-carbon chain; to do this, you first add the strong base (NaNH$_2$) to deprotonate the alkyne and then add a three-carbon halide (I use a bromide here, but you could choose a chloride or iodide as well).

21. Outline a multistep synthesis to convert the shown starting material into the desired product.

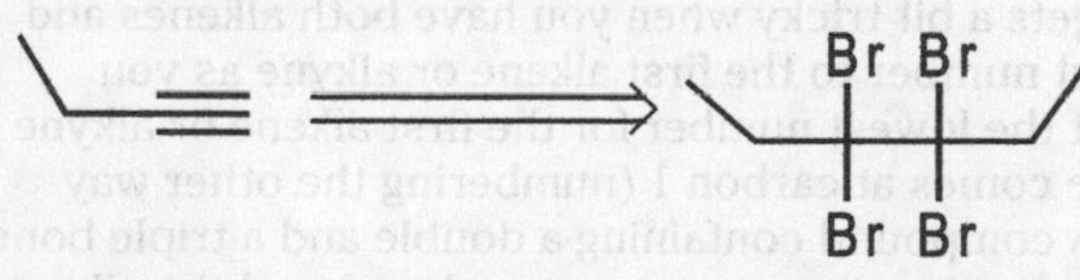

Solve It

22. Outline a multistep synthesis to convert the shown starting material into the desired product.

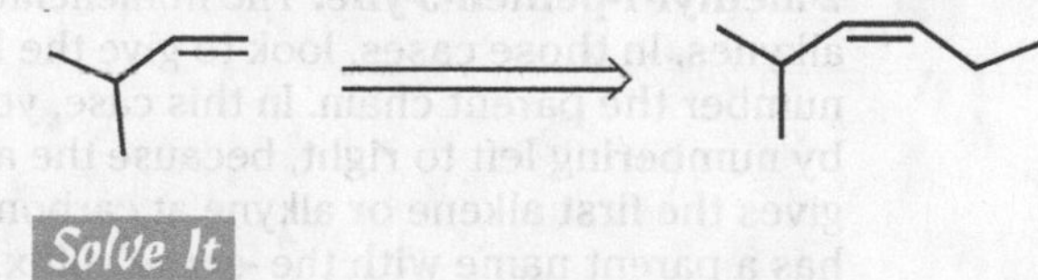

Solve It

Answer Key

The following are the answers to the practice questions presented in this chapter.

1. **2,2-dimethyl-3-hexyne.** First, number the parent chain to give the alkyne the lowest possible number. Numbering right to left or left to right gives the alkyne the same number, so the tie goes to the side that gives the first substituent the lowest number. In this case, that occurs by numbering from right to left. The alkyne comes at carbon 3 and the parent chain is six carbons long, so the parent name is 3-hexyne. Cluster the two identical methyl groups to make a dimethyl unit in the name and then place it in front of the parent name with two numbers to indicate the position of the two methyl groups.

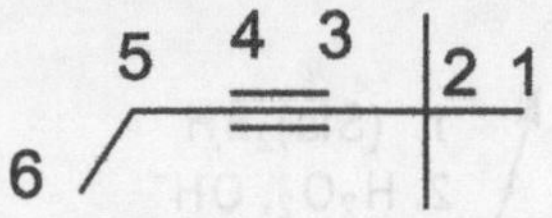

2. **3,3-dimethyl-1-cyclooctyne.** When an alkyne is in a ring (unusual in small rings), you number the alkyne so the first two carbons encompass the carbon-carbon triple bond. In this case, number toward the substituents to give them the lower number. Because the ring is eight carbons, the parent name is 1-cyclooctyne. Adding the two methyl substituents with numbers to indicate their position provides the full name.

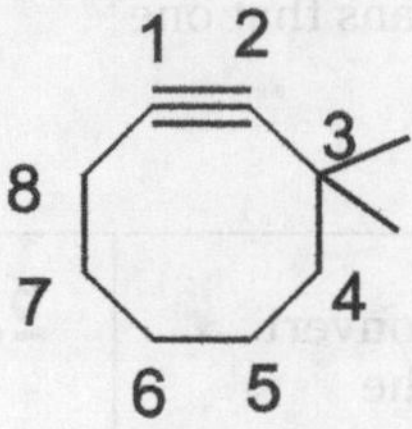

3. **2-methyl-1-penten-3-yne.** The nomenclature gets a bit tricky when you have both alkenes and alkynes. In those cases, look to give the lowest number to the first alkene or alkyne as you number the parent chain. In this case, you get the lowest number for the first alkene or alkyne by numbering left to right, because the alkene comes at carbon 1 (numbering the other way gives the first alkene or alkyne at carbon 3). A compound containing a double and a triple bond has a parent name with the *-enyne* suffix. Because the alkene comes at carbon 1 and the alkyne comes at carbon 3 and the parent chain is five carbons long, the parent name is 1-penten-3-yne. Adding the methyl substituent provides the full name.

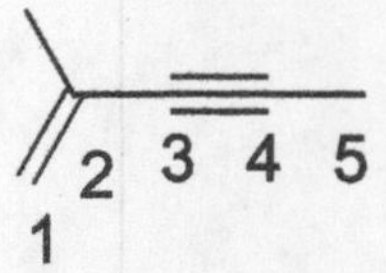

4 ***cis*-3-hexen-1,5-diyne.** This problem is challenging. Numbering is straightforward because it doesn't matter whether you number left to right or right to left. A molecule that contains one alkene and two alkynes is called an *-endiyne* (one *-ene* two *-ynes*). In this case, the alkynes are at positions 1 and 5, the alkene is at position 3, and the parent chain is six carbons long, so the parent name is 3-hexen-1,5-diyne. The double bond is of *cis* (or *Z*) stereochemistry, so add the stereochemistry as a prefix of the parent name.

1 2 5 6

3 4

5

H H

H_2, Lindlar Catalyst

cis alkene

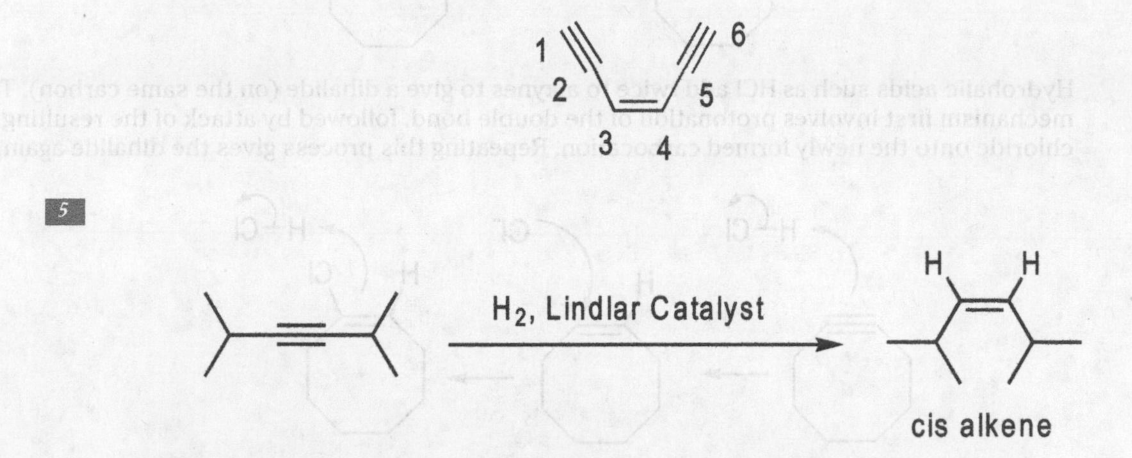

Lindlar's catalyst adds hydrogen to the alkyne to yield a *cis* alkene.

6

Na, NH_3

H

H

trans alkene

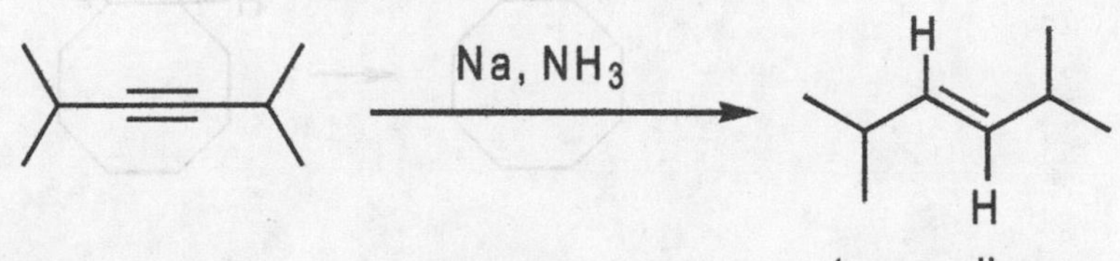

Sodium (Na) in ammonia (NH_3) adds hydrogen to the alkyne to produce a *trans* alkene.

7

H_2, Lindlar Catalyst

H H

Hydrogen and Lindlar's catalyst accomplish the conversion to make the *cis* alkene.

8

H_2, Pd/C

Hydrogen and palladium on carbon reduces double and triple bonds to the corresponding alkanes, not stopping at the alkene stage.

9

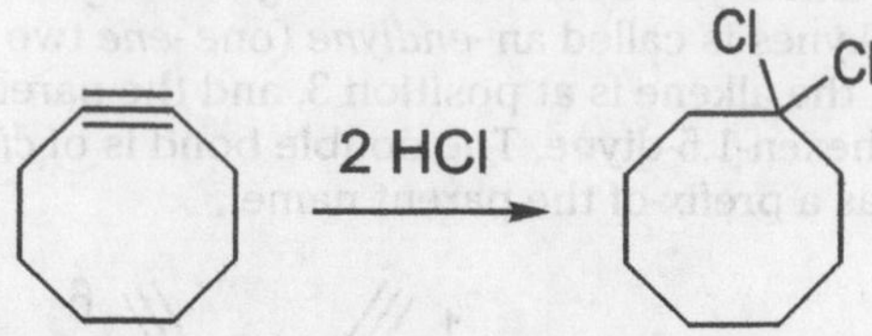

Hydrohalic acids such as HCl add twice to alkynes to give a dihalide (on the same carbon). The mechanism first involves protonation of the double bond, followed by attack of the resulting chloride onto the newly formed carbocation. Repeating this process gives the dihalide again.

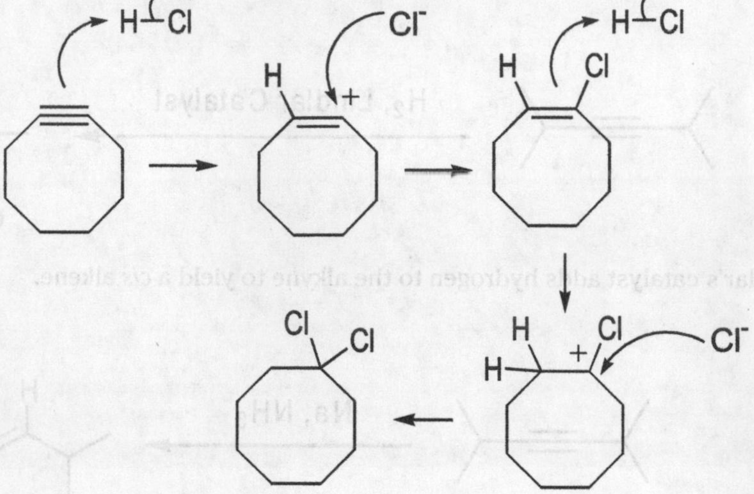

10

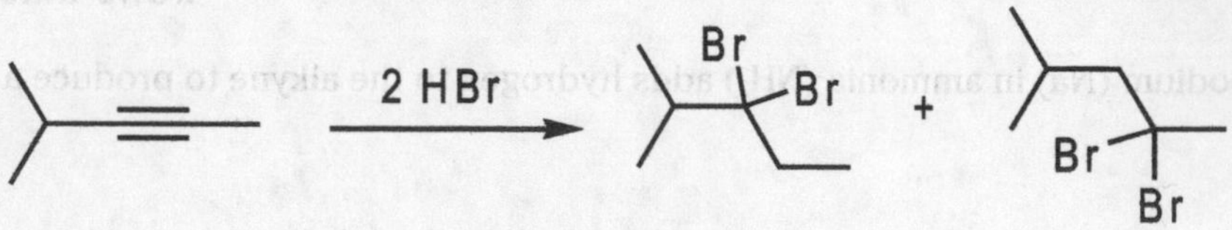

In this case, you can get two dibrominated products because both sides of the alkyne are equally substituted (thus, there's no preference for addition to the left or right side). You get a mixture of HBr addition to the left and right side of the alkyne. Note, though, that the two halides always add to the same carbon.

11

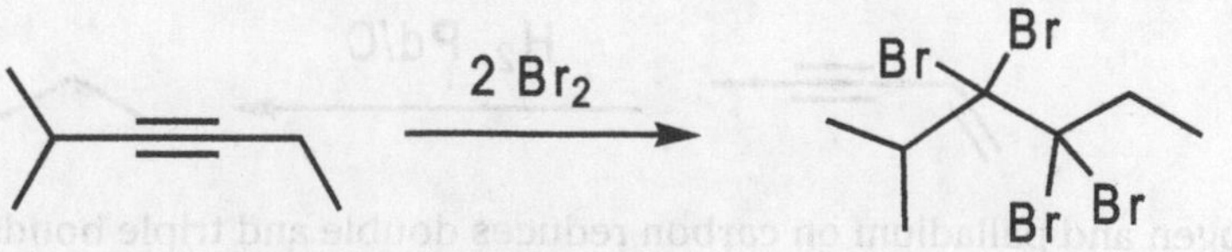

Halogens add to alkynes twice to give the tetrahalogenated product (product with four halogens).

The mechanism goes through a bromonium ion following attack of the alkyne on the bromine. Addition of bromide leads to the dibrominated alkene. Addition of a second molecule of Br_2 through another bromonium ion intermediate produces the tetrabrominated product.

Br—Br Br+ Br⁻ Br Br—Br Br Br Br Br Br Br Br Br Br⁻

12

2 HCl Cl Cl

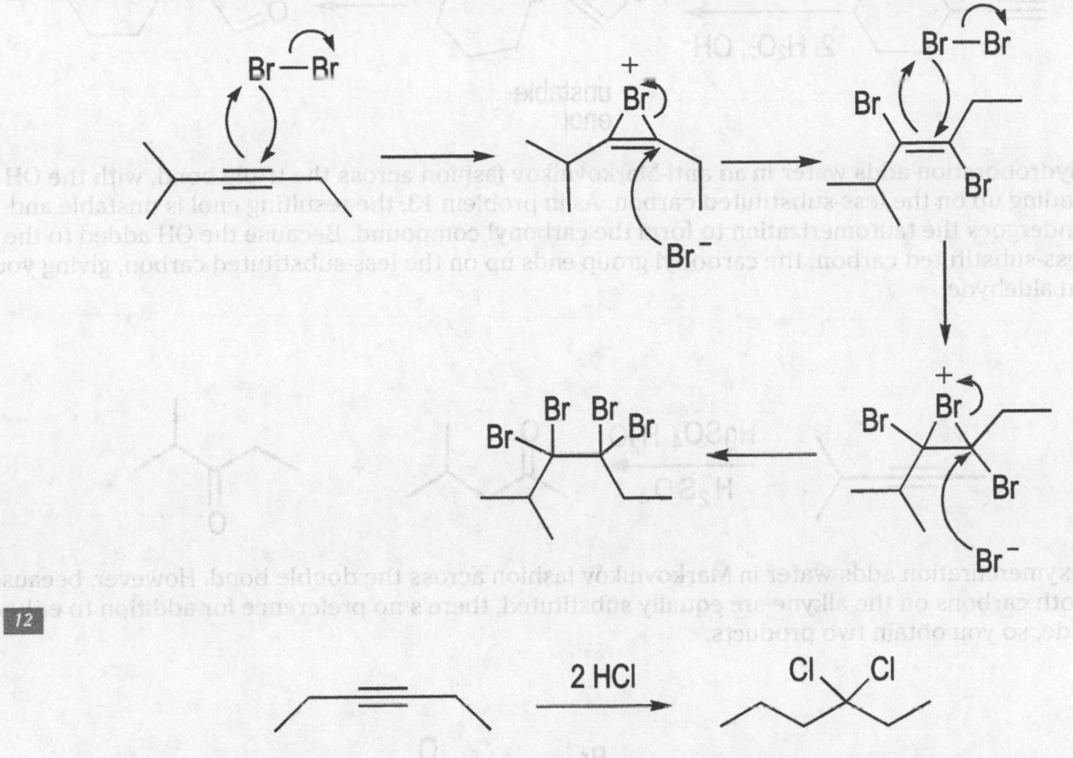

Because the alkyne is symmetrical, each side of the triple bond is identical, so you get only a single product. In other words, regardless of whether the two chlorine atoms add to the left carbon or the right carbon of the triple bond, you obtain the same product.

13

$HgSO_4$, H_2O H_2SO_4 OH O

unstable enol

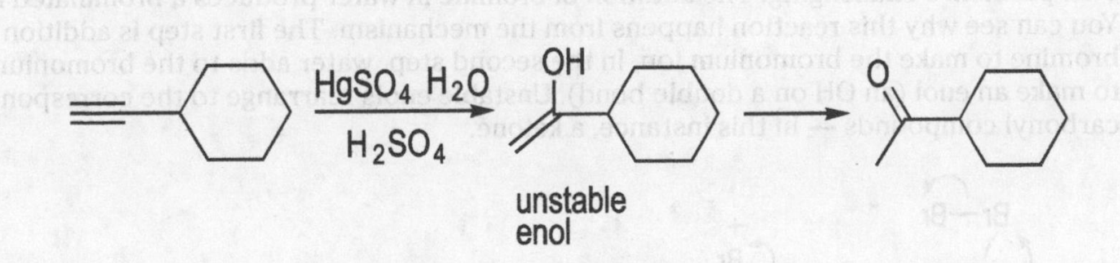

Mercury sulfate ($HgSO_4$) and sulfuric acid add water in a Markovnikov fashion (in which the OH adds to the more-substituted carbon) to make an enol (an OH on a double bond). Enols are unstable and undergo a rearrangement (tautomerization) to give the carbonyl compound. Because the OH adds to the more-substituted carbon, the carbonyl (C=O) ends up on the more-substituted carbon, producing a ketone.

14

1. $(Sia)_2BH$

2. H_2O_2, OH^-

HO

unstable enol

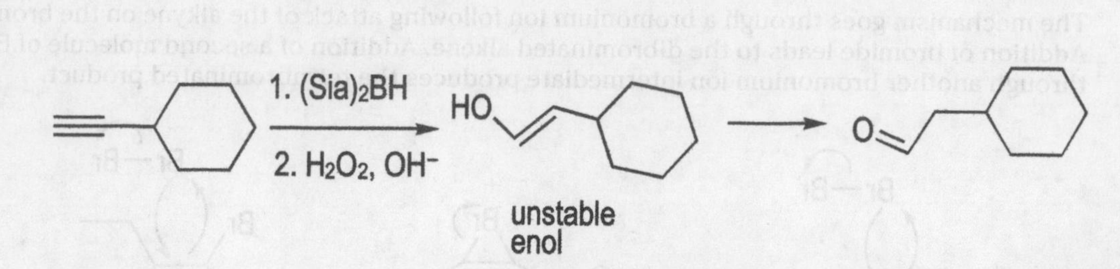

Hydroboration adds water in an anti-Markovnikov fashion across the triple bond, with the OH ending up on the less-substituted carbon. As in problem 13, the resulting enol is unstable and undergoes the tautomerization to form the carbonyl compound. Because the OH added to the less-substituted carbon, the carbonyl group ends up on the less-substituted carbon, giving you an aldehyde.

15

$HgSO_4$ H_2O

H_2SO_4

O

\+

O

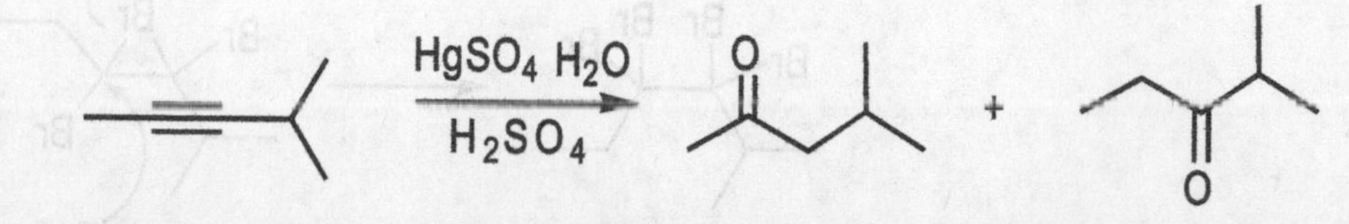

Oxymercuration adds water in Markovnikov fashion across the double bond. However, because both carbons on the alkyne are equally substituted, there's no preference for addition to either side, so you obtain two products.

16

Br_2

H_2O

O

Br

This problem's challenging. The addition of bromine in water produces a brominated ketone. You can see why this reaction happens from the mechanism. The first step is addition of bromine to make the bromonium ion. In the second step, water adds to the bromonium ion to make an enol (an OH on a double bond). Unstable enols rearrange to the corresponding carbonyl compounds — in this instance, a ketone.

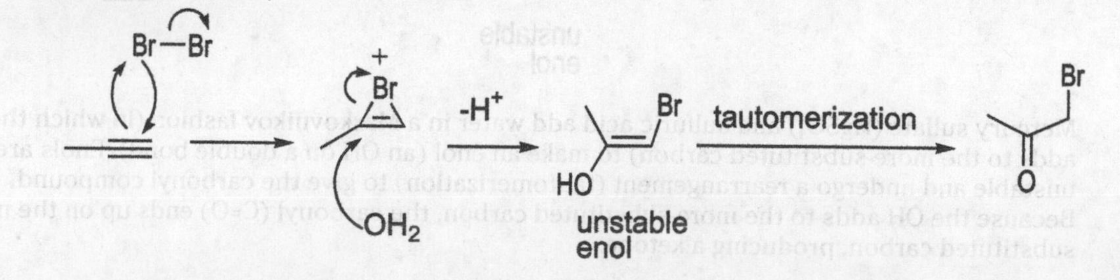

Br—Br

Br^+

$-H^+$

Br

tautomerization

Br

HO

OH_2

unstable enol

O

17

Cl Cl 2 NaOH

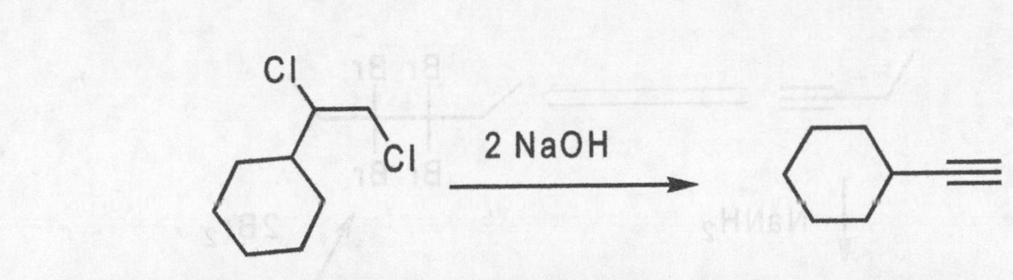

Two equivalents of base eliminate two molecules of HCl to make the alkyne.

18

Br Br 2 NaNH$_2$

Two equivalents of strong base eliminate two molecules of HBr to form the alkyne.

19

1. NaNH$_2$ 2. CH$_3$Br CH$_3$

$\overline{:}$NH$_2$ Na$^+$ C–H C$\overline{:}$ Na$^+$ CH$_3$-Br C–CH$_3$

This reaction is an acetylide reaction. First, the base deprotonates the alkyne to form the acetylide; second, the acetylide undergoes an S_N2 reaction with the alkyl halide. In this case, the halide is one carbon long, so the alkyne is extended by one carbon.

20

NaNH$_2$ Br

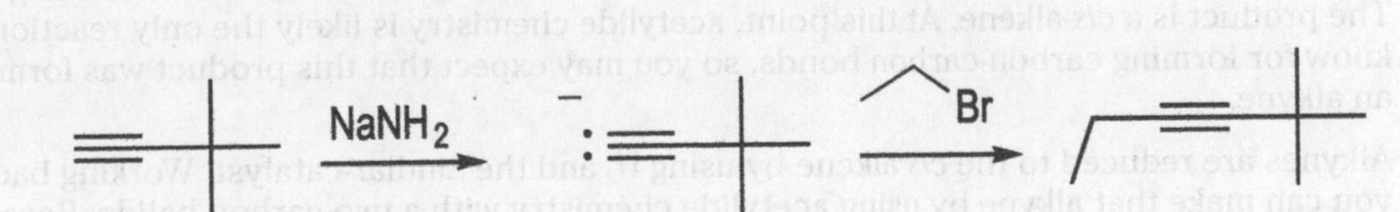

The key to this problem is to remember that acetylide chemistry works only for primary alkyl halides. Therefore, the carbon chain that you add to the alkyne has to have an initial CH_2 unit. The only way to make this product is to cause a reaction between the t-butyl alkyne and $NaNH_2$ and then add a two-carbon halide to give the product. The other possibility — starting with an ethyl alkyne and adding a t-butyl halide — doesn't work because the halide wouldn't be primary.

21

Br Br
Br Br
$NaNH_2$
$2Br_2$
Na^+
Br

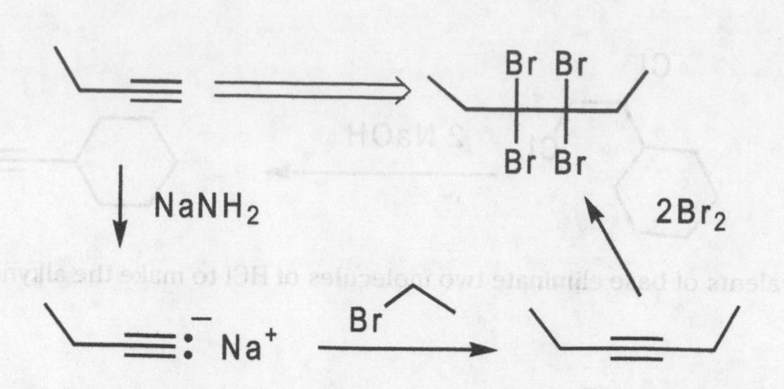

After comparing the number of carbons in the starting material and product, work backward. Here, you notice that four carbons are in the starting material and six carbons are in the product, so somewhere along the way, you need to add two carbons. The product is a tetrabromide. Working backward, this can come from an alkyne by addition of two molecules of Br_2. Next, think about how to make that alkyne. Because you need to add two carbons, use the acetylide chemistry in which you add a two-carbon halide.

22

Br_2
H_2, Lindlar Catalyst
Br
Br
2. $NaNH_2$
1. $NaNH_2$
2. Br

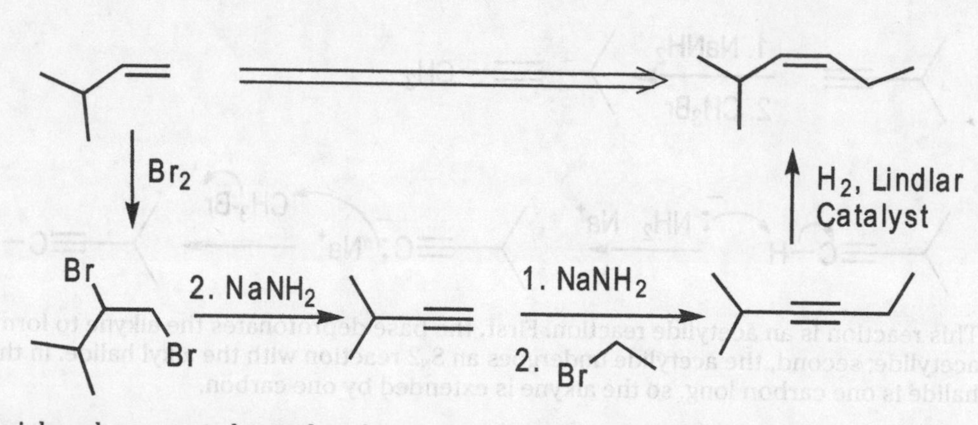

A quick carbon count shows that the product has two more carbons than the starting material. The product is a *cis*-alkene. At this point, acetylide chemistry is likely the only reaction you know for forming carbon-carbon bonds, so you may expect that this product was formed from an alkyne.

Alkynes are reduced to the *cis* alkene by using H_2 and the Lindlar-catalyst. Working backward, you can make that alkyne by using acetylide chemistry with a two-carbon halide. Because the starting material isn't an alkyne, the way to prepare this functional group is through double elimination of halides (such as Br) by using $NaNH_2$. You can make this dibromide from the starting material by reaction with bromine (Br_2).

Part III
Functional Groups and Their Reactions

"I always get a good night's sleep the day before an Orgo test so I'm relaxed and alert the next morning. Then I grab my pen, eat a banana and I'm on my way."

In this part . . .

This is the part where things really get interesting because this is where you get to the heart of organic chemistry: the reactions of organic molecules. These reactions show you how to transform one molecule into another.

The cool thing about functional groups is that you can predict their reactivity. Functional groups tend to react in the same way from one molecule to the next, so if you uncover the reactions of a single functional group, you essentially discover the reactions of many thousands of specific compounds that contain that functional group. The functional groups included in this part are the alkyl halides, the alcohols, and the aromatic compounds, each of which has its own interesting and unique chemistry.

Chapter 10

The Leaving Group Boogie: Substitution and Elimination of Alkyl Halides

In This Chapter

- Looking at S_N1 and S_N2 reactions
- Comparing E1 and E2 reactions
- Predicting products of substitution and elimination reactions

The substitution and elimination reactions are two of the most important classes of reactions that you cover in organic chemistry. *Substitution reactions* are reactions in which an incoming group takes the place of a portion of a molecule that's given the boot, called the *leaving group. Elimination reactions* are reactions in which a leaving group is expelled from a molecule. In relationship terms, an elimination reaction is akin to dumping your significant other. A substitution reaction is when you dump your significant other and replace him or her with someone else.

In this chapter — which you can think of in terms of a molecular soap opera if it helps you get into the spirit of the thing — you see how to distinguish among the different types of substitution and elimination reactions, both of which come in two flavors. Finally, you put everything you know together and predict the product of reactions in which all four reactions are possible.

The Replacements: Comparing S_N1 and S_N2 Reactions

Substitution reactions come in two flavors, called the S_N1 and S_N2 substitution reactions. These two reactions replace one group (the leaving group) with another (a *nucleophile*). Although these two reactions accomplish essentially the same task, the reactions occur by different mechanisms.

Substitution reactions are very important in organic chemistry because they provide a way to synthesize a large variety of molecules. Because these reactions are so important to the organic chemist, most orgo classes cover the particulars of these reactions in all their gory detail. The general mechanism of the S_N1 and S_N2 reactions are shown in Figure 10-1. In the S_N1 reaction, the first step is loss of the leaving group to form an intermediate carbocation. In the second step, the nucleophile attacks the carbocation to form the product. In the S_N2 mechanism, the reaction occurs in a single step: The nucleophile attacks at the same time that the leaving group is given the boot.

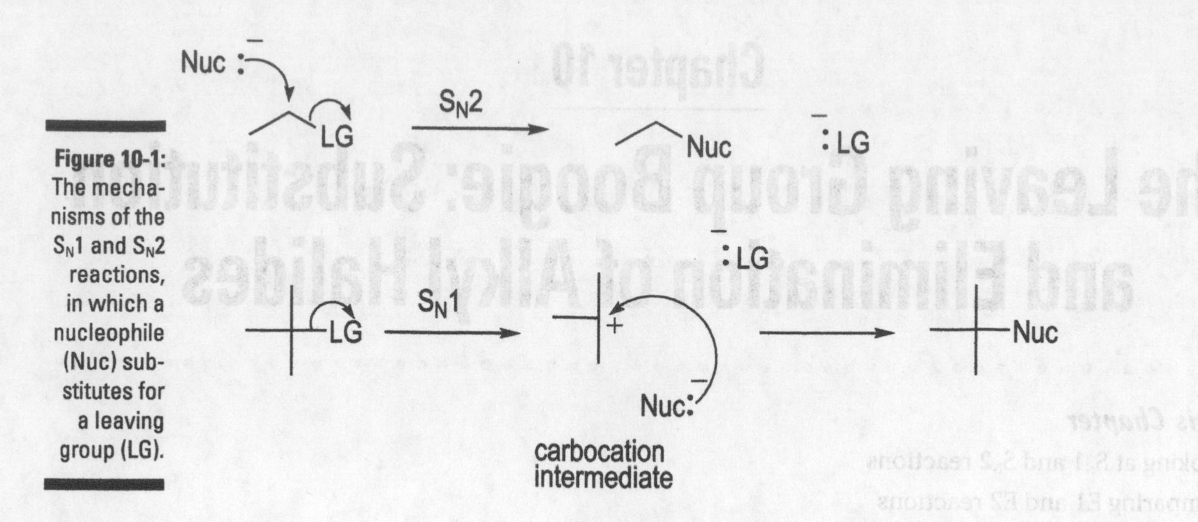

Figure 10-1: The mechanisms of the S_N1 and S_N2 reactions, in which a nucleophile (Nuc) substitutes for a leaving group (LG).

The structure of the molecule and the reaction conditions (solvent, nucleophile, and so on) dictate the mechanism by which the reaction occurs. Table 10-1 shows some of the structural preferences and reaction conditions that favor each of the two mechanisms.

Table 10-1 Structural Preferences for Replacements

Reaction Conditions	***S_N1***	***S_N2***
Substrate	Prefers tertiary (3°); secondary (2°) okay	Prefers methyl, primary (1°) substrates; prefers less steric crowding
Solvent	Polar protic (has NH or OH bonds)	Polar aprotic (no NH or OH bonds)
Nucleophile	Poor nucleophile okay	Good nucleophile needed
Stereochemistry	Mix of stereoisomers (called a *racemic* mix) obtained	Product has inverted stereochemistry
Rate equation	Rate = k[substrate]	Rate = k[substrate][nucleophile]

Here's how to determine the mechanism:

1. **Look at the substrate.**

   If your substrate is tertiary — that is, the leaving group is attached to a tertiary carbon — the substitution has to proceed via the S_N1 mechanism; if the substrate is methyl or primary, the substitution has to proceed via the S_N2 mechanism.

2. **For secondary substrates, look to the solvent and nucleophile.**

   Secondary substrates are slippery ground because both substitutions can occur, so look to the solvent and the nucleophile to see which mechanism will be favored.

The strength of a nucleophile generally parallels basicity, so strong bases (such as OH^-) are usually strong nucleophiles. Generally, negatively charged nucleophiles are strong. A few exceptions: Halides (Cl^-, Br^-, I^-), thiols (*R*-SH), cyanide (CN^-) and azide (N_3^-) are generally considered weak bases but good nucleophiles.

Q. Draw the mechanism for the following reaction.

CI —EtOH→ OEt

A.

CI⁻ :O: H Et carbocation intermediate

-CI⁻ HCl H Et O+ :O Et

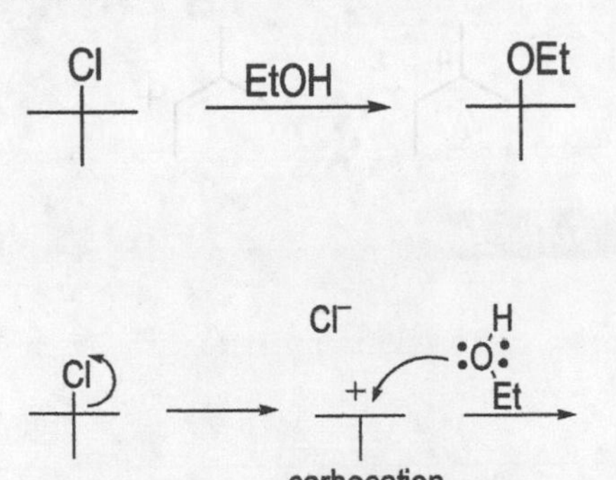

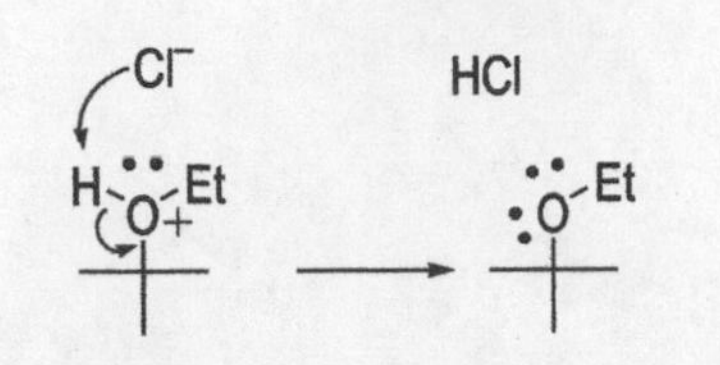

This substitution reaction occurs by an S_N1 mechanism. The big giveaway is that the substrate is a tertiary halide (the halogen is connected to a tertiary carbon, a carbon attached to three other carbons). S_N2 reactions don't occur on tertiary substrates, but S_N1 reactions do. The first step in an S_N1 reaction is loss of the leaving group to form a carbocation intermediate (a positively charged carbon). In the second step, the nucleophile (in this case, CH_3CH_2OH) adds to the carbocation. Loss of a proton forms the product.

Q. Order the following substrates from best to worst for their ability to react in an S_N2 reaction.

$H_3C{-}Br$ Br Br Br

A.

$H_3C{-}Br$ Br Br Br

1 best 3 4 worst 2

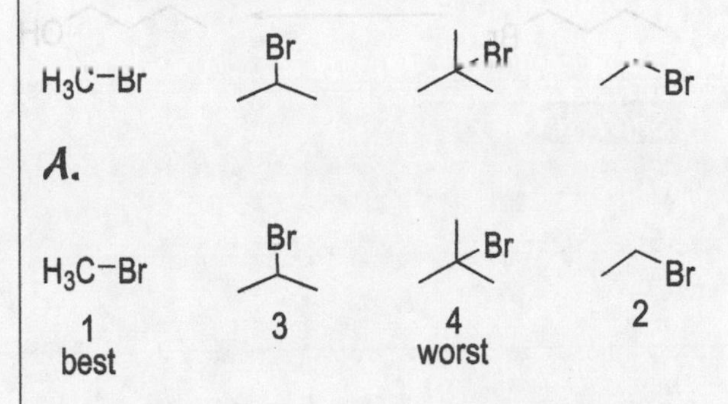

S_N2 reactions work best with the least sterically hindered substrates. In the S_N2 mechanism, the nucleophile attacks the carbon that contains the leaving group, displacing the leaving group. Attacking the carbon with the leaving group is easier if you have less steric crowding around that carbon. Therefore, methyl substrates (CH_3X) react the fastest in an S_N2 reaction, followed by primary substrates (RCH_2X), followed by secondary (R_2CHX). Tertiary substrates don't react via the S_N2 mechanism because there's simply too much crowding around the carbon with the leaving group for the nucleophile to approach.

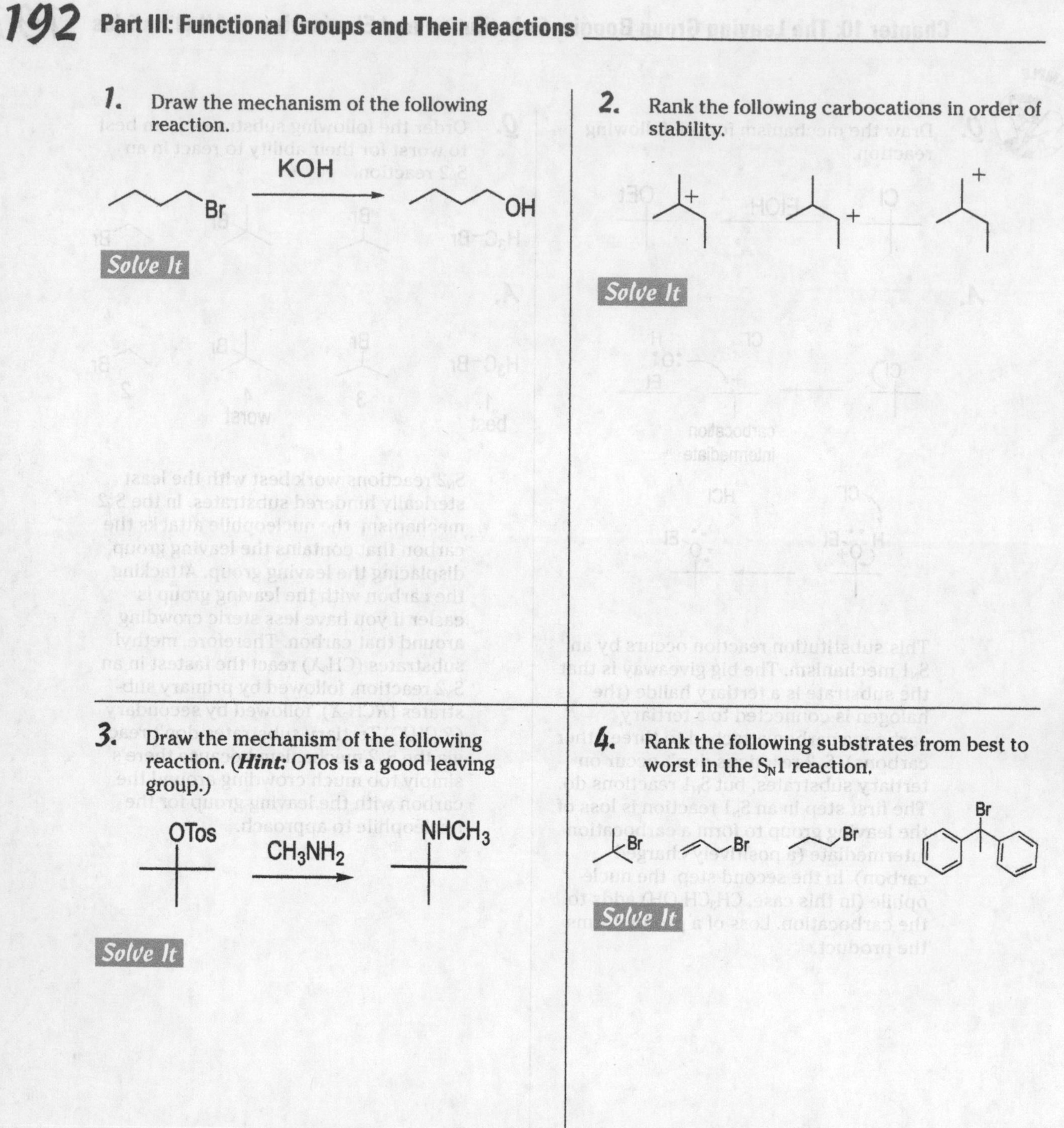

1. **Draw the mechanism of the following reaction.**

KOH

Br → OH

Solve It

2. **Rank the following carbocations in order of stability.**

Solve It

3. **Draw the mechanism of the following reaction. (*Hint:* OTos is a good leaving group.)**

OTos → CH_3NH_2 → $NHCH_3$

Solve It

4. **Rank the following substrates from best to worst in the S_N1 reaction.**

Br Br Br Br

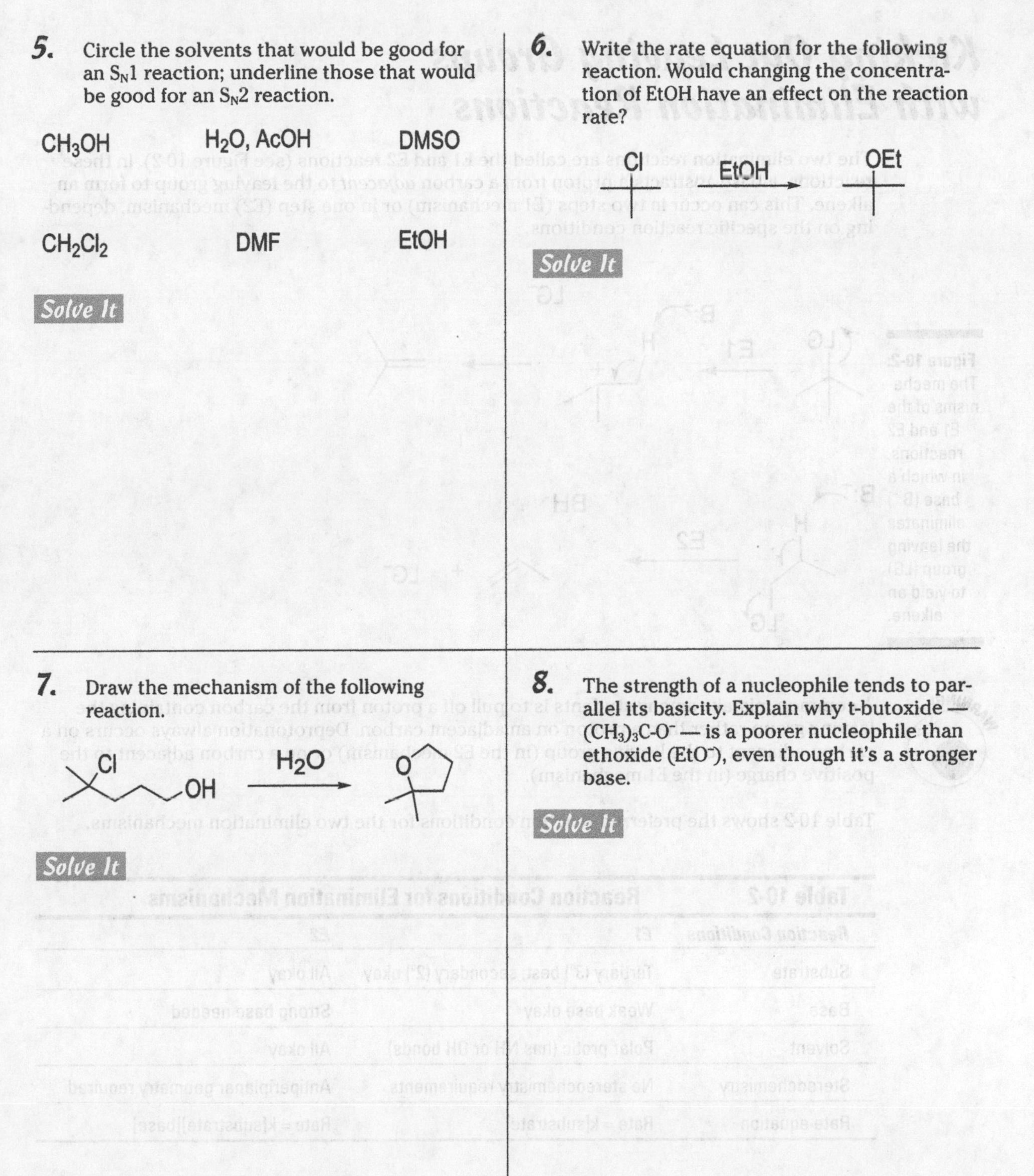

Solve It

5. **Circle the solvents that would be good for an S_N1 reaction; underline those that would be good for an S_N2 reaction.**

CH_3OH H_2O, AcOH   DMSO

CH_2Cl_2 DMF   EtOH

Solve It

6. **Write the rate equation for the following reaction. Would changing the concentration of EtOH have an effect on the reaction rate?**

Cl —(EtOH)→ OEt

Solve It

7. **Draw the mechanism of the following reaction.**

Cl, OH —(H_2O)→ O

Solve It

8. **The strength of a nucleophile tends to parallel its basicity. Explain why t-butoxide — $(CH_3)_3C\text{-}O^-$ — is a poorer nucleophile than ethoxide (EtO^-), even though it's a stronger base.**

Solve It

Kicking Out Leaving Groups with Elimination Reactions

The two elimination reactions are called the E1 and E2 reactions (see Figure 10-2). In these reactions, a base abstracts a proton from a carbon *adjacent* to the leaving group to form an alkene. This can occur in two steps (E1 mechanism) or in one step (E2) mechanism, depending on the specific reaction conditions.

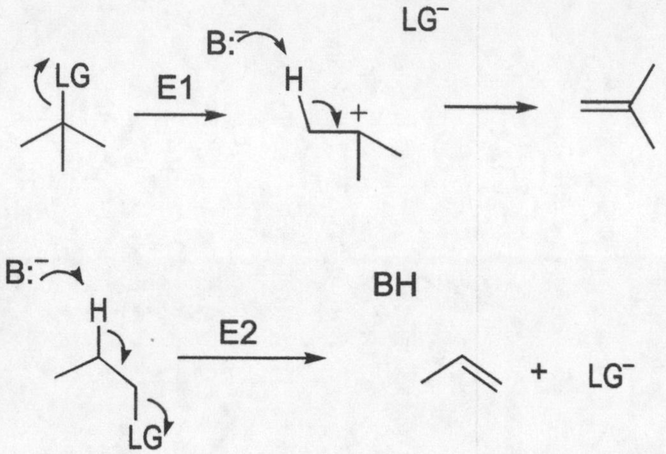

Figure 10-2: The mechanisms of the E1 and E2 reactions, in which a base (B:⁻) eliminates the leaving group (LG) to yield an alkene.

A common mistake among students is to pull off a proton from the carbon containing the leaving group rather than a proton on an adjacent carbon. Deprotonation always occurs on a carbon adjacent to the leaving group (in the E2 mechanism) or on a carbon adjacent to the positive charge (in the E1 mechanism).

Table 10-2 shows the preferred reaction conditions for the two elimination mechanisms.

Table 10-2 Reaction Conditions for Elimination Mechanisms

Reaction Conditions	*E1*	*E2*
Substrate	Tertiary (3°) best; secondary (2°) okay	All okay
Base	Weak base okay	Strong base needed
Solvent	Polar protic (has NH or OH bonds)	All okay
Stereochemistry	No stereochemistry requirements	Antiperiplanar geometry required
Rate equation	Rate = k[substrate]	Rate = k[substrate][base]

E2 mechanisms are more likely when you have a strong base (a base with a negative charge); in terms of stereochemistry, the leaving group and the proton to be pulled off have to be *antiperiplanar* to undergo the E2 reaction — in other words, the leaving group and the proton have to be on opposite sides of the intervening bond (anti stereochemistry) and in the same plane (*planar*). E1 mechanisms occur with weaker bases, secondary or tertiary substrates, and a protic solvent (one containing OH or NH bonds).

When attempting to determine which elimination mechanism is dominant, keep these steps in mind:

1. **Look to the substrate.**

   E1 eliminations occur only with tertiary (3°) substrates or with secondary (2°) substrates. E2 works with all substrate types.

2. **Look to the base.**

   If the base is strong (negatively charged), the E2 mechanism is favored; with weak bases, the E1 mechanism is favored.

3. **If you still don't have a satisfactory answer, look to the solvent.**

   E1 reactions occur only in protic solvents (solvents with NH or OH bonds). Conversely, the E2 reaction can occur in all solvents.

Q. Draw the mechanism of the following reaction.

Cl, MeOH

A.

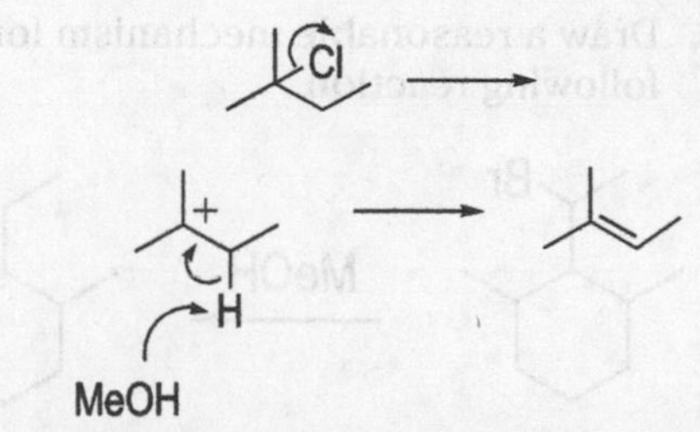

Cl, +, H, MeOH

This elimination reaction can occur by either the E1 or the E2 mechanism. For the E2 mechanism, you need a strong base. Here, the base (MeOH) is weak. E1 eliminations require either a tertiary or secondary substrate, as well as a protic solvent. Here, the substrate is tertiary and the solvent (MeOH) is protic because it contains an OH bond, perfect for the E1 mechanism. The first step is loss of the leaving group to form a carbocation. In the second step, the solvent also acts as the base to pull off an adjacent proton to form the double bond.

Q. Rank the following leaving groups in order of leaving group ability.

F^- Cl^- Br^- I^- $OTos^-$

A. The leaving group ability of the halides increases as you go down the periodic table. Thus $I^- > Br^- > Cl^- > F^-$. Tosylate ($OTos^-$) is a better leaving group still than iodide, so this is the best leaving group of the five.

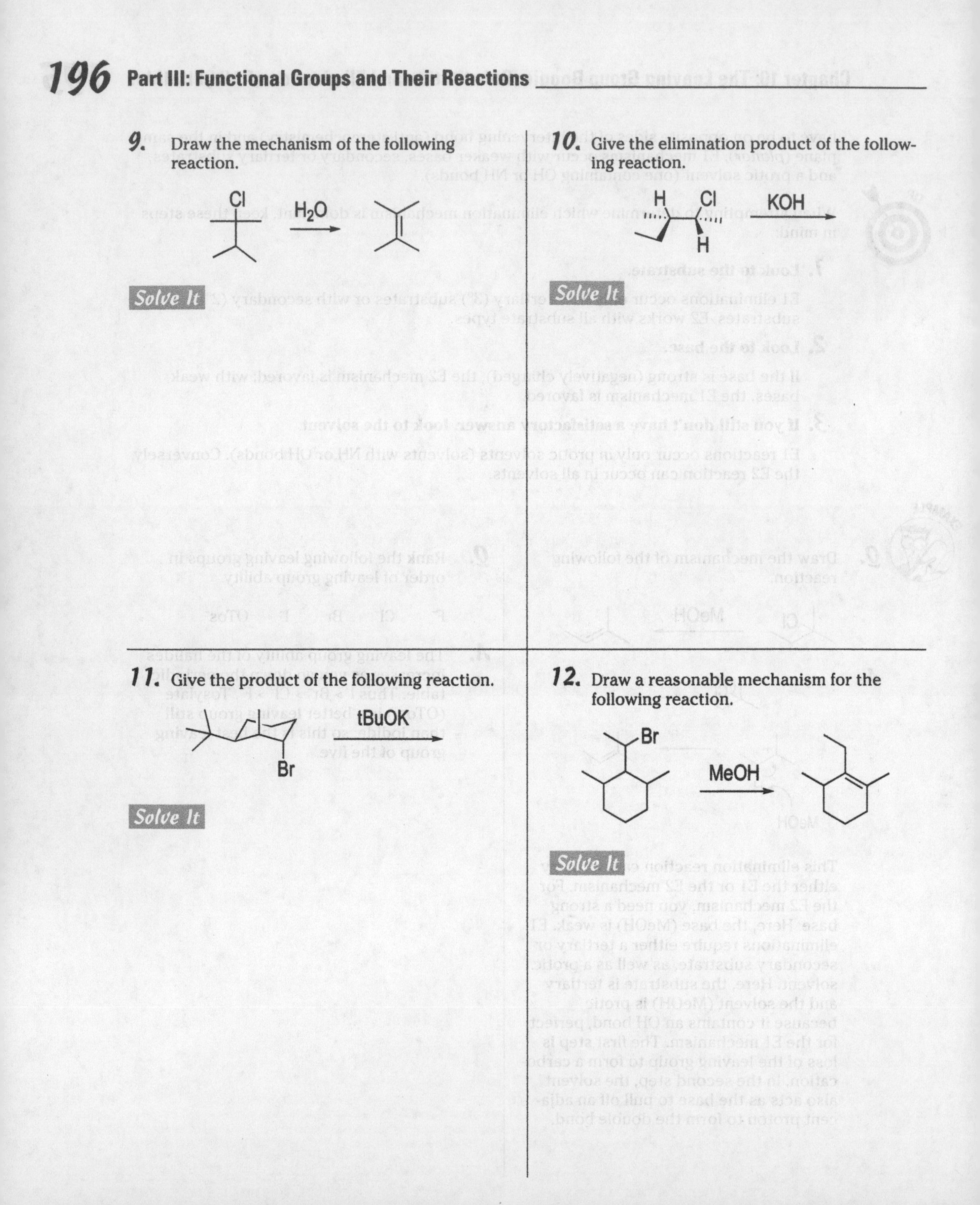

9. Draw the mechanism of the following reaction.

Cl

H_2O

Solve It

10. Give the elimination product of the following reaction.

H Cl

H

KOH

Solve It

11. Give the product of the following reaction.

tBuOK

Br

Solve It

12. Draw a reasonable mechanism for the following reaction.

Br

MeOH

Solve It

Putting It All Together: Substitution and Elimination

Here you put the different types of reactions together. Because E1, E2, S_N1, and S_N2 reactions can occur on very similar substrates, you need to be able to predict which mechanism and what product will form. Figure 10-3 provides a rule of thumb reaction tree for determining what mechanism will occur given the substrate and reaction conditions. You can use this as a guide until you get the feel for the conditions that lead to different products.

To use this reaction tree, start at the top and work down the branches until you've identified the mechanism. The first step is to identify the substrate as methyl, primary (1°), secondary (2°), or tertiary (3°). Each of these choices leads down a different branch, which asks further questions until you've identified the reaction type. As you can see from the reaction tree, secondary substrates are harder to predict than the others, and so this branch has the most questions to identify the reaction mechanism. Under some conditions, two mechanisms may be operating to give a mixture of products.

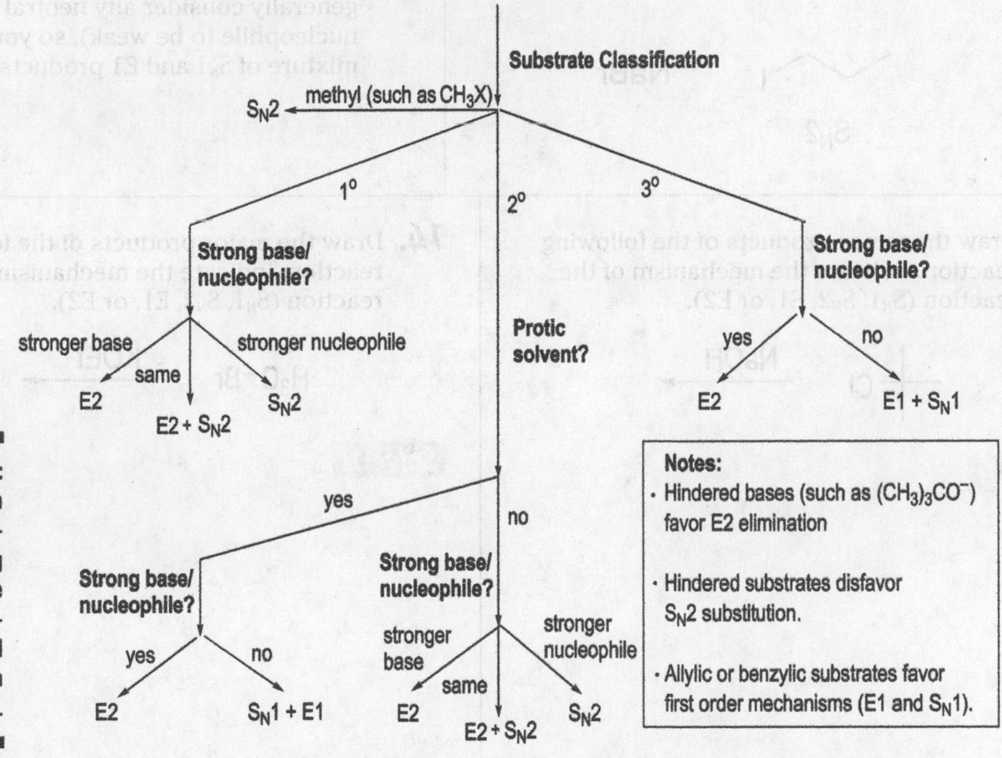

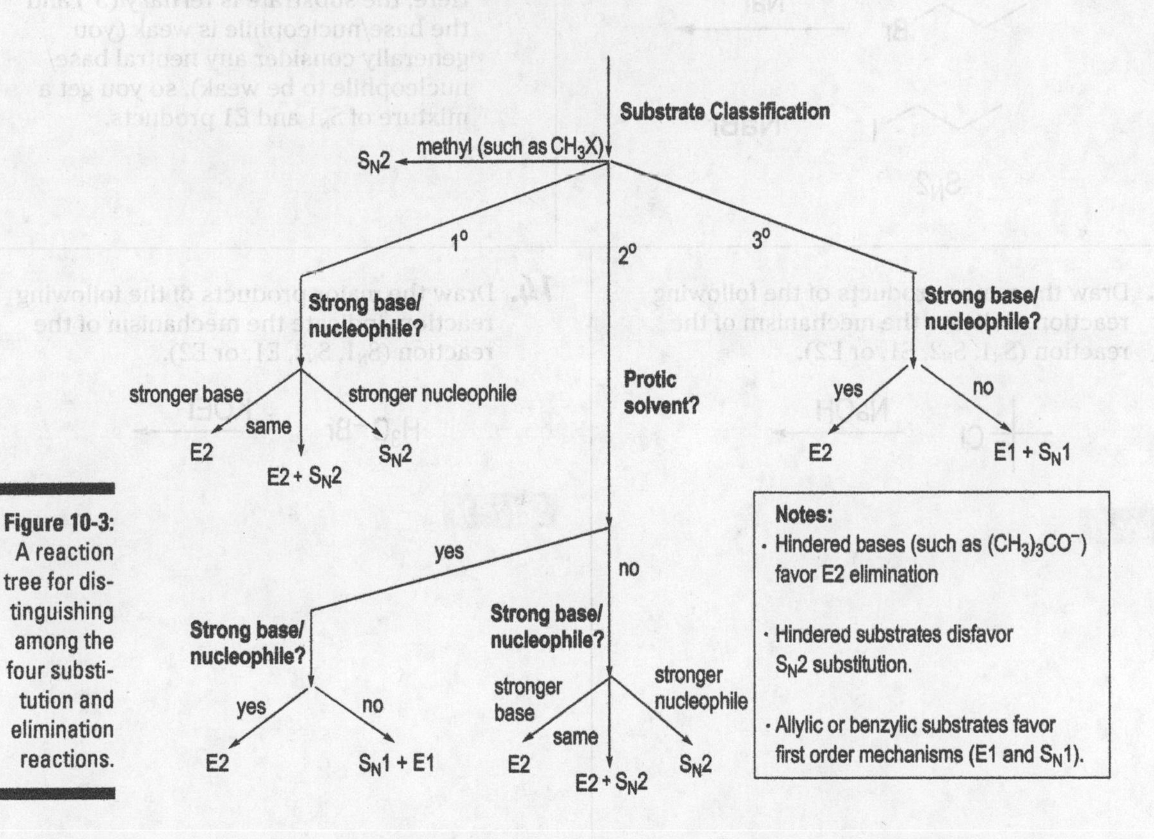

Figure 10-3: A reaction tree for distinguishing among the four substitution and elimination reactions.

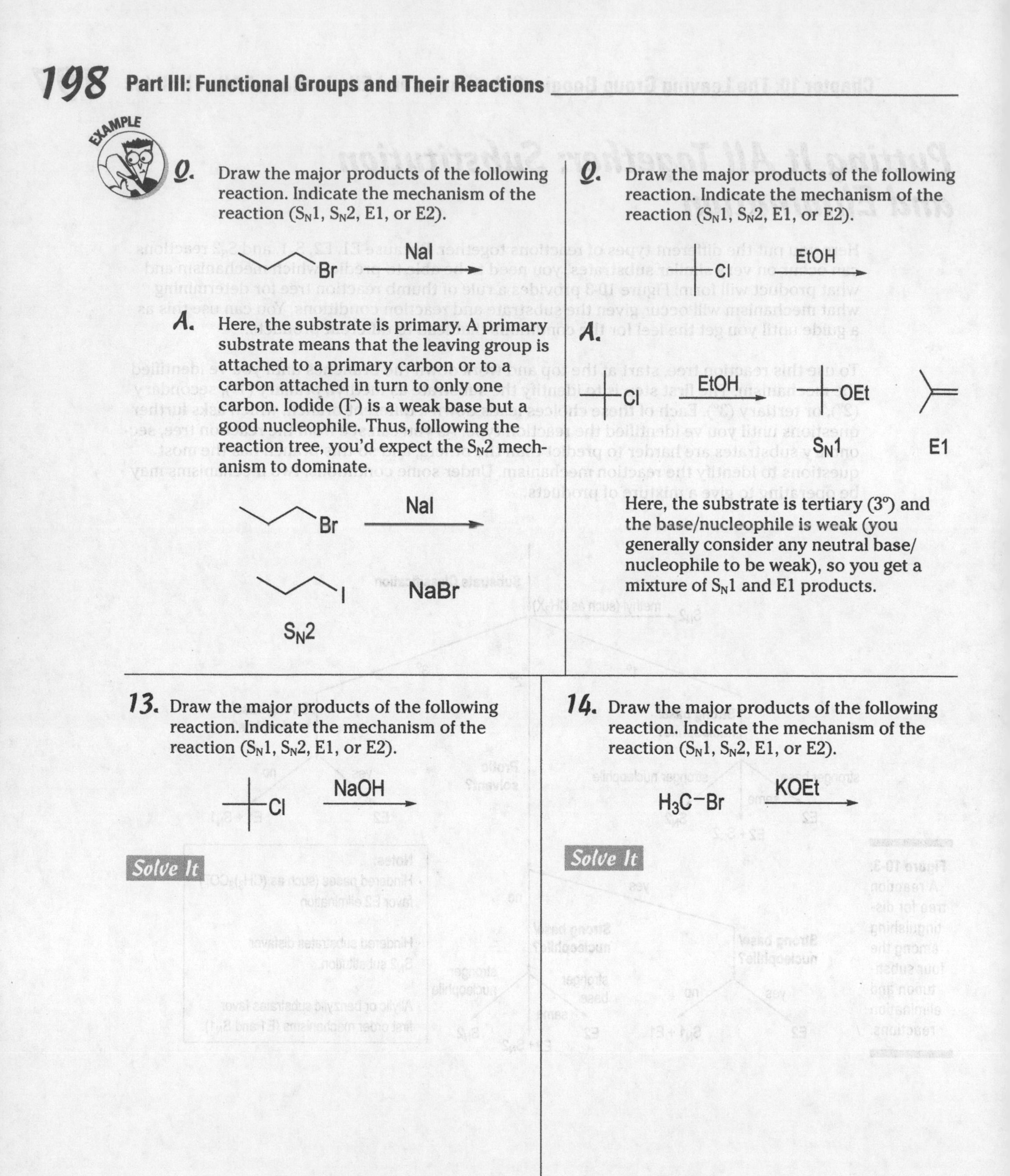

Q. Draw the major products of the following reaction. Indicate the mechanism of the reaction (S_N1, S_N2, E1, or E2).

(1-bromopropane) $\xrightarrow{NaI}$

A. Here, the substrate is primary. A primary substrate means that the leaving group is attached to a primary carbon or to a carbon attached in turn to only one carbon. Iodide (I^-) is a weak base but a good nucleophile. Thus, following the reaction tree, you'd expect the S_N2 mechanism to dominate.

(1-bromopropane) $\xrightarrow{NaI}$ (1-iodopropane) + NaBr

S_N2

Q. Draw the major products of the following reaction. Indicate the mechanism of the reaction (S_N1, S_N2, E1, or E2).

(tert-butyl chloride) $\xrightarrow{EtOH}$

A. (tert-butyl chloride) $\xrightarrow{EtOH}$ (tert-butyl ethyl ether, –OEt) + (2-methylpropene)

S_N1 E1

Here, the substrate is tertiary (3°) and the base/nucleophile is weak (you generally consider any neutral base/nucleophile to be weak), so you get a mixture of S_N1 and E1 products.

13. Draw the major products of the following reaction. Indicate the mechanism of the reaction (S_N1, S_N2, E1, or E2).

(tert-butyl chloride) $\xrightarrow{NaOH}$

Solve It

14. Draw the major products of the following reaction. Indicate the mechanism of the reaction (S_N1, S_N2, E1, or E2).

$H_3C{-}Br \xrightarrow{KOEt}$

Solve It

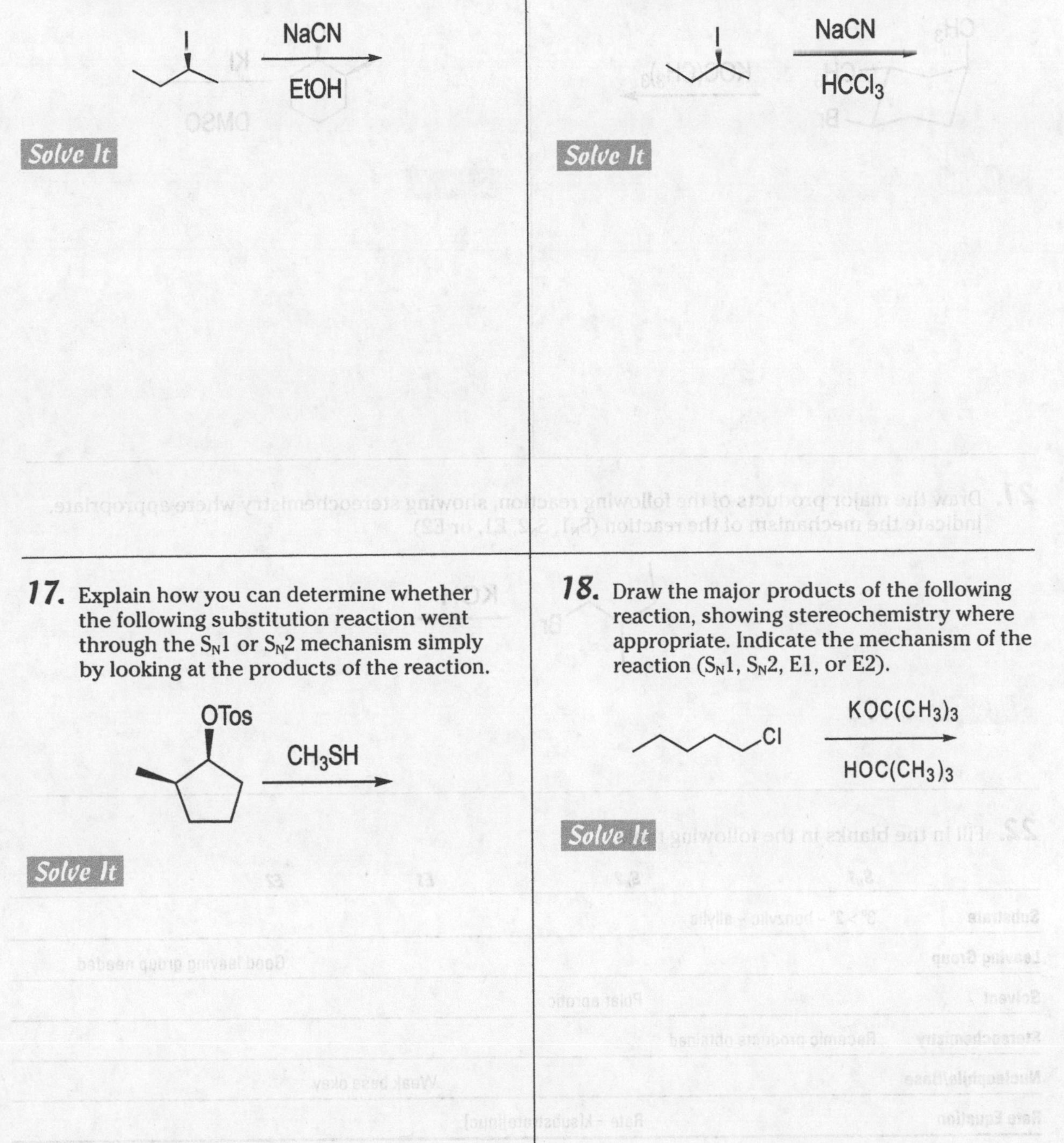

15. Draw the major products of the following reaction, showing stereochemistry where appropriate. Indicate the mechanism of the reaction (S_N1, S_N2, E1, or E2).

I

NaCN

EtOH

Solve It

16. Draw the major products of the following reaction, showing stereochemistry where appropriate. Indicate the mechanism of the reaction (S_N1, S_N2, E1, or E2).

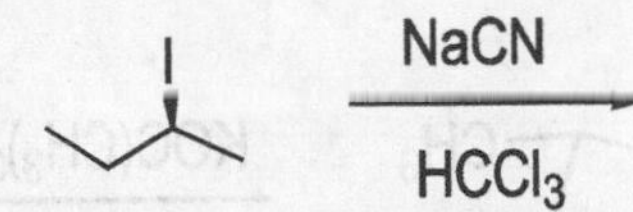

Solve It

17. Explain how you can determine whether the following substitution reaction went through the S_N1 or S_N2 mechanism simply by looking at the products of the reaction.

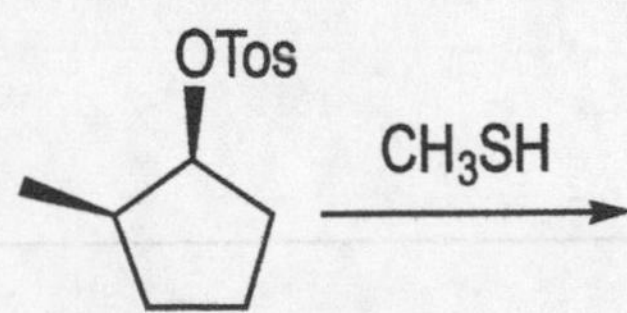

Solve It

18. Draw the major products of the following reaction, showing stereochemistry where appropriate. Indicate the mechanism of the reaction (S_N1, S_N2, E1, or E2).

Cl

$KOC(CH_3)_3$

$HOC(CH_3)_3$

Solve It

19. Draw the major products of the following reaction, showing stereochemistry where appropriate. Indicate the mechanism of the reaction (S_N1, S_N2, E1, or E2).

CH_3 CH_3 Br $KOC(CH_3)_3$

Solve It

20. Draw the major products of the following reaction, showing stereochemistry where appropriate. Indicate the mechanism of the reaction (S_N1, S_N2, E1, or E2).

Cl KI DMSO

Solve It

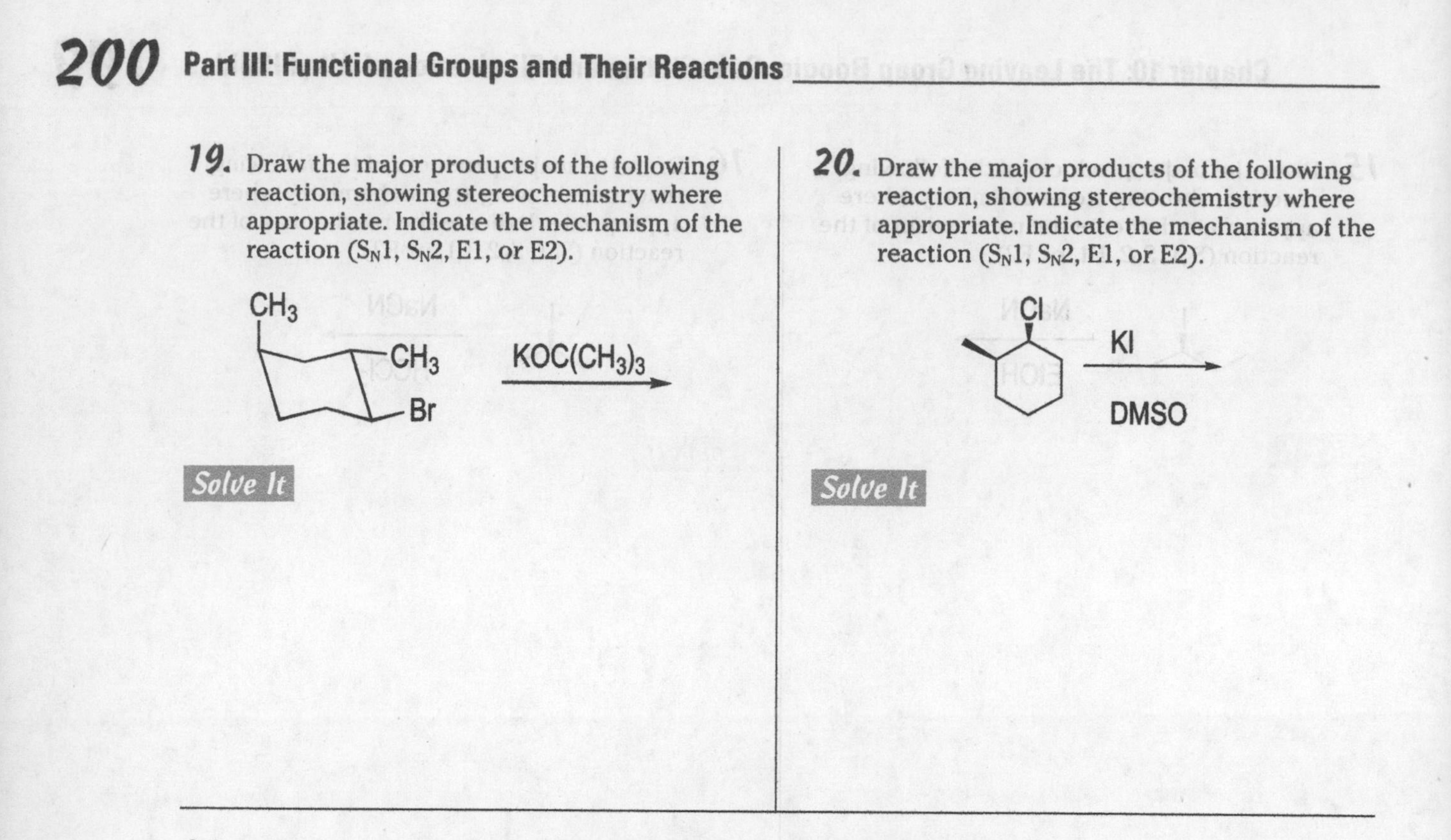

21. Draw the major products of the following reaction, showing stereochemistry where appropriate. Indicate the mechanism of the reaction (S_N1, S_N2, E1, or E2).

Br KOH

Solve It

22. Fill in the blanks in the following table.

	S_N1	S_N2	E1	E2
Substrate	3° > 2° ~ benzylic ~ allylic			
Leaving Group				Good leaving group needed
Solvent		Polar aprotic		
Stereochemistry	Racemic products obtained			
Nucleophile/Base			Weak base okay	
Rate Equation		Rate = k[substrate][nuc]		

Solve It

Answer Key

The following are the answers to the practice questions presented in this chapter.

1

HO⁻ Br → OH

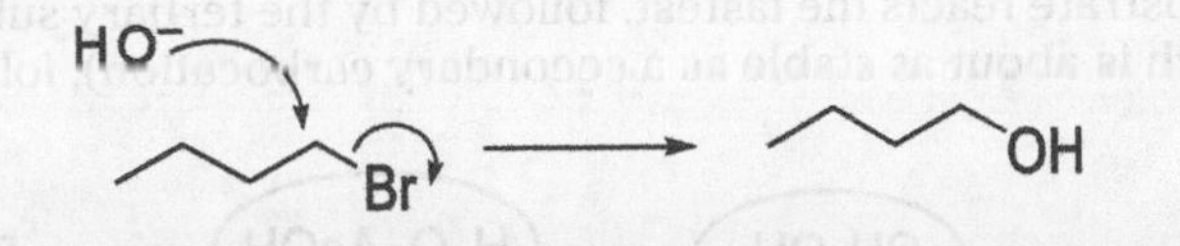

The substrate here is primary, meaning that the leaving group is attached to a primary carbon (a carbon attached to just one other carbon), so this reaction goes through the S_N2 mechanism. In general, primary substrates don't go through the S_N1 mechanism. In the S_N2 mechanism, the nucleophile attacks the carbon from the backside of the carbon that contains the leaving group, displacing the leaving group in a single step.

2

most stable (tertiary) (secondary) least stable (primary)

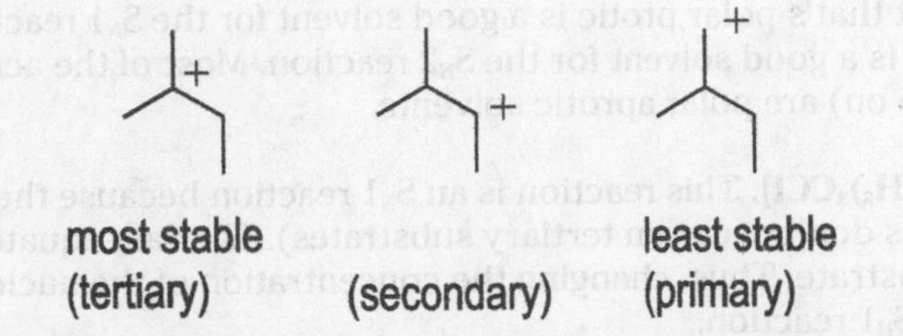

Tertiary carbocations, or positive charges on tertiary carbons (carbons attached to three other carbons), are more stable than secondary carbocations. *Secondary carbocations* are positive charges on secondary carbons (carbons attached to two other carbons), which in turn are more stable than *primary carbocations* (positive charges on a primary carbon — see Chapter 8 for more on carbocations). This stability comes as a result of the electron-donating ability of adjacent *R* groups. These groups can donate some electron density to the carbocation, thereby stabilizing it by delocalizing the charge across more of the molecule. Therefore, the more *R* groups attached to the cationic carbon, the more stable the charge.

3

OTos → (CH_3NH_2) → + → ($TosO^-$) → $H-N^+(H)-CH_3$ → (TosOH) → $:N(H)-CH_3$

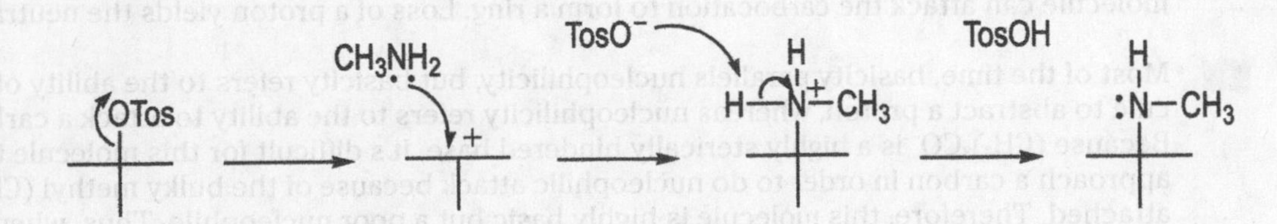

This is a tertiary substrate, so the substitution is via the S_N1 mechanism. The first step is loss of the leaving group ($OTos^-$) to form the carbocation. In the second step, the nucleophile adds to the carbocation. Loss of a proton gives the product.

4

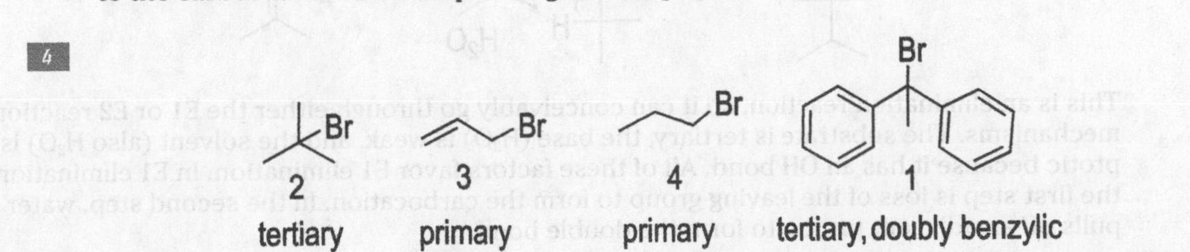

Br 2 tertiary; Br 3 primary allylic; Br 4 primary; Br 1 tertiary, doubly benzylic

The best substrates for the S_N1 reaction are those that form stable carbocations after the leaving group leaves. Tertiary (3°) carbocations are more stable than secondary (2°) carbocations, and secondary carbocations are more stable than primary (1°) carbocations, so the order is tertiary > secondary > primary. Additionally, adjacent double bonds or aromatic rings stabilize the carbocation through resonance (for more on resonance, see Chapter 3). Thus, the tertiary double benzylic substrate reacts the fastest, followed by the tertiary substrate, followed by the primary allylic (which is about as stable as a secondary carbocation), followed by the primary substrate.

5

Any solvent that's polar protic is a good solvent for the S_N1 reaction. Any solvent that's polar but aprotic is a good solvent for the S_N2 reaction. Most of the acronym solvents (DMSO, DMF, THF, and so on) are polar aprotic solvents.

6 **Rate = k[$(CH_3)_3CCl$].** This reaction is an S_N1 reaction because the substrate is tertiary (S_N2 mechanisms don't occur on tertiary substrates). The rate equation of an S_N1 reaction includes just the substrate. Thus, changing the concentration of the nucleophile has no effect on the rate of the S_N1 reaction.

7

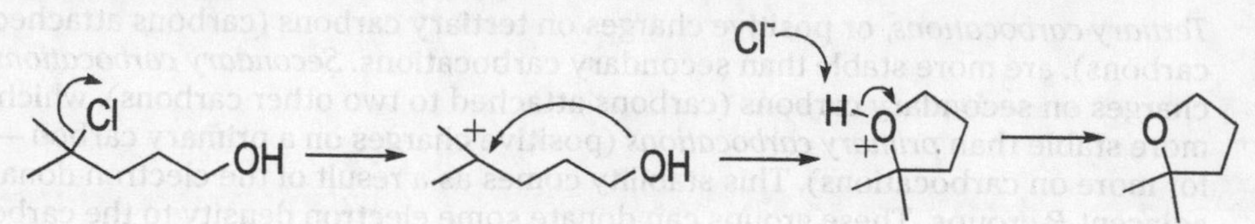

In this mechanism, you have a leaving group (Cl) on a tertiary carbon. Therefore, the first step is loss of the leaving group to form a carbocation. At this point, the OH group at the end of the molecule can attack the carbocation to form a ring. Loss of a proton yields the neutral product.

8 Most of the time, basicity parallels nucleophilicity, but basicity refers to the ability of a molecule to abstract a proton, whereas nucleophilicity refers to the ability to attack a carbon. Because $(CH_3)_3CO^-$ is a highly sterically hindered base, it's difficult for this molecule to approach a carbon in order to do nucleophilic attack because of the bulky methyl (CH_3) groups attached. Therefore, this molecule is highly basic but a poor nucleophile. Thus, whenever you see this ion in a reaction, its function is usually as a base and seldom as a nucleophile.

9

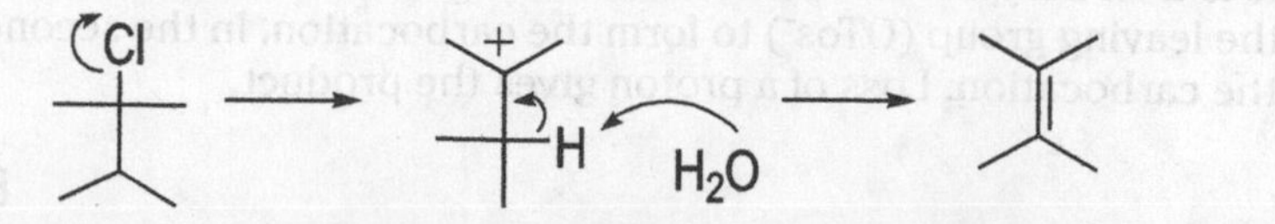

This is an elimination reaction, so it can conceivably go through either the E1 or E2 reaction mechanisms. The substrate is tertiary, the base (H_2O) is weak, and the solvent (also H_2O) is protic because it has an OH bond. All of these factors favor E1 elimination. In E1 elimination, the first step is loss of the leaving group to form the carbocation. In the second step, water pulls off an adjacent proton to form the double bond.

10

HO⁻ H Cl H H H Cl H

antiperiplanar geometry

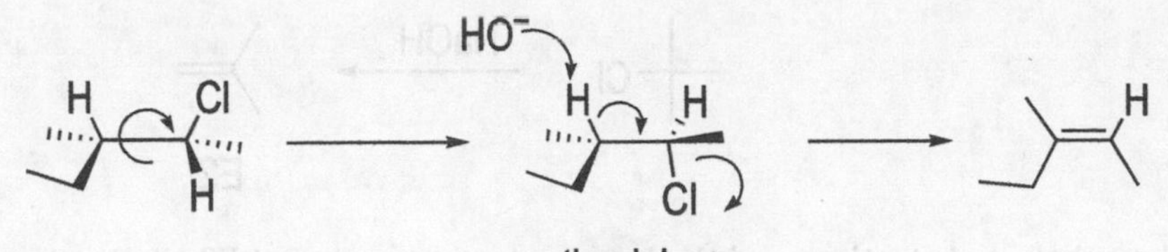

This problem is challenging. The elimination is E2 because the base (OH^-) is very strong (E1 eliminations occur with weak bases). In the E2 reaction, the stereochemical requirements are that the leaving group and the proton to be pulled off have to be *antiperiplanar*. As drawn, however, the leaving group and the adjacent proton aren't anti to each other. Thus, you have to rotate around the carbon-carbon bond to make them anti. Doing the elimination from this geometry leads to the two methyl groups appearing on opposite sides of the double bond.

11

⁻O H H H Br H

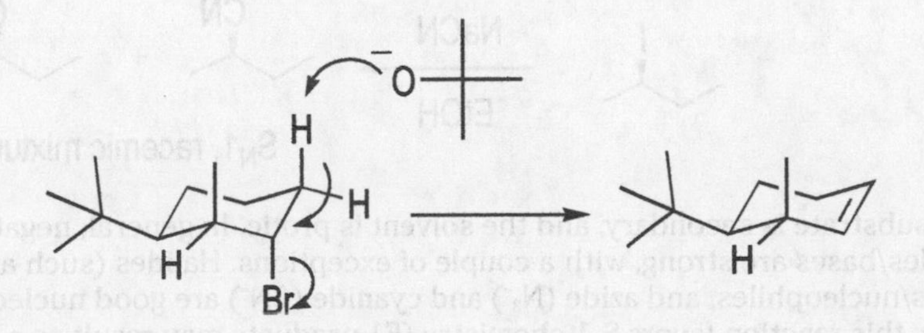

Potassium tert-butoxide — tBuOK — is a powerful base. Whenever you see this reagent, you can almost bet that you'll see E2 elimination as a result of its high basicity and poor nucleophilicity. For an E2 elimination to occur, the leaving group and proton have to be antiperiplanar. On a cyclohexane ring, the only time such a geometry occurs is when both the leaving group and the adjacent hydrogen are axial. Thus, the proton to the left can't be eliminated because it's equatorial, but the proton to the right can be eliminated because it's axial. Pulling off this proton leads to the alkene product.

12

Br + H hydride shift H CH_3 O H + H

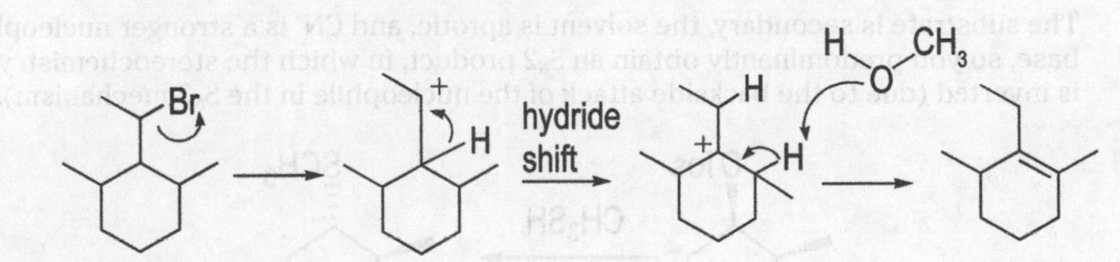

This problem is tough because the product's double bond is in a location that you wouldn't expect. The reaction conditions are ripe for E1 elimination because the base is weak (E2 reactions generally require strong, negatively charged bases). So the first step is loss of the leaving group to form the carbocation. If you perform the elimination at this point, however, you don't obtain the shown product. However, you have a secondary carbocation adjacent to a tertiary carbon, so the cation badly wants to get to that tertiary carbon, where it's more stable. Thus, the molecule performs a hydride shift so that the secondary cation becomes tertiary. Elimination of a hydrogen adjacent to the cation leads to the correct product. (See Chapter 8 for problems on carbocations and hydride shifts.)

13

NaOH

E2

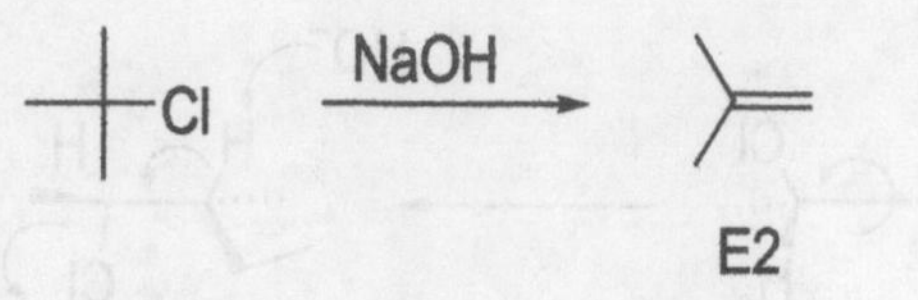

The substrate is tertiary and the base is strong, so you get E2 elimination.

14

$H_3C{-}Br \xrightarrow{KOEt} CH_3OEt$

S_N2

Methyl substrates always give S_N2 products because they can't do elimination reactions (no adjacent protons!) and methyl cations are too unstable to form.

15

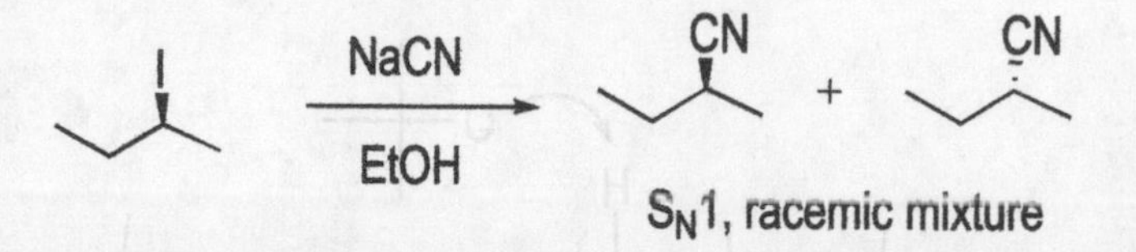

NaCN / EtOH

CN + CN

S_N1, racemic mixture

The substrate is secondary, and the solvent is protic. In general, negatively charged nucleophiles/bases are strong, with a couple of exceptions. Halides (such as Cl^-, Br^-, and I^-) are weak bases/nucleophiles, and azide (N_3^-) and cyanide (CN^-) are good nucleopiles but weak bases. Thus, this reaction favors S_N1 chemistry (E1 products may result as a minor product). S_N1 reactions lead to racemic products — a rough 50:50 mixture of the two stereoisomers.

16

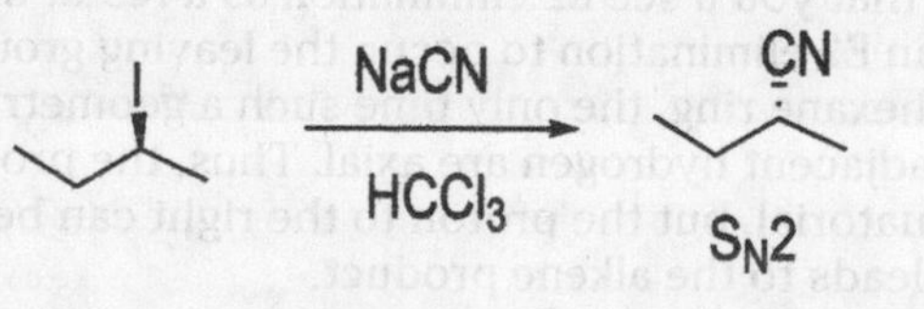

NaCN / $HCCl_3$

CN

S_N2

The substrate is secondary, the solvent is aprotic, and CN^- is a stronger nucleophile than a base, so you predominantly obtain an S_N2 product, in which the stereochemistry of the product is inverted (due to the backside attack of the nucleophile in the S_N2 mechanism).

17

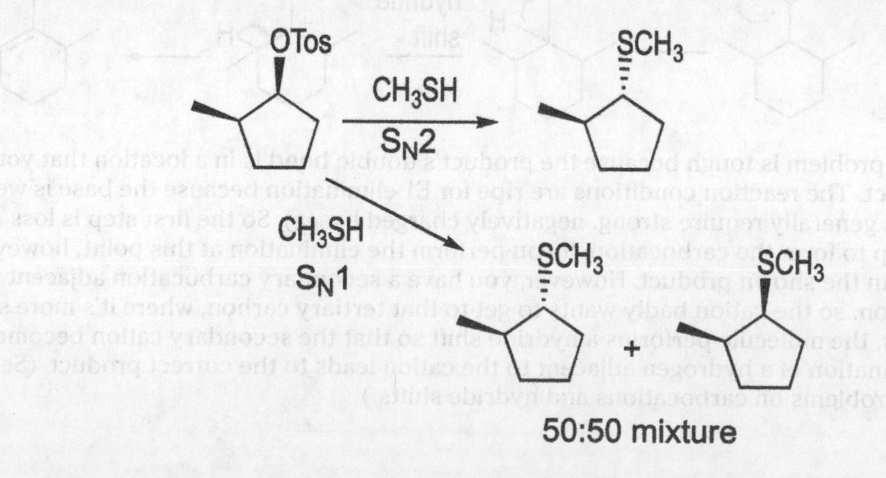

OTos

CH_3SH / S_N2

SCH_3

CH_3SH / S_N1

SCH_3 + SCH_3

50:50 mixture

S_N2 reactions lead exclusively to inversion of stereochemistry in the product. In S_N1 reactions, you get a roughly 50:50 mixture of inversion and retention of the stereochemistry.

18

Cl $\xrightarrow[HOC(CH_3)_3]{KOC(CH_3)_3}$

E2

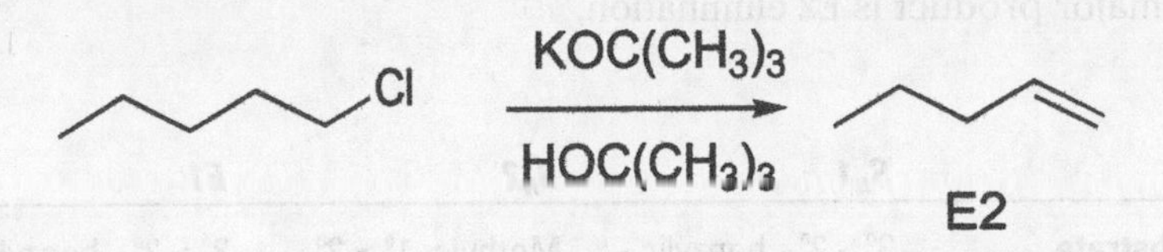

The substrate is primary, but the powerful bulky base tert-butoxide forces the reaction to undergo an E2 reaction (tert-butoxide is generally too bulky to efficiently undergo S_N2 reactions).

19

CH_3 H CH_3 H Br H — ring flip → Br H H_3C H H $^-OC(CH_3)_3$ — E2 → H H_3C H

only H antiperiplanar to Br!

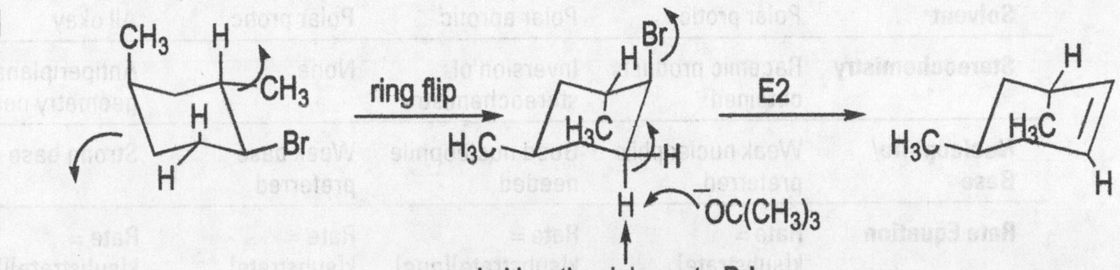

This problem is challenging. The substrate is secondary, and the base is the powerful tert-butoxide. This reagent almost always forces E2 elimination. For the molecule to undergo E2 elimination, however, the hydrogen and the leaving group have to have an *antiperiplanar* geometry (they need to be on opposite sides of the connecting bond [anti] and in the same plane). In a cyclohexane ring, the only way for such an antiperiplanar geometry to occur is if both the hydrogen and the leaving group are axial. Because the leaving group as shown is equatorial, you have to perform a ring flip to make it axial (Chapter 6 explains ring flips). After you do that, only one adjacent hydrogen is also axial and can undergo this elimination. Thus, this hydrogen undergoes the elimination to give the double bond via an E2 mechanism.

20

Cl $\xrightarrow[DMSO]{KI}$ I

S_N2

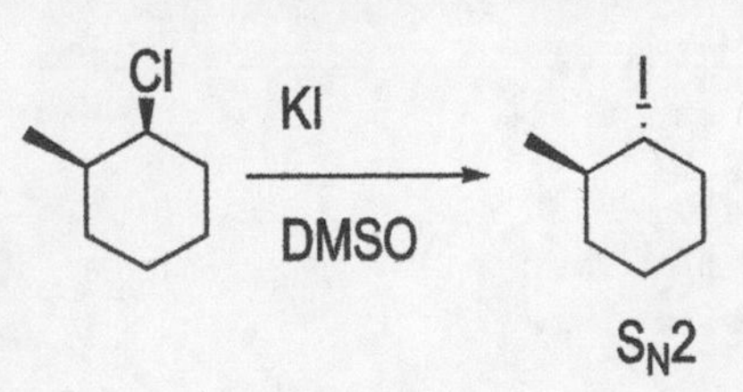

This substrate is secondary, and the solvent is polar aprotic (most of the acronym solvents, such as DMSO, THF, and DMF, are polar aprotic). Generally, negatively charged species are strong nucleophiles and strong bases, but the halides are good nucleophiles but poor bases. Thus, this is an S_N2 reaction, which gives inversion of stereochemistry.

21

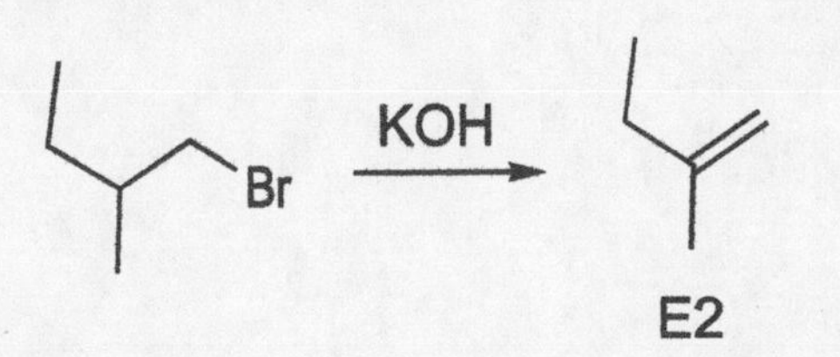

Br $\xrightarrow{KOH}$

E2

The substrate is primary and OH^- is a strong base/nucleophile, so you may expect a mix of S_N2 and E2 products. However, the substrate is bulky, which favors elimination over substitution, so the major product is E2 elimination.

22

	S_N1	S_N2	E1	E2
Substrate	3° > 2° ~ benzylic ~ allylic	Methyl > 1° > 2°	3° > 2° ~ benzylic ~ allylic	All okay
Leaving Group	Good leaving group needed	Good leaving group needed	Good leaving group needed	Good leaving group needed
Solvent	Polar protic	Polar aprotic	Polar protic	All okay
Stereochemistry	Racemic products obtained	Inversion of stereochemistry	None	Antiperiplanar geometry needed
Nucleophile/ Base	Weak nucleophile preferred	Good nucleophile needed	Weak base preferred	Strong base needed
Rate Equation	Rate = k[substrate]	Rate = k[substrate][nuc]	Rate = k[substrate]	Rate = k[substrate][base]

Chapter 11

Not as Thunk as You Drink I Am: The Alcohols

In This Chapter

- Naming alcohols
- Synthesizing alcohols
- Seeing the reactions of alcohols

Most people — and in particular, college students — are familiar with the alcohol in beer, wine, and other strong drinks. This alcohol is ethanol (CH_3CH_2OH), which is a byproduct of fermentation, the breakdown of sugars into alcohol by yeast enzymes. You may also know some other alcohols, such as wood alcohol, also called methanol (CH_3OH), which is the alcohol that makes people go blind from drinking poorly distilled moonshine. There's also rubbing alcohol, or isopropyl alcohol — $(CH_3)_2CHOH$ — which is often used to sterilize skin before you get an injection.

But there are many thousands of alcohols that have been made, and in fact, *alcohols,* or molecules that contain a hydroxyl (OH) group, are one of the most versatile and important functional groups in organic chemistry. Not only are alcohols found in many useful end products such as drugs and organic materials, but they're also useful intermediates in synthesis because they're easily made and transformed into other functional groups. The chemistry of alcohols, therefore, plays a key role in synthetic organic chemistry.

This chapter covers the most important reactions of alcohols, with a wee bit of nomenclature thrown in so you can be clear about the structure of a molecule indicated by a given name.

Name Your Poison: Alcohol Nomenclature

Naming alcohols follows along a similar line to the nomenclature of other organic molecules, except that the names of alcohols end with the suffix *-ol.* The primary difference between naming alcohols and naming alkanes (Chapter 6) is in the numbering of the parent chain. For alcohols, you start numbering from the end that gives the alcohol the lowest possible number rather than the end that reaches the first substituent soonest.

To name an alcohol, follow these steps:

1. **Identify the parent chain.**

   The parent chain of an alcohol is the longest chain of carbons that contains the hydroxyl (OH) group.

2. **Number the parent chain.**

   Start at the end that reaches the hydroxyl group sooner to give the alcohol the lowest possible number.

3. **Name the substituents and place them alphabetically in front of the parent chain.**

   Be sure to use a number to indicate the position of any substituents along the parent chain.

4. **Assign stereochemistry (if applicable).**

   If a chiral center is shown (see Chapter 5), assign the stereochemistry of the chiral center and place this designation in the front of the name. Include a number to indicate the location of the chiral centers if the molecule has more than one.

Q. Name the following alcohol.

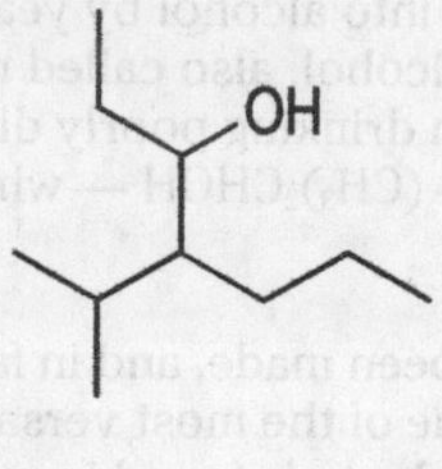

OH

A. **4-isopropyl-3-heptanol.**

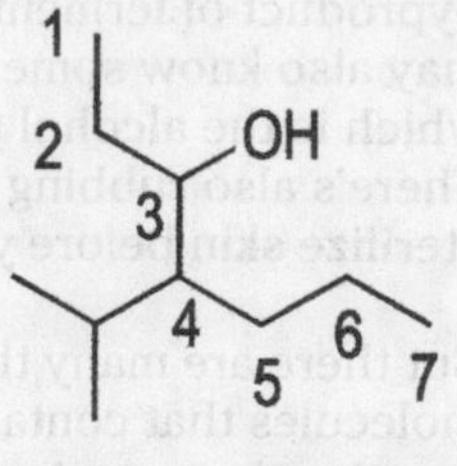

1 2 OH 3 4 5 6 7

In this case, the parent chain is seven carbons. Numbering from the top down reaches the OH group first, at carbon 3. The names of alcohols end with the suffix *-ol*, so the parent name of this molecule is 3-heptanol (or heptan-3-ol if you prefer). In this case, the only substituent is the isopropyl group (which looks like a snake's tongue) at the 4 carbon. Thus, the full name is 4-isopropyl-3-heptanol.

Q. Name the following alcohol.

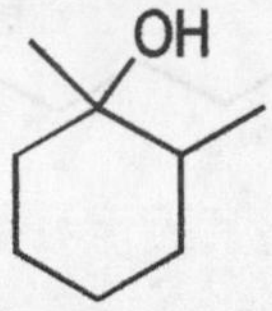

A. **1,2-dimethyl-1-cyclohexanol.**

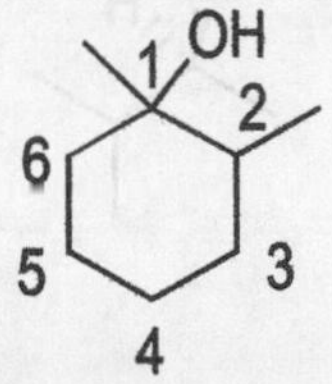

The ring is the parent chain here, so you number the ring so as to give the hydroxyl group the lowest number (1). The ring is six carbons, so the parent name is 1-cyclohexanol (you can omit the 1 if you like because it's assumed that the hydroxyl group is at the first position). Methyl (CH_3) groups come at carbons 1 and 2, and these are clustered together as dimethyl in the name.

1. Name the following alcohol.

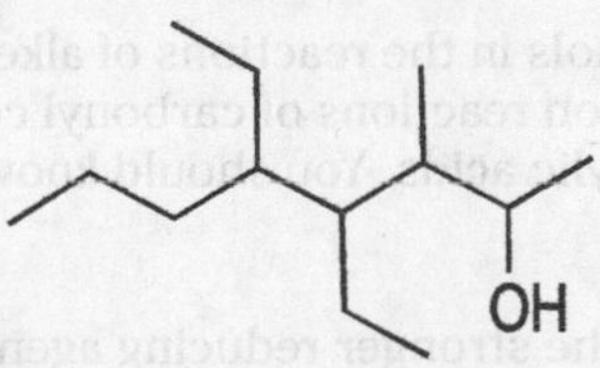

Solve It

2. Name the following alcohol.

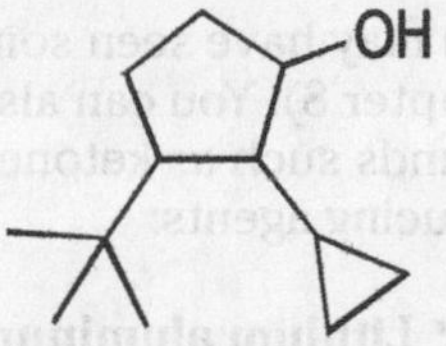

Solve It

3. Name the following alcohol.

HO H

Solve It

4. Name the following alcohol.

OH OH

Solve It

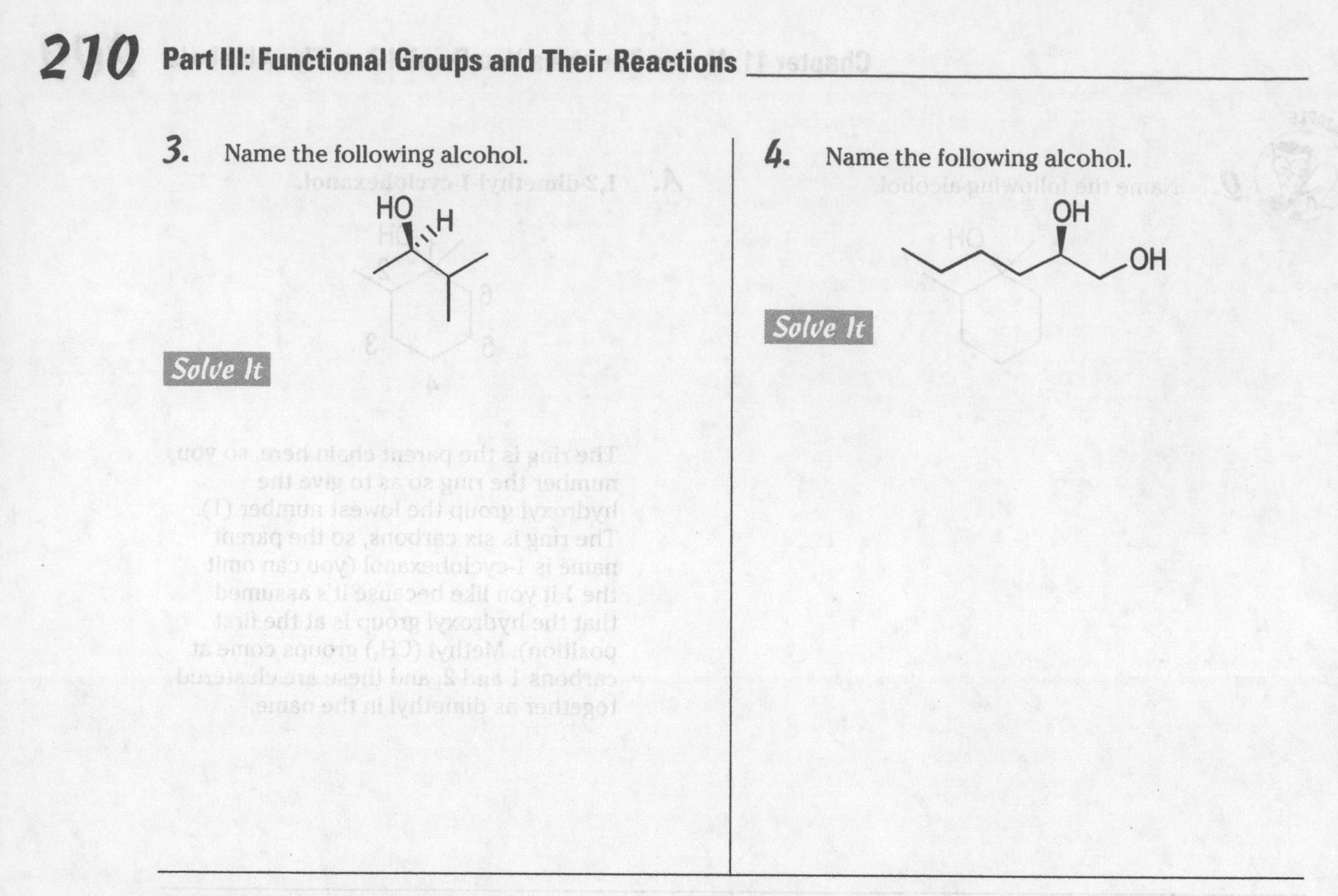

Beyond Homebrew: Making Alcohols

You may have seen some of the reactions to make alcohols in the reactions of alkenes (see Chapter 8). You can also make alcohols through reduction reactions of carbonyl compounds such as ketones, aldehydes, esters, and carboxylic acids. You should know two reducing agents:

- **Lithium aluminum hydride ($LiAlH_4$):** This one is the stronger reducing agent of the two. It reduces esters (*R*COO*R*), carboxylic acids (*R*COOH), ketones (*R*CO*R*), and aldehydes (*R*CHO) to alcohols.
- **Sodium borohydride ($NaBH_4$):** Sodium borohydride is a less powerful reducing agent. It can only reduce ketones and aldehydes to alcohols. This reagent leaves esters and carboxylic acids intact.

Figure 11-1 shows how these two agents work.

ester (RCOOR') $\xrightarrow{LiAlH_4}$ R–CH_2OH + HO-R' (primary alcohols)

carboxylic acid (RCOOH) $\xrightarrow{LiAlH_4}$ R–CH_2OH + H_2O (primary alcohol)

ketone/aldehyde (RCOR') $\xrightarrow{LiAlH_4}$ RCH(OH)R' (primary or secondary alcohol)

ester (RCOOR') $\xrightarrow{NaBH_4}$ No Reaction

carboxylic acid (RCOOH) $\xrightarrow{NaBH_4}$ No Reaction

ketone/aldehyde (RCOR') $\xrightarrow{NaBH_4}$ RCH(OH)R' (primary or secondary alcohol)

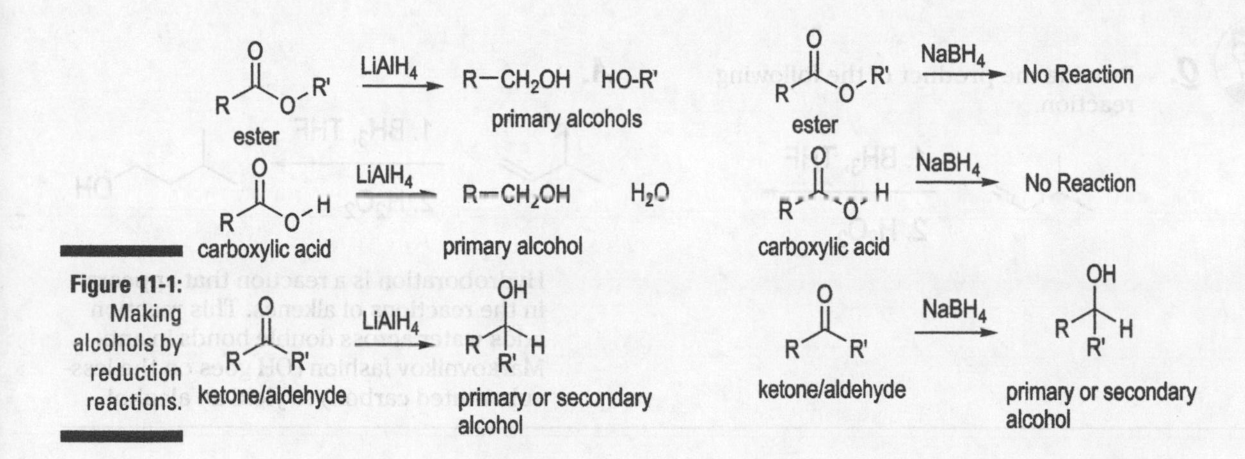

Figure 11-1: Making alcohols by reduction reactions.

The biggest reaction newcomer, though, is the *Grignard reaction,* an important alcohol-making reaction that forms a carbon-carbon bond in the process. When working with Grignard reactions, you start with an alkyl halide (usually bromide or chloride). This alkyl halide reacts with magnesium metal to form the Grignard reagent.

Grignard reagents react like carbanions. When predicting products of Grignard reactions, you may want to cross off the Mg*X* portion of the Grignard reagent and make it an anionic carbon, as shown in Figure 11-2, to make the reactivity clearer.

R–Br $\xrightarrow[\text{ether}]{Mg}$ R–MgBr ("Grignard reagent") $\xrightarrow{\text{"R}^-\text{" + R'C(=O)R'}}$ R–C(O^-)(R')–R' $\xrightarrow{H^+}$ R–C(OH)(R')–R'

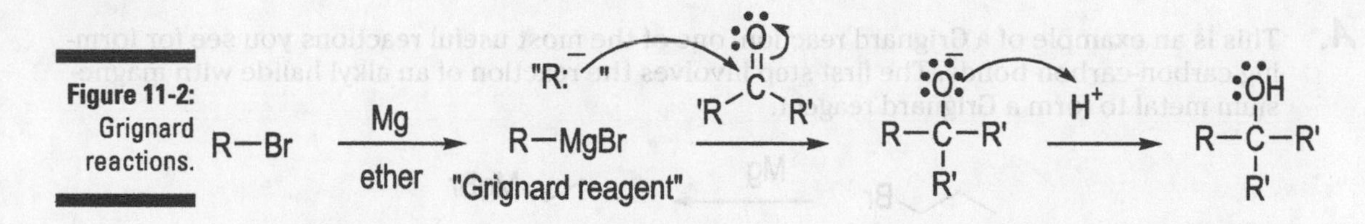

Figure 11-2: Grignard reactions.

The next step is to react the Grignard reagent with a carbonyl (C=O) compound. Because Grignard reagents act like disguised carbanions, you can draw the Grignard reagent acting as a nucleophile by attacking the carbonyl carbon and pushing the C=O double bond onto the oxygen as a lone pair. This leads to a negatively charged oxygen, which is protonated in the final step by shaking the reaction with dilute acid to yield the alcohol product.

Reactions of Grignard reagents with ketones yield tertiary alcohol products (the OH group on a tertiary, thrice-substituted carbon). With aldehydes, they yield secondary alcohols (an OH group attached to a twice-substituted carbon). Reactions with formaldehyde (H_2C=O) yield primary alcohols (OH group on a monosubstituted carbon).

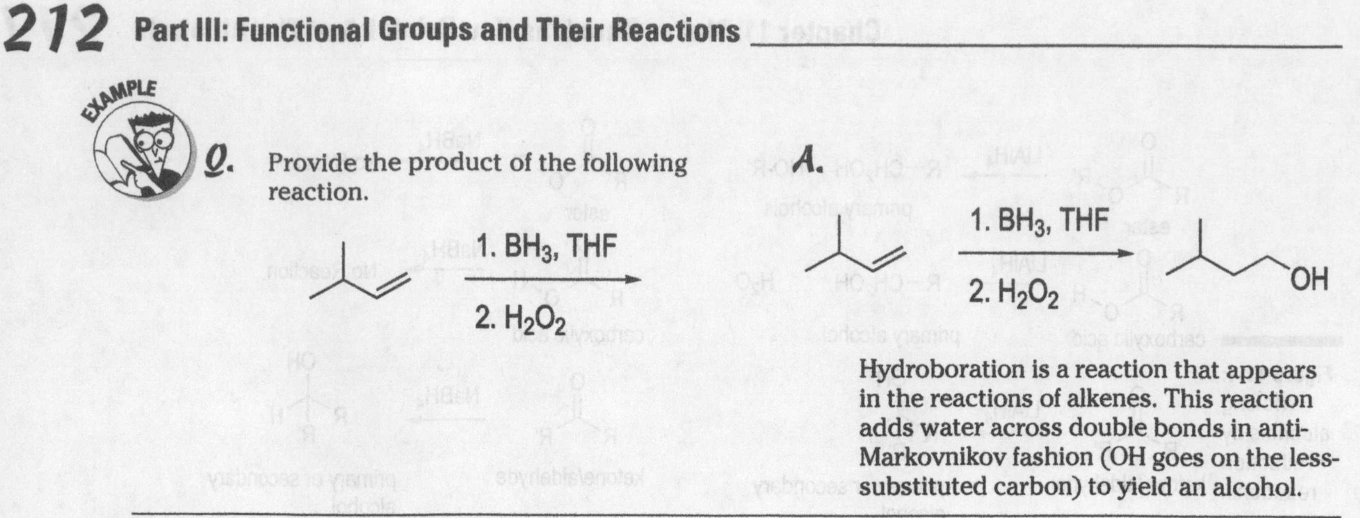

Q. Provide the product of the following reaction.

1. BH_3, THF
2. H_2O_2

A.

1. BH_3, THF
2. H_2O_2

OH

Hydroboration is a reaction that appears in the reactions of alkenes. This reaction adds water across double bonds in anti-Markovnikov fashion (OH goes on the less-substituted carbon) to yield an alcohol.

Q. Provide the product of the following reaction.

Br

1. Mg, Et_2O
2. (O)
3. H^+

A. This is an example of a Grignard reaction, one of the most useful reactions you see for forming carbon-carbon bonds. The first step involves the reaction of an alkyl halide with magnesium metal to form a Grignard reagent.

Br —Mg→ MgBr

Grignard reagent

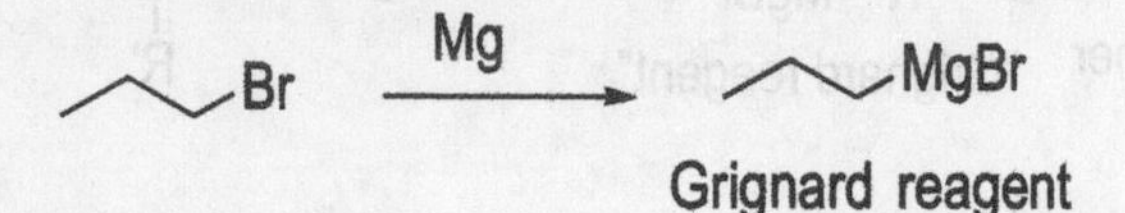

Grignard reagents react like carbanions in disguise. These "disguised anions" are great nucleophiles and add to carbonyl (C=O) compounds by attacking the carbonyl carbon and kicking the double-bond electrons onto the oxygen as a lone pair. In the final step, add the acid to protonate the oxygen and give the alcohol product.

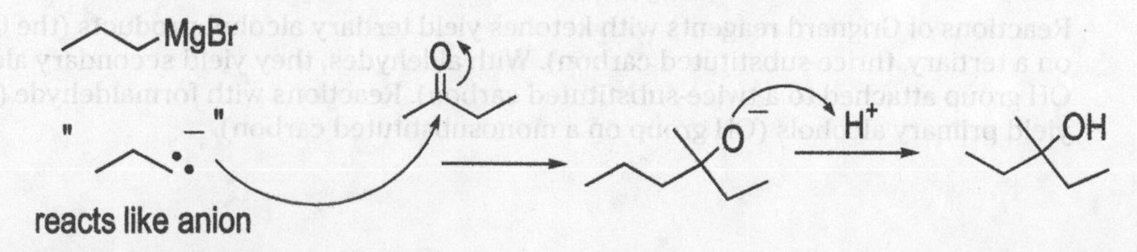

MgBr " = " O → O⁻ —H^+→ OH

reacts like anion

5. Provide the product of the following reaction.

1. BH_3, THF
2. H_2O_2

Solve It

6. Provide the product of the following reaction.

1. $Hg(OAc)_2$
2. $NaBH_4$

Solve It

7. Provide the product of the following reaction.

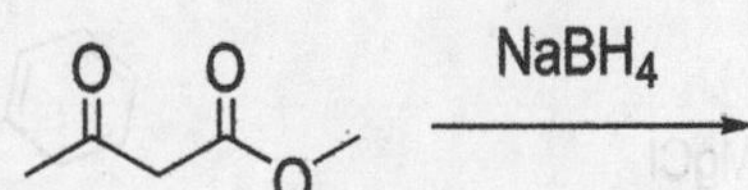

Solve It

8. Provide the product of the following reaction.

O O O $LiAlH_4$

Solve It

9. Fill in the boxes.

Br — 1. Mg, Ether →

H⁺

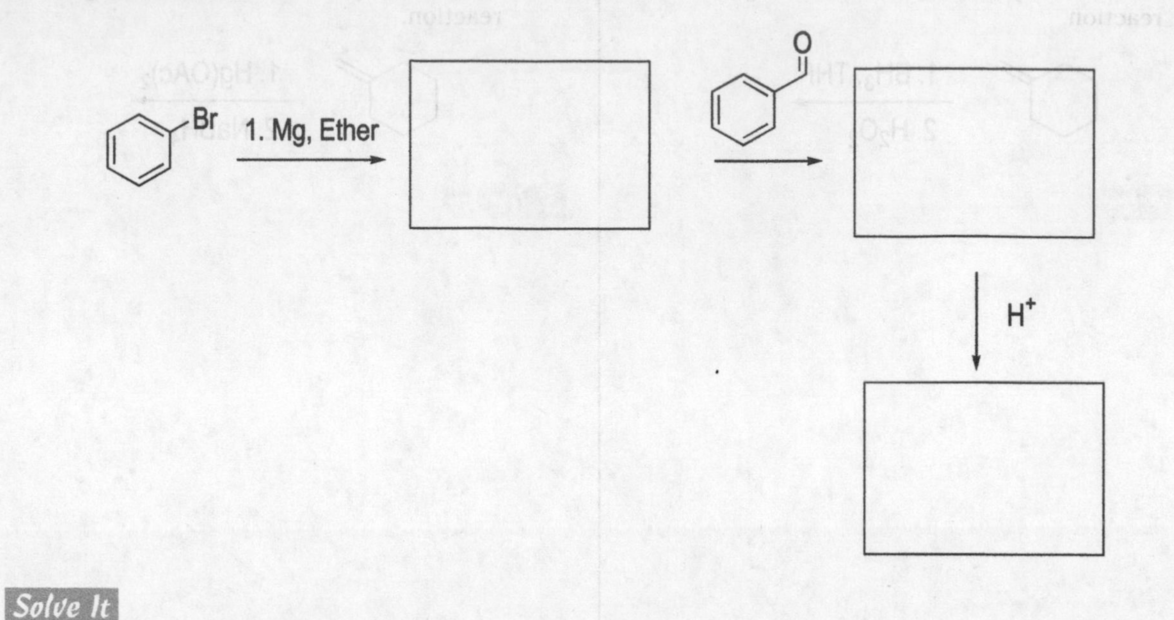

Solve It

10. Provide the product of the following reaction sequence.

Cl — 1. Mg, Ether; 2. (acetone, O); 3. H^+ →

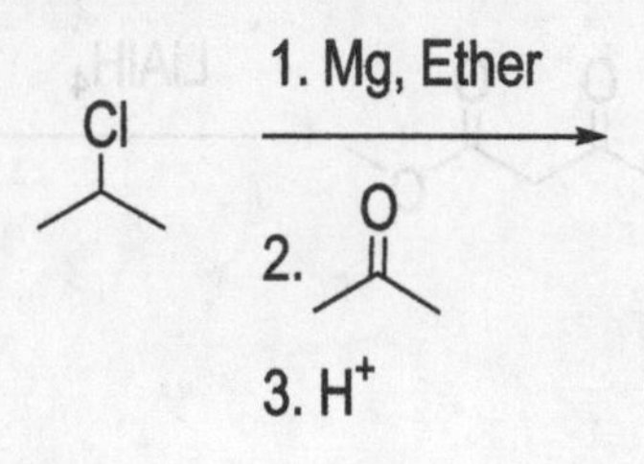

Solve It

11. Provide the missing reactant that would accomplish this reaction.

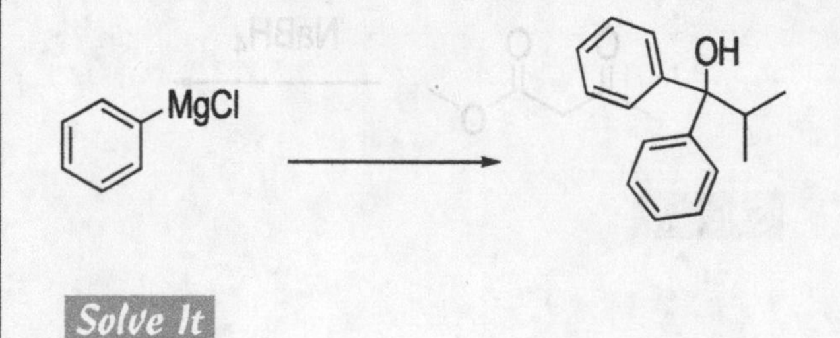

MgCl → OH

Solve It

Transforming Alcohols (without Committing a Party Foul)

The alcohol functional group has some amazing transformational powers — besides turning your usually shy roommate into a die-hard disco enthusiast. Reactions can easily convert alcohols into other functional groups, including ethers and carbonyl compounds. This transformational power makes alcohols a versatile and important functional group for performing the synthesis of useful molecules.

One of the most important functional group transformations of alcohols is their oxidation to carbonyl (C=O) compounds using oxidizing agents. You should know two oxidizing agents:

- PCC (short for *pyridinium chlorochromate*) converts alcohols into carbonyl (C=O) compounds. PCC is milder; it oxidizes primary alcohols to aldehydes.
- Jones' reagent (CrO_3, H_2SO_4) and other oxidizing agents (in some textbooks you see $Na_2Cr_2O_7$ or $KmnO_4$, which accomplish the same transformation as Jones' reagent) oxidize primary alcohols to carboxylic acids.

Secondary alcohols are oxidized to ketones with all the oxidizing agents.

Tertiary alcohols can't be oxidized.

Figure 11-3 compares these two oxidizing agents.

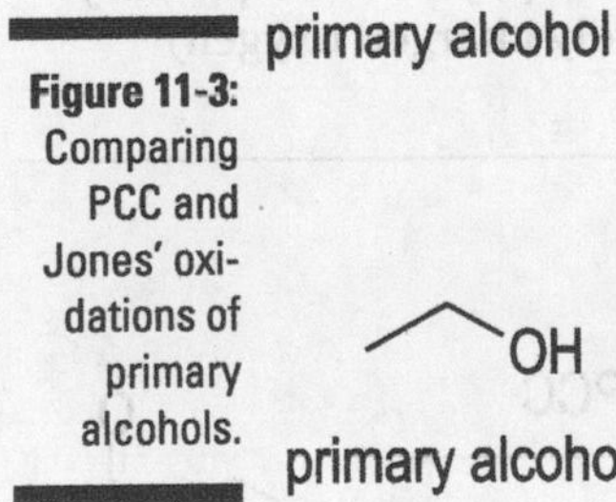

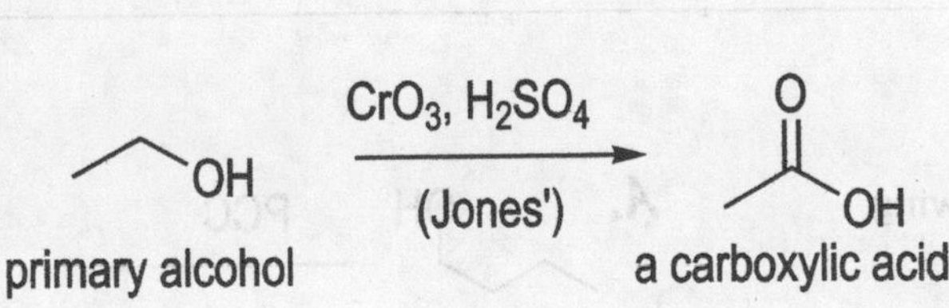

Figure 11-3: Comparing PCC and Jones' oxidations of primary alcohols.

Another important reaction of alcohols is the *Williamson ether synthesis,* which converts alcohols (*R*-OH group) into ether functional groups (*R*-O-*R*). This reaction works by first turning the weakly nucleophilic alcohol (OH) group into a strong nucleophile by deprotonating it to make an *alkoxide* (negatively charged oxygen). This deprotonation is accomplished using sodium metal (Na) or sodium hydride (NaH). In both cases, hydrogen gas is the byproduct of the deprotonation reaction (meaning you have to careful with these reactions in real life or you'll have an explosion on your hands). In the next step, a primary alkyl halide is added to the solution and undergoes an S_N2 substitution reaction to give the ether product.

One requirement of Williamson ether synthesis is that the alkyl halide be primary, meaning that the halide has to be attached to a primary carbon (a *primary carbon* is a carbon attached to only one other carbon). This requirement exists because S_N2 substitutions work best with primary alkyl halides. See Figure 11-4.

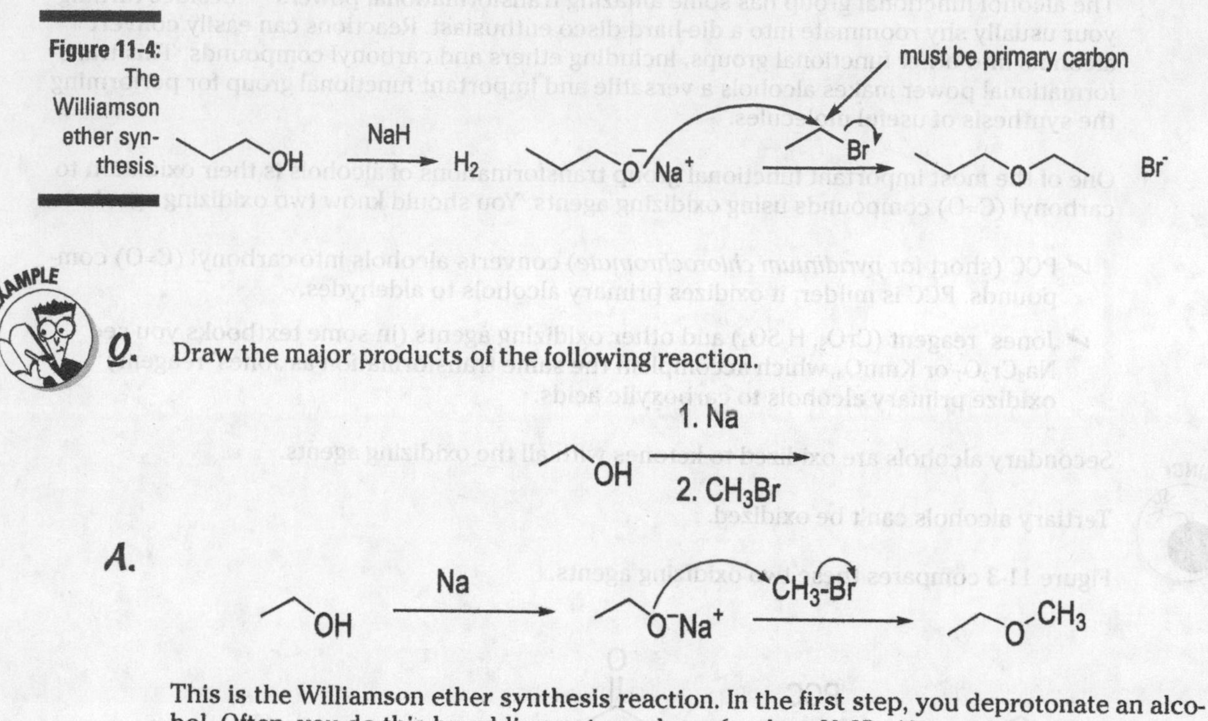

Figure 11-4: The Williamson ether synthesis.

$$\text{CH}_3\text{CH}_2\text{CH}_2\text{OH} \xrightarrow{\text{NaH}} \text{H}_2 + \text{CH}_3\text{CH}_2\text{CH}_2\text{O}^-\,\text{Na}^+ \xrightarrow{\text{CH}_3\text{CH}_2\text{CH}_2\text{Br}} \text{CH}_3\text{CH}_2\text{CH}_2\text{OCH}_2\text{CH}_2\text{CH}_3 + \text{Br}^-$$

must be primary carbon

Q. **Draw the major products of the following reaction.**

$$\text{CH}_3\text{CH}_2\text{OH} \xrightarrow[\text{2. CH}_3\text{Br}]{\text{1. Na}}$$

A.

$$\text{CH}_3\text{CH}_2\text{OH} \xrightarrow{\text{Na}} \text{CH}_3\text{CH}_2\text{O}^-\,\text{Na}^+ \xrightarrow{\text{CH}_3\text{-Br}} \text{CH}_3\text{CH}_2\text{O-CH}_3$$

This is the Williamson ether synthesis reaction. In the first step, you deprotonate an alcohol. Often, you do this by adding a strong base (such as NaH). Alternatively, you can add sodium metal as shown to deprotonate the alcohol. In the second step, you add a primary alkyl halide and get an S_N2 reaction in which the alkoxide (negatively charged oxygen) undergoes nucleophilic attack to displace the halide.

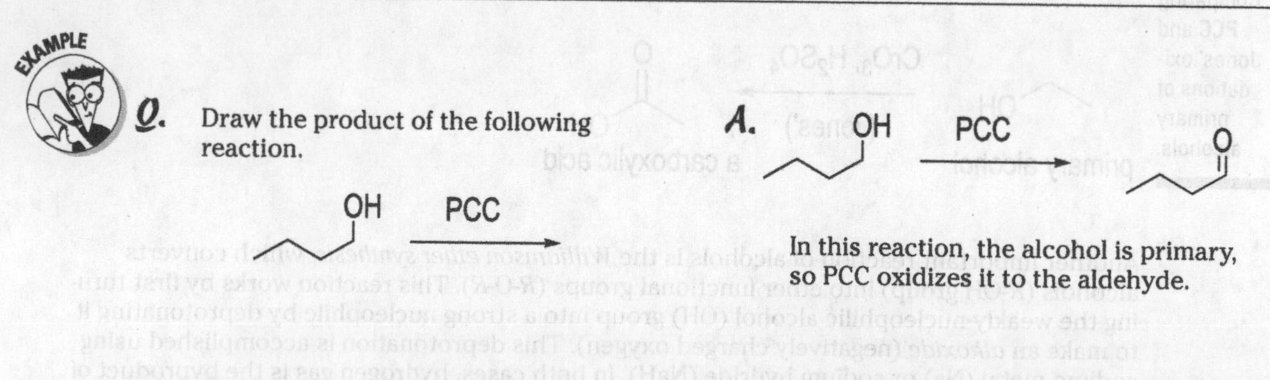

Q. **Draw the product of the following reaction.**

$$\text{CH}_3\text{CH}_2\text{CH}_2\text{OH} \xrightarrow{\text{PCC}}$$

A.

$$\text{CH}_3\text{CH}_2\text{CH}_2\text{OH} \xrightarrow{\text{PCC}} \text{CH}_3\text{CH}_2\text{CHO}$$

In this reaction, the alcohol is primary, so PCC oxidizes it to the aldehyde.

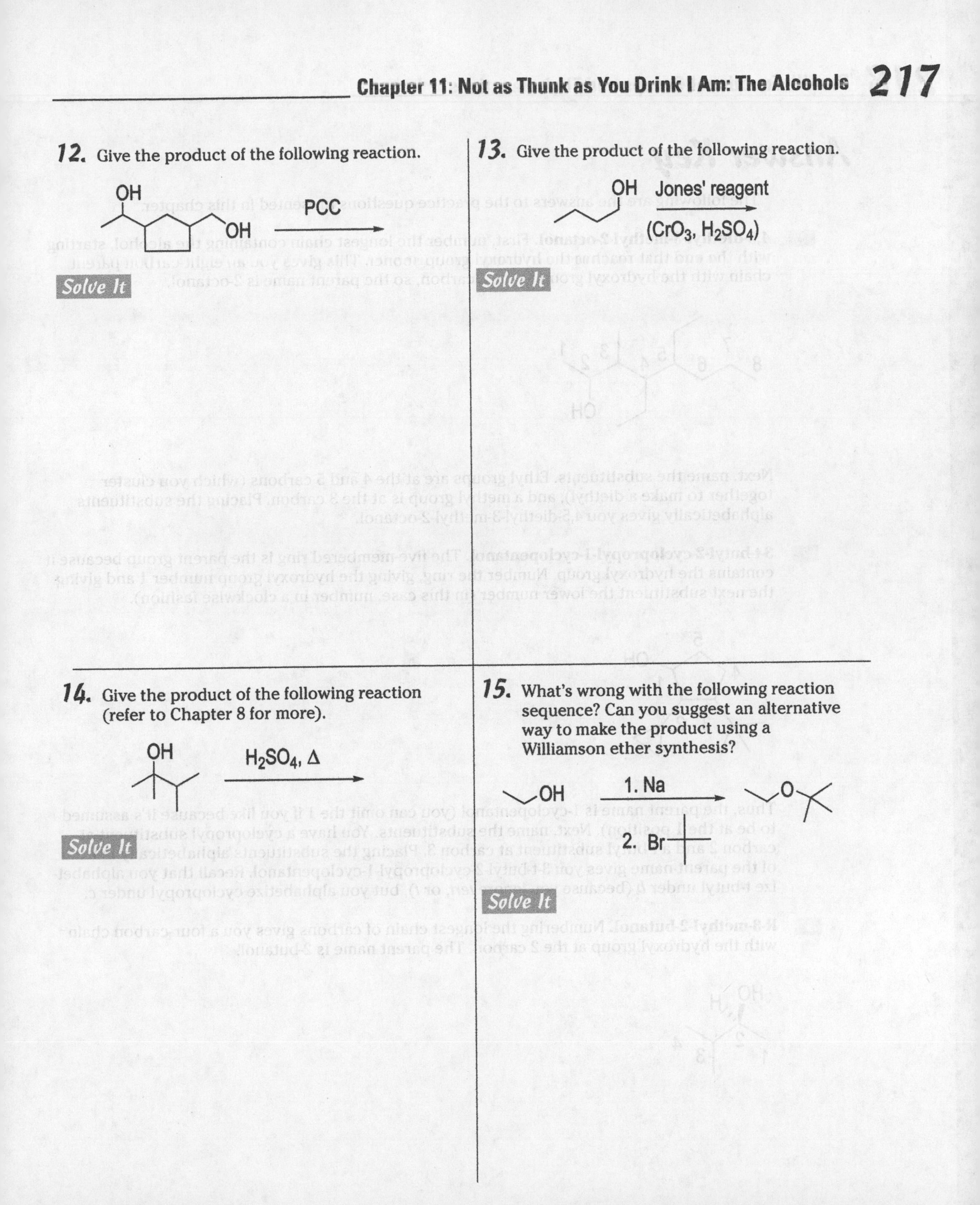

12. Give the product of the following reaction.

OH OH PCC →

Solve It

13. Give the product of the following reaction.

OH Jones' reagent (CrO_3, H_2SO_4) →

Solve It

14. Give the product of the following reaction (refer to Chapter 8 for more).

OH H_2SO_4, Δ →

Solve It

15. What's wrong with the following reaction sequence? Can you suggest an alternative way to make the product using a Williamson ether synthesis?

OH 1. Na 2. Br → O

Solve It

Answer Key

The following are the answers to the practice questions presented in this chapter.

1 **4,5-diethyl-3-methyl-2-octanol.** First, number the longest chain containing the alcohol, starting with the end that reaches the hydroxyl group sooner. This gives you an eight-carbon parent chain with the hydroxyl group at the 2 carbon, so the parent name is 2-octanol.

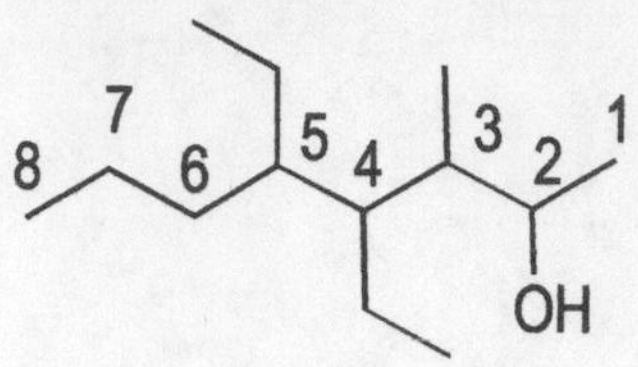

Next, name the substituents. Ethyl groups are at the 4 and 5 carbons (which you cluster together to make a diethyl), and a methyl group is at the 3 carbon. Placing the substituents alphabetically gives you 4,5-diethyl-3-methyl-2-octanol.

2 **3-t-butyl-2-cyclopropyl-1-cyclopentanol.** The five-membered ring is the parent group because it contains the hydroxyl group. Number the ring, giving the hydroxyl group number 1 and giving the next substituent the lower number (in this case, number in a clockwise fashion).

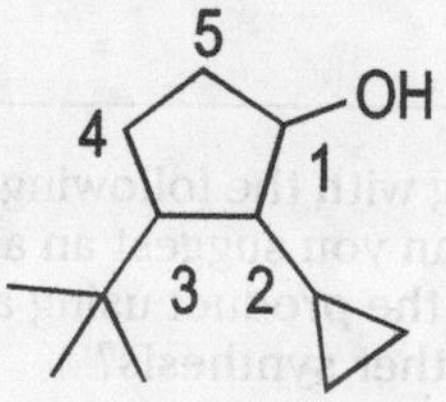

Thus, the parent name is 1-cyclopentanol (you can omit the 1 if you like because it's assumed to be at the 1 position). Next, name the substituents. You have a cyclopropyl substituent at carbon 2 and a t-butyl substituent at carbon 3. Placing the substituents alphabetically in front of the parent name gives you 3-t-butyl-2-cyclopropyl-1-cyclopentanol. Recall that you alphabetize t-butyl under *b* (because you ignore *tert*, or *t*), but you alphabetize cyclopropyl under *c*.

3 **R-3-methyl-2-butanol.** Numbering the longest chain of carbons gives you a four-carbon chain with the hydroxyl group at the 2 carbon. The parent name is 2-butanol.

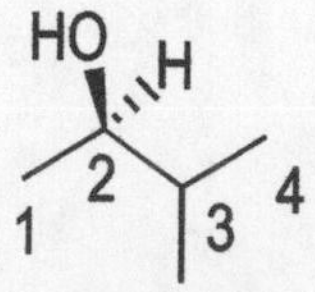

The only substituent is a methyl at carbon 3. Of course, you can't forget the stereochemistry (see Chapter 7 for a refresher). To assign the R/S designation of carbon 2, you first prioritize the groups. First priority goes to the OH group because O has a higher atomic number than C. The right carbon gets second priority because this carbon is attached to two other carbons (the left carbon is attached only to hydrogens). The left carbon gets third priority, and hydrogen (of course) is fourth priority. Because the hydrogen is in the back, you can draw a curve from first- to second- to third priority substituents. Because the curves are clockwise, the configuration is R. Thus, the total name is R-3-methyl-2-butanol (or 2R-3-methyl-2-butanol).

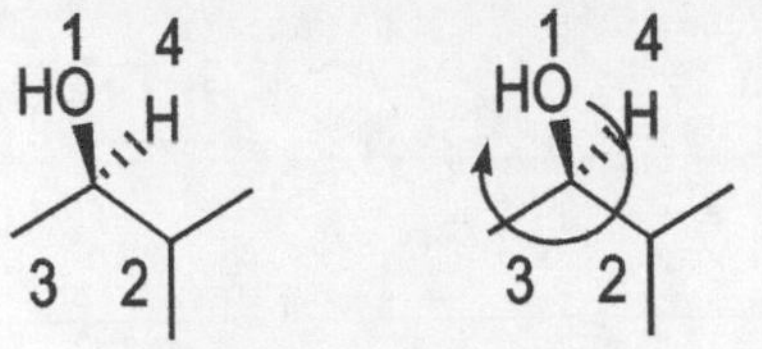

4 **2R-1,2-hexanediol.** This molecule has two hydroxyl groups and is therefore considered a diol. Numbering the parent chain to give the lower number to the hydroxyl groups gives you the following:

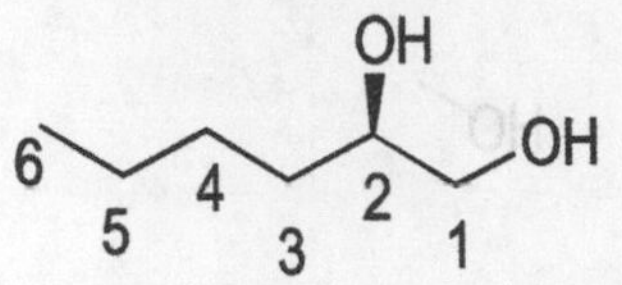

Because the chain is six carbons long, the hydroxyl groups come at carbons 1 and 2; the parent name is thus 1,2-hexanediol.

Next, determine the stereochemistry. Prioritizing the substituents and drawing the curve yields R stereochemistry for the chiral center because the curve goes clockwise (the fourth-priority substituent is in the back, so you don't need to rotate the molecule).

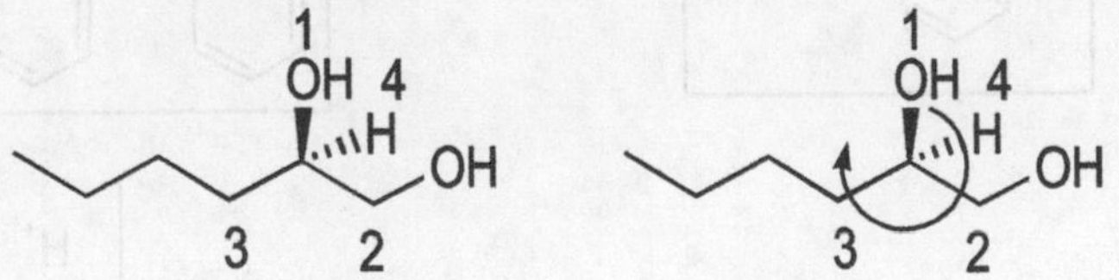

The name is therefore 2R-1,2-hexanediol (or simply R-1,2-hexanediol, because there's only one chiral center).

5 1. BH_3, THF

2. H_2O_2

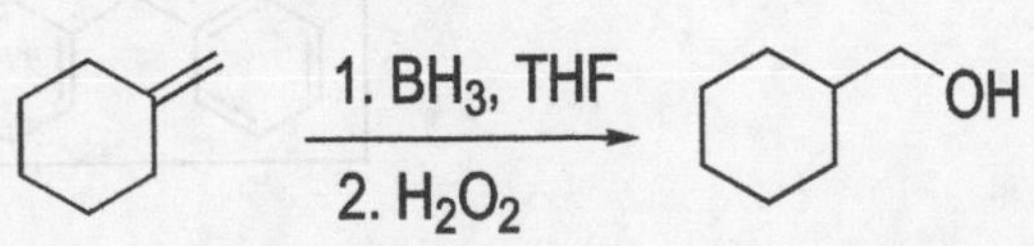

Hydroboration adds water across the double bond in an anti-Markovnikov fashion, with the hydroxyl group adding to the less-substituted carbon.

6

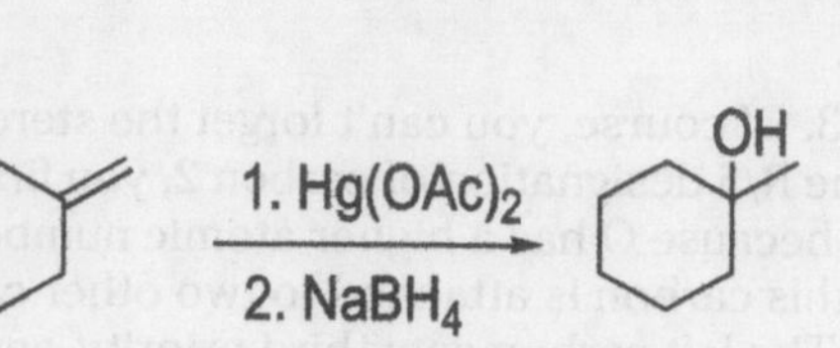

1. $Hg(OAc)_2$
2. $NaBH_4$

OH

Oxymercuration adds water across the double bond in a Markovnikov fashion, with the hydroxyl group winding up on the more-substituted carbon.

7

O O — $NaBH_4$ → OH O

Sodium borohydride ($NaBH_4$) is a weaker reducing agent than lithium aluminum hydride ($LiAlH_4$). This reagent can reduce ketones and aldehydes to alcohols, but it's not strong enough to reduce more-difficult carbonyl compounds such as esters or carboxylic acids. Thus, in this molecule, the $NaBH_4$ can only reduce the ketone to the alcohol, leaving the ester untouched.

8

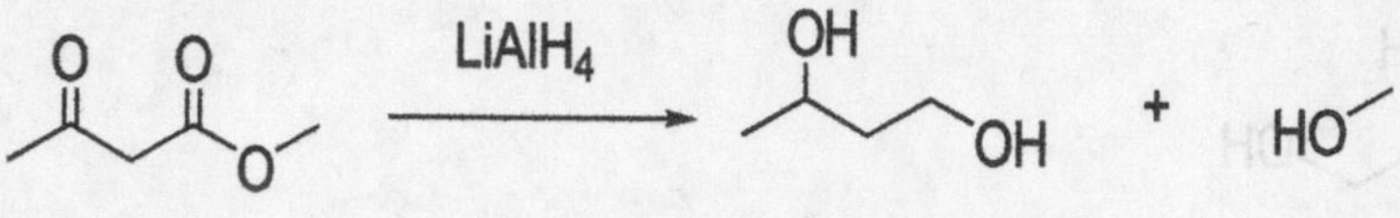

O O — $LiAlH_4$ → OH, OH + HO

Lithium aluminum hydride ($LiAlH_4$) is a strong reducing agent that can reduce all the carbonyl compounds to alcohols. Esters and carboxylic acids are reduced to primary alcohols.

9

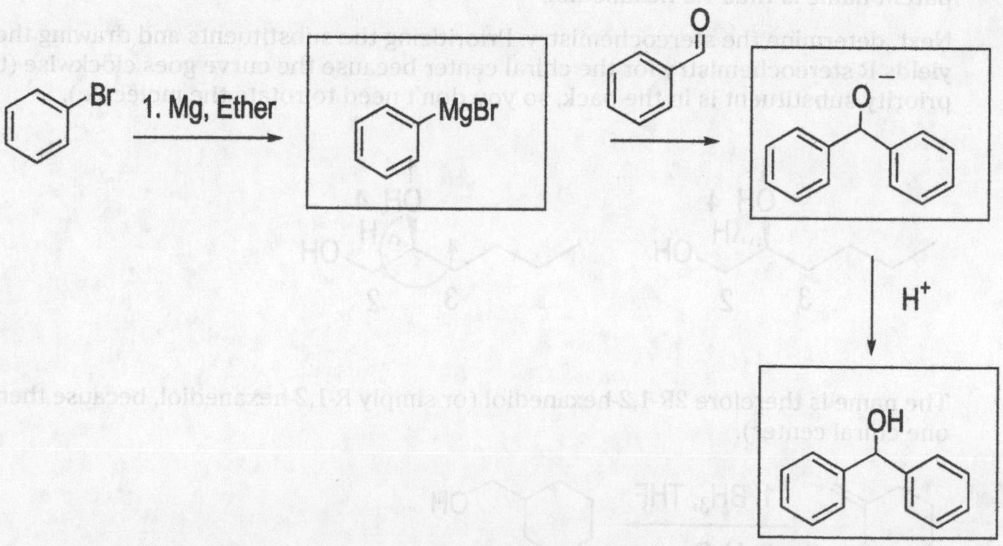

Br — 1. Mg, Ether → MgBr — O → O⁻ — H^+ → OH

10

Cl → Mg, ether → MgCl ≡ ... → H+ → OH

Addition of magnesium (in ether solvent) forms a Grignard reagent. In the second step, the Grignard reagent adds to the carbonyl carbon, kicking the double-bond electrons onto the oxygen as a lone pair. In the final step, dilute acid is added to protonate the oxygen and give the alcohol product.

11

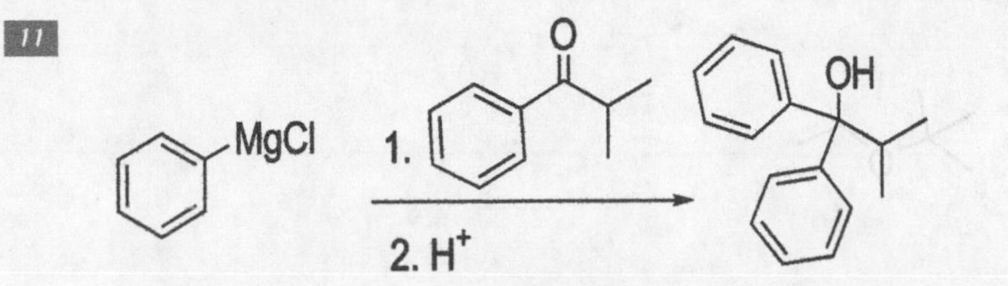

MgCl 1. O 2. H^+ → OH

12

OH OH → PCC → O O

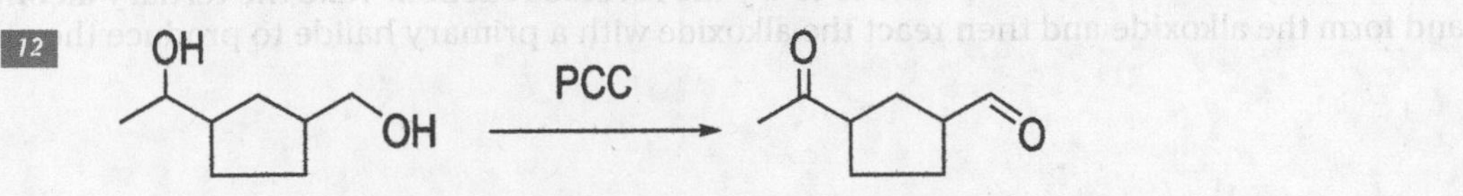

PCC oxidizes alcohols to carbonyl compounds. PCC is a mild oxidizing agent, so primary alcohols are oxidized to aldehydes and not to carboxylic acids.

13

OH → Jones' reagent (CrO_3, H_2SO_4) → O OH

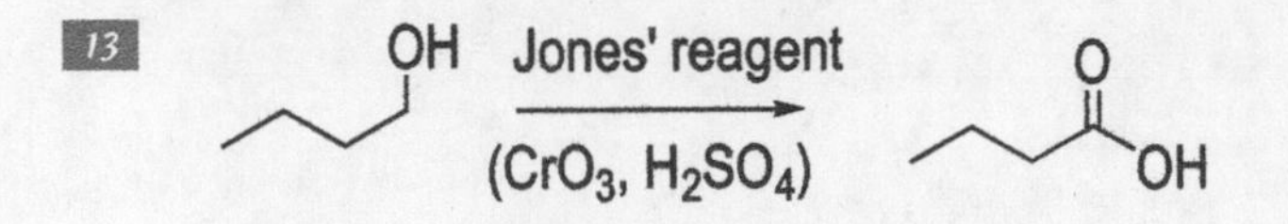

Jones' reagent is a strong oxidizing agent, and it converts primary alcohols into carboxylic acids.

14

OH → H_2SO_4, Δ → E1

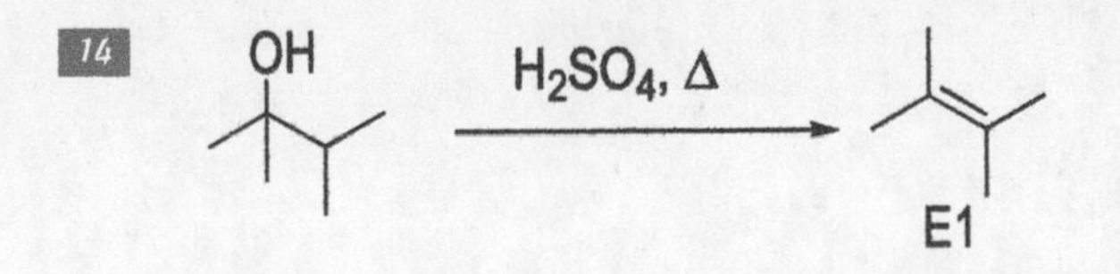

This problem is an oldie but a goodie. Treatment of tertiary alcohols with sulfuric acid and heat leads to an E1 elimination to produce the most-substituted alkene. (See Chapter 10 for more on E1 elimination.) The Greek letter delta, Δ, indicates that heat was added to the system.

15

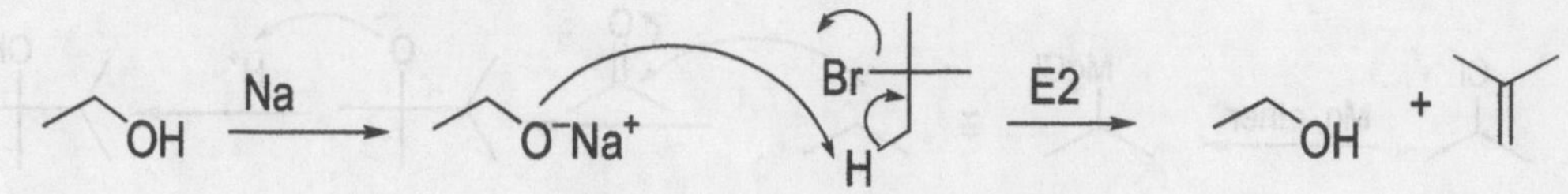

Williamson ether synthesis reactions work only when you react an *alkoxide* (deprotonated alcohol) with a primary halide (a halide attached to a primary carbon). When you react an alkoxide with a secondary or tertiary halide, you get an E2 elimination instead (see Chapter 10) because alkoxides are good bases in addition to being good nucleophiles.

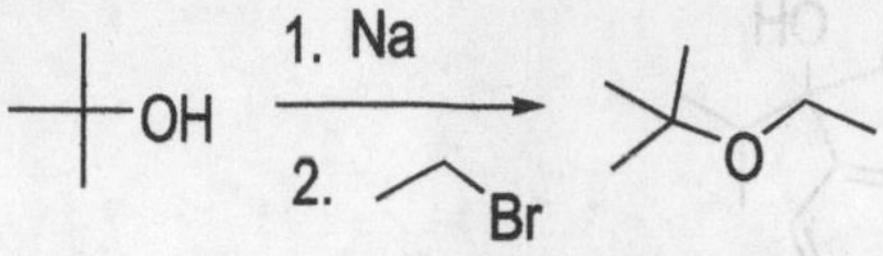

Another way to make the compound is to try the reverse reaction. Take the tertiary alcohol and form the alkoxide and then react the alkoxide with a primary halide to produce the ether.

Chapter 12

Conjugated Dienes and the Diels-Alder Reaction

In This Chapter

- Adding hydrohalic acids to conjugated dienes
- Finding the products of the Diels-Alder reaction
- Working back to the starting materials of a Diels-Alder reaction

Isolated functional groups each have a unique set of reactions, and the chapters on alkenes (Chapter 8), alkynes (Chapter 9), and alcohols (Chapter 11) cover these for some important functional groups. But sometimes when you put two functional groups next to each other, you get different reactivity entirely than when the functional group is by itself. That's the case with *conjugated dienes,* or *alternating double bonds* (two double bonds separated by a single carbon-carbon bond). Some very important reactions that don't occur with isolated double bonds — or double bonds separated by more than one single bond — occur with molecules containing conjugated double bonds. In this chapter, you see two interesting reactions of conjugated dienes:

- **The addition of hydrohalic acids (such as HBr and HCl) to conjugated dienes:** This reaction is similar to the reactions of isolated alkenes (Chapter 8) — but with a bit of a twist, now that two double bonds are involved.
- **The Diels-Alder reaction:** This reaction is a most interesting and unusual reaction, and it's one of the most powerful reactions you see in organic chemistry for forming carbon-carbon bonds.

This chapter provides you plenty of help in getting you up to speed on working with both of these reactions.

Seeing 1,2- and 1,4-Addition Reactions to Conjugated Dienes

One type of reaction of conjugated dienes is the addition of hydrohalic acids. Adding HBr (or HCl) to a conjugated diene leads to one of two products:

- **The 1,2-addition product:** Hydrogen adds to the first carbon, and the halide adds to the second carbon.

- **The 1,4-addition product:** Hydrogen adds to the first carbon, and the halide adds to the fourth carbon.

Figure 12-1 shows examples of these two reactions.

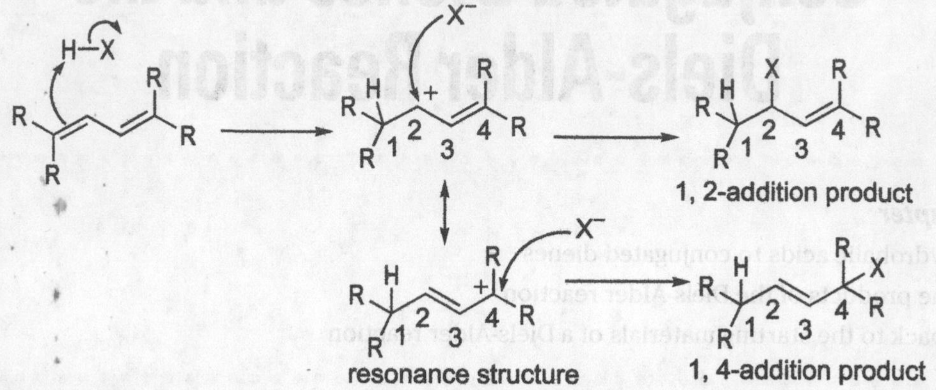

Figure 12-1: The 1,2- and 1,4-addition reactions of a hydrogen halide (H-*X*) to a conjugated diene.

To draw the mechanism of this addition reaction, follow these steps:

1. **Protonate one of the double bonds.**

   After you protonate the double bond by taking the electrons from the double bond to form a new bond to hydrogen (from the acid H-*X*), a positive charge forms on one of the carbons, creating a *carbocation intermediate.* The positive charge prefers to reside on the carbon closer to the other double bond so resonance can stabilize the charge (see Chapter 3 to brush up on resonance structures). Thus, the hydrogen goes on the 1 carbon, and the positive charge goes on the 2 carbon.

2. **Add the halide.**

   - **To obtain the 1,2-addition product:** The halide adds to the carbocation formed on carbon 2 to yield the 1,2-addition product. This is the usual product you see from the reaction of isolated alkenes (Chapter 8).
   - **To obtain the 1,4-addition product:** Draw the resonance structure for the carbocation by moving the double bond toward the cation, which reforms the double bond and moves the charge onto the number 4 carbon. The halide then attacks the carbocation to yield the 1,4-addition product. You don't see this product in reactions of isolated alkenes.

Often, one of these products, called the *thermodynamic product,* is lower in energy. The other, called the *kinetic product,* forms more quickly as a result of having to climb a smaller barrier, or *energy hill,* to get to the product. In such cases, you can change the product distributions by altering the reaction temperature:

- **To obtain a kinetic product, lower the reaction temperature so that the reaction has only enough energy to go over small energy hills.**
- **To obtain the thermodynamic product, heat the reaction so it has enough energy to go over large hills. At equilibrium, the thermodynamic product is favored because it's lower in energy.**

Very generally speaking, 1,4-additions tend to yield the thermodynamic product (lower energy) and 1,2-additions yield the kinetic product (faster forming product with the lower energy barrier to formation). Thus, in most cases, you lower the temp to obtain a 1,2-addition product and raise the temp to obtain a 1,4-addition product.

A useful way of visualizing the energetics of the 1,2- and 1,4-addition reactions is through a *reaction coordinate diagram,* a plot of the reaction progress versus the energy. In Figure 12-2, the first step in this reaction is the same for both the 1,2- and 1,4-addition reactions, and it yields the carbocation intermediate. Because two resonance structures describe a single intermediate ion, going from one resonance structure to another isn't a reaction but merely a way of showing that in the actual structure, half the charge resides on carbon 2 and half is on carbon 4 (the actual molecule looks like an average of all the resonance structures). In the second step, the reaction with the lower barrier leads to the kinetic product, whereas the product that yields a lower-energy product is the thermodynamic product.

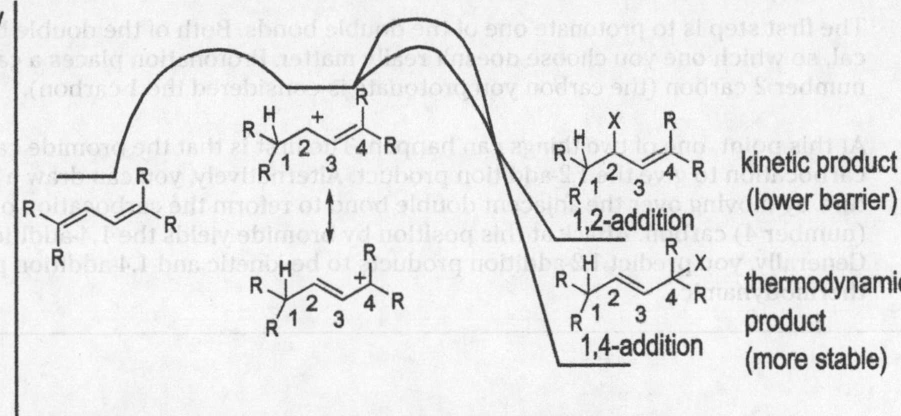

Figure 12-2: Reaction coordinate diagram showing kinetic and thermodynamic products.

Q. **Draw the two products of the following reaction. Then draw the mechanism leading to those two products. Finally, state which product will likely be the kinetic product and which will likely be the thermodynamic product.**

H—Br

A.

H—Br Br- 1,2-addition

Br

kinetic product (formed quicker)

Br- 1,4-addition Br

thermodynamic product (more stable)

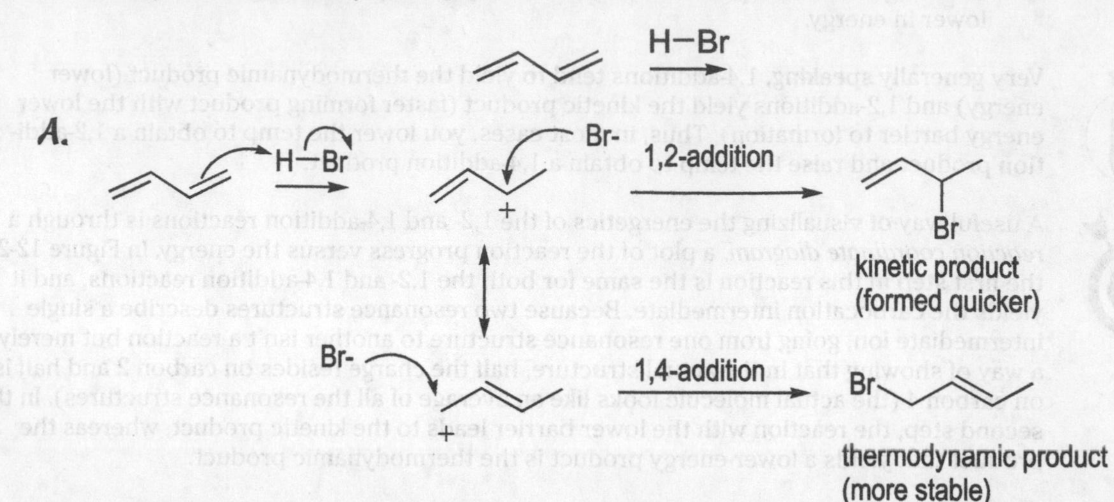

The first step is to protonate one of the double bonds. Both of the double bonds are identical, so which one you choose doesn't really matter. Protonation places a carbocation on the number 2 carbon (the carbon you protonate is considered the 1 carbon).

At this point, one of two things can happen. The first is that the bromide can attack the carbocation to give the 1,2-addition product. Alternatively, you can draw a resonance structure by moving over the adjacent double bond to reform the carbocation on the primary (number 4) carbon. Attack at this position by bromide yields the 1,4-addition product. Generally, you predict 1,2-addition products to be kinetic and 1,4-addition products to be thermodynamic.

Q. **Why might the 1,4-addition product from the previous example be more stable than the 1,2-addition product? That is, why is it the thermodynamic product?**

A. **Attaching substituents stabilizes double bonds. Often, 1,4-additions yield double bonds that are more highly substituted than 1,2-additions (more *R* groups are attached to them). More highly substituted double bonds are more stable than less-substituted double bonds, so 1,4-additions tend to yield more stable alkenes than 1,2-addition reactions. In the first example, the 1,4-addition product double bond is disubstituted, whereas the 1,2-addition product is monosubstituted. As a result, the 1,4-addition product is more stable than the 1,2-addition product, so the 1,4-addition product is the thermodynamic product.**

1. Draw the mechanism leading to the 1,2- and 1,4-addition products of the following reaction. Then, assuming the 1,2-addition product is the kinetic product and the 1,4-addition product is the thermodynamic product, draw the reaction coordinate diagram of the reaction leading to these two products.

H-Cl

Energy

Reaction Progress

Solve It

2. Which product from question 1 would be favored at low temperatures? At high temperatures?

Solve It

3. What would the reaction coordinate diagram look like for a hypothetical addition reaction that led to a 1,4-addition product that was both the kinetic and thermodynamic product, as compared to the 1,2-addition product?

Solve It

Energy

Reaction Progress

4. For the reaction in question 3, which product would form at low temperatures? At high temperatures?

Solve It

Dienes and Their Lovers: Working Forward in the Diels-Alder Reaction

The Diels-Alder reaction is one of the most interesting reactions in organic chemistry, but it's also somewhat difficult to visualize initially. In this reaction, a conjugated *diene* (a molecule with two alternating double bonds) reacts with a *dienophile* (a "diene-lover" — an alkene, with one double bond). Two new bonds form simultaneously in this reaction, and the result is a six-membered ring. See Figure 12-3 for an example.

new bond

new bond

diene dienophile

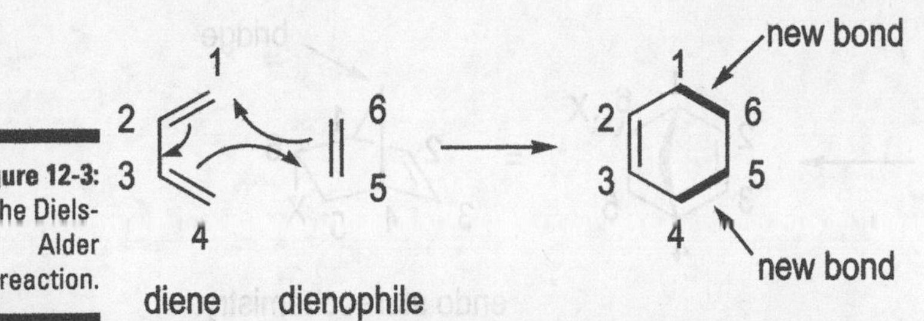

Figure 12-3: The Diels-Alder reaction.

To work these types of problems, follow these steps:

1. **Orient the diene and the dienophile correctly.**

   Dienes should be in the s-*cis* orientation, with both double bonds sticking off the same side of the bond connecting them (dienes in rings react quickly because they're locked in this s-*cis* conformation). The two double bonds should point toward the dienophile.

2. **Number the diene and dienophile.**

   I generally begin at the top of the diene and go counterclockwise, numbering its first four carbons. Thus, 1-2 and 3-4 are the double bonds, and 2-3 is a single bond. I continue counterclockwise, labeling two carbons on the dienophile so that its double bond lies at 5-6. Numbering helps you keep track of all the atoms and provides a visual aid so you can see where bonds form and break. It's particularly crucial when the diene is in a ring.

3. **Work the reaction by arrow-pushing.**

   Simultaneously make three arrows, starting by forming bonds between carbons 4-5 and 6-1 by moving the double bonds; also move the double bond at 1-2 between carbons 2-3. The result is new bonds between carbons 4-5 and 6-1 as well as a new double bond between carbons 2-3. All the arrows should move in the same direction, either all clockwise or all counterclockwise.

4. **Make sure you have the correct stereochemistry.**

   When you have two substituents on the dienophile, if the substituents have *cis* orientation on the double bond, they end up with *cis* orientation in the product; if they start *trans,* they end *trans* in the product.

   If the diene starts in a ring, you get a *bicyclic product* — a fused ring system that takes some practice to draw correctly. In this case, any substituent on the dienophile prefers to orient down relative to the bridge, giving you what's called the *endo* product (in the *exo* product, the substituent sticks up relative to the bridge). Figure 12-4 shows an example. Here, when you have a substituent *(X)* on a dienophile, the down orientation relative to the bridgehead is preferred to give the endo product.

One more note: Diels-Alder reactions work best when the dienophile is substituted with electron-withdrawing groups (such as CN, NO_2, COR) and the diene is substituted with electron-donating groups (such as OR, NH_2, NR_2, alkyl). ***Remember:*** Any substituent attached to the diene by an atom with a lone pair will be electron-donating.

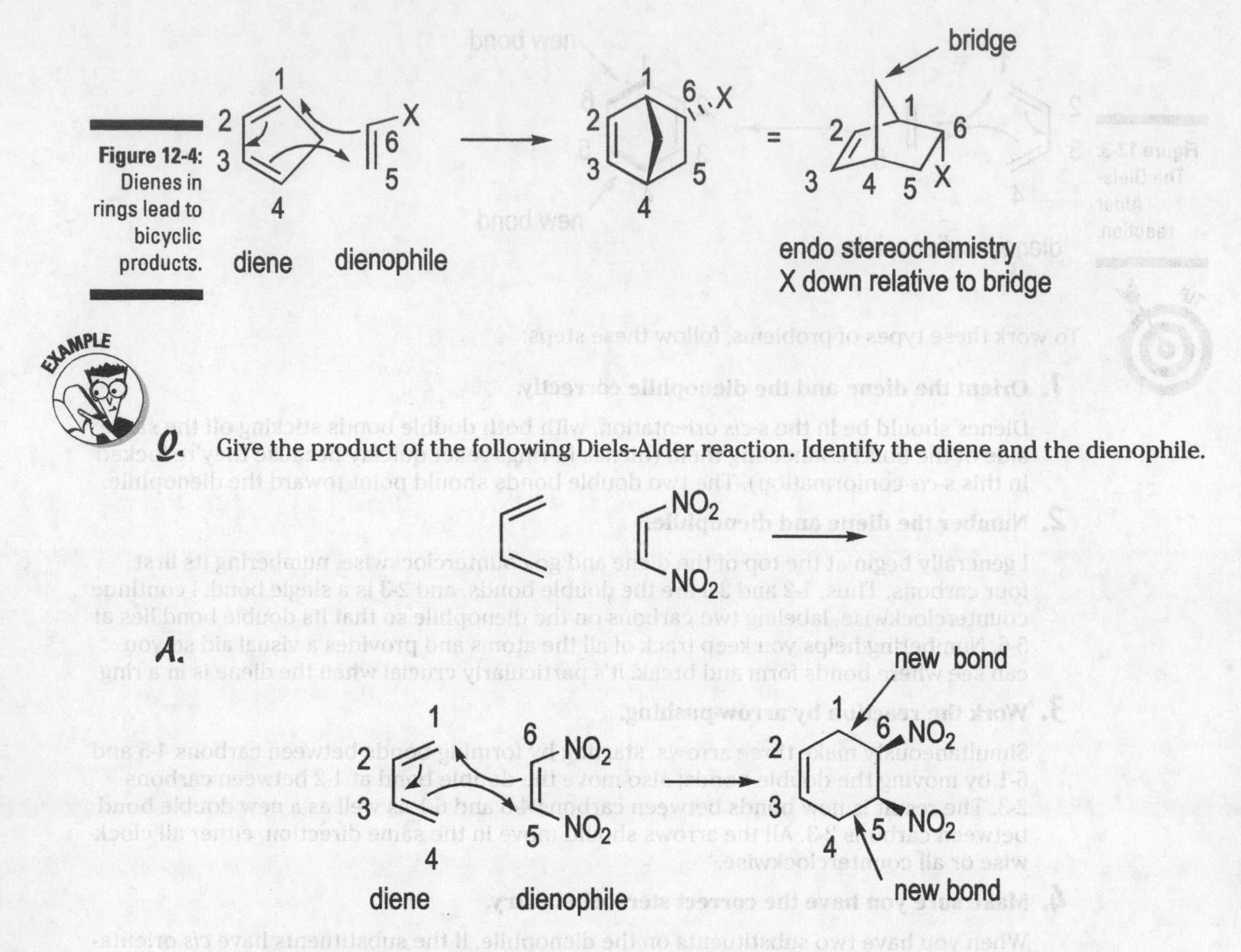

Figure 12-4: Dienes in rings lead to bicyclic products.

Q. Give the product of the following Diels-Alder reaction. Identify the diene and the dienophile.

A.

The diene is in s-*cis* orientation, and both double bonds are pointing toward the dienophile, so the orientation is fine. Number the diene and dienophile. I recommend that you number the four diene carbons 1 through 4 and then number the two dienophile carbons 5 and 6. To work the reaction, take a deep breath and push the arrows. One double bond on the diene attacks the dienophile, which in turn attacks the other side of the diene. Bonds form between carbons 4-5 and 1-6, and a new double bond forms between carbons 2-3. As a result, you make a six-membered ring. Now check the stereochemistry. Because the substituents (NO_2 groups) on the dienophile started *cis* in the starting material, they have to be *cis* in the product.

Q. Give the product of the following Diels-Alder reaction.

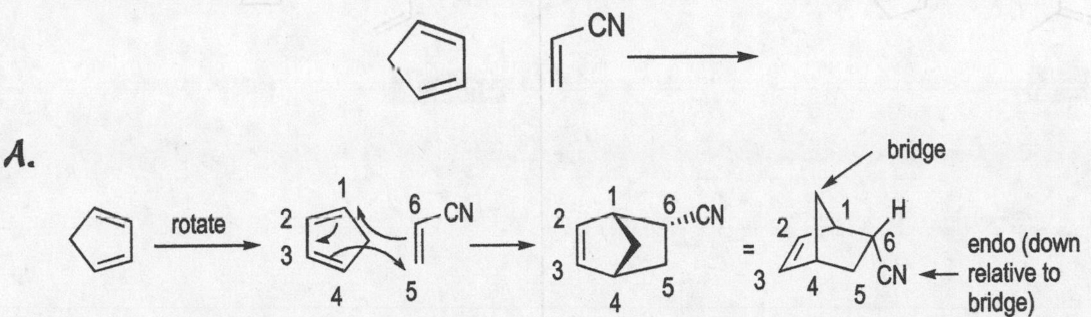

CN

A.

rotate 1 2 3 4 5 6 CN 1 2 3 4 5 6 CN = bridge 1 2 3 4 5 6 H CN endo (down relative to bridge)

Things get a bit hairy when the diene is incorporated into a ring because you get a bicyclic product. But these problems aren't too hard if you take them stepwise. In this case, the diene is pointing away from the dienophile, so orient the diene so that its two double bonds point out toward the dienophile. Then number the diene and the dienophile. Push the arrows (three of them) so that bonds form between carbons 4-5 and 6-1 and a new double bond forms between 2-3. Finally, make sure you have the correct stereochemistry. The product is a bicyclic ring, so the cyano (CN) substituent prefers to orient down from the bridge, giving you the endo product.

5. Give the product of the following reaction.

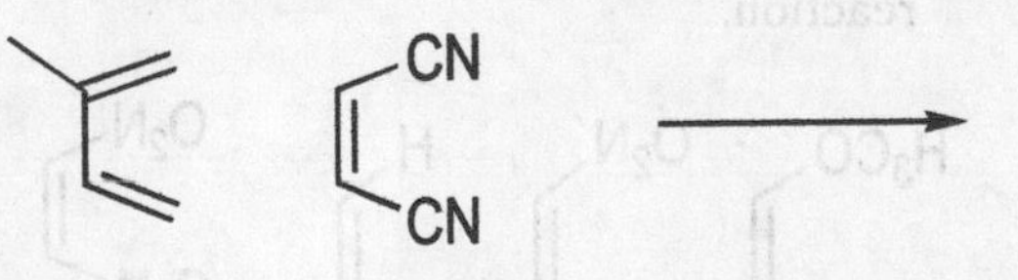

Solve It

6. Give the product of the following reaction.

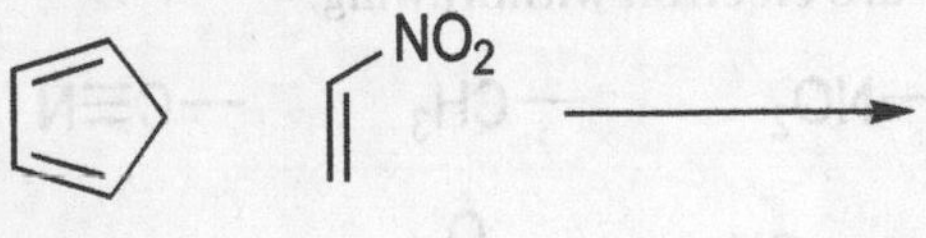

Solve It

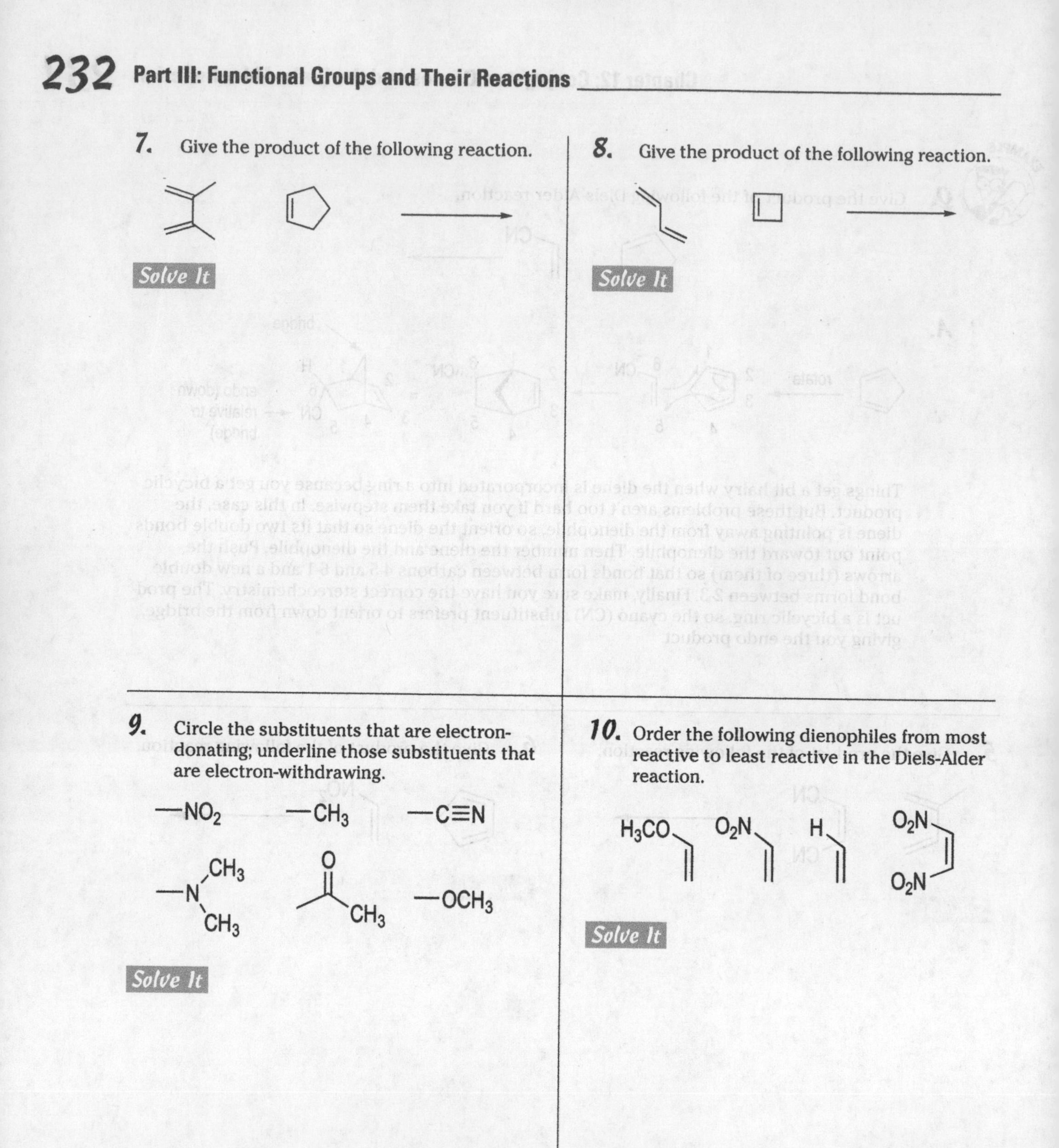

7. Give the product of the following reaction.

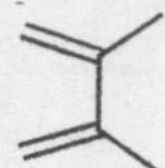 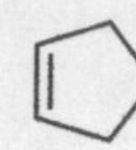 ⟶

Solve It

8. Give the product of the following reaction.

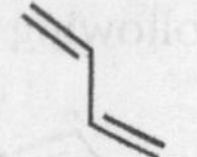  ⟶

Solve It

9. Circle the substituents that are electron-donating; underline those substituents that are electron-withdrawing.

$-NO_2$ $-CH_3$ $-C\equiv N$

$-N(CH_3)_2$ $-C(=O)CH_3$ $-OCH_3$

Solve It

10. Order the following dienophiles from most reactive to least reactive in the Diels-Alder reaction.

H_3CO O_2N H 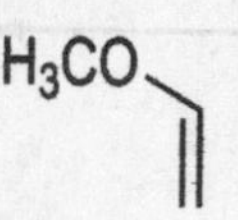

Solve It

11. Which of the following dienes can react in the Diels-Alder reaction?

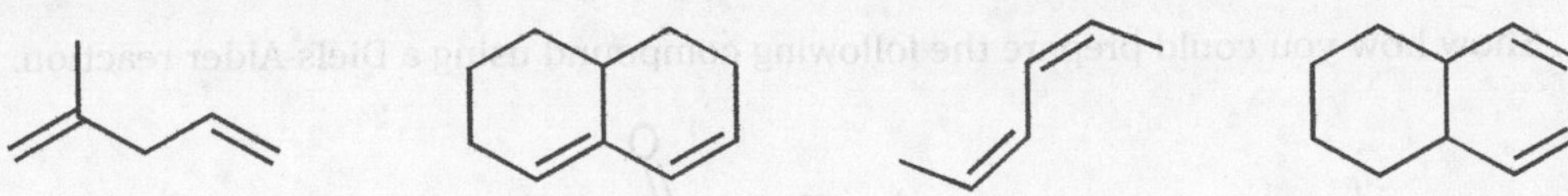

Solve It

12. Order the following dienes from most reactive to least reactive in the Diels-Alder reaction.

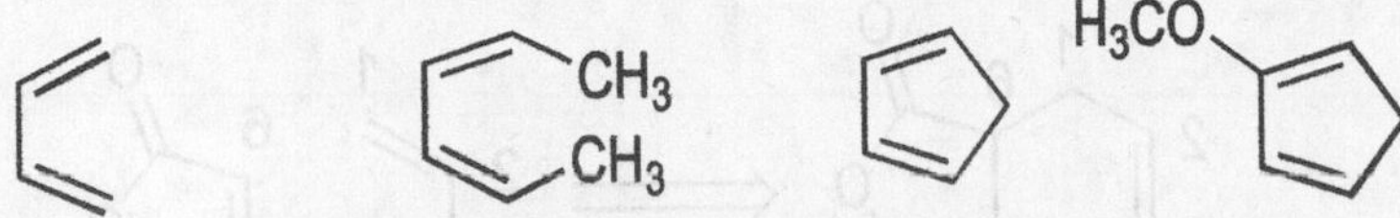

Solve It

Reverse Engineering: Working Backward in the Diels-Alder Reaction

One challenge of working with Diels-Alder reactions is working backward, or using *retrosynthesis,* to show how a Diels-Alder reaction can make a given product. The tough bit is seeing where the two pieces connected together to give the product, a problem that I like to solve by using a simple numbering scheme.

The one foothold that you have when working backward in the Diels-Alder reaction is the double bond in the product. From there, you can back out what the starting materials must have been because the double bond ends up between the 2 and 3 carbons of the diene. In order to work backward in the Diels-Alder reaction, follow these steps:

1. **Number the carbons.**

   Number the product carbons from 1 through 6, starting by giving the double bond the 2-3 position.

2. **Break the bonds in the reverse Diels-Alder reaction.**

   Break the bonds between 4-5 and 1-6, the two bonds formed in the Diels-Alder reaction.

3. **Add the double bonds to the diene and dienophile.**

   The diene's double bonds go between carbons 1-2 and 3-4, and in the dienophile, they go between 5-6.

4. **Check the stereochemistry of the dienophile.**

   If the substituents are *trans* in the product, the starting dienophile must've likewise been *trans;* if they're *cis,* they must've been *cis.*

Q. **Show how you could prepare the following compound using a Diels-Alder reaction.**

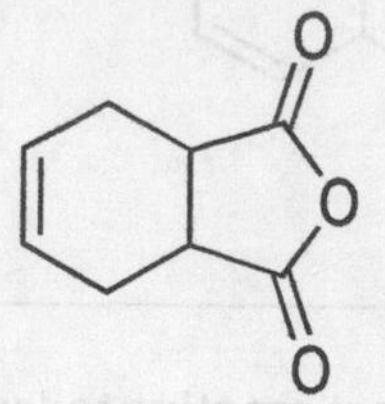

A.

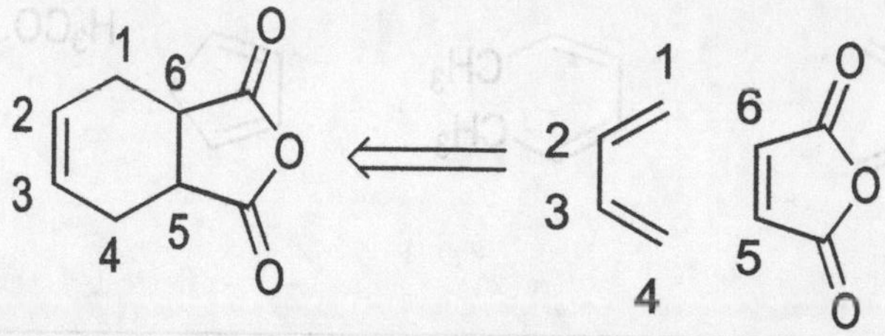

Start with the numbering, putting the 2-3 carbons on the double bond. Then break the bonds between 4-5 and 1-6. Finally, add double bonds between carbons 1-2, 3-4, and 5-6.

Q. **Show how you could prepare the following compound using a Diels-Alder reaction.**

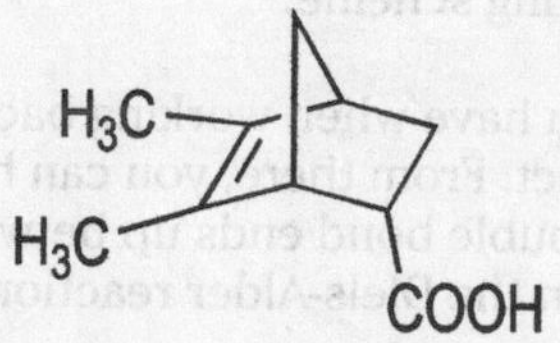

A.

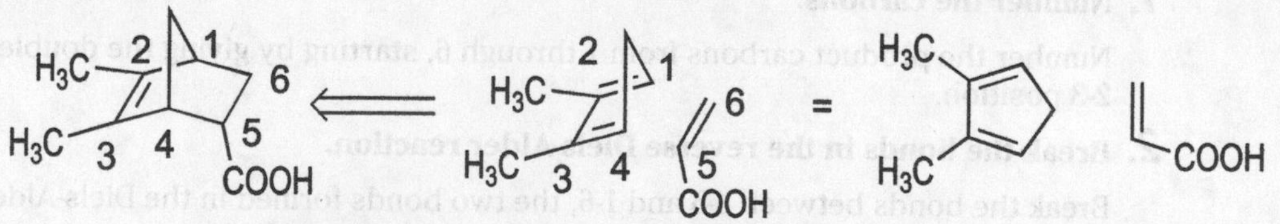

Make sure that numbers 2-3 are the double-bond carbons. You then make the breaks between carbons 1-6 and 4-5. In the starting material, the double bonds start between carbons 1-2, 3-4, and 5-6.

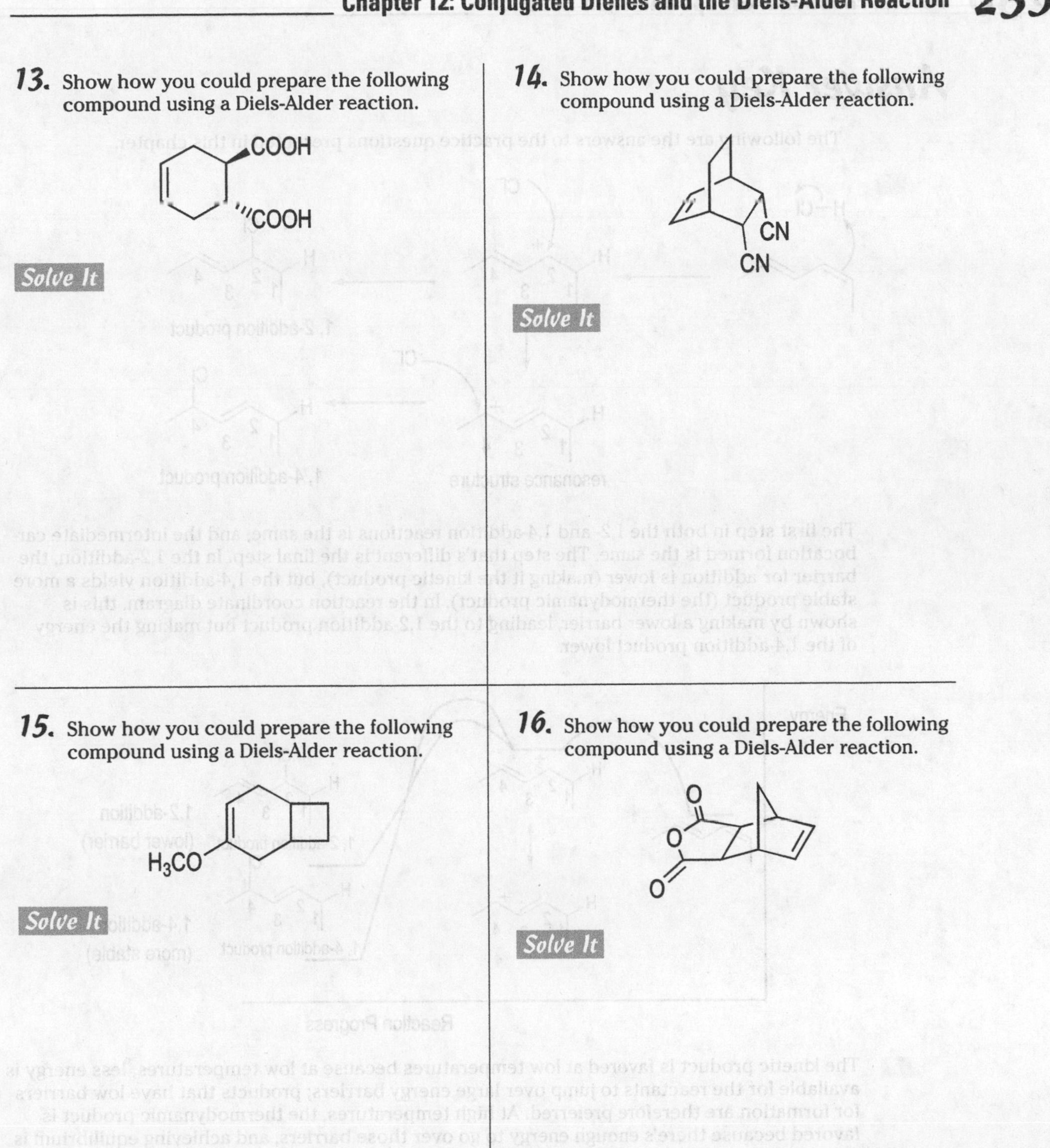

13. Show how you could prepare the following compound using a Diels-Alder reaction.

Solve It

14. Show how you could prepare the following compound using a Diels-Alder reaction.

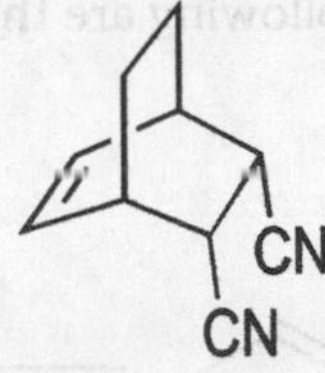

Solve It

15. Show how you could prepare the following compound using a Diels-Alder reaction.

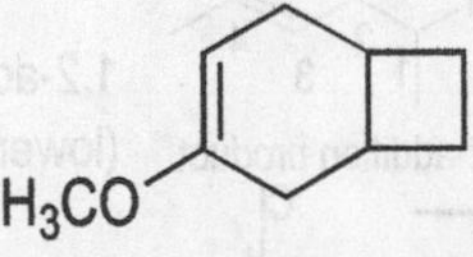

Solve It

16. Show how you could prepare the following compound using a Diels-Alder reaction.

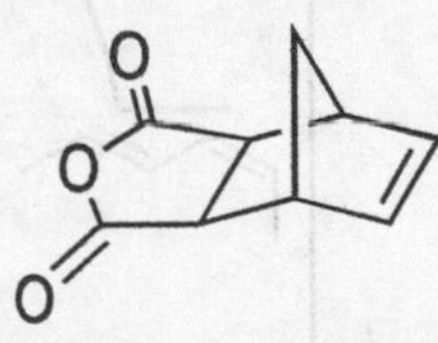

Solve It

Answer Key

The following are the answers to the practice questions presented in this chapter.

1

H–Cl

Cl⁻

1, 2-addition product

Cl⁻

resonance structure

1, 4-addition product

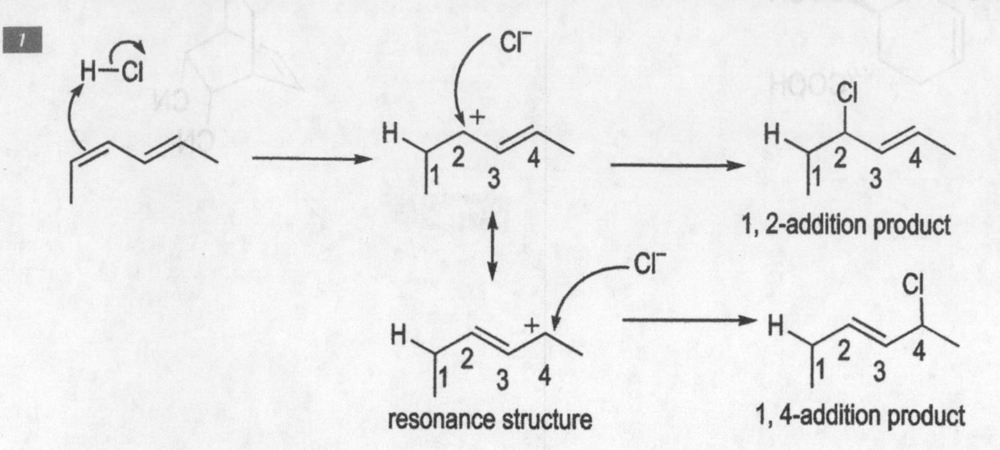

The first step in both the 1,2- and 1,4-addition reactions is the same, and the intermediate carbocation formed is the same. The step that's different is the final step. In the 1,2-addition, the barrier for addition is lower (making it the kinetic product), but the 1,4-addition yields a more stable product (the thermodynamic product). In the reaction coordinate diagram, this is shown by making a lower barrier, leading to the 1,2-addition product but making the energy of the 1,4-addition product lower.

Energy

1, 2-addition product

1,2-addition (lower barrier)

1, 4-addition product

1,4-addition (more stable)

Reaction Progress

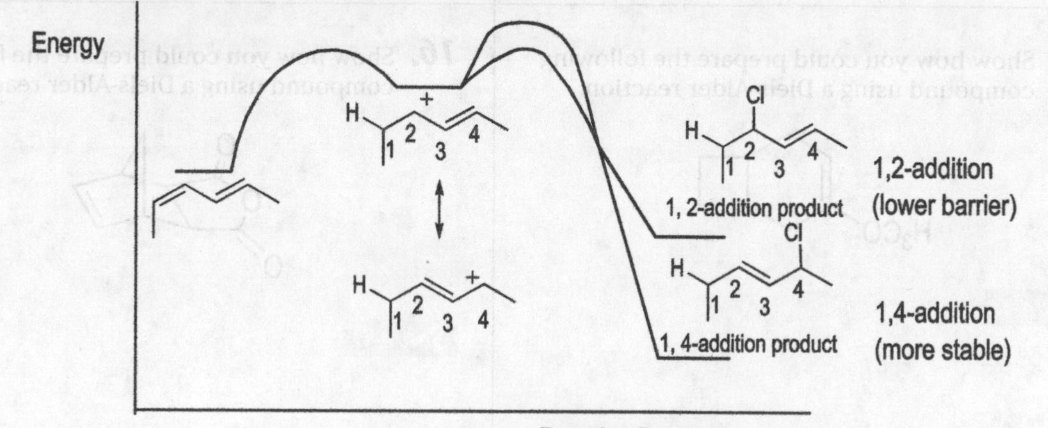

2 The kinetic product is favored at low temperatures because at low temperatures, less energy is available for the reactants to jump over large energy barriers; products that have low barriers for formation are therefore preferred. At high temperatures, the thermodynamic product is favored because there's enough energy to go over those barriers, and achieving equilibrium is easier for the system. After the system reaches equilibrium, the lowest-energy product is favored.

3

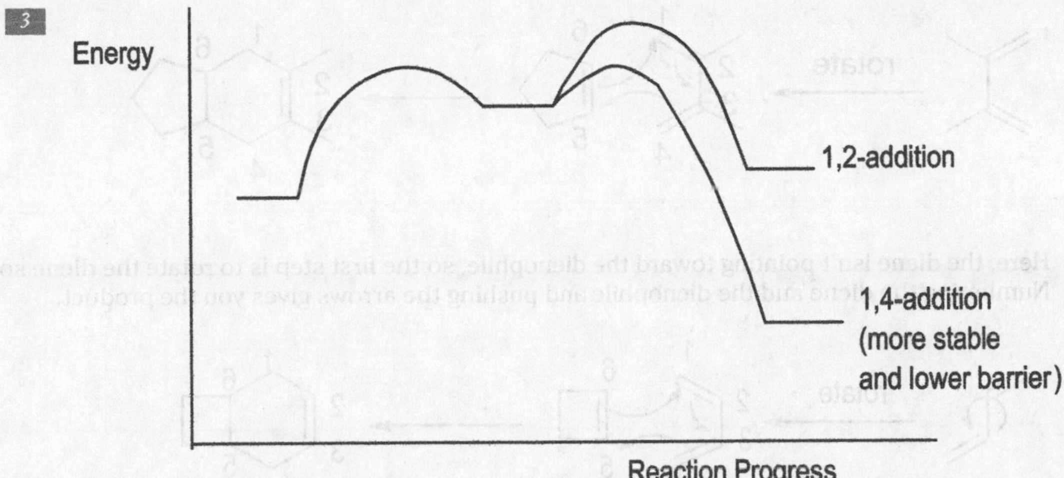

In such a reaction, the barrier leading to 1,4-addition is lower, and it also gives you a product that's lower in energy.

4 The 1,4-addition product is favored at both low temperatures and high temperatures if it's both the thermodynamic and kinetic product.

5

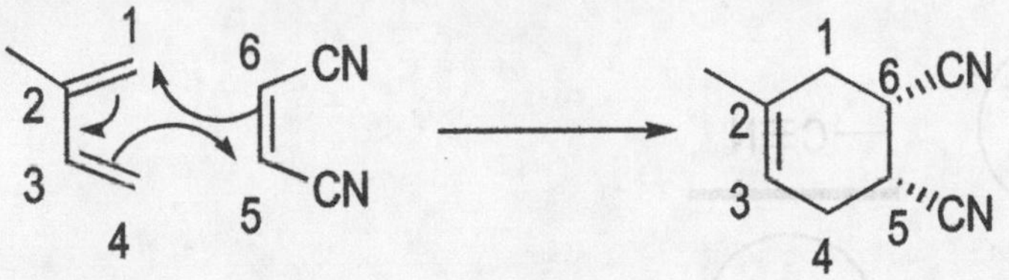

Because the diene and dienophile are lined up correctly, you number them and draw the arrows. New bonds form between carbons 4-5 and 1-6, and the double bond moves between carbons 2-3. Because the cyano groups started *cis* on the dienophile, they end up *cis* in the product (you can draw them as both sticking up or both sticking down).

6

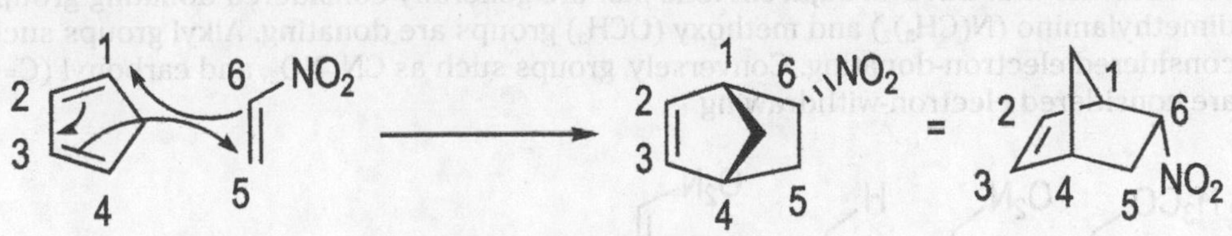

The diene starts out in a ring, so you get a bicyclic product. Still, you can perform the same steps to determine the product — numbering the diene and the dienophile and pushing the arrows. The stereochemistry is also important. When forming a bicyclic product, the endo product is preferred, in which the dienophile substituent sticks down relative to the bridge. Admittedly, getting used to drawing these bicyclic structures correctly takes some practice.

7

rotate

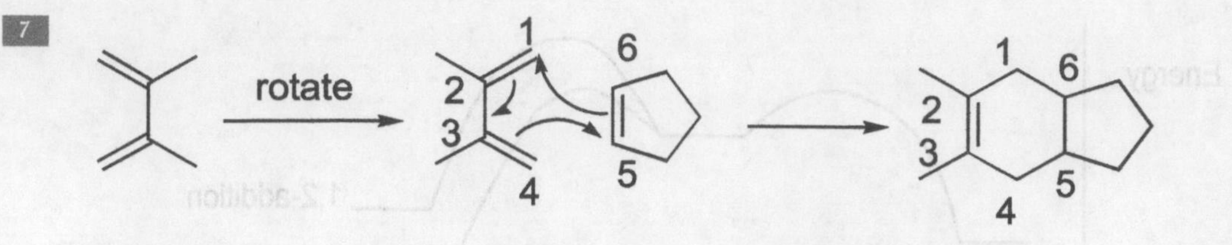

Here, the diene isn't pointing toward the dienophile, so the first step is to rotate the diene so that it is. Numbering the diene and the dienophile and pushing the arrows gives you the product.

8

rotate

s-trans s-cis

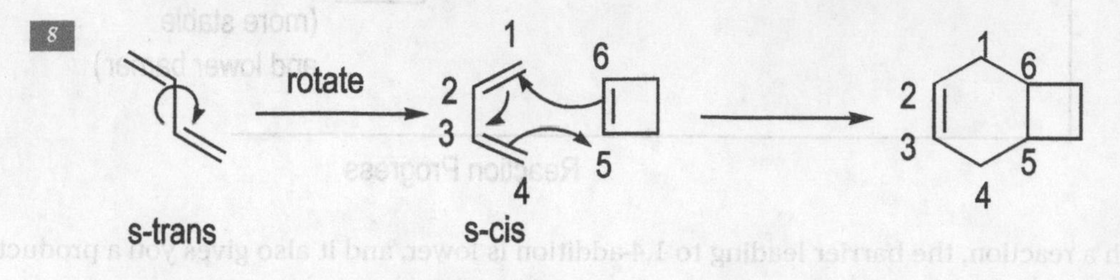

The diene here is drawn in the *s-trans* form (the double bonds stick off opposite sides of the connecting bond. Thus, you need to rotate around the single bond to put the diene in the *s-cis* conformation, from which all Diels-Alder reactions occur. Numbering and arrow-pushing then leads to the product.

9

$—NO_2$ $—CH_3$ —C≡N $—N(CH_3)_2$ $—C(=O)CH_3$ $—OCH_3$

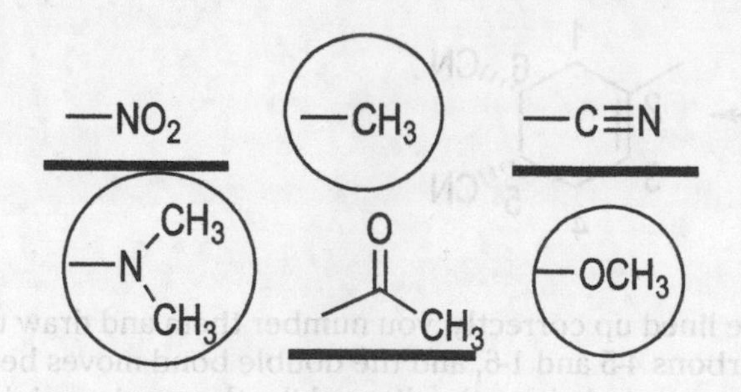

Substituents that have an adjacent lone pair are generally considered donating groups; thus, the dimethylamino ($N(CH_3)_2$) and methoxy (OCH_3) groups are donating. Alkyl groups such as CH_3 are also considered electron-donating. Conversely, groups such as CN, NO_2, and carbonyl (C=O) compounds are considered electron-withdrawing.

10

H_3CO 4 O_2N 2 H 3 O_2N O_2N 1

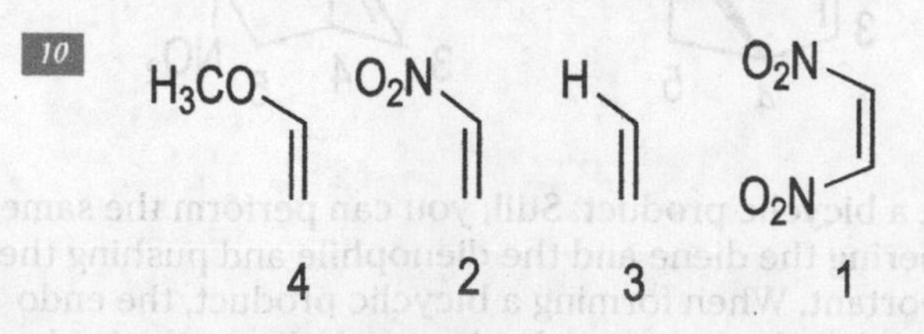

Diels-Alder reactions work best when the dienophile is substituted with electron-withdrawing groups. The nitro group (NO_2) is an electron-withdrawing group, so the dienophile with this group as two of its substituents is the best. The methoxy (OCH_3) group is an electron-donating group, so this dienophile is the least reactive.

11 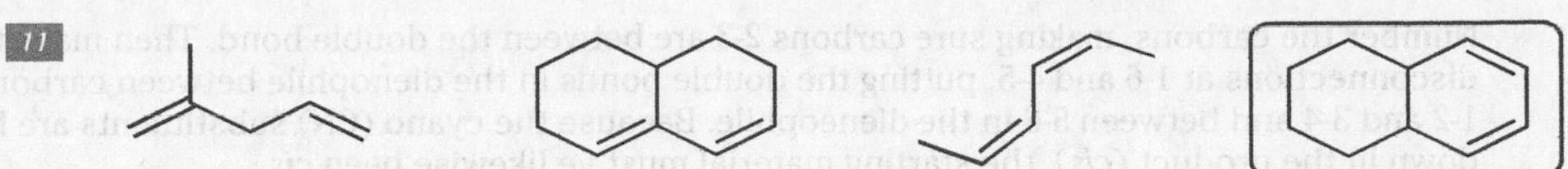

The first diene isn't conjugated (with alternating double bonds), so it can't react as a diene in the Diels-Alder reaction. The second diene in the fused rings is locked in the s-*trans* conformation, so it can't react, either (the Diels-Alder reaction requires the diene to be in the s-*cis* conformation). The third one is in the s-*trans* conformation and can't be rotated to the s-*cis* conformation because the two methyl groups would be sticking in the same space (steric strain). The final structure can react because it's a conjugated diene locked in the preferred s-*cis* conformation.

12

3 4 2 1

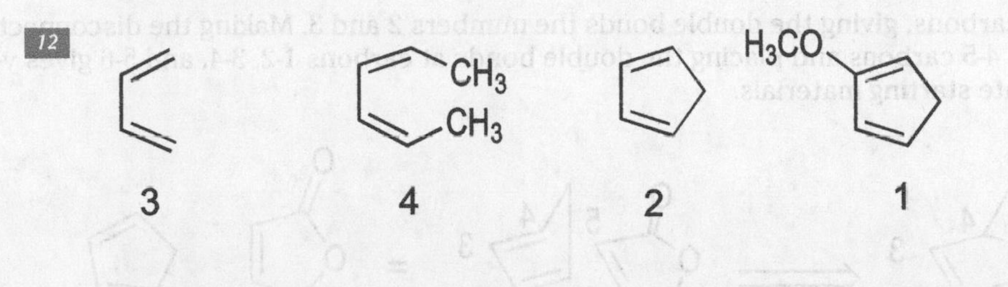

Dienes that are in rings are generally more reactive than their non-cyclic counterparts because these dienes are locked into the s-*cis* conformation. Moreover, the most reactive dienes are substituted with electron-donating groups, such as the methoxy (OCH_3) group. Thus, the most reactive diene is the one locked in the ring that also contains the electron-donating group. Dienes that can't easily rotate into the s-*cis* conformation — such as the second diene, in which two methyl groups suffer from *steric strain* (trying to share the same space) in the s-*cis* conformation — are generally unreactive.

13

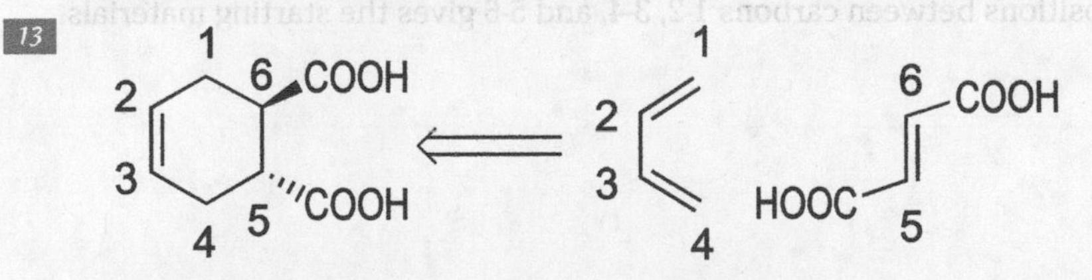

The first step is to number the carbons, making sure numbers 2-3 are the double-bond carbons. Then make the breaks between carbons 1-6 and 4-5. In the starting material, the double bonds start between carbons 1-2, 3-4, and 5-6. Additionally, because the substituents are *trans* in the product, the starting dienophile must've likewise been *trans*.

14

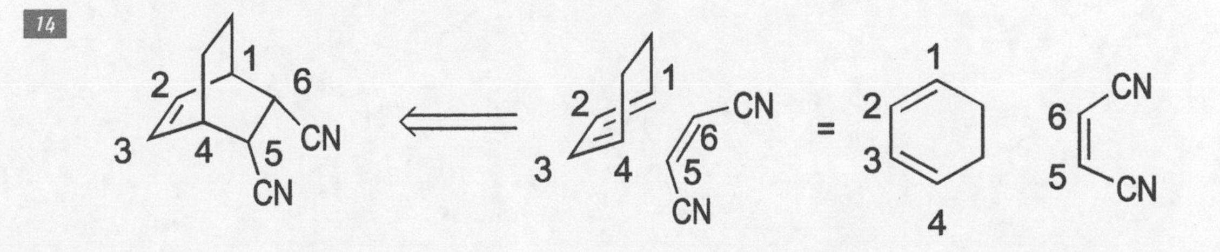

Number the carbons, making sure carbons 2-3 are between the double bond. Then make the disconnections at 1-6 and 4-5, putting the double bonds in the dienophile between carbons 1-2 and 3-4 and between 5-6 in the dieneophile. Because the cyano (CN) substituents are both down in the product *(cis)*, the starting material must've likewise been *cis*.

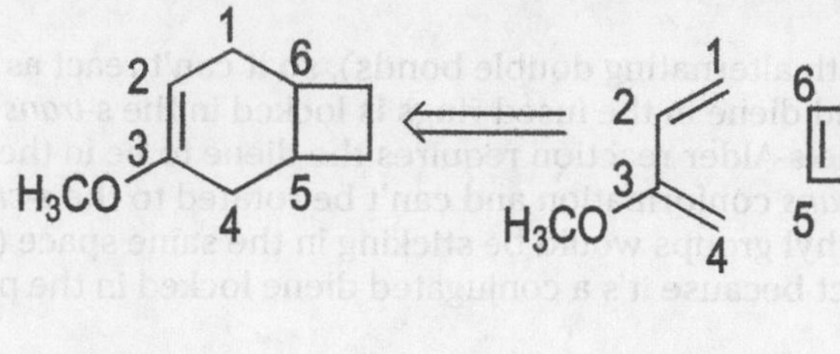

Number the carbons, giving the double bonds the numbers 2 and 3. Making the disconnects at the 1-6 and 4-5 carbons and placing the double bonds at carbons 1-2, 3-4, and 5-6 gives you the appropriate starting materials.

16

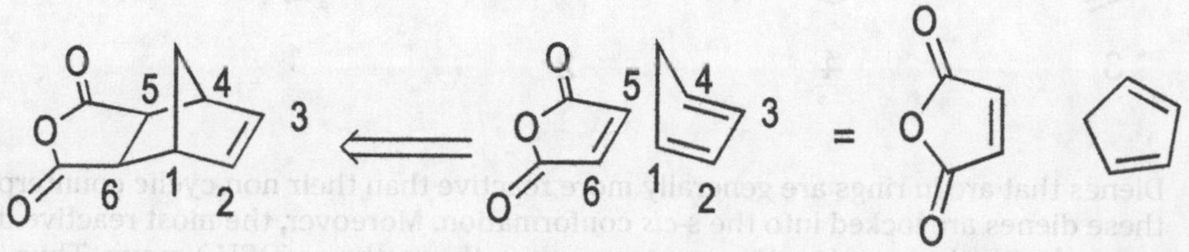

The difficult part of this problem is simply seeing the product molecule drawn in a different orientation. However, you can tackle this problem in normal fashion by giving the numbers 2-3 to the double bond and making the disconnects at carbons 1-6 and 4-5. Placing the double bonds in their original positions between carbons 1-2, 3-4, and 5-6 gives the starting materials.

Chapter 13

The Power of the Ring: Aromatic Compounds

In This Chapter

- Determining whether a ring is aromatic, anti-aromatic, or nonaromatic
- Constructing MO diagrams of aromatics
- Working with the reactions of aromatics
- Tackling multistep syntheses of polysubstituted aromatic compounds

One of the most unusual features of organic compounds is that they can form stable ring systems, a feature that's relatively uncommon with elements other than carbon. Interestingly, experimental observation has shown that some rings with alternating double bonds are exceptionally stable, whereas other rings are exceptionally unstable.

Those that are exceptionally stable are called *aromatic* compounds, the name referring to the first isolated stable ring systems, which had a pungent odor. The aromatic compounds make up the backbone of a lot of pharmaceuticals, perfumes, polymers, and plastics, to name just a few practical applications. (In fact, organic chemistry research only really started in earnest in the mid 19th century after William Perkin fortuitously discovered that he could make a fortune manufacturing the dye *mauve*, a brilliantly colored organic compound that contains an aromatic ring.) Exceptionally unstable ring systems are called, naturally, *anti-aromatic*.

This chapter gives you practice working with ring systems. First, you see how to determine whether ring systems are aromatic, anti-aromatic, or nonaromatic. You then see how making molecular orbital (MO) diagrams of these ring compounds can be a helpful tool in explaining the stability of aromatic ring systems and the instability of anti-aromatic ring systems. Then you get to the meat and potatoes of aromatic chemistry — the reactions. After you see the individual reactions, you string together everything you know about the reactivity of aromatic compounds to perform multistep synthesis of polysubstituted aromatic compounds.

Determining Aromaticity, Anti-aromaticity, or Nonaromaticity of Rings

Being able to predict which ring systems will be stable (and aromatic) and which will be unstable (and anti-aromatic) is an important part of orgo. For a molecule to be stable and *aromatic,* it needs to adhere to these four rules:

- It has to be a ring.
- It has to be *planar* (flat).
- Each atom of the ring needs to have a *p* orbital that's at a right angle to the plane of the ring. In other words, no atom on the ring can be sp^3 hybridized. (See Chapter 1 for more on hybridization.)
- The ring needs a *Hückel number* of pi electrons, following the $4n + 2$ rule (a *pi electron* is an electron occupying an orbital made from the side-by-side overlap of *p* orbitals). That is, if you set $4n + 2$ equal to the number of pi electrons in the ring, *n* must be an integer. If *n* isn't, the ring system has a non-Hückel number of electrons.

  If the ring has a heteroatom (such as O, S, or N), you have to determine which lone pairs are part of the pi system. First draw in any of the lone pairs on the heteroatoms, and then follow these guidelines:

  - If the heteroatom is part of a double bond, it can't contribute its lone pairs to the pi system.
  - If the heteroatom is connected to the ring with only single bonds, one of the lone pairs (but only one) goes into the pi system.

  Thus, no heteroatom can contribute more than two electrons to the pi system.

If the ring meets the first three criteria but has a non-Hückel number of pi electrons following the equation $4n$, then the ring is considered *anti-aromatic*. However, if the molecule doesn't meet the first three criteria, the ring is simply *nonaromatic* — it's neither exceptionally stable nor unstable.

Q. Determine whether the following ring system is aromatic, anti-aromatic, or nonaromatic.

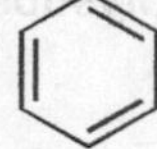

A. **Aromatic.** The molecule is a ring, is flat (you usually assume that any ring system with no sp^3 atoms is flat), has all sp^2-hybridized carbons, and has six pi electrons (two from each double bond). Six pi electrons is a Huckel number of electrons following $4n + 2$ (where $n = 1$), so this molecule meets all four criteria and is therefore aromatic.

Q. Determine whether the following ring system is aromatic, anti-aromatic, or nonaromatic.

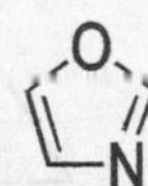

A. **Aromatic.** Rings with heteroatoms are somewhat trickier, but you can still determine whether they're aromatic in the same fashion. The first thing I recommend you do is draw in any of the lone-pairs on the heteroatoms, nitrogen and oxygen.

Now go through the four criteria: Must be a ring? Check! Is the ring planar? You may think the oxygen is sp^3 hybridized because this atom is attached to four different substituents (lone pairs count as substituents). However, the oxygen is actually sp^2 hybridized. The rule for counting substituents to determine atom hybridization (Chapter 1) breaks down when an atom with a lone pair is adjacent to a double or triple bond. In these cases, the atom rehybridizes from sp^3 to sp^2 to place one of the lone pairs in a *p* orbital. Doing so places the lone pair in conjugation (resonance) with the adjacent double or triple bond, which is a stabilizing interaction. Thus, all the atoms in this ring are actually sp^2 hybridized, so you can assume that the ring is planar (flat) and that each atom of the ring has a *p* orbital that's orthogonal to the plane of the ring. Check off the next two criteria.

Does the molecule have a Hückel number of pi electrons? This is the tricky part because you have to determine which lone pairs are in the pi system. In this case, the nitrogen is part of a double bond, so the lone pair on the nitrogen isn't a part of the pi system. The oxygen is connected with single bonds, so one of its lone pairs is in the pi system. Thus, the total number of pi electrons in the molecule is six — four pi electrons come from the double bonds (two electrons from each) and two come from the lone pair on oxygen. Six pi electrons is a Hückel number following the $4n + 2$ rule (where $n = 1$), so this molecule is aromatic.

1. Determine whether the following ring system is aromatic, anti-aromatic, or nonaromatic.

Solve It

2. Determine whether the following ring system is aromatic, anti-aromatic, or nonaromatic.

Solve It

3. **Determine whether the following ring system is aromatic, anti-aromatic, or nonaromatic.**

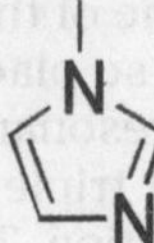

Solve It

4. **Determine whether the following ring system is aromatic, anti-aromatic, or nonaromatic.**

Solve It

5. **Determine whether the following ring system is aromatic, anti-aromatic, or nonaromatic.**

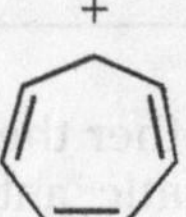

Solve It

6. **Determine whether the following ring system is aromatic, anti-aromatic, or nonaromatic.**

Solve It

7. Determine whether the following ring system is aromatic, anti-aromatic, or nonaromatic.

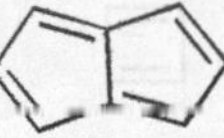

Solve It

8. Explain why [10]-annulene isn't aromatic but naphthalene is aromatic, even though they both have the same number of pi electrons. (***Hint:*** Draw out the hydrogens.)

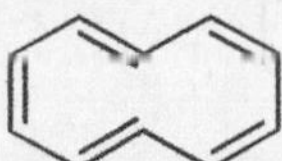

[10]-annulene
not aromatic

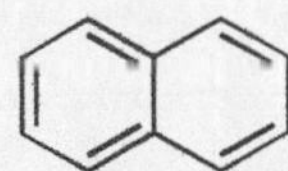

naphthalene
aromatic

Solve It

Figuring Out a Ring System's MO Diagram

Perhaps the easiest way to see why aromatic ring systems are so stable and why anti-aromatic ring systems are so unstable is to look at their molecular orbital (MO) diagrams. An *MO diagram* depicts the molecular orbitals of the pi electrons, their electron occupation, and the relative energies of those orbitals.

To generate the MO diagram of a single ring system, follow these steps:

1. **Place the ring point-down inside a circle.**

   Any place where the points of the ring touch the circle represents an MO at that energy level; therefore, the point of the ring touching the bottom of the circle would be the lowest-energy MO. This shortcut for quickly determining the number and energy levels of the MOs in a ring is called a *Frost circle.* You can use Frost circles for all single-ring systems with alternating double bonds around the ring.

2. **Redraw those diagrams to the right of the circle to make viewing the MO diagram clearer.**

3. **Fill the orbitals by placing the electrons in the lowest-energy MO and building up.**

   You do this just like you build the atomic electron configuration using the Aufbau principle, making sure to follow Hund's rule (see Chapter 1).

Q. Generate the MO diagram of the pi orbitals of benzene.

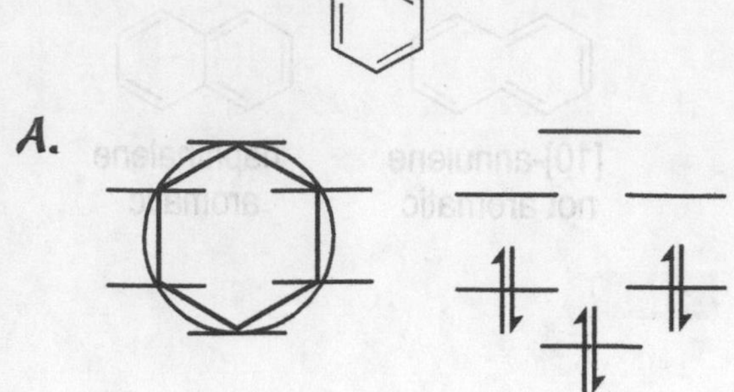

A. Because benzene has six pi electrons (two from each double bond), you fill the MOs with six electrons. A general feature of aromatic compounds such as benzene is that all the low-energy MOs are completely filled with electrons — a very stable electron configuration.

Q. Generate the MO diagram of the pi orbitals of 1,3-cyclobutadiene.

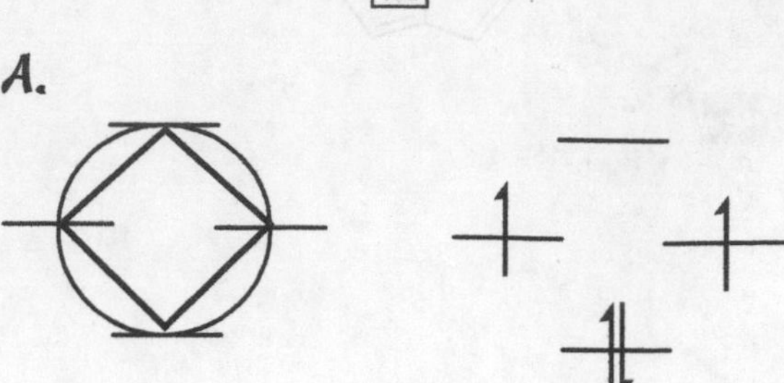

A. Start making the Frost circle by placing the four-membered ring point-down in the circle. Wherever the points of the ring touch the circle, you place an MO. Then fill the MOs with electrons. 1,3-cyclobutadiene has four pi electrons, so you start by filling the lowest-energy MO with two electrons. The next two electrons split up, with one going into each of the degenerate orbitals (following Hund's rule), and these electrons get the same spin. These incompletely filled MOs are characteristic of anti-aromatic rings. Because of these incompletely filled orbitals, anti-aromatic rings are quite unstable.

9. Generate the MO diagram of the pi orbitals of 1,3-cyclopentadienyl cation. Based on the resulting MO diagram, state whether you'd expect the ring to be aromatic or anti-aromatic.

Solve It

10. Generate the MO diagram of the pi orbitals of 1,3-cyclopentadienyl anion. Based on the resulting MO diagram, state whether you'd expect the ring to be aromatic or anti-aromatic.

Solve It

11. Generate the MO diagram of the pi orbitals of cyclopropenyl cation. Based on the resulting MO diagram, state whether you'd expect the ring to be aromatic or anti-aromatic.

Solve It

12. Generate the MO diagram of the pi orbitals of 1,3, 5,7,-cyclooctatetraene. Based on the resulting MO diagram, state whether you'd expect the ring to be aromatic or anti-aromatic.

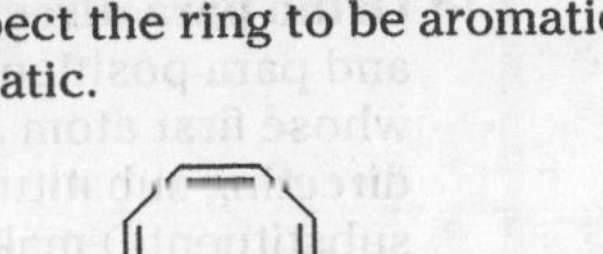

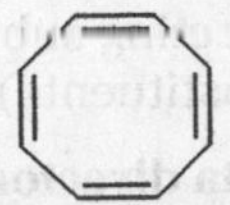

Solve It

Dealing with Directors: Reactions of Aromatic Compounds

Here you practice some of the basic reactions of aromatic compounds. Perhaps the most important reaction type, called *electrophilic aromatic substitution,* replaces a hydrogen on benzene with an electrophile (a Lewis acid, which can accept a lone pair of electrons — see Chapter 4). These reactions become more interesting when the benzene ring is already substituted because then you can make three different possible products by adding another group. See Figure 13-1.

X Y ortho

X Y meta

X Y para

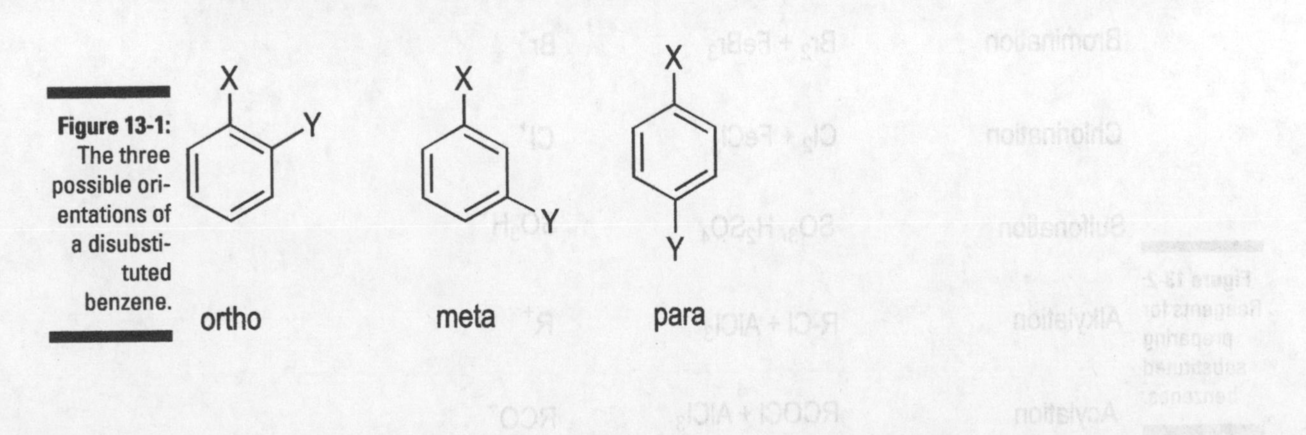

Figure 13-1: The three possible orientations of a disubstituted benzene.

The nature of the substituent attached to the benzene ring determines where any incoming electrophiles add, of which there are two types. Here's a shortcut for remembering which is which:

- **Ortho-para directors:** These substituents direct incoming electrophiles to the ortho and para positions of the ring, so you get a mixture of two products. Any substituent whose first atom attached to the ring has a lone pair of electrons is an ortho-para directing substituent. These substituents (in addition to alkyl groups and aromatic ring substituents) make up the vast majority of ortho-para directors.
- **Meta directors:** These substituents direct incoming electrophiles to the meta position. If the substituent doesn't have a lone pair on the first atom and isn't an alkyl group or an aryl group, you may strongly suspect that the substituent is a meta director. Generally, meta directors are considered to be electron-withdrawing substituents (such as CN, NO_2, and COR).

Figure 13-2 shows the generic electrophilic aromatic substitution reaction. In this reaction, a double bond from the benzene ring attacks the powerful electrophile (abbreviated E^+) to give you an intermediate carbocation. A base then pulls off the adjacent proton to produce the substituted benzene. Because powerful electrophiles (E^+) are unstable and can't be stored in a jar, they have to be made in the reaction mixture on the fly *(in situ)* by mixing together different reagents. For example, to brominate a ring, you mix bromine (Br_2) and iron tribromide ($FeBr_3$) in benzene. These reagents generate the unstable Br^+ electrophile in solution, which then adds to the ring to give the substituted product. Memorize these reagents for preparing the different electrophiles.

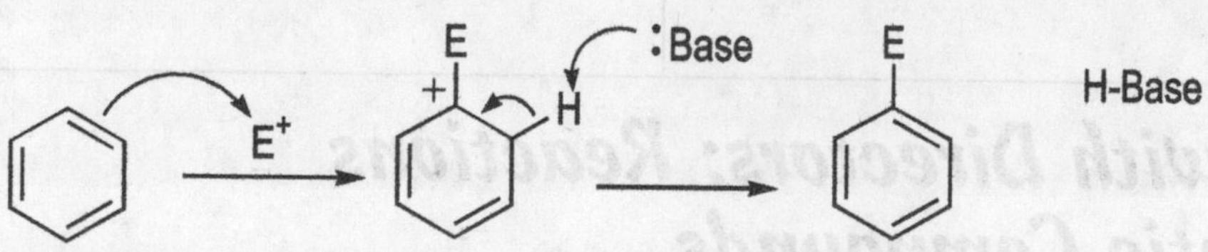

Reaction	Reagents	Electrophile (E^+)
Nitration	HNO_3, H_2SO_4	$O=\overset{+}{N}=O$
Bromination	$Br_2 + FeBr_3$	Br^+
Chlorination	$Cl_2 + FeCl_3$	Cl^+
Sulfonation	SO_3, H_2SO_4	SO_3H^+
Alkylation	$R\text{-}Cl + AlCl_3$	R^+
Acylation	$RCOCl + AlCl_3$	RCO^+

Figure 13-2: Reagents for preparing substituted benzenes.

Figure 13-3 shows a few other general aromatic reactions that you should know for interchanging functional groups attached to aromatic rings. Keep in mind that these reactions apply only to molecules containing a benzene ring.

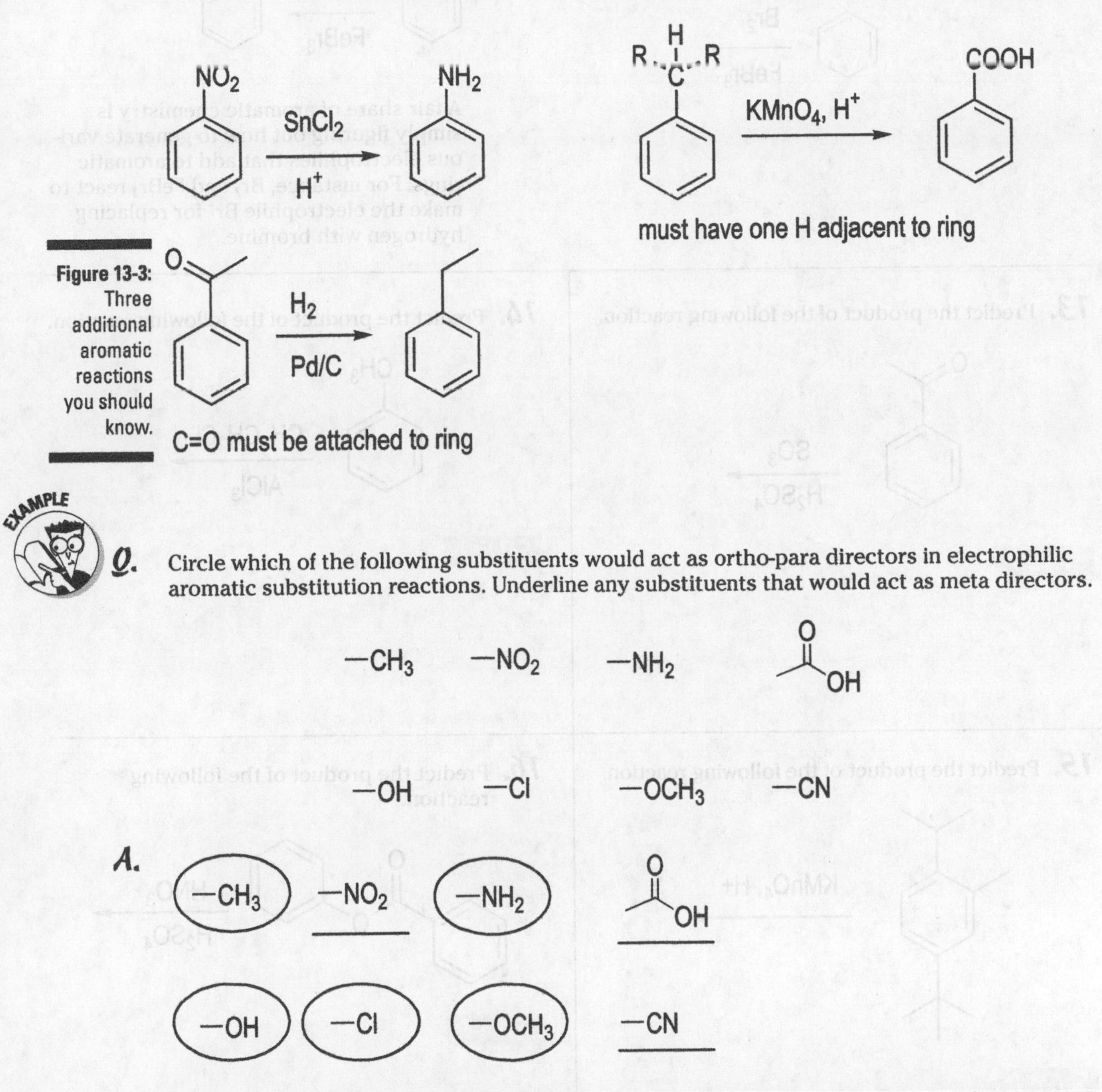

Figure 13-3: Three additional aromatic reactions you should know.

Q. Circle which of the following substituents would act as ortho-para directors in electrophilic aromatic substitution reactions. Underline any substituents that would act as meta directors.

$-CH_3$ $-NO_2$ $-NH_2$ $-COOH$

$-OH$ $-Cl$ $-OCH_3$ $-CN$

A. Circled (ortho-para): $-CH_3$, $-NH_2$, $-OH$, $-Cl$, $-OCH_3$. Underlined (meta): $-NO_2$, $-COOH$, $-CN$.

Any substituent with a first-atom lone pair, such as O*R* or NH_2, is an ortho-para director. Alkyl substituents and aryl substituents (other benzene rings) are also ortho-para directors. Electron-withdrawing substituents without first-atom lone pairs, such as CN, NO_2, and CO*R* groups, are meta directing.

Q. Predict the product of the following reaction.

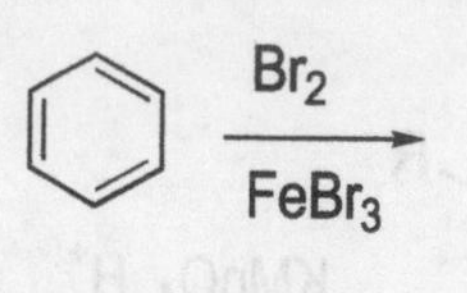

$$\text{benzene} \xrightarrow[\text{FeBr}_3]{\text{Br}_2}$$

A. benzene $\xrightarrow[\text{FeBr}_3]{\text{Br}_2}$ bromobenzene (Br)

A fair share of aromatic chemistry is simply figuring out how to generate various electrophiles that add to aromatic rings. For instance, Br_2 and $FeBr_3$ react to make the electrophile Br^+ for replacing hydrogen with bromine.

13. Predict the product of the following reaction.

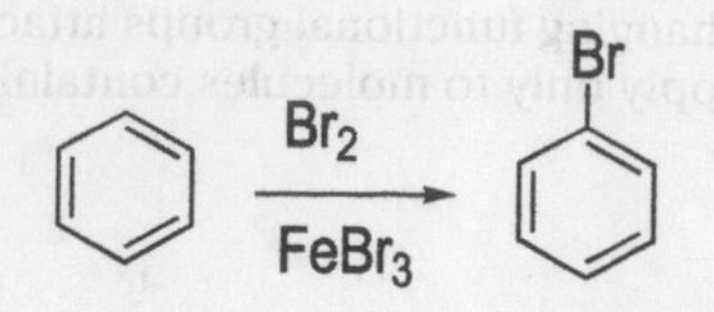

acetophenone $\xrightarrow[\text{H}_2\text{SO}_4]{\text{SO}_3}$

Solve It

14. Predict the product of the following reaction.

toluene (CH_3) $\xrightarrow[\text{AlCl}_3]{\text{CH}_3\text{CH}_2\text{Cl}}$

Solve It

15. Predict the product of the following reaction.

$\xrightarrow{\text{KMnO}_4\text{, H+}}$

Solve It

16. Predict the product of the following reaction.

phenyl benzoate (O, O) $\xrightarrow[\text{H}_2\text{SO}_4]{\text{HNO}_3}$

Solve It

17. Give the product of the following reactions and draw the mechanism of the reaction.

Solve It

18. Give the product of the following reactions and draw the mechanism of the reaction.

Solve It

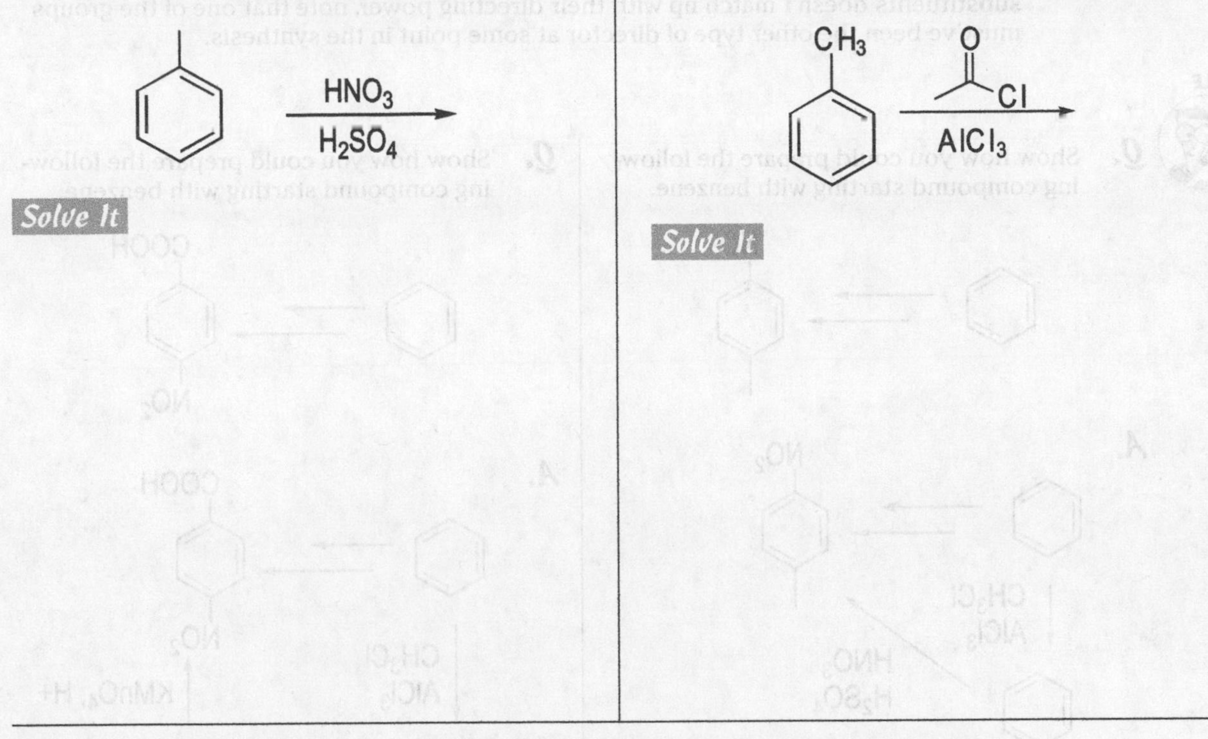

Order! Tackling Multistep Synthesis of Polysubstituted Aromatic Compounds

Making complex aromatic compounds using a multistep synthesis is actually fun. Of crucial importance is the order in which you add the substituents because these groups need to be added in such a way that the first group added directs subsequent additions to the correct location on the ring.

To solve these types of problems, keep these steps in mind:

1. **Look at the product and determine whether each of the substituents is ortho-para directing or meta directing.**

   See the preceding section for details.

2. **Note the location of the substituents in the product.**

   Determine whether they're *ortho* (on adjacent carbons), *meta* (separated by one carbon), or *para* (separated by two carbons).

3. **Add the substituents in logical order.**

   The first substituent should direct placement of the second substituent, and the first and second together should direct placement of the third (if needed). Thus, if the product is a para-disubstituted ring and one of the substituents is ortho-para directing and the other is meta directing, you'd add the ortho-para directing substituent *first* in order to direct the second substituent to the para position. ***Tip:*** If the final location of the substituents doesn't match up with their directing power, note that one of the groups must've been the other type of director at some point in the synthesis.

Q. Show how you could prepare the following compound starting with benzene.

Benzene ⟶ (two steps) ⟶ p-nitrotoluene (NO_2 para to CH_3)

A.

Benzene —(CH_3Cl, $AlCl_3$)→ toluene —(HNO_3, H_2SO_4)→ p-nitrotoluene (NO_2)

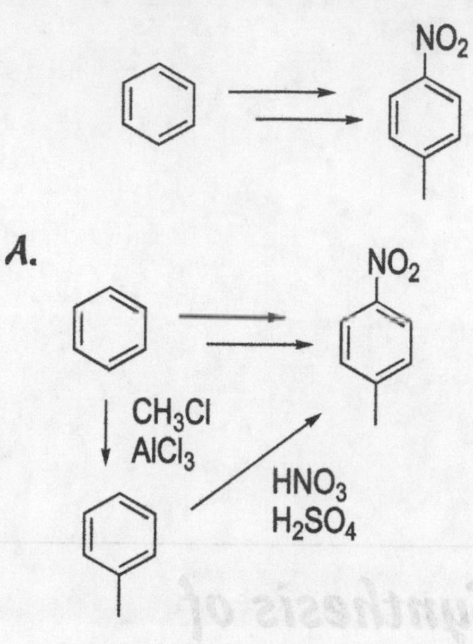

The product has both an ortho-para directing substituent (the methyl group) and a meta-directing substituent (the nitro group). Because the product is para-substituted, you thus need to add the ortho-para director first — the methyl group — in order to direct the nitro group to the para position. If you were to add the nitro group first, the next addition would be to the meta position, which you don't want.

Q. Show how you could prepare the following compound starting with benzene.

Benzene ⟶ (two steps) ⟶ p-nitrobenzoic acid (COOH, NO_2)

A.

Benzene —(CH_3Cl, $AlCl_3$)→ toluene (CH_3) —(HNO_3, H_2SO_4)→ p-nitrotoluene (CH_3, NO_2) —($KMnO_4$, H+)→ p-nitrobenzoic acid (COOH, NO_2)

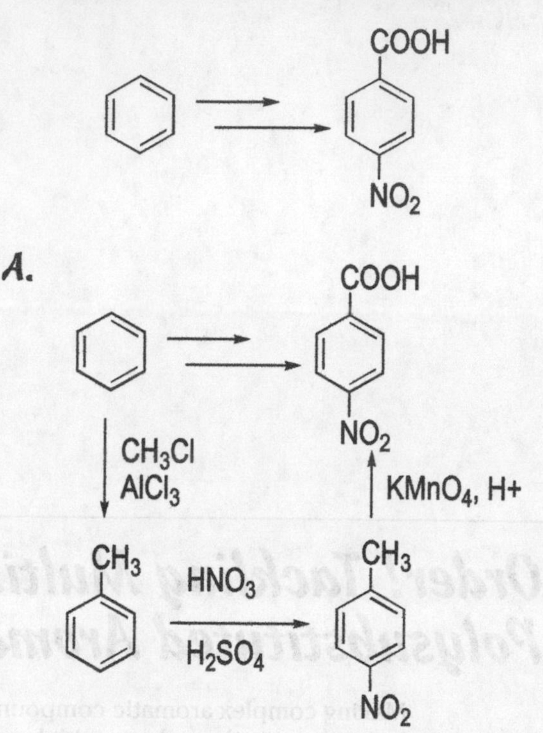

Looking at the product shows that both substituents are meta directors, but the substituents are para-substituted. Although this seems like a real boondoggle, it just means that at some point in the synthesis, one of the groups had to have been an ortho-para director to get the second substituent in the right location. Carboxylic acids on benzene rings are made by oxidation of alkyl groups attached to the aromatic ring. Because alkyl groups are ortho-para directors, the correct sequence is to first alkylate the benzene ring and then use the ortho-para directing ability of the alkyl group to put nitrate in the para position. Finally, oxidize the alkyl group to the carboxylic acid using potassium permanganate ($KMnO_4$).

19. Show how you could prepare the following compound starting with benzene.

Br

NO_2

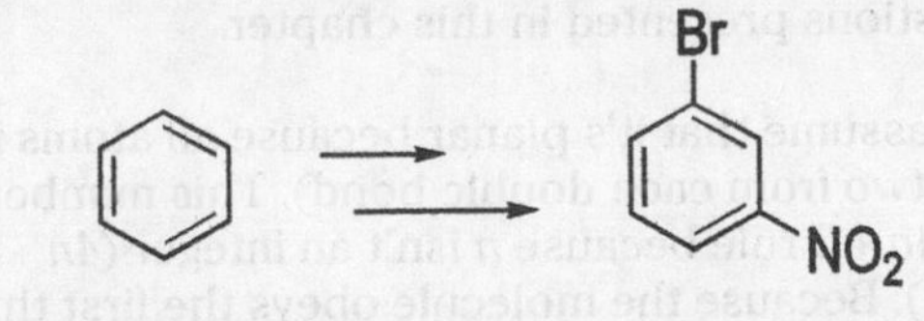

Solve It

20. Show how you could prepare the following compound starting with benzene.

COOH

Br

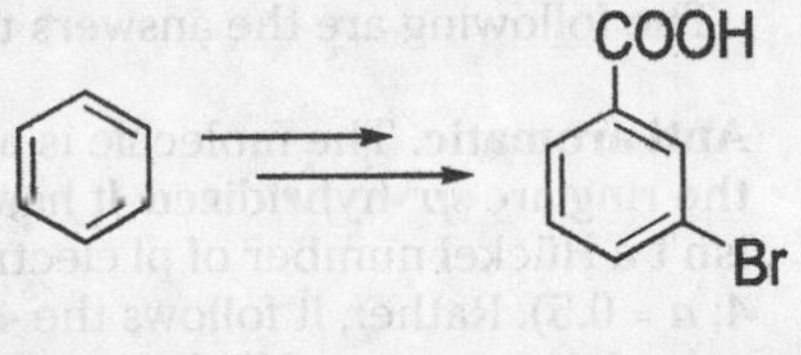

Solve It

21. Show how you could prepare the following compound starting with benzene.

Cl

Br

NO_2

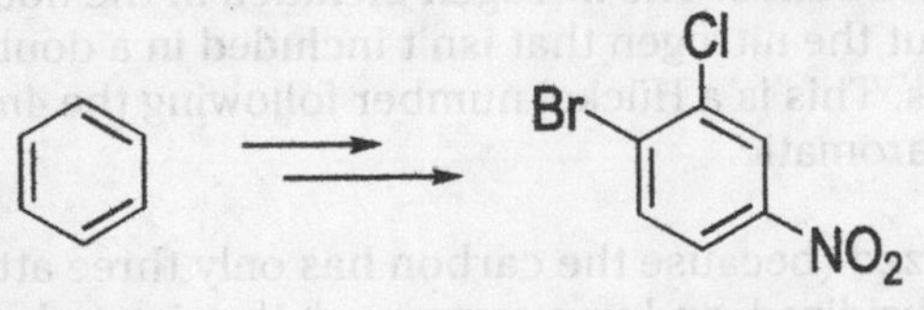

Solve It

22. Show how you could prepare the following compound starting with benzene.

Br

SO_3H

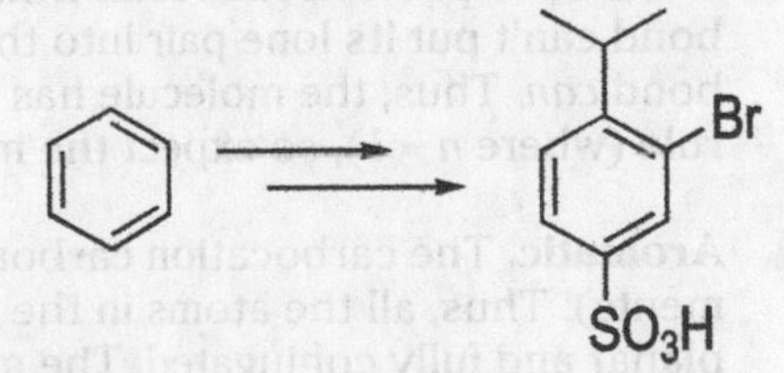

Solve It

Answer Key

The following are the answers to the practice questions presented in this chapter.

1 **Anti-aromatic.** The molecule is a ring, and you can assume that it's planar because all atoms in the ring are sp^2-hybridized. It has four pi electrons (two from each double bond). This number isn't a Hückel number of pi electrons following the $4n + 2$ rule because n isn't an integer ($4n + 2 = 4$; $n = 0.5$). Rather, it follows the $4n$ rule (where $n = 1$). Because the molecule obeys the first three rules but has a non-Hückel number of electrons, you predict that it's anti-aromatic.

2 **Anti-aromatic.** This molecule is cyclic and flat because all atoms are sp^2-hybridized (the carbocation carbon is sp^2-hybridized because the carbon is attached to three other atoms). Because every atom is sp^2-hybridized, you can assume that the ring is planar. The ring has four pi electrons, or two pi electrons per double bond. However, four isn't a Hückel number of pi electrons ($4n + 2 = 4$; $n = 0.5$). Instead, it follows the non-Hückel ($4n$) number of pi electrons, making this molecule anti-aromatic.

3 **Aromatic.** Anytime you have heteroatom in a ring (such as nitrogen or oxygen), draw out all the lone pairs so you can more easily figure out how many pi electrons are there.

N
N

All the atoms are sp^2 hybridized. The nitrogen with the methyl substituent rehybridizes to sp^2 because this atom has a lone pair adjacent to a double bond. Because all the atoms in the ring are sp^2-hybridized, you expect the molecule to be planar. Next, count the number of pi electrons. Four pi electrons result from the two double bonds. The nitrogen included in the double bond can't put its lone pair into the pi system, but the nitrogen that isn't included in a double bond *can.* Thus, the molecule has six pi electrons. This is a Hückel number following the $4n + 2$ rule (where $n = 1$), so expect the molecule to be aromatic.

4 **Aromatic.** The carbocation carbon is sp^2 hybridized (because the carbon has only three attachments). Thus, all the atoms in the ring are sp^2 hybridized, and you can expect the ring to be planar and fully conjugated. The molecule has only two pi electrons, but this is a Hückel number of pi electrons ($4n + 2 = 2$, where $n = 0$). Thus, this molecule is likely aromatic.

5 **Aromatic.** All the carbons are sp^2 (the cationic carbon has only three attachments), so you can assume that the ring is flat. The molecule has six pi electrons, a Hückel number (where $n = 1$), so predict the ring to be aromatic.

6 sp^3 carbon

H H

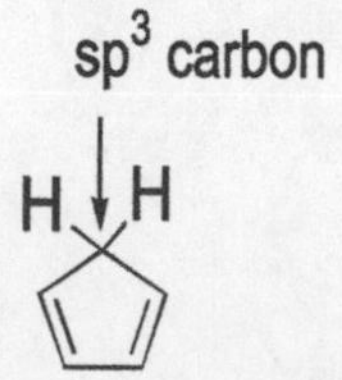

Nonaromatic. This ring has an sp^3-hybridized carbon. Thus, the ring is a nonaromatic system.

7 **Anti-aromatic.** This molecule has every carbon sp^2 hybridized, so you presume it's flat. It has eight pi electrons, a non-Hückel ($4n$) number. Thus, the molecule is likely anti-aromatic. (As an aside, it turns out experimentally that this molecule puckers out of planarity into a tub-shaped ring to avoid the instability of being anti-aromatic, but the prediction, at least, is that this molecule would be anti-aromatic if it were planar.)

8 hydrogens clash, making the molecule become nonplanar

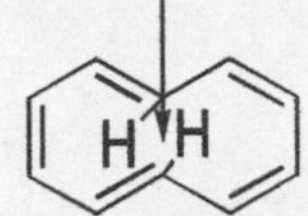

[10]-annulene
nonaromatic

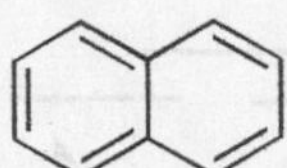

naphthalene
aromatic

The [10]-annulene molecule has two *trans* double bonds in the ring. If you draw out the hydrogens, you see that in a planar geometry, they're trying to occupy the same space. To avoid this steric hindrance, the molecule twists and breaks the planarity. Because planarity is a requirement for aromaticity, the molecule is nonaromatic. On the other hand, naphthalene has a fused ring junction that forces the rings to be flat and is therefore aromatic.

9

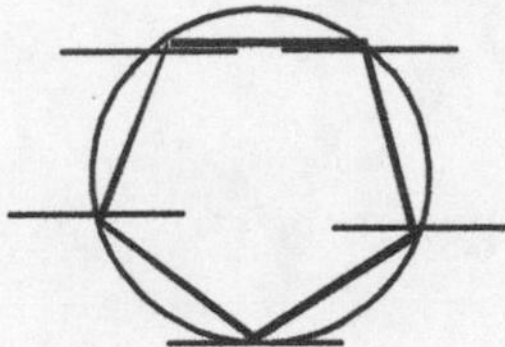

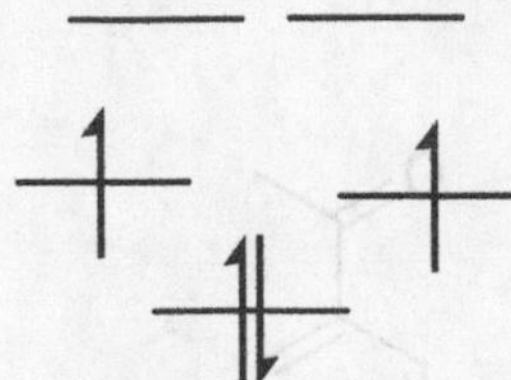

Anti-aromatic. Here, you have four pi electrons. Adding them to the MOs generated from the Frost circle and following Hund's rule gives you two unfilled orbitals. This electron configuration indicates an anti-aromatic ring system.

10

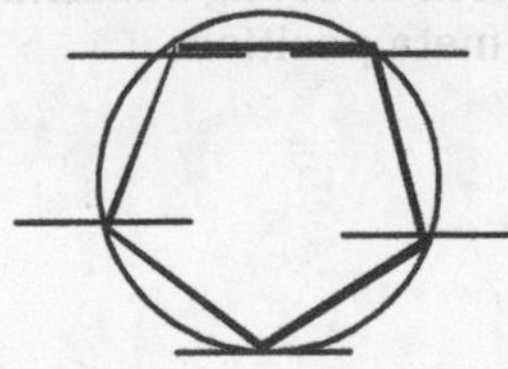

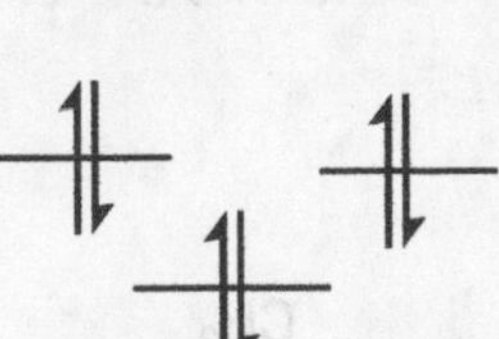

Aromatic. Here, you have six pi electrons (four from the two double bonds and two from the lone pair). Adding these to the MOs shows that all the low-energy orbitals are completely filled with electrons, a configuration that indicates an aromatic ring system.

11

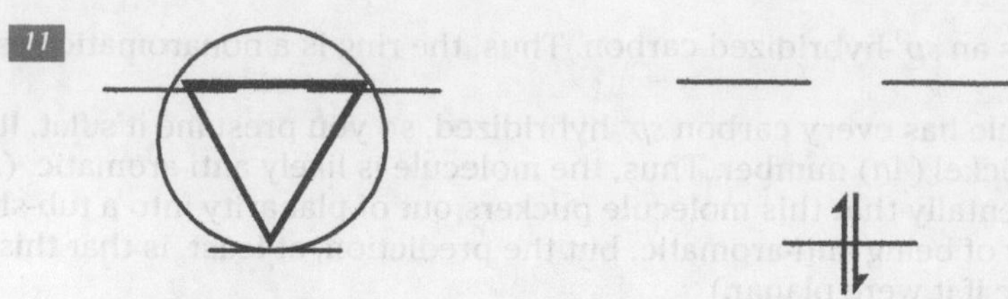

Aromatic. In this three-membered ring system, you have two pi electrons. Making the Frost circle ar filling the lowest-energy orbital with the two pi electrons results in an electron configuration that in cates an aromatic ring system.

12

Anti-aromatic. The hardest part of this problem is drawing the eight-membered ring with the point down in the Frost circle. After you do that, you get the MOs and can fill them with the eight pi electrons, following Hund's rule for placement of the last two electrons into the degenerate orbitals. The last two orbitals remain unfilled, characteristic of an anti-aromatic ring system.

13 meta directing

O

SO_3

H_2SO_4

O

SO_3H

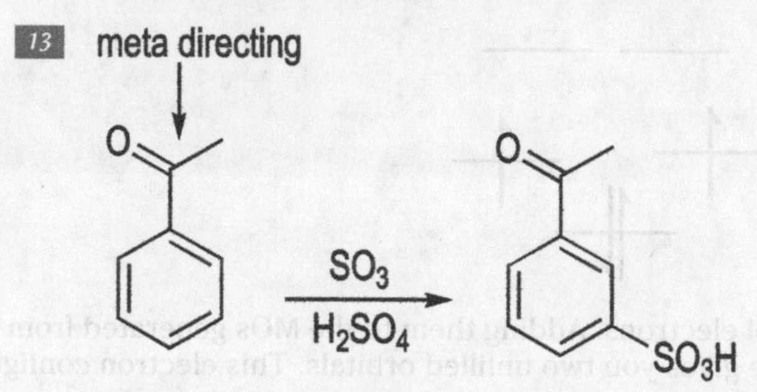

The carbonyl (C=O) group attached to the benzene ring is a meta directing substituent. Therefore, when you add the sulfonic acid (SO_3H) group, it goes into the meta position.

14 ortho-para directing

CH_3

CH_3CH_2Cl

$AlCl_3$

CH_3

CH_2CH_3

+

CH_3

CH_2CH_3

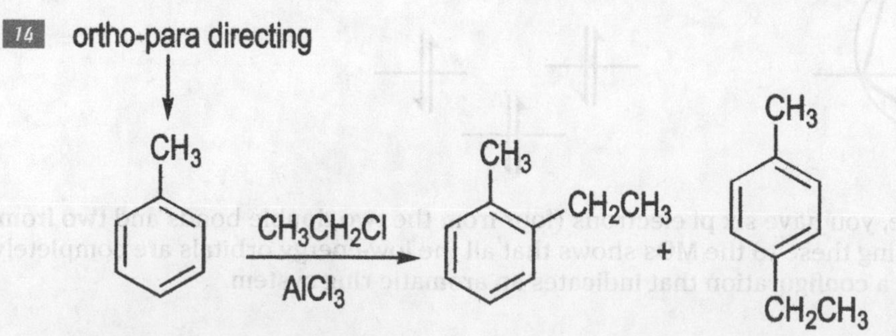

The methyl group is ortho-para directing. Therefore, you get a mix of the ortho- and para-substituted products.

15

KMnO$_4$, H+

HO_2C — CO_2H

quaternary carbons untouched!

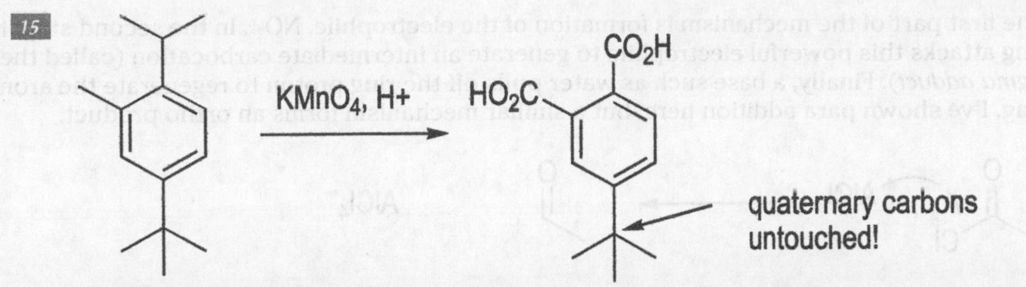

Permanganate oxidations essentially chew all alkyl substituents into carboxylic acid groups. The one exception is that quaternary carbons adjacent to the ring remain unscathed.

16

deactivated ring

HNO_3 / H_2SO_4

activated ring

NO_2 + NO_2

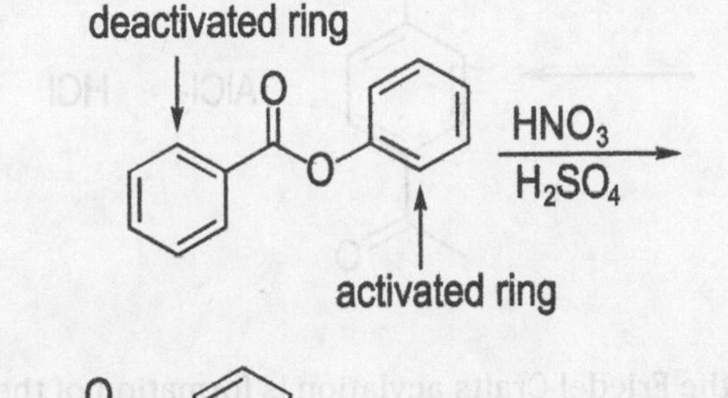

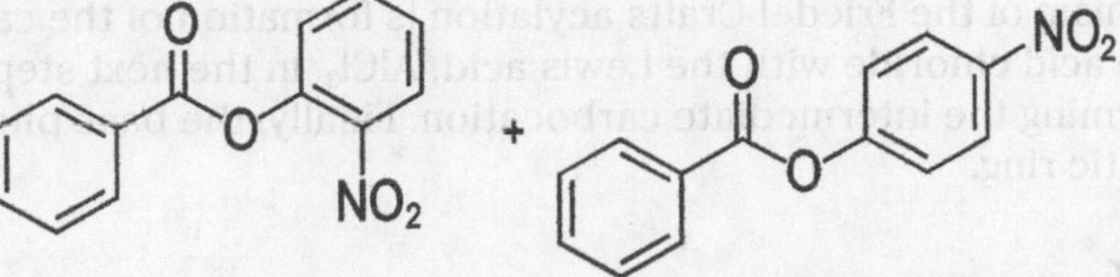

Here, you have two possible benzene rings that can be nitrated. The one adjacent to the carbonyl (C=O) group is deactivated because carbonyls are electron-withdrawing, meta directing, and deactivating. On the other hand, any atom that has a lone pair on the atom attached to the aromatic ring is activating and is an ortho-para director (the one exception is the halides, which are ortho-para directing but deactivating). Thus, the activated ring is the ring that gets nitrated at the ortho and para positions.

17

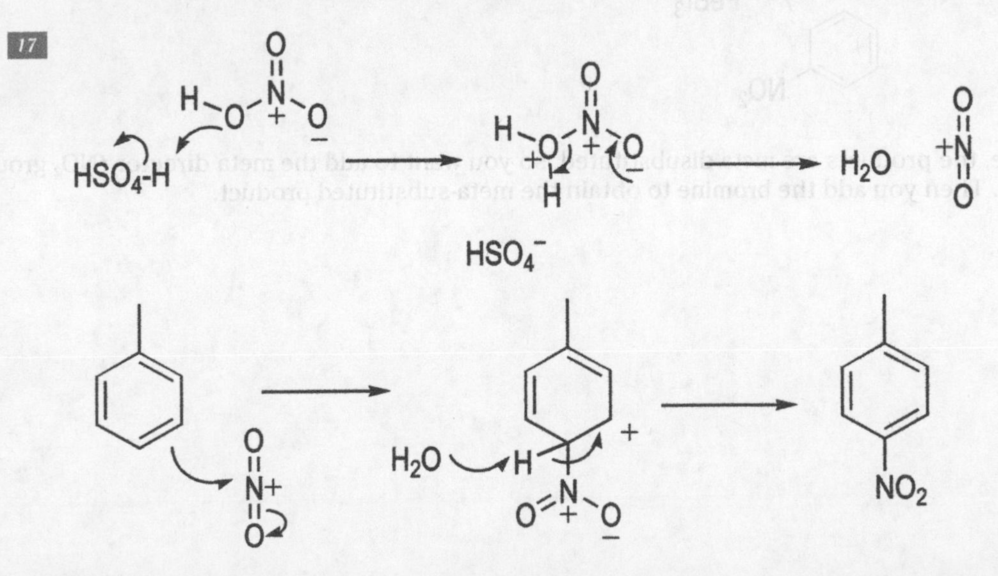

HSO_4-H

HSO_4^-

H_2O

H_2O

NO_2

The first part of the mechanism is formation of the electrophile, NO_2^+. In the second step, the ring attacks this powerful electrophile to generate an intermediate carbocation (called the *sigma adduct*). Finally, a base such as water pulls off the ring proton to regenerate the aromatic ring. I've shown para addition here, but a similar mechanism forms an ortho product.

18

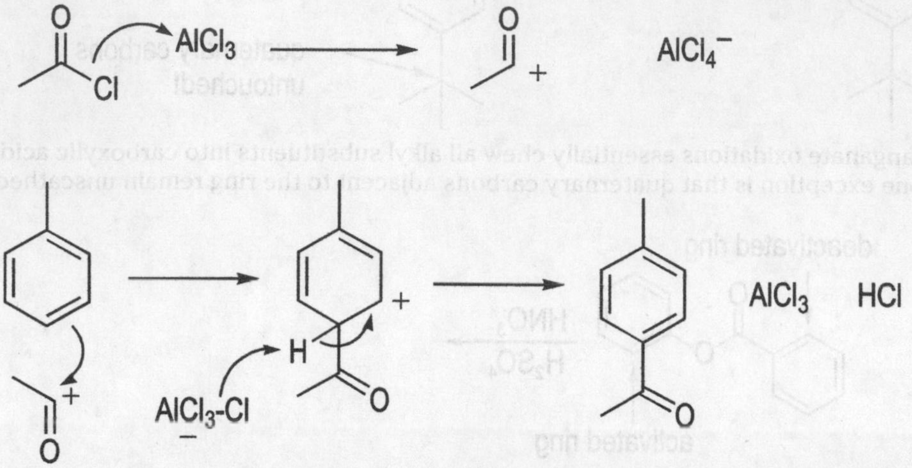

O Cl $AlCl_3$ → O + $AlCl_4^-$

→ + → $AlCl_3$ HCl

H $AlCl_3$-Cl O O O

The first step in the mechanism of the Friedel-Crafts acylation is formation of the carbocation electrophile by reacting the acid chloride with the Lewis acid, $AlCl_3$. In the next step, the ring attacks the electrophile, forming the intermediate carbocation. Finally, the base plucks off the proton to reform the aromatic ring.

19

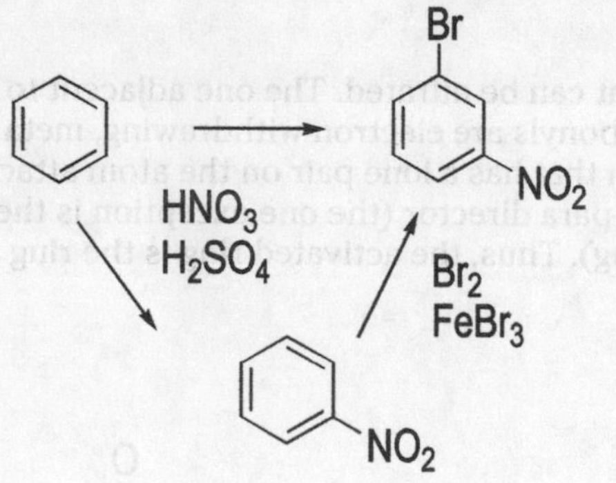

Br NO₂ HNO_3 H_2SO_4 Br_2 $FeBr_3$ NO_2

Here, the products are meta-disubstituted, so you want to add the meta director (NO_2 group) first. Then you add the bromine to obtain the meta-substituted product.

20

COOH

Br

CH_3Cl
$AlCl_3$

Br_2, $FeBr_3$

COOH

$KMnO_4$, H+

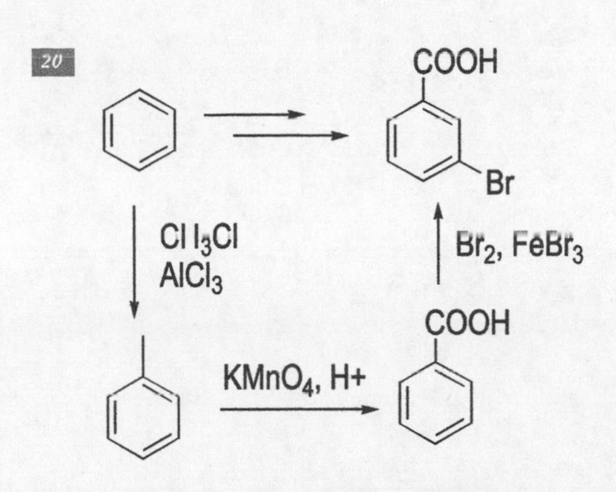

The order of adding substituents is of primary importance. The substituents are located meta to one another, so you want to add the meta director (COOH) first to direct the second substituent (Br) into the meta position. You can accomplish this by methylating the ring using Friedel-Crafts alkylation and then oxidizing the methyl group to the carboxylic acid with potassium permanganate ($KMnO_4$). Finally, add the bromine with the bromine and $FeBr_3$ catalyst.

21

Cl

Br

NO_2

Br_2
$FeBr_3$

Cl_2
$FeCl_3$

HNO_3
H_2SO_4

Br

Br

NO_2

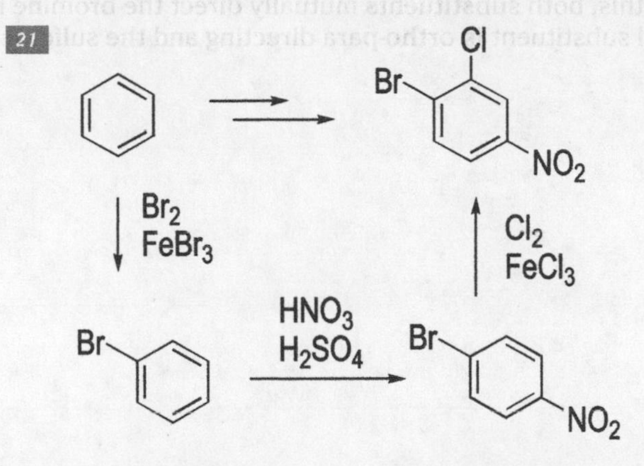

The situation becomes more complex with three substituents because the first substituent you add should direct the second to the correct place, but then both the first and second substituents influence where the third adds. Sometimes figuring out the correct order comes down to trying different possibilities of adding substituents. Here, adding the bromine first and the nitro group second sets up the last addition in the correct place. You get this result because the nitro is meta directing and the bromine is ortho-para directing, both mutually directing where you want the chlorine to go. ***Note:*** Other reasonable alternatives for ordering the addition of the substituents may exist. Keep in mind that whenever you have a reaction that generates both ortho and para products, it's generally assumed that you can separate your desired ortho or para product out from this mixture and continue with the synthesis.

22

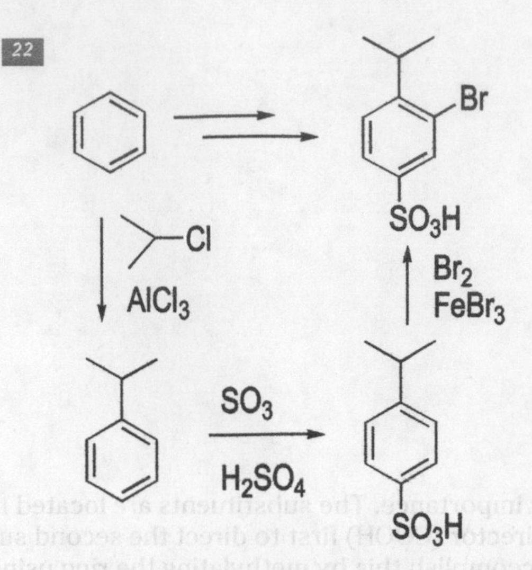

This is a challenging problem because you have to get three substituents in the right spots on the ring. If you add the alkyl group first (an ortho-para director), you can then add the sulfonic acid group in the para position. After this, both substituents mutually direct the bromine into the proper location (because the alkyl substituent is ortho-para directing and the sulfonic acid group is meta directing).

Part IV

Detective Work: Spectroscopy and Spectrometry

"Dr. Duncan's spectroscopy results show the absorption bands to be medium, broad, bold, but fruity with a long robust finish."

In this part . . .

In this part, you become a makeshift Sherlock Holmes and see how to gather clues from the different spectroscopic analyses of organic compounds to deduce the structure of organic molecules. Essentially, spectroscopy measures molecules' absorption of light, and the wavelengths of light that an unknown molecule absorbs give you info about the molecule itself. In spectrometry, the unknown molecule is smashed into wee little bits and then weighed (sort of). The weights of the bits give you clues to the nature of the substance.

Chapter 14

Breaking Up (Isn't Hard to Do): Mass Spectrometry

In This Chapter

- Identifying molecules and their fragments with mass spectrometry
- Using a mass spectrum to determine a structure

Mass spectrometry (or *mass spec,* as it's called by the cool crowd) is a tool chemists use to get clues about the identity of an unknown molecule. A *mass spectrometer* is an instrument that essentially smashes up a molecule into little bits and then weighs each of the bits. The plot of the weights of the fragments versus the abundance of those fragments detected is called the *mass spectrum.* The way in which an unknown molecule fragments in a mass spectrometer gives you clues about its structure.

This chapter covers the basics of working with mass spectrometry, from using mass spec for distinguishing compounds to assigning peaks in the mass spectrum. In addition, you use mass spectrometry to identify simple organic structures.

Identifying Fragments in the Mass Spectrum

One way of distinguishing one molecule from another is by seeing how the molecule falls apart in a mass spectrometer. Mass spectrometry ionizes molecules through collisions with high-energy electrons to form radical cations. These radical cations usually fragment to give you two pieces:

- A cation piece, which is observed in the mass spectrum
- A radical piece, which isn't observed

Figure 4-1 shows you an example.

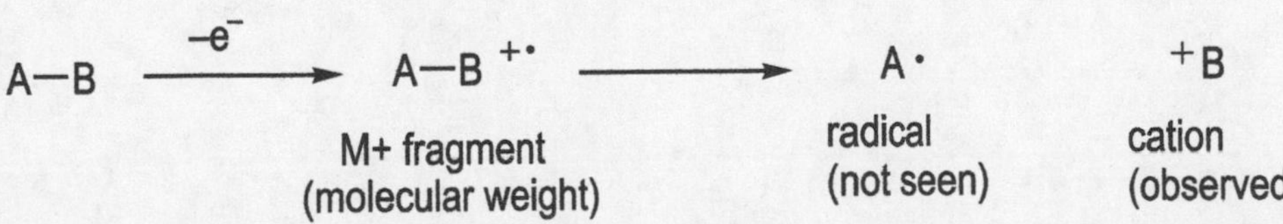

Figure 14-1: How molecules break apart in a mass spectrometer.

The mass spectrum plots the mass-to-charge-ratio *(m/z)* of all the fragments versus their intensity (abundance). However, because the fragments you talk about in an introductory course have charges (z) of +1, most people speak of the m/z value as simply the *mass* (or molecular weight [MW]) of the fragment.

Because the ionizing electrons are very high-energy electrons, the ionized molecules have enough energy to break apart in many different ways, which is why you get so many peaks in the mass spectrum. In general, though, fragmentations tend to occur in places that give the *most stable* resulting carbocation, so the peaks with highest intensity usually correspond to the weights (m/z) of cationic fragments that are the most stable. Thus, for mass spectrometry, being able to predict the relative stability of carbocations is important.

The *M+ peak,* or the *molecular ion,* is the peak that represents the molecular weight of the molecule. To determine the structure of fragments that correspond to a given peak, find the difference between the M+ weight and the weight of the peak under consideration; doing so tells you the weight of the fragment that was lost. Here are the molecular weights of some common fragments you should probably memorize:

Molecular Weight	***Fragment***
15 mass units	Methyl group (CH_3)
29 mass units	Ethyl group (CH_3CH_2)
43 mass units	Propyl group (C_3H_7)

Suppose you have an M+ ion at m/z = 85 and the peak you're interested in is at m/z = 70. Subtract 70 from 85 to get 15 mass units. A methyl (CH_3) group corresponds to 15 mass units, so you'd expect that the structure of the fragment is the original structure minus a methyl group. ***Note:*** Any fragment you draw representing a peak in the mass spectrum must be positively charged because mass specs don't detect neutral fragments.

Aside from fragmentations of the cation radical that lead to a cation and radical fragment, cation radicals can also undergo one rearrangement that you should know: the *McLafferty rearrangement* (see Figure 14-2). This rearrangement occurs with carbonyl (C=O) compounds that have a hydrogen on a *gamma* (γ) hydrogen, which is a hydrogen on the third carbon away from the carbonyl group. In this rearrangement, the carbonyl abstracts the γ hydrogen to yield an enol cation radical fragment that's observed in the mass spec, plus a neutral alkene that isn't.

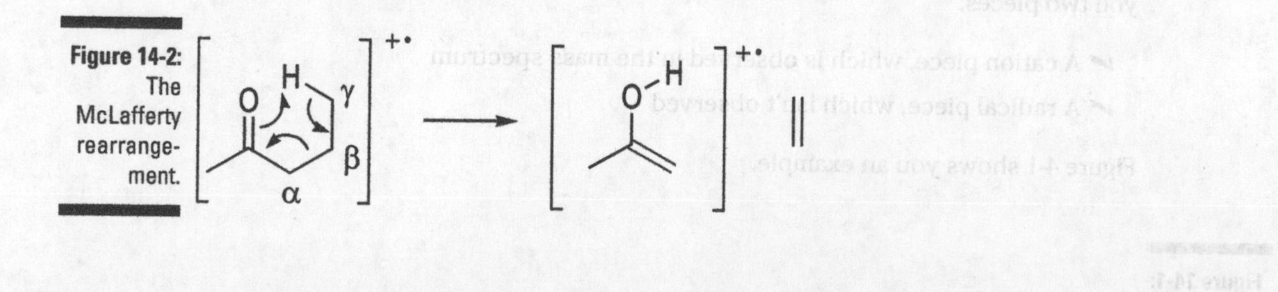

Figure 14-2: The McLafferty rearrangement.

Q. **Look at the mass spectrum of 2-methylhexane (MW = 100 g/mol). What's the m/z value of the base peak? The M+ peak? Additionally, give possible structures of the fragments giving rise to the large peaks at m/z = 85, 57, and 43.**

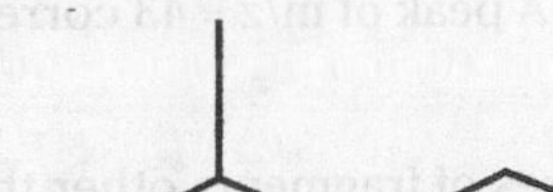

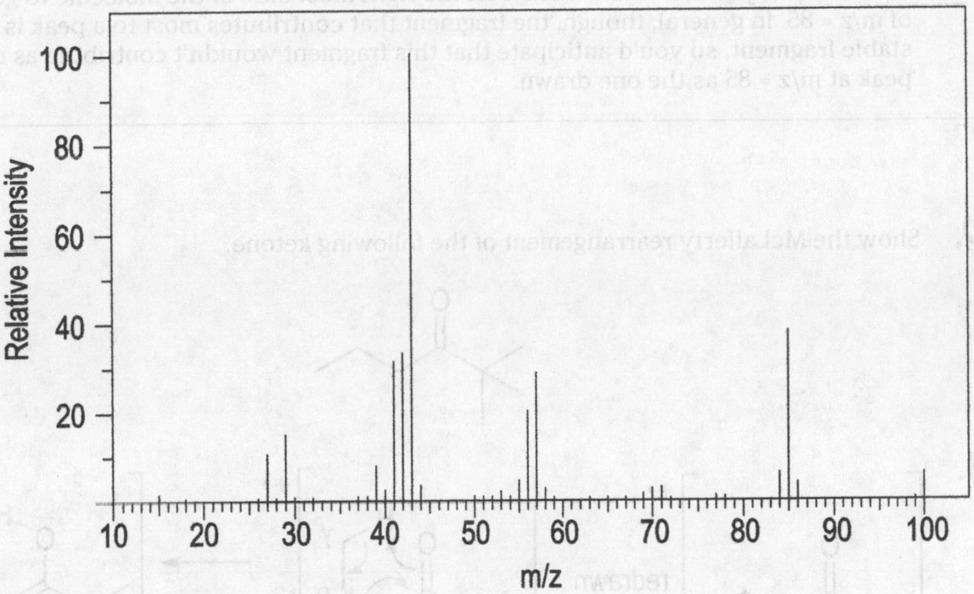

Relative Intensity

m/z

A.

m/z = 85

m/z = 57

m/z = 43

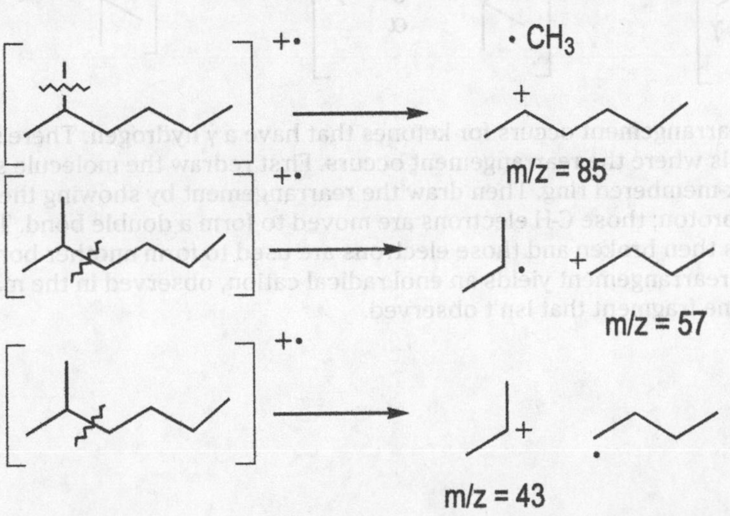

The base peak, the tallest peak in the spectrum, is at 43 m/z. The M+ peak is the 100 m/z peak that represents the molecule's molecular weight, representing a molecule that didn't fragment in the mass spec. A peak representing m/z = 85 corresponds to a fragment that has

lost 15 mass units — the weight of a methyl group (CH_3) — from the original molecule (100 – 15). The most likely place to lose a methyl group is in a location that gives you a secondary carbocation because secondary cations are more stable than primary cations (you can't get a tertiary cation from loss of a methyl group on this molecule). The peak at m/z = 57 indicates a loss of a unit with a mass of 43 (100 – 57), which corresponds to loss of a propyl (three-carbon) group. A peak of m/z = 43 corresponds to a three-carbon (propyl) fragment.

Note: You may provide structures of fragments other than the ones shown with identical weights, which is fine so long as the mass is correct and the fragment is positively charged. For example, you can lose a methyl off the right-most side of the molecule to get a fragment of m/z = 85. In general, though, the fragment that contributes most to a peak is the most stable fragment, so you'd anticipate that this fragment wouldn't contribute as much to the peak at m/z = 85 as the one drawn.

Q. Show the McLafferty rearrangement of the following ketone.

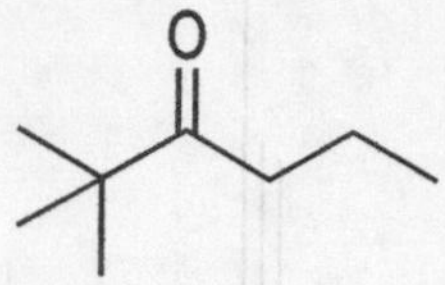

A.

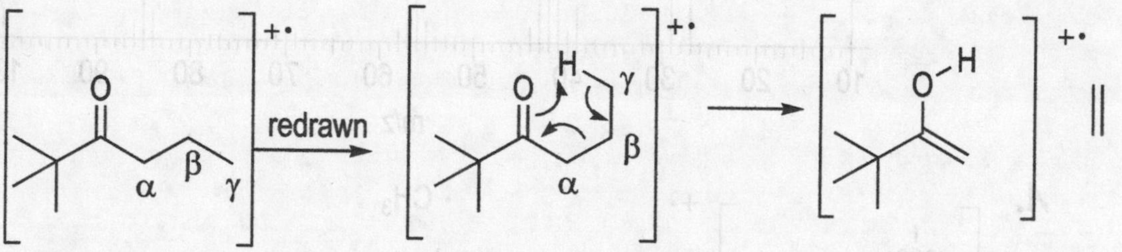

The McLafferty rearrangement occurs for ketones that have a γ hydrogen. There's only one γ hydrogen, so this is where the rearrangement occurs. First redraw the molecule so that it almost forms a six-membered ring. Then draw the rearrangement by showing the oxygen abstracting the γ proton; those C-H electrons are moved to form a double bond. The neighboring C-C bond is then broken and those electrons are used to form another bond with the C=O carbon. This rearrangement yields an enol radical cation, observed in the mass spec, and a neutral alkene fragment that isn't observed.

1. A lab assistant, Daniel Dumkopf, forgot to label the bottles for 2-methylpentane and 3-methylpentane. He took a mass spec on a sample from each bottle. Help Daniel decide the chemical identities in the two bottles.

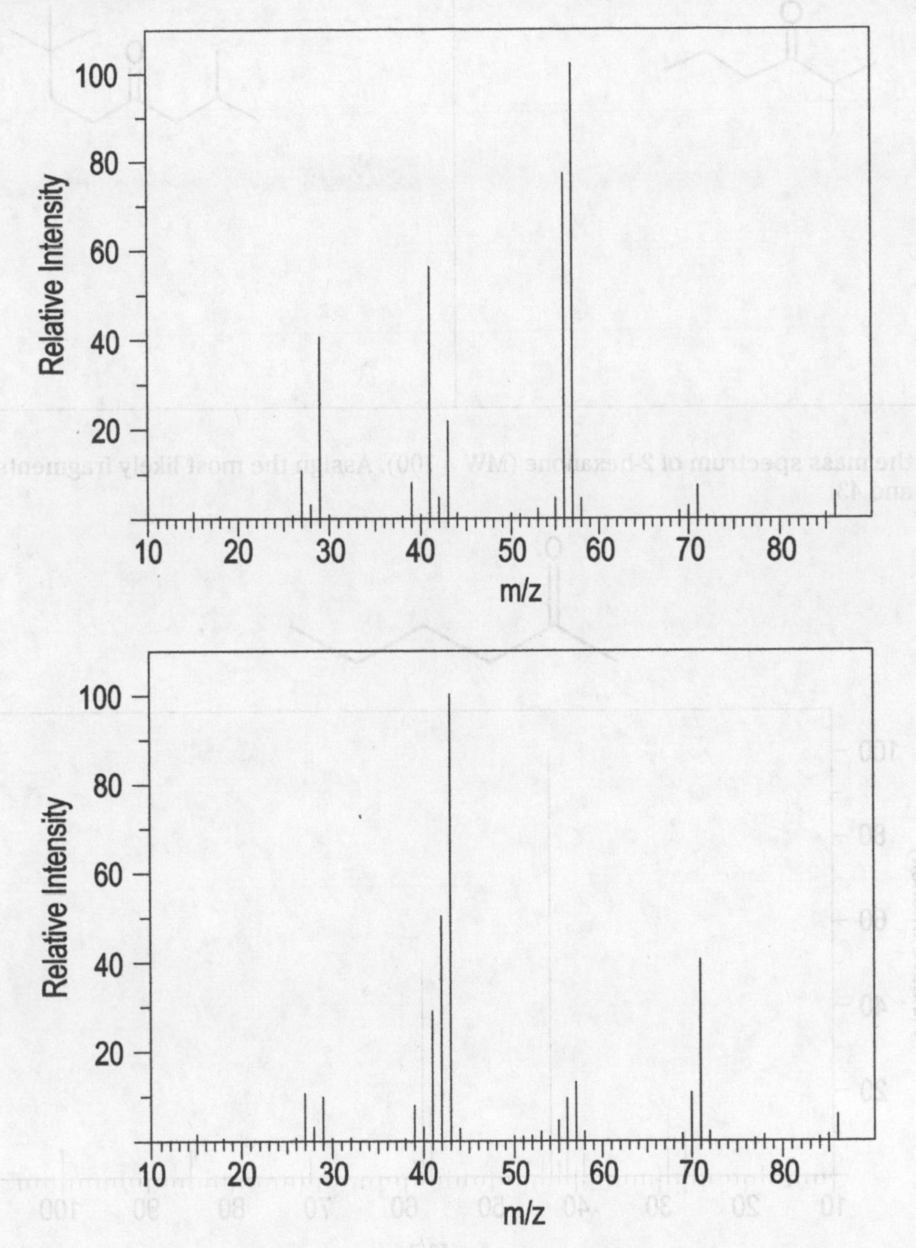

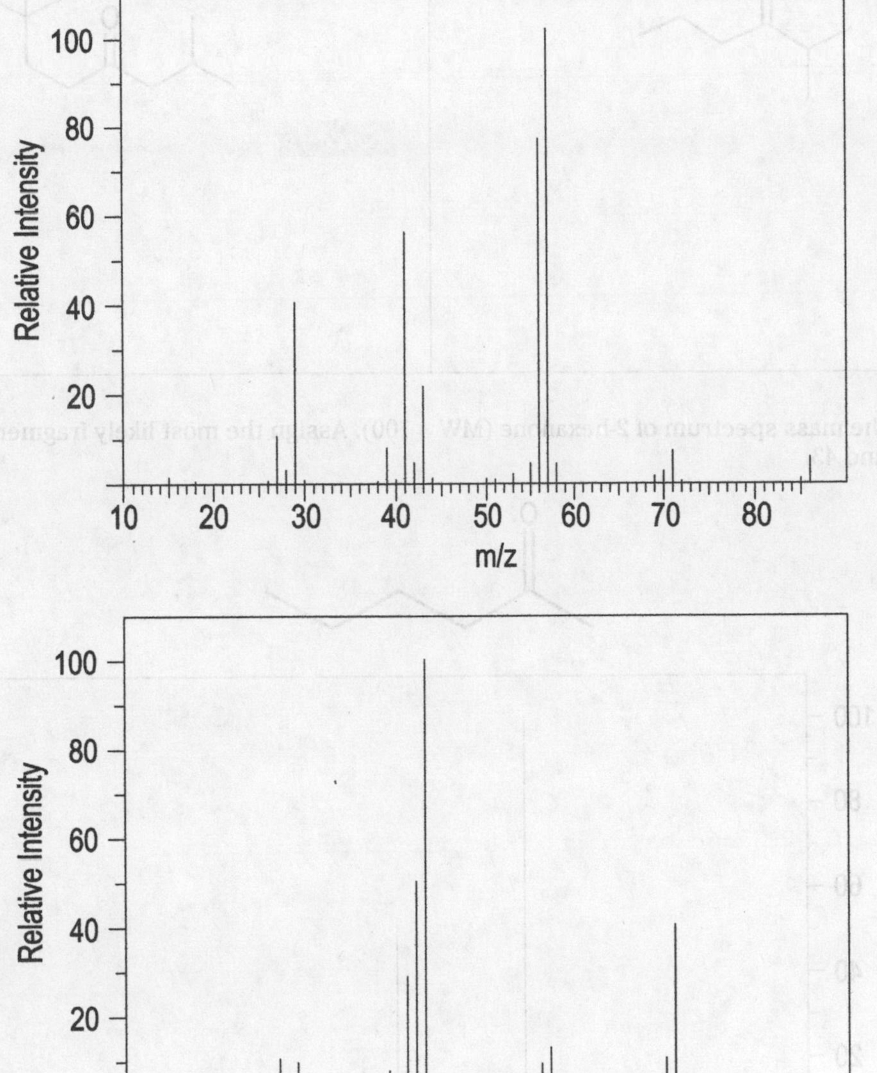

Solve It

2. Show the McLafferty rearrangement fragment for the following ketone.

O

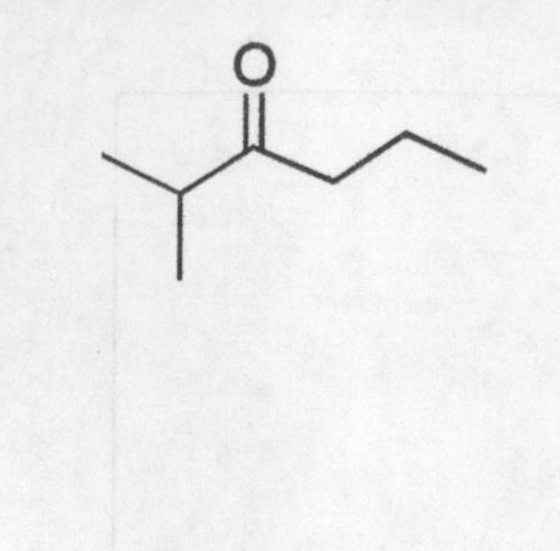

Solve It

3. Show the McLafferty rearrangement fragment for the following ketone.

O

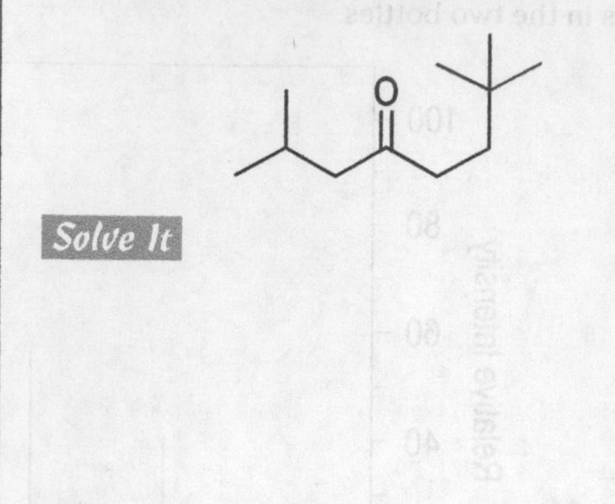

Solve It

4. Here's the mass spectrum of 2-hexanone (MW = 100). Assign the most likely fragments for m/z = 85, 58, and 43.

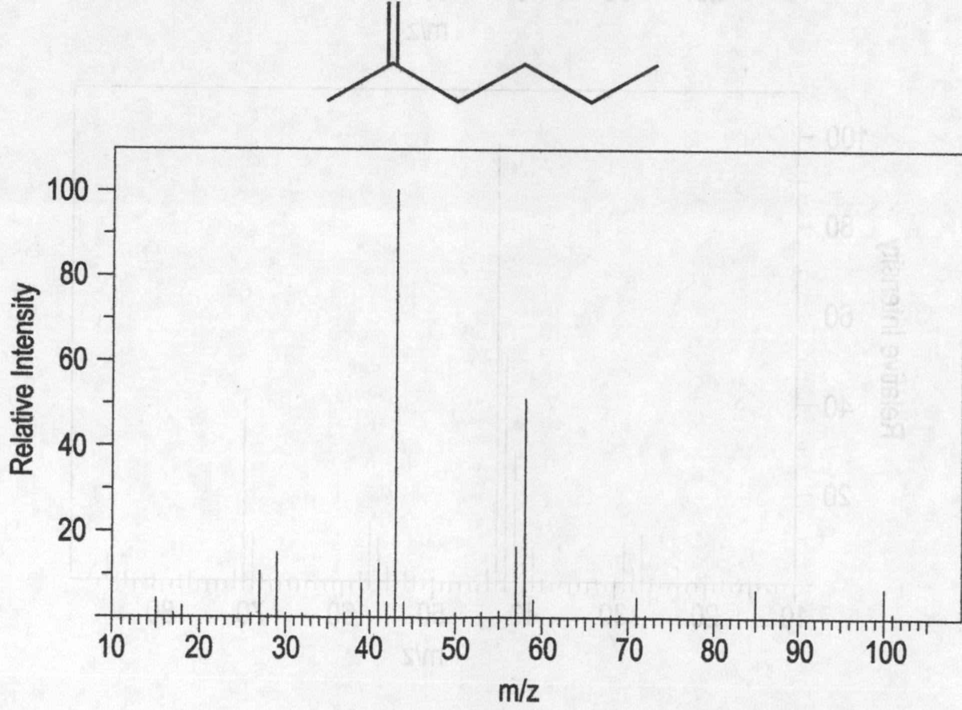

Solve It

5. A molecule has the molecular ion shown and a curious peak at 2 mass units higher with one-third of the intensity. What atom is most likely in this molecule? Explain the peak at 2 higher mass units (M+2).

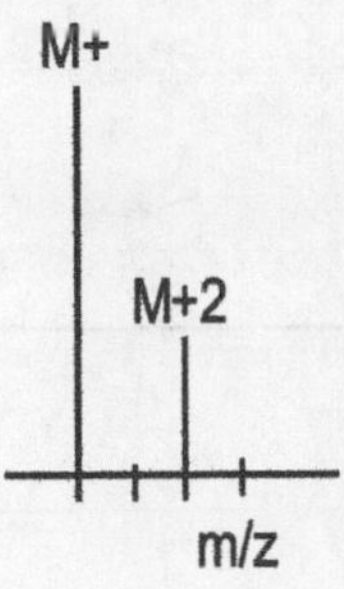

Solve It

6. A molecule has the molecular ion (M+ peak) shown and a peak at 2 mass units higher that has the same intensity as the M+ peak. What atom is most likely in the molecule? Explain the existence of the M+2 peak.

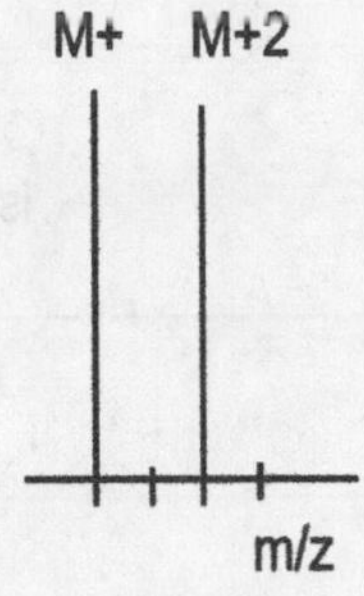

Solve It

7. One of these mass spectra corresponds to propylbenzene, and the other, to isopropyl benzene. Which is which?

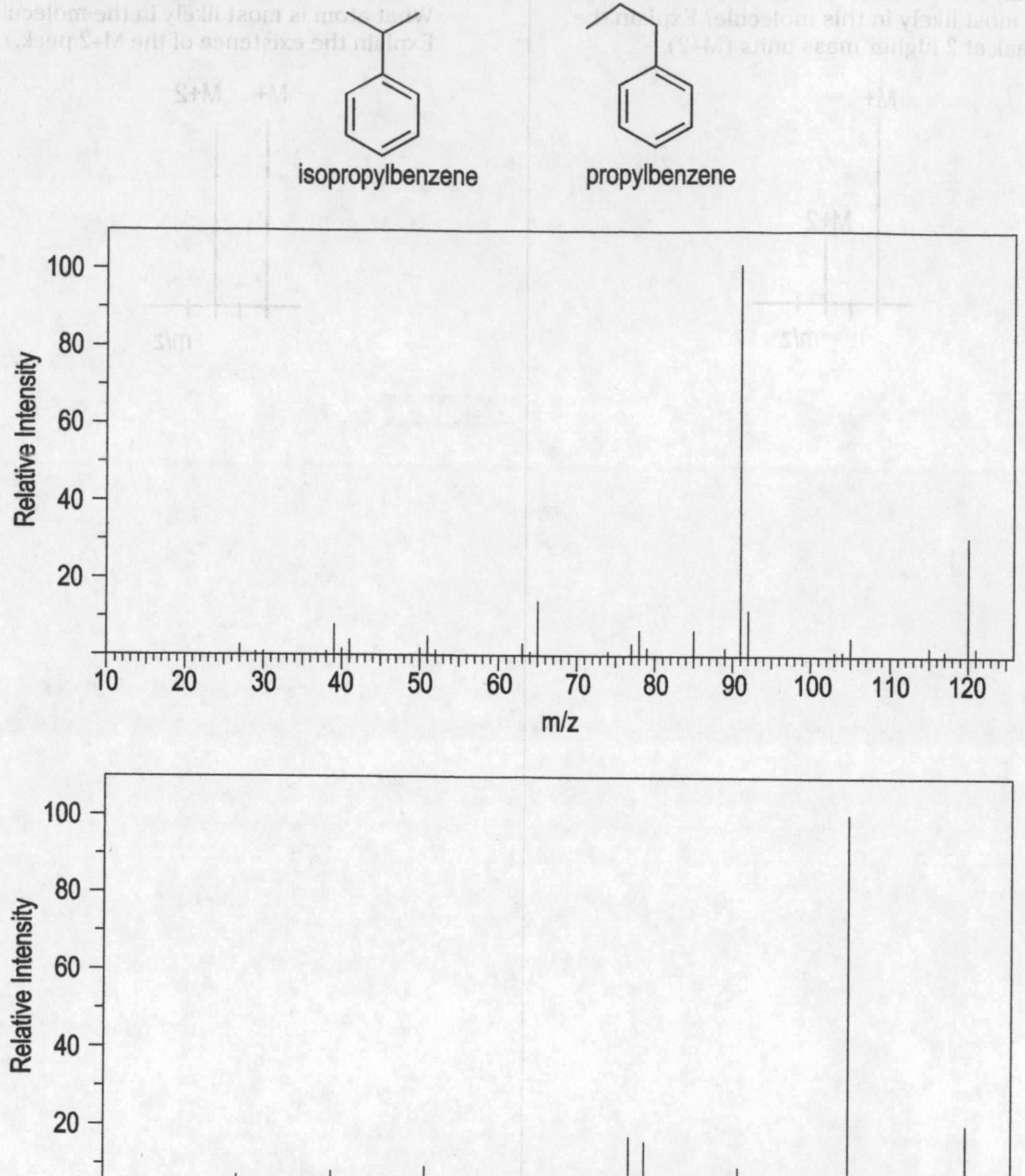

Solve It

8. Give a possible molecular formula for a hydrocarbon that has an M+ at 100 m/z.

Solve It

Predicting a Structure Given a Mass Spectrum

Determining structures from a mass spec by itself is usually possible only for the simplest of molecules. Still, these problems offer a great way to really hone your critical thinking skills and sharpen your understanding of mass spec.

To predict a structure given its mass spectrum, follow these steps:

1. **Identify any atoms that have common isotopes.**

   For example, bromine has two nearly equally abundant isotopes with ^{79}Br and ^{81}Br, giving rise to M+ and M+2 peaks of nearly equal intensity. Chlorine has two isotopes, ^{35}Cl and ^{37}Cl, in roughly a 3:1 ratio, giving rise to M+ and M+2 peaks in roughly a 3:1 ratio. Iodine has a peak at 129 corresponding to the I^+ fragment.

2. **If there are any halogens, subtract the weight of this atom from the M+ ion to obtain the weight of the rest of the molecule.**

   Memorize the weights of simple hydrocarbon fragments such as CH_3 (15), CH_2CH_3 (29), and C_3H_7 (43). If you reach an unknown weight, a common trick for determining how many carbons are in a given weight is to divide by 14, the mass of a methylene (CH_2) unit. For example, say you have a mass of 43. You take $43 \div 14$, which is approximately equal to 3. Three carbons weigh 36 mass units (3×12), so the remaining 7 mass units would be hydrogens, and you'd expect the formula of this 43 weight to correspond to a C_3H_7 unit.

3. **Create a list of all possible structures.**

4. **Use the process of elimination to determine the correct structure by looking for the fragmentation pattern you'd expect for each molecule.**

   Look for which of the possible structures most closely matches what you'd predict the actual spectrum to look like.

Q. Give the structure of the compound that has the following mass spectrum.

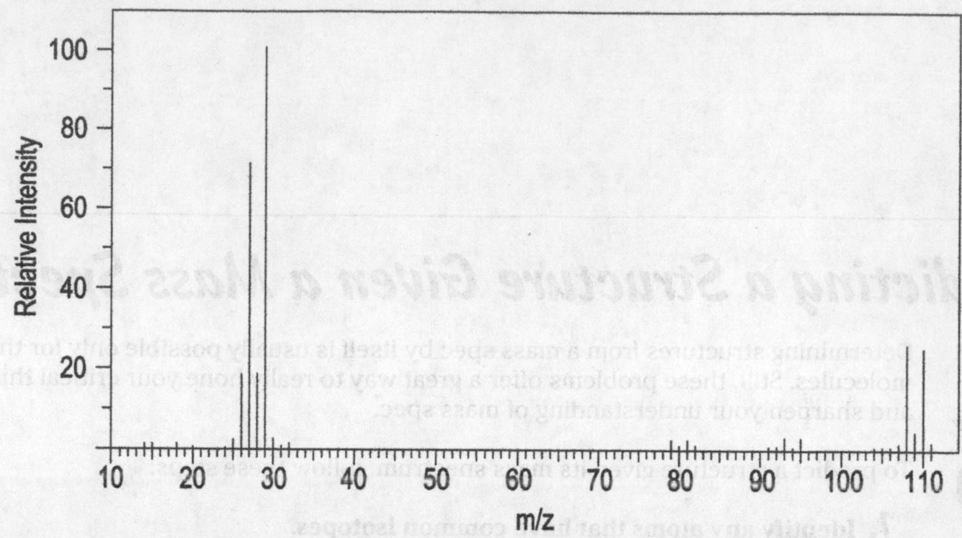

A. **Bromoethane (CH_3CH_2Br). The big tipoff here are the two peaks of nearly equal intensity for the M+ and M+2 ions. These peaks scream out "bromine!" because bromine has the two nearly equally abundant isotopes of ^{79}Br and ^{81}Br. Taking the M+2 peak at m/z = 110 and subtracting the mass of bromine (81) leaves 29 mass units, which you may have craftily memorized as the weight of an ethyl (CH_3CH_2) group. Therefore, this molecule is bromoethane (CH_3CH_2Br).**

---

9. **Give the structure of the molecule that gives the following mass spectrum.**

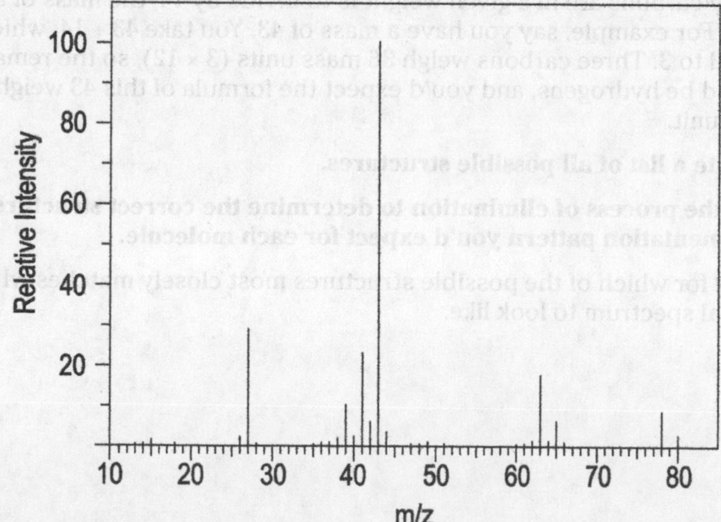

Solve It

10. Estella Stumblefoot found two vials in the stockroom whose labels had been worn to the point of being unreadable. Both vials smell like carbonyl compounds, which she confirmed by chemical tests. She also obtained the following two mass spectra. Help Estella identify what's in the vials by assigning each mass spectrum to a structure.

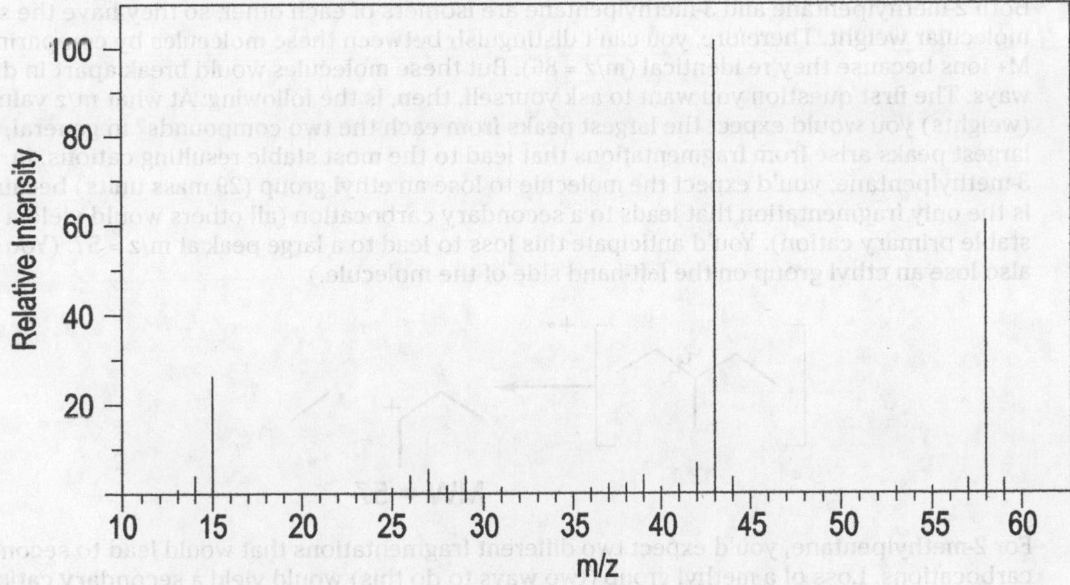

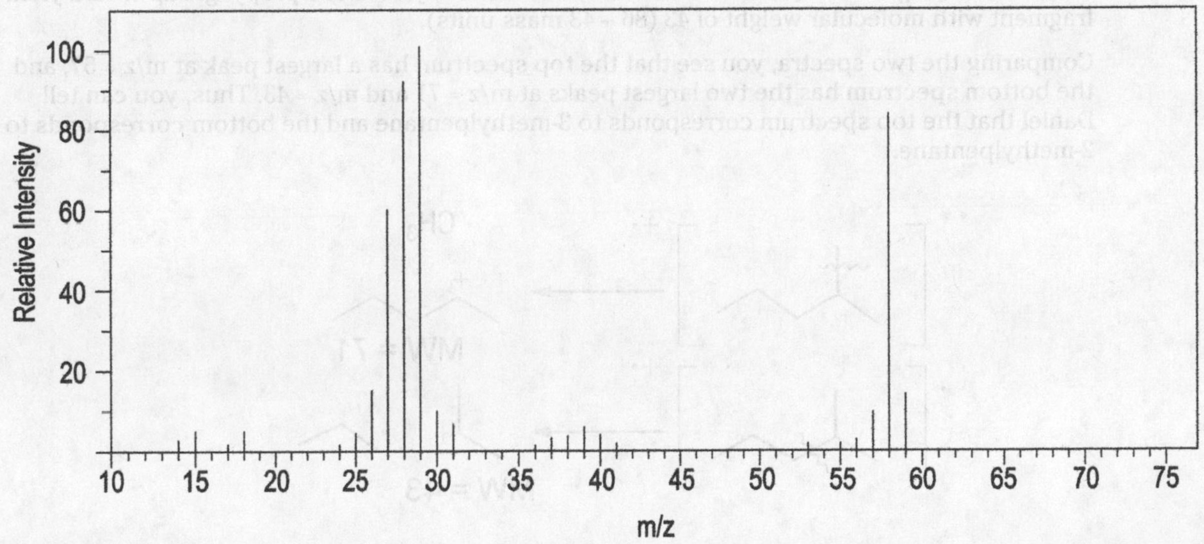

Solve It

Answer Key

The following are the answers to the practice questions presented in this chapter.

1 Both 2-methylpentane and 3-methylpentane are isomers of each other, so they have the same molecular weight. Therefore, you can't distinguish between these molecules by comparing the M+ ions because they're identical (m/z = 86). But these molecules would break apart in different ways. The first question you want to ask yourself, then, is the following: At what m/z values (weights) you would expect the largest peaks from each the two compounds? In general, the largest peaks arise from fragmentations that lead to the most stable resulting cations. In 3-methylpentane, you'd expect the molecule to lose an ethyl group (29 mass units) because this is the only fragmentation that leads to a secondary carbocation (all others would yield a less stable primary cation). You'd anticipate this loss to lead to a large peak at m/z = 57. (You could also lose an ethyl group on the left-hand side of the molecule.)

+•

+

MW = 57

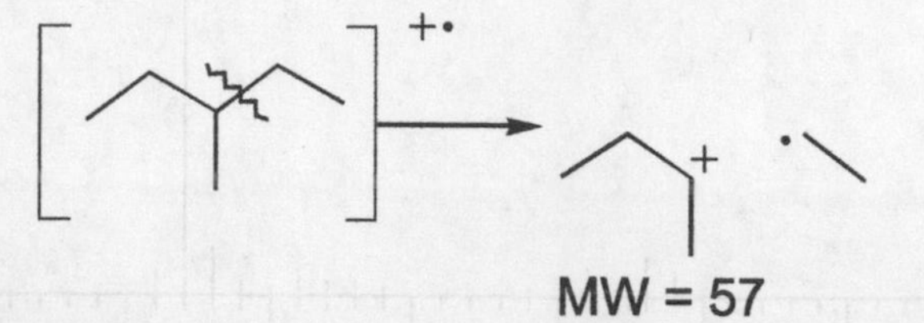

For 2-methylpentane, you'd expect two different fragmentations that would lead to secondary carbocations. Loss of a methyl group (two ways to do this) would yield a secondary cation with a molecular weight of 71 (86 – 15 mass units). Alternatively, loss of a propyl group would yield a fragment with molecular weight of 43 (86 – 43 mass units).

Comparing the two spectra, you see that the top spectrum has a largest peak at m/z = 57, and the bottom spectrum has the two largest peaks at m/z = 71 and m/z = 43. Thus, you can tell Daniel that the top spectrum corresponds to 3-methylpentane and the bottom corresponds to 2-methylpentane.

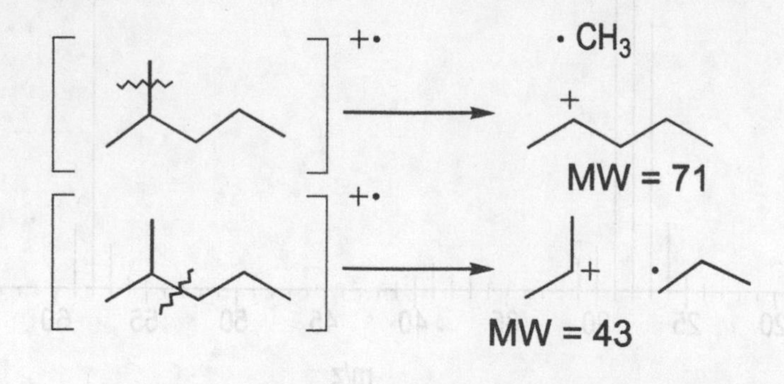

+•

$\cdot CH_3$

+

MW = 71

+•

+

MW = 43

2

+•

O

α β γ

redrawn

+•

O H γ

β

α

O–H

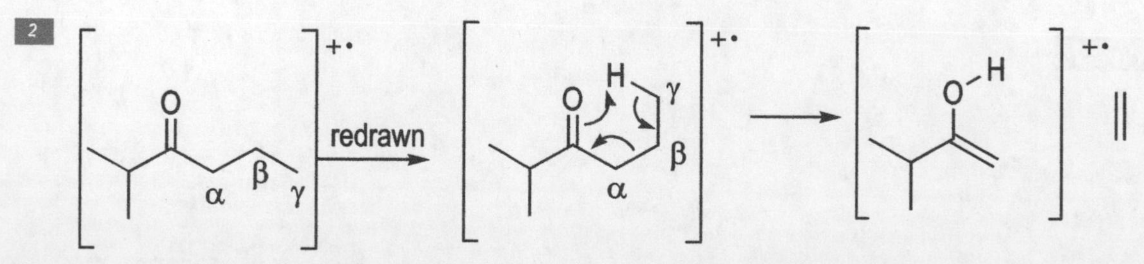

+•

When performing the McLafferty rearrangement, I recommend you label the α, β, and γ carbons first; then add the γ hydrogen and arrange the carbons into a ring-like arrangement. Finally, push the arrows as shown to give the McLafferty product.

3

γ O γ β β α α no H's! γ H O β α OH

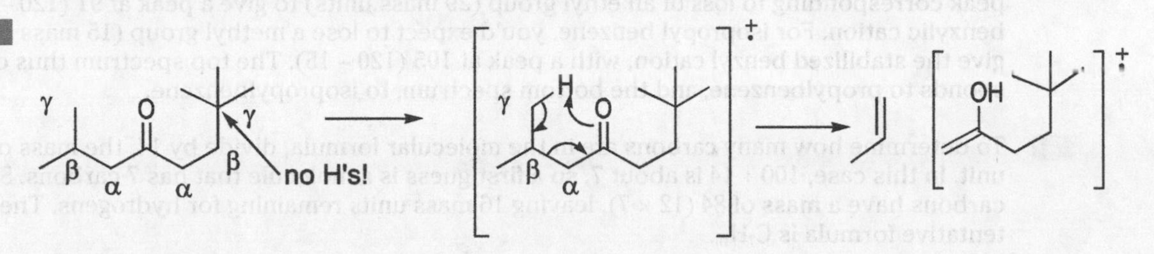

On this compound, you have two γ carbons. However, only one of the γ carbons has an attached hydrogen that can undergo the McLafferty rearrangement. Drawing the rearrangement for this hydrogen gives the McLafferty product.

4

O • CH_3 + O MW = 85

O O + MW = 43

H γ O β α OH MW = 58

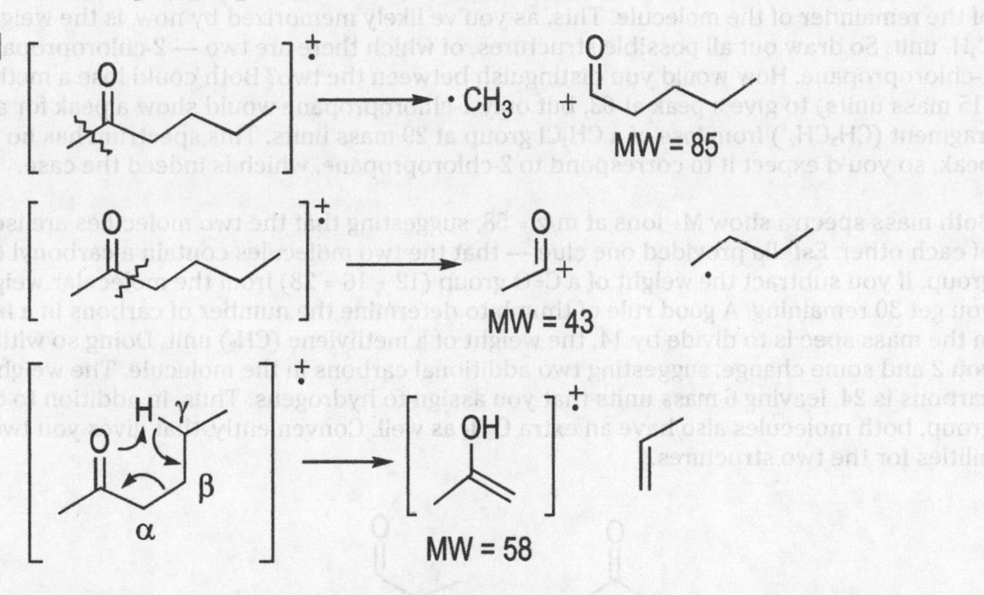

In problems such as this, look for the most likely breaks in the molecule. Because the molecule is a ketone, the most obvious break is α cleavage. Additionally, you have the possibility of getting a McLafferty rearrangement.

5 **This molecule likely contains chlorine. Chlorine has two abundant isotopes, ^{35}Cl and ^{37}Cl, in about a 3:1 ratio. Thus, roughly ¼ of the chlorine-containing fragments that hit the detector in the mass spec have the isotope of chlorine that's 2 mass units heavier, giving rise to this heavier isotope peak.**

6 **This molecule likely contains bromine. Bromine has two abundant isotopes, ^{79}Br and ^{81}Br, in nearly equal abundance. Thus, half the fragments that hit the detector in the mass spec that contain bromine have the heavier isotope.**

7 Molecules tend to fragment to give the most stable cation. For molecules that contain aromatic rings (see Chapter 13), cleavage tends to occur in the benzylic position so as to make a resonance-stabilized benzylic cation (C_6H_5-CH_2^+). Thus, for propylbenzene, you'd expect a large peak corresponding to loss of an ethyl group (29 mass units) to give a peak at 91 (120 – 29) for a benzylic cation. For isopropyl benzene, you'd expect to lose a methyl group (15 mass units) to give the stabilized benzyl cation, with a peak at 105 (120 – 15). The top spectrum thus corresponds to propylbenzene, and the bottom spectrum, to isopropylbenzene.

8 To determine how many carbons are in the molecular formula, divide by 14, the mass of a CH_2 unit. In this case, 100 ÷ 14 is about 7, so a first guess is a molecule that has 7 carbons. Seven carbons have a mass of 84 (12 × 7), leaving 16 mass units remaining for hydrogens. Therefore, a tentative formula is C_7H_{16}.

9 The first thing to notice is that this mass spec shows an M+2 peak that's roughly ⅓ as large as the M+ peak. This indicates a chlorine atom, which has two isotopes, ^{35}Cl and ^{37}Cl, in a 3:1 ratio. Thus, you subtract the weight of chlorine (35) from the M+ peak (78) to get 43, the weight of the remainder of the molecule. This, as you've likely memorized by now, is the weight of a C_3H_7 unit. So draw out all possible structures, of which there are two — 2-chloropropane and 1-chloropropane. How would you distinguish between the two? Both could lose a methyl group (15 mass units) to give a peak at 63, but only 1-chloropropane would show a peak for an ethyl fragment ($CH_3CH_2^+$) from loss of a CH_2Cl group at 29 mass units. This spectrum has no such peak, so you'd expect it to correspond to 2-chloropropane, which is indeed the case.

10 Both mass spectra show M+ ions at m/z = 58, suggesting that the two molecules are isomers of each other. Estella provided one clue — that the two molecules contain a carbonyl (C=O) group. If you subtract the weight of a C=O group (12 + 16 = 28) from the molecular weight of 58, you get 30 remaining. A good rule of thumb to determine the number of carbons in a fragment in the mass spec is to divide by 14, the weight of a methylene (CH_2) unit. Doing so with 30 gives you 2 and some change, suggesting two additional carbons in the molecule. The weight of two carbons is 24, leaving 6 mass units that you assign to hydrogens. Thus, in addition to the C=O group, both molecules also have an extra C_2H_6 as well. Conveniently, that gives you two possibilities for the two structures.

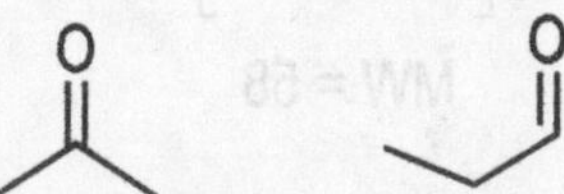

Now, to determine which is which: You'd expect the ketone on the left (acetone) to have a large peak corresponding to loss of a methyl group (15 mass units) from the α cleavage (to give you a peak at m/z = 43). On the other hand, you'd expect the aldehyde on the right to have a large peak corresponding to loss of an ethyl group (29 mass units) from α cleavage (to give you a peak with m/z = 29). The top spectrum has a large peak at m/z = 43, which fits the bill for the ketone on the left, and the bottom spectrum has a large peak at m/z = 29, which matches the aldehyde on the right.

Chapter 15

Cool Vibrations: IR Spectroscopy

In This Chapter

- Differentiating between molecules using IR spectroscopy
- Eyeing functional groups in an IR spectrum

Infrared (IR) spectroscopy is one of the most common tools organic chemists use to deduce the structure of unknown compounds. This instrumental method is particularly useful for determining which functional groups are present.

The quick 'n' dirty version of how IR spectroscopy works is that infrared light is irradiated onto a sample of an unknown compound. Different bond types (functional groups) in the unknown molecule absorb different frequencies of the infrared light. For example, an O-H bond absorbs infrared light of a different frequency than a C-H bond. The IR spectrometer feeds this data into a computer that then plots the frequencies of the light that isn't absorbed by the sample (the transmittance of light) versus the amount of transmittance to give the IR spectrum. It's up to you, the chemist, to look at the spectrum and figure out which functional groups are present in the unknown molecule.

This chapter gives you the lowdown on IR spectroscopy and helps you identify functional groups in an IR spectrum.

Distinguishing between Molecules Using IR Spectroscopy

Before you can understand how to use IR spectroscopy to determine the functional groups in an unknown molecule, you need to know a few strange conventions in IR spectroscopy. First, the IR spectrum plots on its *y* axis the transmittance of light through a sample, not the absorption of light by the molecule. Essentially, this has the effect of making the spectrum look like it's been turned upside down. In other words, the peaks look like stalactites hanging down from the ceiling of the IR spectrum rather than your typical stalagmite variety of peaks that you see in other self-respecting spectroscopies that plot absorbances. In spite of this convention, peaks are referred to as *absorbances,* not transmittances! To add to the confusion, the frequency of IR light is given in bizarre units, reciprocal centimeters (cm^{-1}), often referred to as *wavenumbers.*

One useful feature of IR spectroscopy is that it can distinguish between two molecules. For example, in court cases involving drug possession, IR spectroscopy can distinguish between a sample of cocaine and one of crack cocaine. On a more positive note, you can see whether your reaction worked using IR spectroscopy. For instance, if you run a reaction that converts an alcohol (OH) functional group into a ketone (C=O), IR spectroscopy shows the disappearance of the O-H absorbance in the starting material and the growth of the C=O absorbance in your product.

In order to distinguish between molecules using IR spectroscopy, you first need to know where in the infrared spectrum different bond types absorb light. Table 15-1 lists the absorption locations of the common functional groups in organic molecules.

Table 15-1 IR Absorption of Common Functional Groups

Functional Group	*Absorption Location (cm^{-1})*	*Absorption Intensity*
Alcohol (hydroxyl O–H)	3,400–3,700	Strong, broad
Amines (N–H)	3,300–3,350	Medium ***Note:*** Primary amines (*R*-NH_2) give two peaks in this region, and secondary amines (*R*NH*R*) give a single peak.
Nitrile (C≡N)	2,200–2,250	Medium
Alkyne (C≡C) (C≡C–H)	 2,100–2,250 3,300	 Medium Strong
Carbonyls (C=O): Aldehyde (CHO) Ketone (*R*CO*R*) Ester (*R*COO*R*) Carboxylic acid (*R*COOH)	 1,720–1,740 1,715 1,735–1,750 1,700–1,725	Strong
Aromatics (C=C) Ar-H	1,650–2,000 3,000–3,100	Weak Weak
Alkene (C=C) (C=C–H)	1,640–1,680 3,020–3,100	Weak to medium Medium

You may have glanced at an IR spectrum and thought, "Yikes, they want me to interpret that?!" If you said that, here's some good news: You can generally ignore two major regions in the IR spectrum:

- **The fingerprint region below 1,500 cm^{-1}:** This is generally a disgusting haystack of peaks that are difficult to interpret and aren't typically helpful for giving structural information.
- **From 2,800 cm^{-1} to 3,000 cm^{-1}:** This region is where saturated C-H absorptions show up, and because virtually all organic molecules contain saturated C-H bonds, you almost always see peaks in this region that aren't particularly informative for figuring out the structure of a molecule.

With those parts of the spectrum in the waste bin, you're left looking for peaks between 1,500 cm^{-1} and 2,800 cm^{-1} and above 3,000 cm^{-1} on any IR spectrum. I recommend you draw a line straight up from 3,000 cm^{-1} on your spectrum and note the following:

- Anything just to the left the line at higher wavenumbers is an unsaturated C-H bond; this indicates a double bond or aromatic ring in your molecule.
- In alcohols, the hydroxyl (OH) group is easy to spot as a fat belly that hangs down to the left of the saturated C-H stretches at around 3,400 cm^{-1}.
- Carbonyl (C=O) peaks are also hard to miss as a large finger-like absorption that sticks down around 1,700 cm^{-1}.

Until you memorize the locations of the other functional group peaks, you can use Table 15-1 as a reference (you can find more-comprehensive tables in any good organic text).

The intensity of IR light absorption by a bond in a molecule depends on the dipole moment of that bond (refer to Chapter 1). Bonds with large dipole moments (such as OH bonds) have intense absorption. Without a dipole, the bond is *IR inactive,* and no peaks show up in the IR spectrum for that bond.

Q. How can you tell H-Cl apart from Cl-Cl using IR spectroscopy?

A. Although H-Cl has a bond dipole because of the difference in the electronegativity of the two atoms, you get no bond dipole between two identical atoms, such as in Cl-Cl. Thus, H-Cl absorbs IR light, but Cl-Cl doesn't, and this bond is IR inactive (and shows no peaks in the IR spectrum).

Q. Match the two spectra to the shown alkynes.

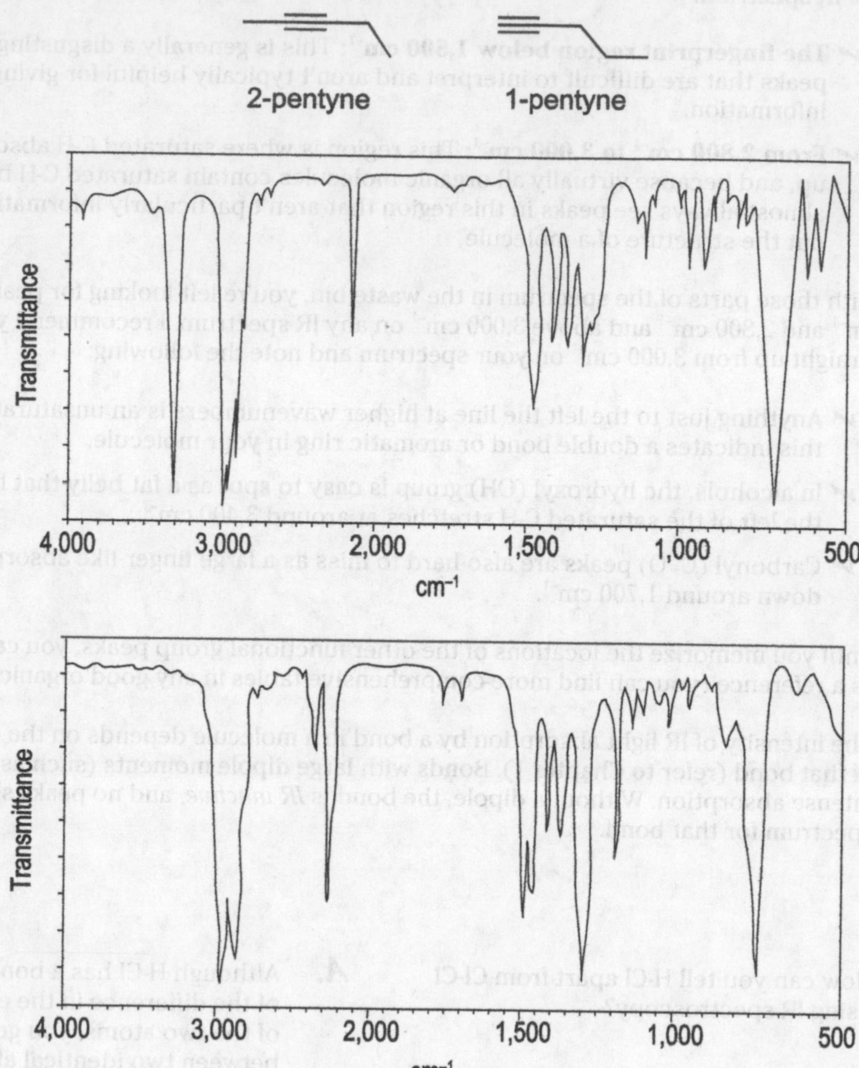

A. **The top spectrum is 1-pentyne; the bottom is 2-pentyne.** In this case, you're comparing two alkynes. The major difference between them is that one is a terminal alkyne (a triple bond is at the end of the molecule) and one is an internal alkyne (an alkyne is in the middle of a molecule). Thus, although they both have triple-bond alkyne stretches at about 2,200 cm^{-1}, the terminal alkyne should also have an alkyne C-H stretch to the left of the saturated C-H stretches at around 3,300 cm^{-1}. The top spectrum has this additional stretch, so it's the terminal alkyne.

1. Explain how you can distinguish between the following two molecules using IR spectroscopy.

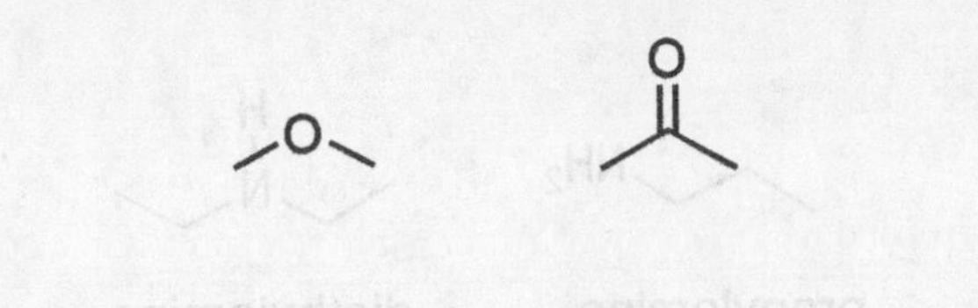

Solve It

---

2. Match each spectrum with the correct molecule. Assign any important functional group bands in the two IR spectra.

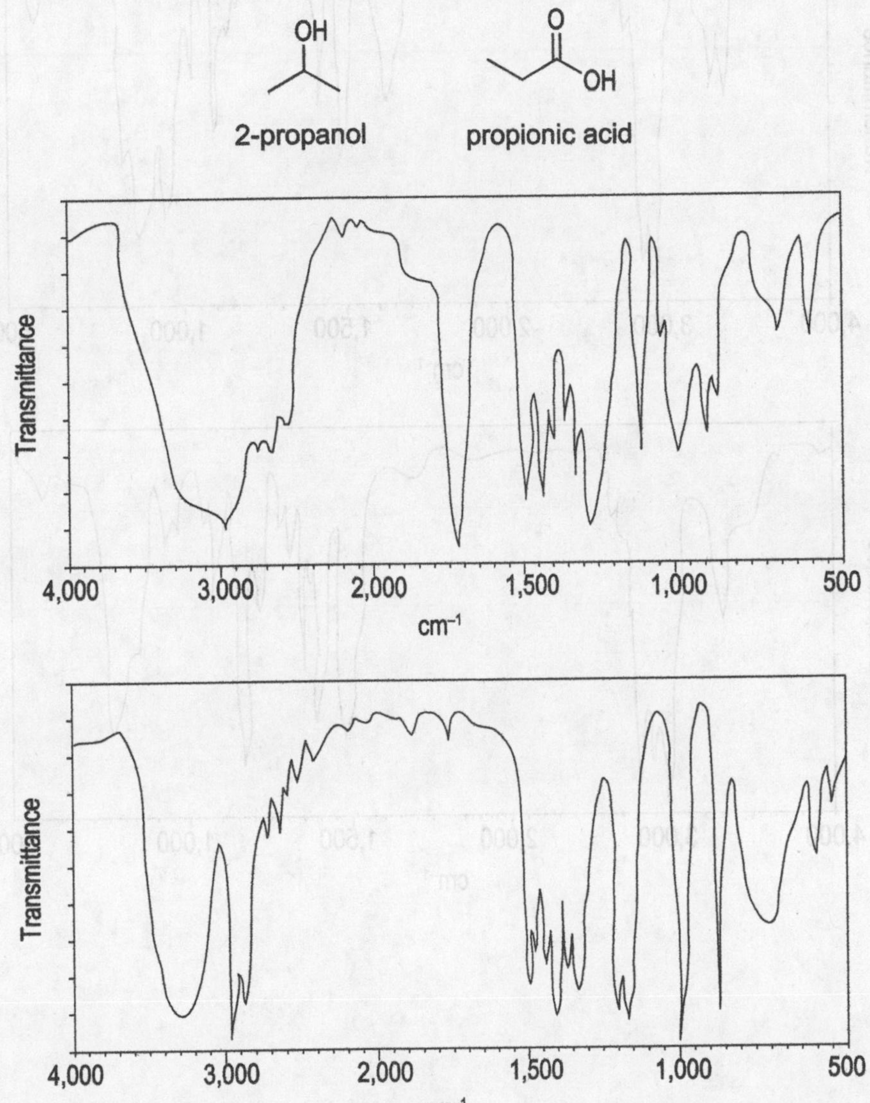

Solve It

3. Match each spectrum with the correct molecule. Assign any important functional group bands in the two IR spectra.

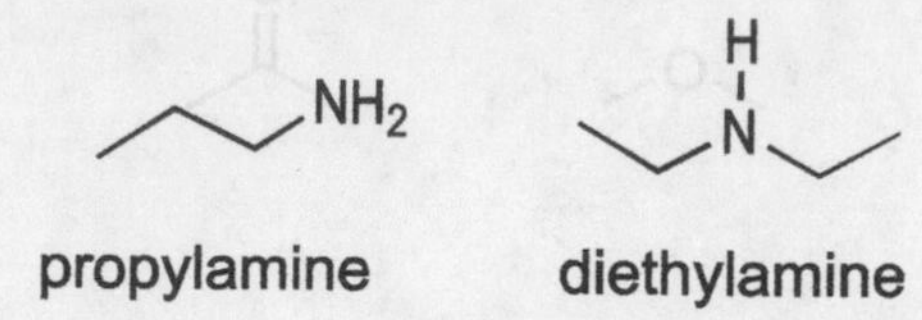

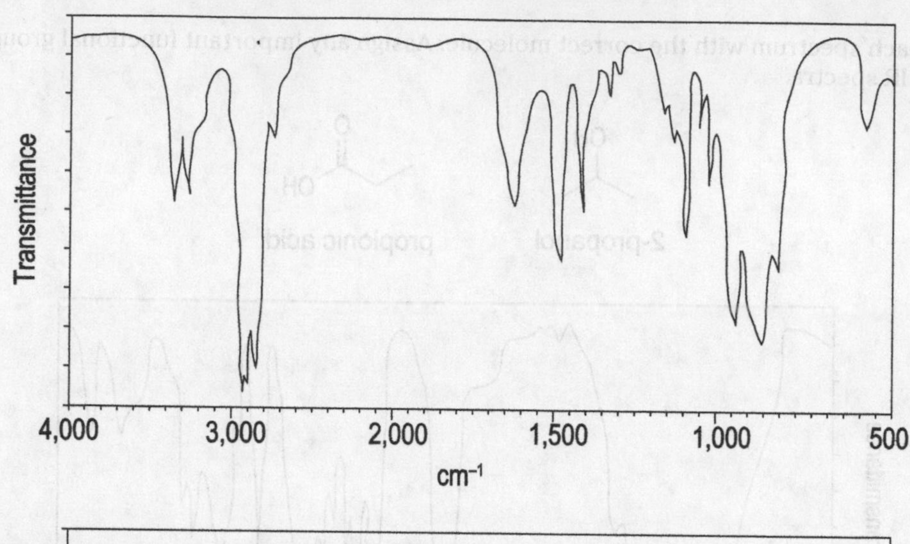

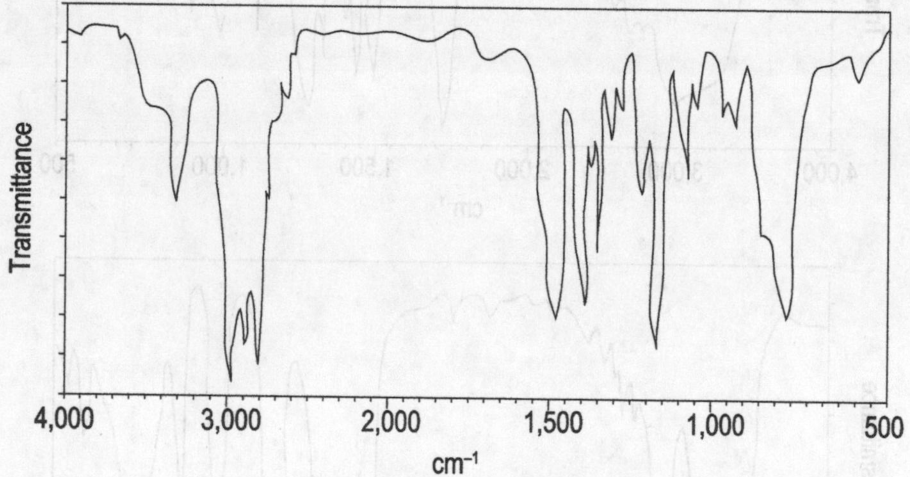

Solve It

4. Match each spectrum with the correct molecule. Assign any important functional group bands in the two IR spectra.

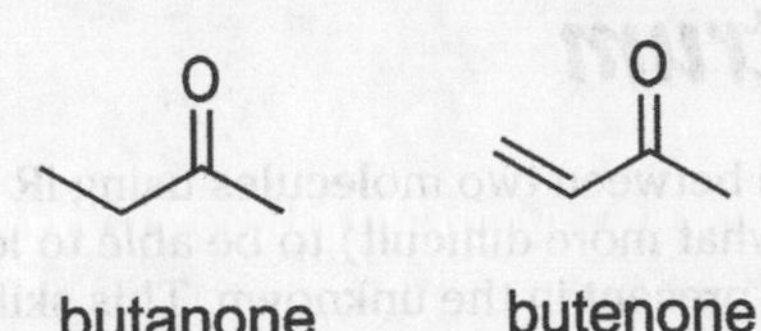

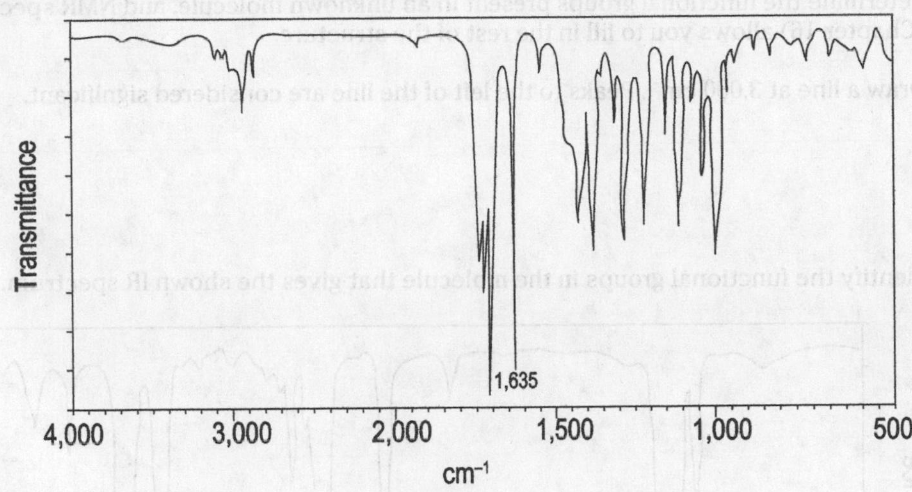

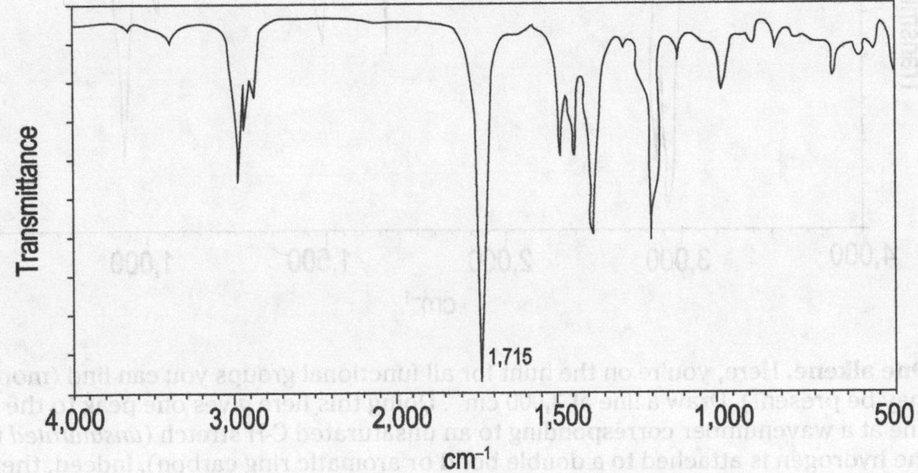

Solve It

Identifying Functional Groups from an IR Spectrum

Being able to distinguish between two molecules using IR spectroscopy is important, but it's also important (if somewhat more difficult) to be able to look at a spectrum and determine all the functional groups present in the unknown. This skill is important when you have no idea of the molecule's structure and want to determine it. IR spectroscopy allows you to determine the functional groups present in an unknown molecule, and NMR spectroscopy (Chapter 16) allows you to fill in the rest of the structure.

Draw a line at 3,000 cm^{-1}. Peaks to the left of the line are considered significant.

Q. Identify the functional groups in the molecule that gives the shown IR spectrum.

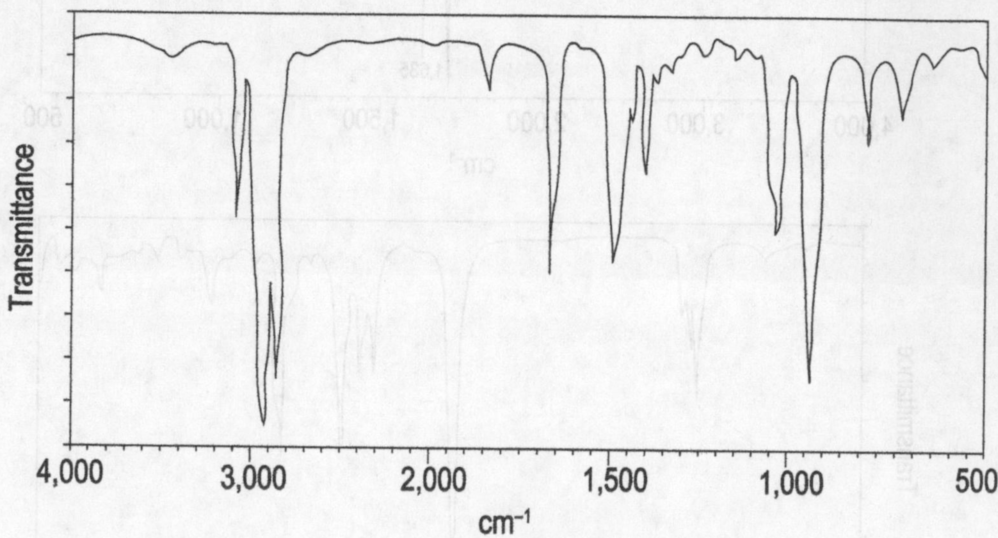

A. **One alkene.** Here, you're on the hunt for all functional groups you can find (more than one may be present). Draw a line at 3,000 cm^{-1}. Doing this here gives one peak to the left of the line at a wavenumber corresponding to an unsaturated C-H stretch (*unsaturated* means that the hydrogen is attached to a double bond or aromatic ring carbon). Indeed, there's a double-bond stretch at ~1,640 cm^{-1}.

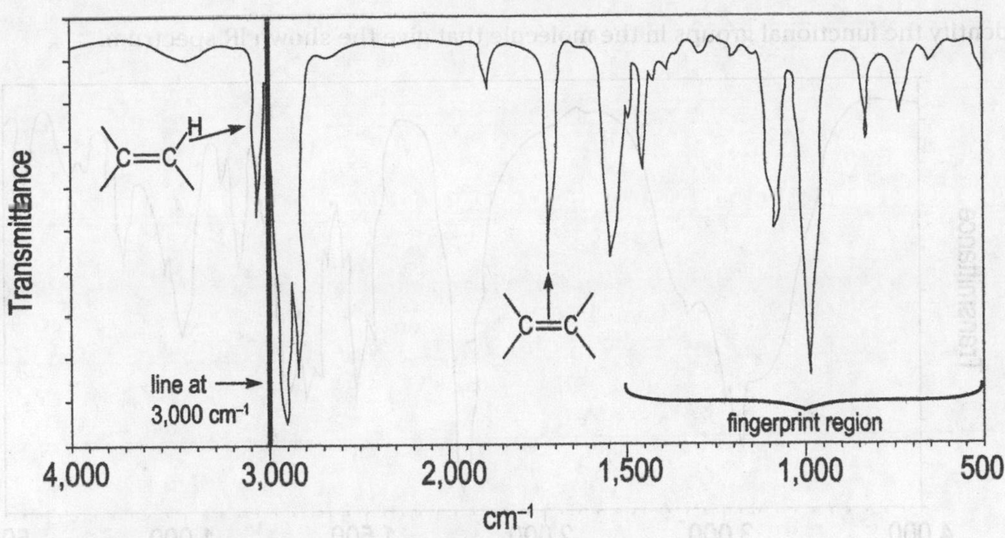

5. Identify the functional groups in the molecule that give the shown IR spectrum.

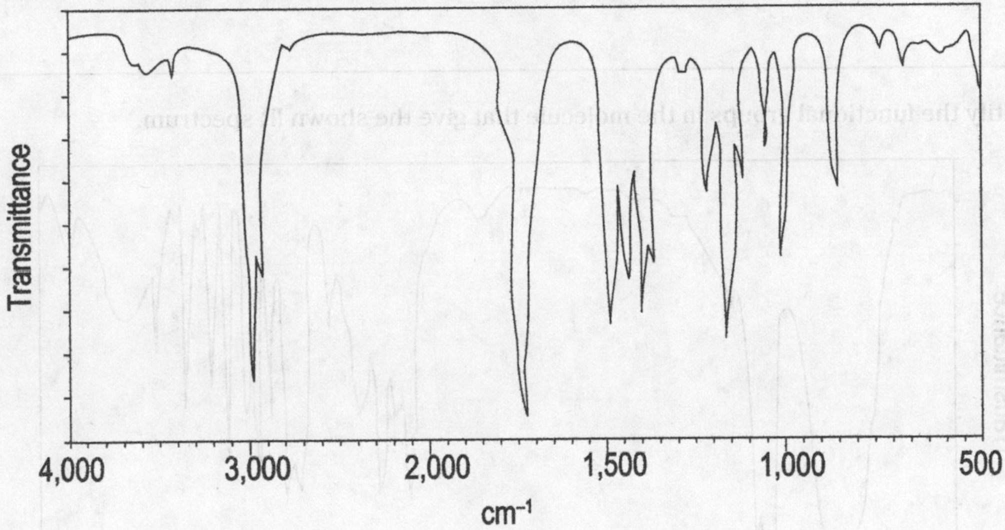

Solve It

6. Identify the functional groups in the molecule that give the shown IR spectrum.

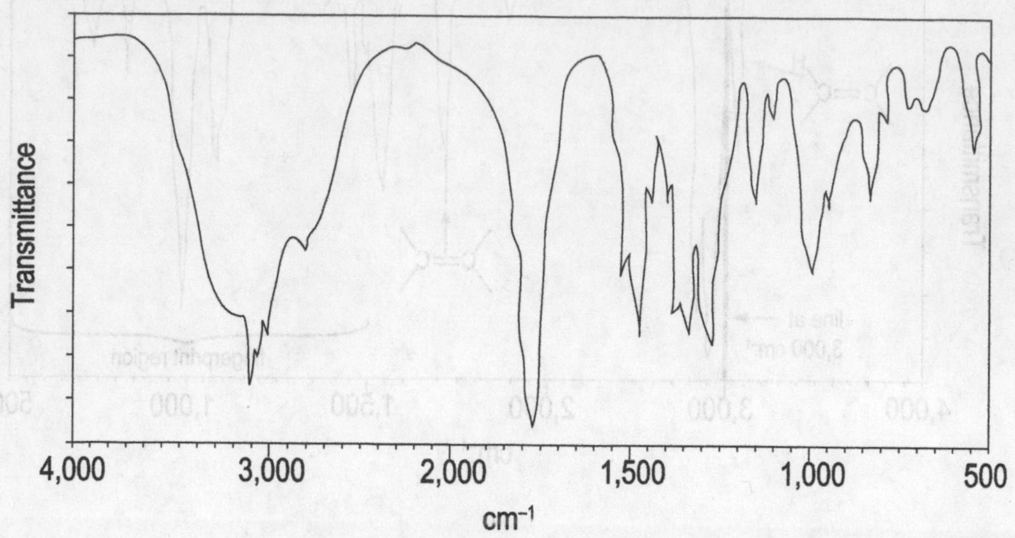

Solve It

7. Identify the functional groups in the molecule that give the shown IR spectrum.

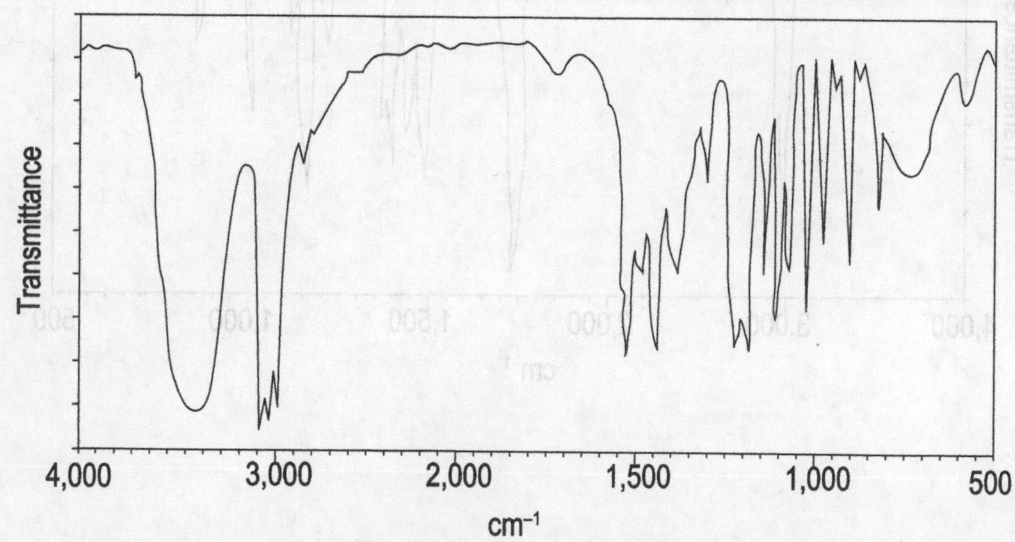

Solve It

8. Identify the functional groups in the molecule that give the shown IR spectrum.

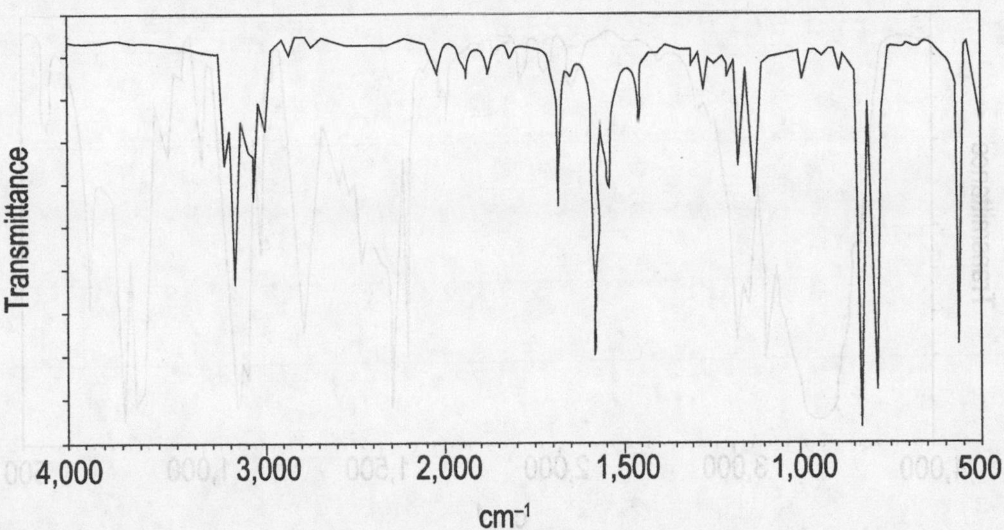

Solve It

9. Identify the functional groups in the molecule that give the shown IR spectrum.

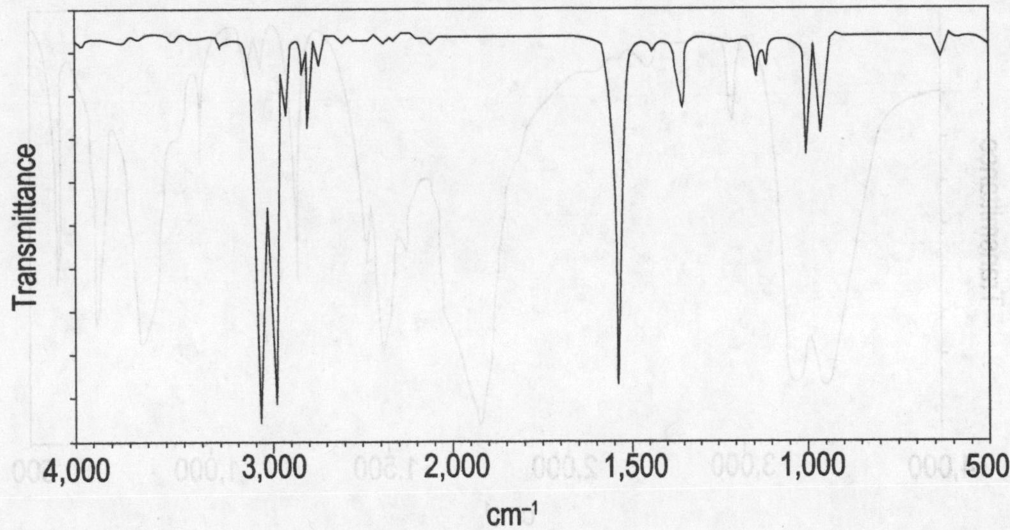

Solve It

10. Identify the functional groups in the molecule that give the shown IR spectrum.

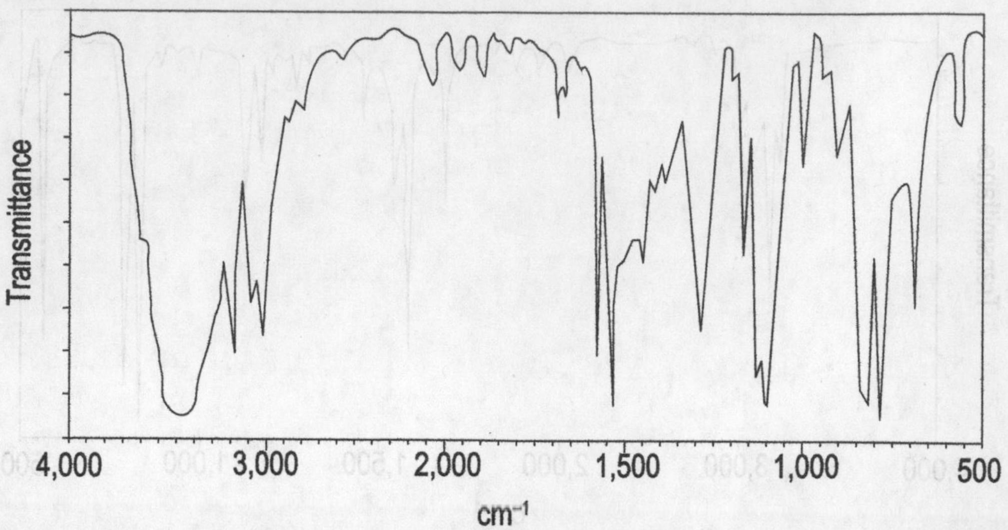

Solve It

11. Identify the functional groups in the molecule that gives the shown IR spectrum.

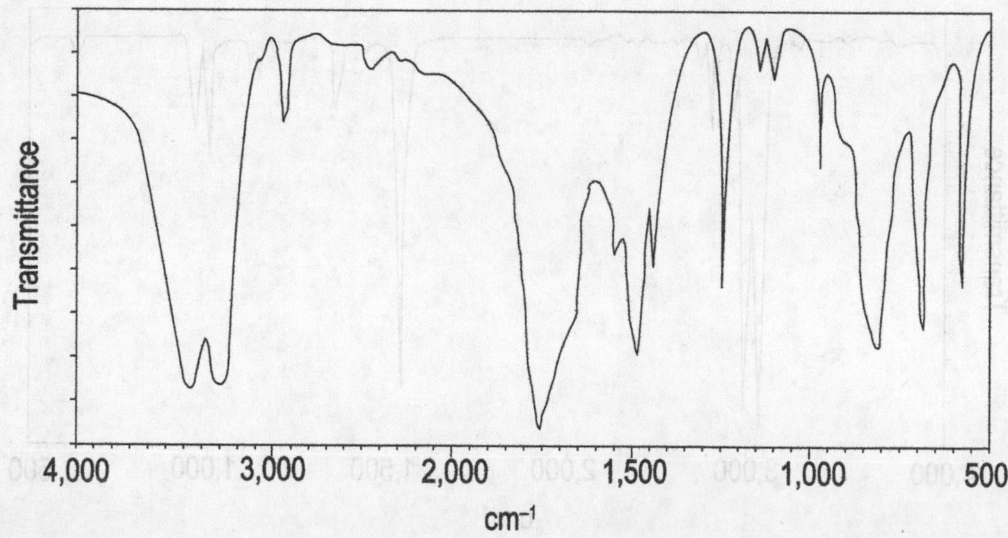

Solve It

12. How can you distinguish among the following four molecules using IR spectroscopy? Explain what would be different in the IR spectra for each of the compounds.

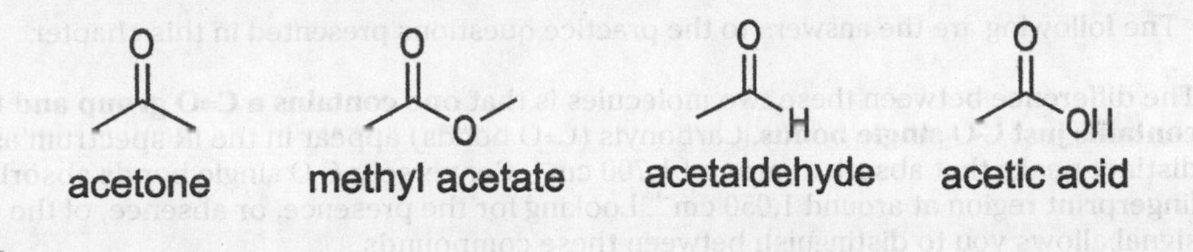

acetone methyl acetate acetaldehyde acetic acid

Solve It

Answer Key

The following are the answers to the practice questions presented in this chapter.

1 The difference between these two molecules is that **one contains a C=O group and the other contains just C-O single bonds.** Carbonyls (C=O bonds) appear in the IR spectrum as large distinct peaks that absorb at around 1,700 cm^{-1}. Conversely, C-O single bonds absorb in the fingerprint region at around 1,050 cm^{-1}. Looking for the presence, or absence, of the C=O signal allows you to distinguish between these compounds.

2 **The top spectrum is propionic acid; the bottom spectrum is 2-propanol.** Both of these compounds contain an alcohol functional group, which have hydroxyl (O-H) groups. Hydroxyl groups show up as broad bands above 3,000 cm^{-1} in the IR spectrum. For carboxylic acid (COOH) hydroxyl groups, the OH peak is usually much broader still and overlaps with the saturated C-H stretches. Additionally, the carboxylic acid spectrum contains a C=O stretch at around 1,700 cm^{-1}. Both the broad OH peak and the carbonyl peak of the top spectrum indicate a carboxylic acid, but the bottom spectrum contains a peak only for an isolated OH group.

3 **The top spectrum is propylamine; the bottom spectrum is diethylamine.** Both compounds are amines. Amine N-H protons absorb at 3,300–3,350 cm^{-1} in the IR spectrum. The difference between these two amines is that propylamine is a primary amine (R-NH_2, with the nitrogen attached to only one *R* group), whereas diethylamine is a secondary amine (R_2NH, where the nitrogen is attached to two *R* groups). Primary amines give two peaks in the N-H region of the spectrum (that together look like a cow udder), whereas secondary amines only give a single absorption. The top spectrum has two NH peaks, and the bottom spectrum has just a single absorption. Thus, the top spectrum is a primary amine, and the bottom is a secondary amine.

4 **The top spectrum is butenone; the bottom spectrum is butanone.** Here, both molecules have carbonyl (C=O) groups. The difference is that butenone contains a C=C bond as well. This leads to three differences in the spectrum of butenone. First, you should see a C=C stretch in the IR spectrum at roughly 1,640 cm^{-1}. Secondly, you should see unsaturated C-H bonds in the butenone spectrum occurring at 3,000–3,200 cm^{-1}. Lastly, conjugated C=O stretches (C=O groups attached to double bonds or aromatic rings) are shifted to lower wavenumbers (often just below 1,700 cm^{-1}) than isolated C=O groups. All three of these features occur in the top spectrum and not in the bottom one, so the top spectrum is butenone.

5 This molecule contains a **carbonyl (C=O) group and is likely a ketone.** The main standout peak here is the carbonyl C=O stretch at about 1,715 cm^{-1}. If you have a carbonyl, it could make up four different functional groups — a carboxylic acid (COOH), a ketone (*R*CO*R*), an aldehyde (*R*CHO), or an ester (*R*COO*R*). So how do you distinguish among these? A carboxylic acid would also have an OH stretch, which is absent here. An ester carbonyl stretch comes higher than normal at 1,735–1,760 cm^{-1}. So that leaves a ketone or aldehyde. Aldehydes have a hydrogen attached to the carbonyl, which absorbs at 2,800 cm^{-1}. That stretch is absent here, so the most likely alternative is a simple ketone (*R*CO*R*). Note that when you have a strong carbonyl stretch, you often see "overtones" of that stretch at twice the wavenumber. Because carbonyl groups have intense absorptions at about 1,700 cm^{-1}, you often see small overtone bumps at 3,400 cm^{-1}, as is the case here. Just ignore these artifact overtone peaks.

6 This molecule contains a **carboxylic acid (COOH)** functional group. The two main peaks of interest is the large OH absorption band from about 2,500–3,500 cm^{-1}. When the OH band is stretched out so far (often overlapping the C-H absorptions), this is a tip-off for a carboxylic acid. Secondly, you see a large C=O band at 1,715 cm^{-1}. These bands also tend to be fatter than normal in a carboxylic acid.

7 This molecule contains an **alcohol (OH)** functional group. The main peak of interest is the broad OH peak at 3,100–3,600 cm^{-1}. Because this band doesn't significantly overlap with the C-H absorptions, it's likely your run-of-the-mill OH absorption, not a carboxylic acid OH. No other significant peaks are present in the spectrum.

8 This molecule contains an **aromatic benzene ring.** Drawing a line at 3,000 cm^{-1} shows unsaturated C-H stretches just to the left. Additionally, the four peaks between 1,600–1,900 cm^{-1} are indicative of an aromatic ring. ***Note:*** The number of peaks can change, depending on the substitution of the aromatic ring.

9 This molecule is an **alkane, with no significant functional groups.** The only peaks in this molecule are the absorptions in the fingerprint region (below 1,500 cm^{-1}) and in the saturated C-H region between 2,800–3,000 cm^{-1}. Thus, this molecule is likely a saturated alkane with no functional groups of interest.

10 This molecule contains an **aromatic benzene ring and an alcohol (OH)** functional group. This spectrum contains three major peaks of interest. Most noticeable is the large OH stretch at 3,000–3,600 cm^{-1}. Second is the series of four peaks between 1,700–2,000 cm^{-1}, indicating an aromatic benzene ring. Most subtle are the unsaturated C-H peaks at just about 3,000 cm^{-1}, belonging to the aromatic ring C-H absorptions.

11 This molecule contains a **primary amide (*R*CONH$_2$)** functional group. You see two very large peaks of interest here. The first is the enormous carbonyl C=O stretch at 1,690 cm^{-1}. The second is the two cow udder–like peaks at 3,100–3,400 cm^{-1}, which represent an NH_2 group. Although these two peaks could represent separated carbonyl and primary amine functional groups, the broadness of the C=O peak (just as in the case of a carboxylic acid) suggests that the two functional groups are connected to each other to form a primary amide ($CONH_2$).

12 Each of these molecules contains a C=O group, so each should have a stretch at around 1,700 cm^{-1}. However, ester C=O stretches, such as the ester C=O in methyl acetate, absorb at slightly higher wavenumbers than other carbonyls between 1,735–1,750 cm^{-1}, distinguishing this functional group from the others. Carboxylic acids, such as acetic acid, can be distinguished because of the additional OH stretch not present in the other molecules. You can distinguish ketones and aldehydes from each other because aldehydes, such as acetaldehyde, have a C-H attached to the carbonyl that absorbs at about 2,800 cm^{-1}, which ketones lack.

Chapter 16

Putting Molecules under the Magnet: NMR Spectroscopy

In This Chapter

- Looking for molecular symmetry
- Understanding chemical shift, integration, and coupling
- Solving structures using NMR

Nuclear magnetic resonance (NMR) spectroscopy is the most powerful tool in the chemist's arsenal for determining the structure of unknown compounds. NMR can help you figure out the precise molecular structure of whatever organic stuff you're looking at. You've probably already seen a few NMR spectra — lovely little haystacks of peaks and squiggles and numbers that you're supposed to able to convert into a chemical structure on an exam. NMR is an instrumental method that collects information about the chemical environments around each nucleus in a molecule. In *carbon (^{13}C) NMR,* you look at the chemical environments of the carbon nuclei; much more common, however, is *proton (^{1}H) NMR,* in which you look at the chemical environments of the hydrogen nuclei. Seeing the chemical environments of each of the protons in your molecule allows you to piece together the structure.

Because of the complexity of this instrumental method, this chapter takes a gradual approach for getting you up to speed with solving chemical structures using NMR spectroscopy. First, I include problems to help you see how each part of the spectrum works and how you can extract all the information from that part of the spectrum. Next, you go over the meaning of the chemical shift, integration, and coupling. After you see how to extract information about a molecule from those pieces of data, you put it all together to decipher entire unknown structures given a spectrum (or multiple spectra).

Seeing Molecular Symmetry

NMR spectroscopy reports on the different chemical environments of the nuclei. You may expect every nucleus in a molecule to give rise to a different signal in the NMR spectrum. In this case, a molecule with five hydrogens would give rise to five signals in the ^{1}H NMR spectrum, and a molecule with six hydrogens would give rise to six signals, and so forth. This is sometimes the case, but as far as NMR is concerned, nuclei that are in identical chemical environments are indistinguishable.

Identical chemical environments means two nuclei are attached to exactly the same types of atoms in the same order. The simplest example is H_2, in which both hydrogens are attached to one H. A more common example is that, in general, two hydrogens attached to the same carbon are in identical chemical environments. Two (or more) nuclei that exist in identical chemical environments give rise to a single peak in the NMR spectrum, making these nuclei *chemically equivalent.*

A lot of times, chemical equivalence arises because of molecular symmetry. In a molecule that has a plane of symmetry, one side of the molecule is the mirror image of the other side, and any atom on one side of the symmetry plane has a twin on the other side of the plane in exactly the same chemical environment. These two chemically equivalent atoms show up as a single peak in the NMR spectrum.

Spotting planes of symmetry in molecules is important so you can determine how many signals a molecule should show in the NMR spectrum. In many cases, symmetry can distinguish two very similar molecules from each other simply because they give a different number of signals.

To figure out how many signals should appear for molecules with molecular symmetry, follow these steps:

1. **Identify any mirror planes in the molecule.**

   Draw each plane of symmetry into the molecule to make the situation clearer.

2. **Count the number of chemically unique atoms.**

   Each atom on one side of the plane has an identical twin on the other side, so only one of these should be counted.

Q. How many signals would you see in the ^{13}C and ^{1}H NMR spectra of the following compound?

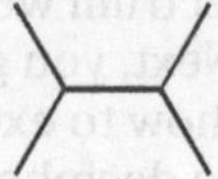

A. **Two signals in both the ^{13}C and ^{1}H NMR.**

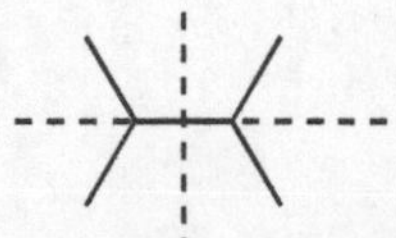

The dotted lines indicate two planes of symmetry in the molecule. Therefore, the top side of the molecule is chemically equivalent to the bottom side, and the right-hand side is chemically equivalent to the left-hand side. Therefore, all the methyl group hydrogens (CH_3's) are chemically equivalent, and the two CH's are also chemically equivalent. Similar reasoning shows that all the methyl carbons and the methine (CH) carbons are also chemically equivalent. Therefore, you see only two signals in the ^{13}C NMR spectrum, one from the methyl carbons and one from the methine carbon, and you likewise see two signals in the ^{1}H NMR spectrum from the methyl hydrogens and the methine hydrogens.

Q. How many signals would you see in the ^{13}C and ^{1}H NMR spectra of the following compound?

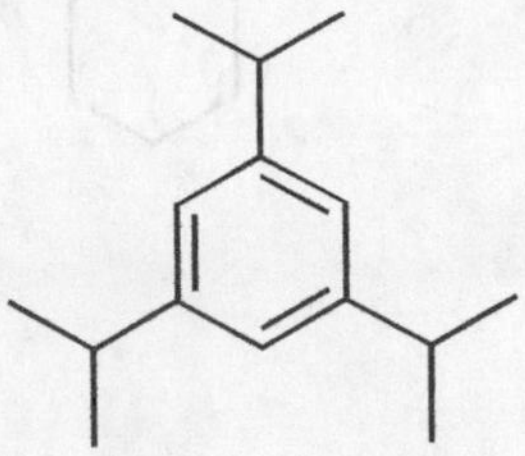

A. **Three hydrogen signals in the ^{1}H NMR and four signals in the ^{13}C NMR.**

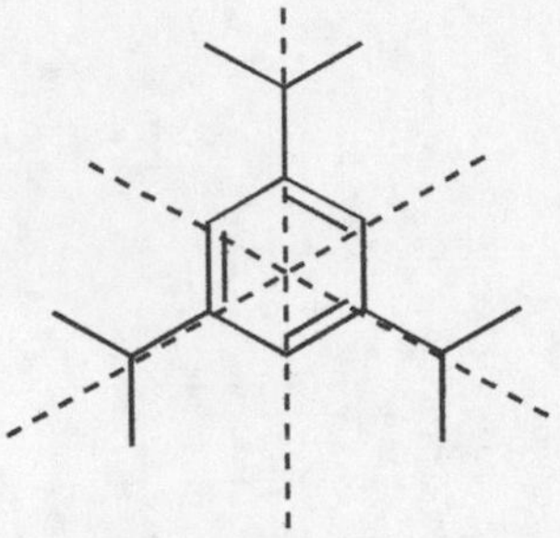

In this molecule, you have three planes of symmetry. As a result, all the methyl (CH_3) hydrogens and methyl carbons are chemically equivalent, and the CH groups to which they attach are also chemically equivalent. Finally, you encounter two types of ring carbons: the carbons to which the isopropyl groups are attached (which have no attached hydrogens) and the ring carbons with a hydrogen. Thus, the molecule has four distinct carbon types — the two ring carbons, the methyl (CH_3) carbons, and the methine (CH) carbons — and three distinct hydrogen types: the methyl hydrogens, the methine hydrogens, and the ring hydrogens.

1. How many signals would you see in the ^{13}C and ^{1}H NMR spectra of the following compound?

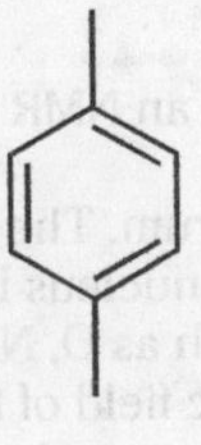

Solve It

2. How many signals would you see in the ^{13}C and ^{1}H NMR spectra of the following compound?

O

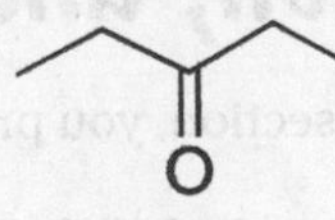

3. How many signals would you see in the ^{13}C and ^{1}H NMR spectra of the following compound?

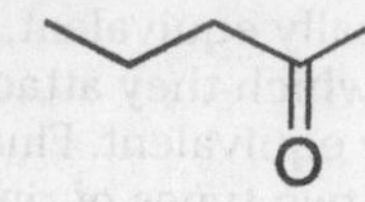

Solve It

4. How many signals would you see in the ^{13}C and ^{1}H NMR spectra of the following compound?

Solve It

Working with Chemical Shifts, Integration, and Coupling

In this section, you practice working with the individual parts of an NMR spectrum:

- **Chemical shift:** This is where the peaks show up on a spectrum. The higher the chemical shift of a peak (larger ppm value), the more deshielded that nucleus is. In most cases, *deshielding* is caused by adjacent electronegative atoms such as O, N, and halogens. Benzene ring hydrogens are also deshielded by the magnetic field of the circulating electrons in the aromatic ring, which is termed *diamagnetic anisotropy* for those of you who like naming things (scary word, though, huh?).
- **Integration:** Integration is those little *s*-shaped curves over the peaks. The height of these curves gives information about the relative number of hydrogens in that peak. A peak that has an integration that's twice as high as the integration of another peak means that the first peak represents twice as many protons as the other peak.
- **Coupling:** This is the way some peaks split into a number of different lines. The number of lines a peak splits into depends on the number of hydrogen neighbors the peak has, following the $n + 1$ rule. Thus, a proton with one hydrogen neighbor splits into two lines (a *doublet*), a proton with two hydrogen neighbors splits into three lines (a *triplet*), and so forth.

When solving a structure, knowing the absolute number of hydrogens each peak represents is a very important piece of information. You can typically assume that *all* the hydrogens in the molecule show up in the ^{1}H NMR. Comparing the heights of the integration curves gives you the *relative* number of hydrogens in each peak. Unfortunately, it doesn't tell you exactly how many hydrogens each peak represents. Here's how to get the actual number of hydrogens:

1. **Determine the heights of the integration curves.**

   Sometimes you can eyeball it, but if that proves difficult, use a ruler. Measure the height of each curve, from where the curve is flat up to where it goes flat again.

2. **Divide each of the heights by the smallest height to get the relative ratio.**

   The ratio of heights, from left to right, represents the ratio of hydrogens in each peak.

3. **Find the sum of the relative ratios.**

4. **Compare this sum to the total number of hydrogens in the molecule.**

   If the numbers match, the relative ratio equals the actual number of hydrogens in each peak. If not, divide the total number of hydrogens by the sum of the ratio; then multiply this answer by each number in the relative ratio to find the actual number of hydrogens.

   For example, suppose you have two peaks in a proton NMR, and one has an integration curve that's twice as large as the other, in a 2:1 height ratio. From the formula, you know there are six hydrogens in the molecule. So you take this number (6) divided by the sum of the ratios (2 + 1 = 3), which gives you 2. Then multiply the relative ratio of 2:1 by 2 to get the absolute number of hydrogens in each peak, which is 4:2. The taller peak represents the signal from four hydrogens, and the smaller peak represents the signal from two hydrogens.

Q. The following integration curves represent four signals in the ^{1}H NMR spectrum (peaks not shown for clarity) of a molecule with formula C_8H_8O. How many hydrogens does each peak represent?

A.

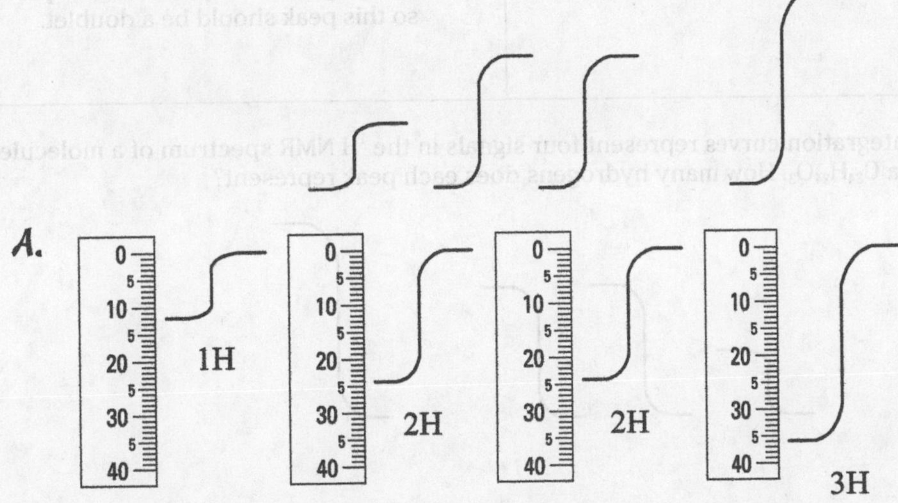

In this problem, the first thing you want to do is get the relative heights of the integration curves. You can use a ruler. In this case, the heights are 12:24:24:36 mm, which, dividing by the smallest height (12), gives you a relative ratio of 1:2:2:3. The sum of these relative ratios is 8 (1 + 2 + 2 + 3). Because the formula includes eight hydrogens, the relative ratio is also the absolute number of hydrogens in each peak. Thus, the left-most peak represents 1H; the second and third peaks, 2H; and the last, 3H.

Q. **Rank the hydrogens in the shown molecule from highest chemical shift (1) to lowest chemical shift (3) in the ^{1}H NMR spectrum. Also, rank the carbons from highest chemical shift to lowest.**

$CH_3CH_2CH_2Cl$

A. **3 2 1**

$CH_3CH_2CH_2Cl$

Hydrogens adjacent to electronegative elements (such as chlorine) come at higher chemical shift than hydrogens farther away from electronegative atoms; likewise for carbons, so the carbon attached to the chlorine comes at highest chemical shift and the carbon farthest away comes at the lowest chemical shift.

Q. **Predict the coupling (number of lines) each peak would split into for the protons *a*, *b*, and *c* using the *n* + 1 rule.**

(structure: $(H_3C)_2CH{-}C(=O){-}H$, methyl protons labeled **a**, methine proton **b**, aldehyde proton **c**)

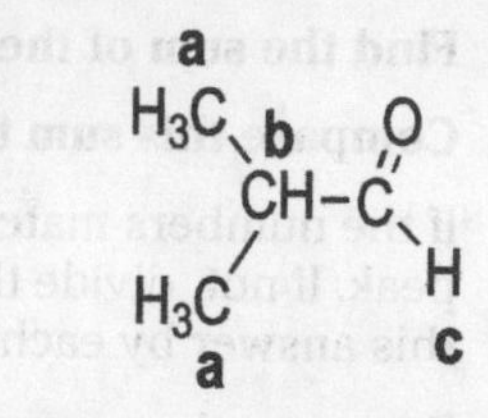

A. **A = doublet, b = septet, c = doublet.** **Follow the *n* + 1 rule. The two methyl (CH_3) group protons, *a*, are in identical chemical environments because both methyls are attached to the same carbon. These protons are adjacent to one hydrogen, so you'd expect the signal for these protons to split into two lines, or a doublet, following the *n* + 1 rule. For the methine (CH) proton *b*, there are six hydrogen neighbors, so you'd expect this peak to split into seven lines, or a septet. The proton *c* is adjacent to one proton, so this peak should be a doublet.**

5. The following integration curves represent four signals in the ^{1}H NMR spectrum of a molecule with the formula $C_{24}H_{24}O_3$. How many hydrogens does each peak represent?

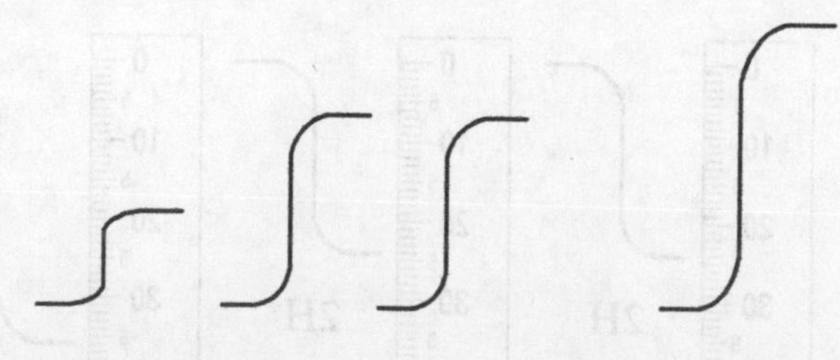

Solve It

6. Assign each of the H's to the approximate region you expect the peak for this H to show up in the ^{1}H NMR spectrum. ***Note:*** Only one structure is present for each region.

X = halogen, O, N

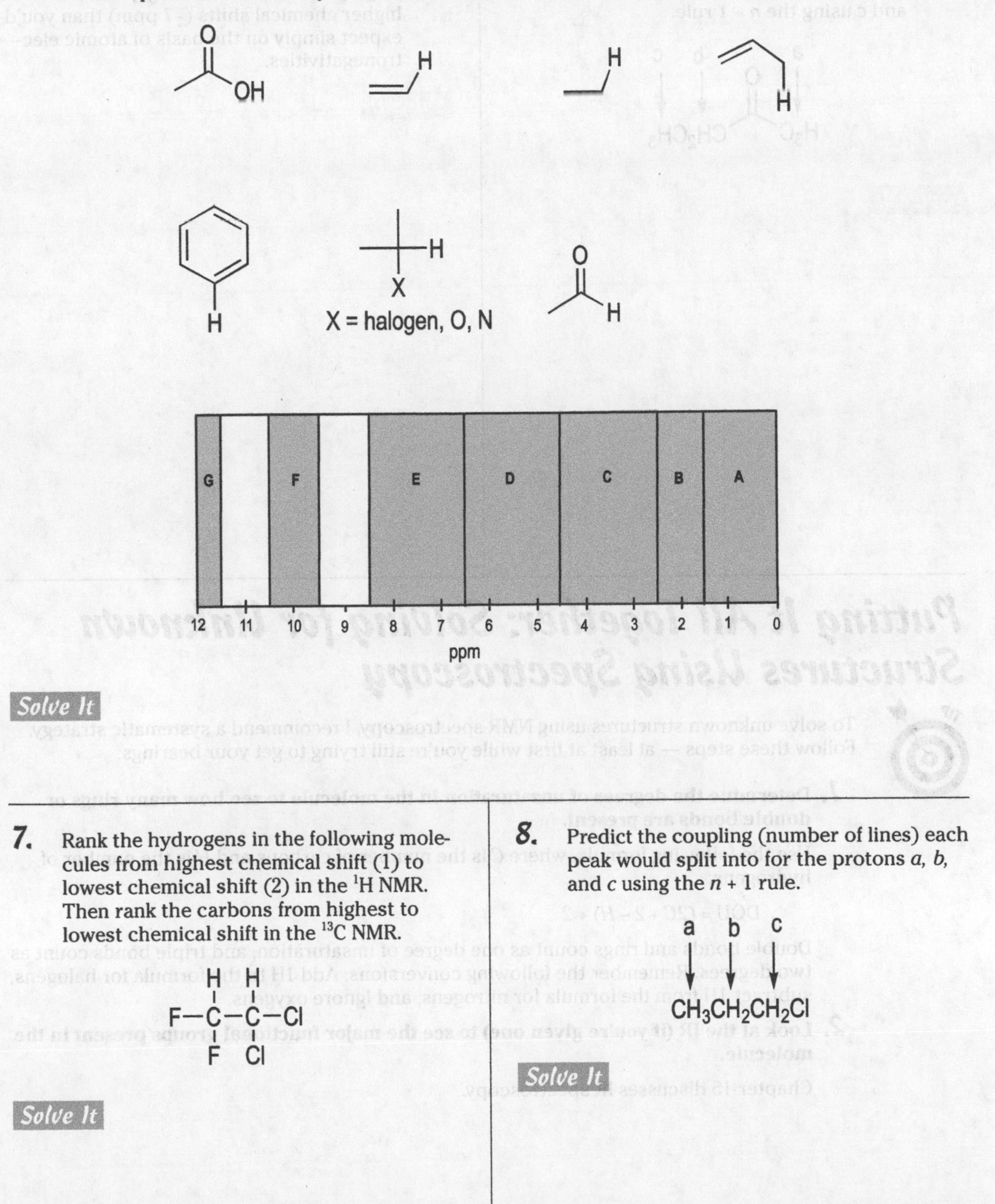

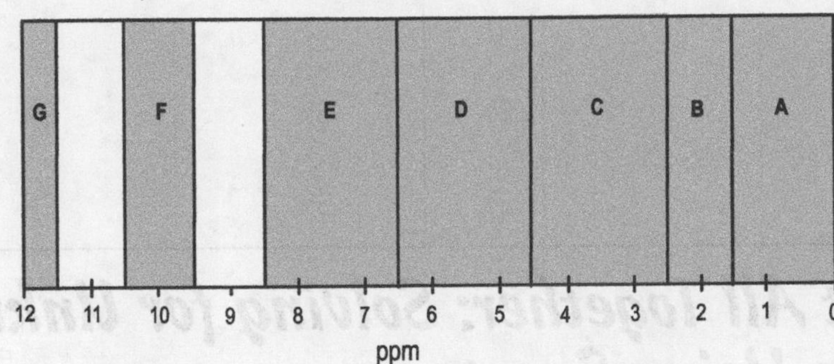

Solve It

7. Rank the hydrogens in the following molecules from highest chemical shift (1) to lowest chemical shift (2) in the ^{1}H NMR. Then rank the carbons from highest to lowest chemical shift in the ^{13}C NMR.

$F{-}CHF{-}CHCl{-}Cl$ (F–C(H)(F)–C(H)(Cl)–Cl)

Solve It

8. Predict the coupling (number of lines) each peak would split into for the protons *a, b,* and *c* using the *n* + 1 rule.

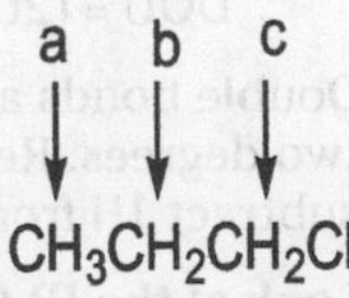

Solve It

9. Predict the coupling (number of lines) each peak would split into for the protons *a*, *b*, and *c* using the *n* + 1 rule.

a b c

H_3C–C(=O)–CH_2CH_3

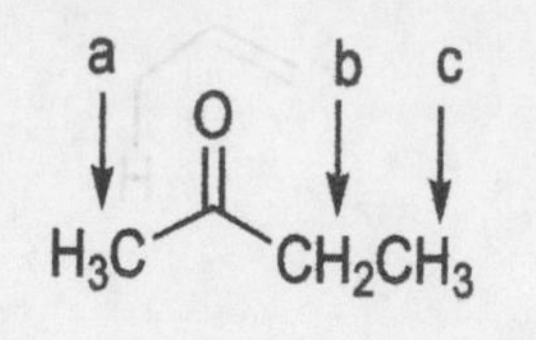

Solve It

10. Explain why protons attached to aromatic rings (such as benzene rings) come at higher chemical shifts (~7 ppm) than you'd expect simply on the basis of atomic electronegativities.

Solve It

Putting It All Together: Solving for Unknown Structures Using Spectroscopy

To solve unknown structures using NMR spectroscopy, I recommend a systematic strategy. Follow these steps — at least at first while you're still trying to get your bearings:

1. **Determine the degrees of unsaturation in the molecule to see how many rings or double bonds are present.**

   Use the following formula, where *C* is the number of carbons and *H* is the number of hydrogens:

   $$DOU = (2C + 2 - H) \div 2$$

   Double bonds and rings count as one degree of unsaturation, and triple bonds count as two degrees. Remember the following conversions: Add 1H to the formula for halogens, subtract 1H from the formula for nitrogens, and ignore oxygens.

2. **Look at the IR (if you're given one) to see the major functional groups present in the molecule.**

   Chapter 15 discusses IR spectroscopy.

3. Determine from the integration how many hydrogens each peak represents, and use that information to determine the structure of all the fragments of the molecule.

Table 16-1 shows a list of the common fragments. (See the preceding section for info on finding the number of hydrogens each peak represents.)

Table 16-1 Common Fragments

<table>
<tr><th>Number of Hs</th><th>Likely Fragment</th><th>Notes</th></tr>
<tr><td>1H</td><td>H
|
—C
|</td><td>Check the IR spectrum to make sure this is not an alcohol (OH), secondary amine (NH), aldehyde (CHO), or acid proton (COOH).</td></tr>
<tr><td>2H</td><td>H
|
—C—H
|</td><td>Rarely: This is two symmetric CH groups.</td></tr>
<tr><td>3H</td><td>H
|
—C—H
|
H</td><td>Rarely: This is three symmetric CH groups.</td></tr>
<tr><td>4H</td><td>2 H
|
—C—H
|</td><td>Usually two symmetric methylene (CH_2) groups.</td></tr>
<tr><td>6H</td><td>2 H
|
—C—H
|
H</td><td>Rarely: This is three symmetric CH_2 groups. Often indicates an isopropyl group.</td></tr>
<tr><td>9H</td><td>3 H
|
—C—H
|
H</td><td>Often indicative of a tertiary butyl group.</td></tr>
</table>

Sum together all the atoms in your fragments to make certain that the sum of the atoms in the fragments matches the molecular formula (to make sure you're not missing any atoms).

4. **Put together the fragments of the molecule that makes sense with the chemical shift and coupling.**

   Generate all possible structures based on the fragments you found in Step 3. Remember that structures you propose almost always have no charges and that each atom obeys the octet rule.

5. **Double check your proposed structure to ensure it makes sense with the chemical shifts, integration, and coupling.**

   Go back and predict what you think your proposed structure's NMR spectrum would look like and see whether it matches the actual spectrum.

Q. Give the structure of the molecule with formula $C_4H_8O_2$ that gives the following ^{1}H NMR and IR spectra.

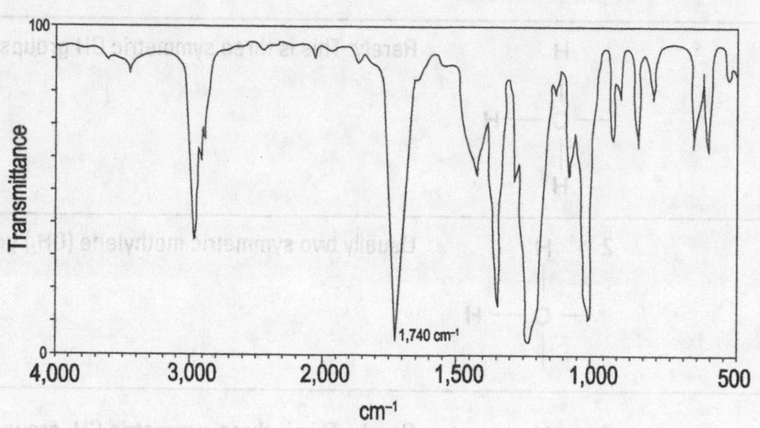

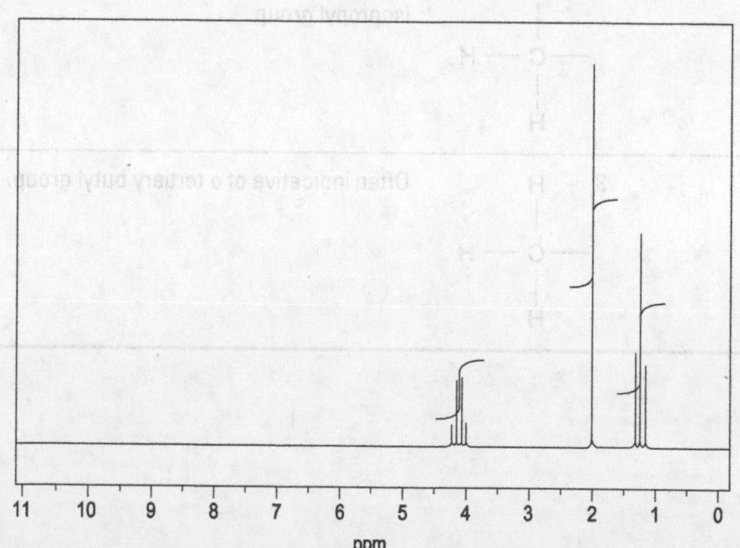

A. Tackle the problem stepwise. First, determine the degrees of unsaturation. You have four carbons and eight hydrogens, so plugging your numbers into the DOU formula gives you DOU = $(2 \times 4 + 2 - 8) \div 2 = (8 + 2 - 8) \div 2 = 1$. There's one degree of unsaturation, suggesting the molecule has either one ring or one double bond.

Next, look at the IR spectrum to determine any major functional groups. The peak that should stand out to you is the large peak at 1,740 cm^{-1}, indicating a carbonyl (C=O) group. Furthermore, 1,740 cm^{-1} is quite high for a carbonyl group (aldehydes and ketone C=O stretches absorb ~1,700–1,715 cm^{-1}). This high carbonyl frequency indicates an ester carbonyl stretch, which absorbs at higher wavenumbers than aldehydes, ketones, or carboxylic acids. Because an ester contains the C=O double bond, this double bond accounts for the one degree of unsaturation in the molecule; thus, the molecule can't contain any other double bonds or rings.

Then determine the fragments of the molecule. Look at the integrations to determine the number of hydrogens each peak represents. The relative integration ratio, from left to right, is 2:3:3. To get the actual number of hydrogens that each peak represents, sum the relative ratios, which gives you 8 (2 + 3 + 3). Because the molecular formula includes eight hydrogens, the relative ratio and the absolute ratio of hydrogens are the same. Thus, from left to right, the peaks represent 2H, 3H, and 3H, respectively. Next, find the fragments. The 2H peak most likely represents a CH_2 fragment, and the two 3H peaks most likely represent two CH_3 fragments (see Table 16-1). You already know the molecule contains the ester fragment. In this case, the sum of the atoms in the fragments matches the molecular formula, so you're good to go.

Now piece together the fragments in a way that makes sense with the chemical shifts and the peak coupling. In this case, you have only two options (A and B) for piecing the fragments together so that each carbon has four bonds. It comes down to distinguishing these two structures. If the structure were A, you'd expect the CH_2 fragment to have the highest chemical shift because it's directly bonded to the oxygen. If the structure were B, you'd expect one of the methyl (CH_3) groups to have the highest chemical shift as a result of being directly bonded to the oxygen. In this case, the highest chemical shift is the CH_2 fragment, so the structure is A.

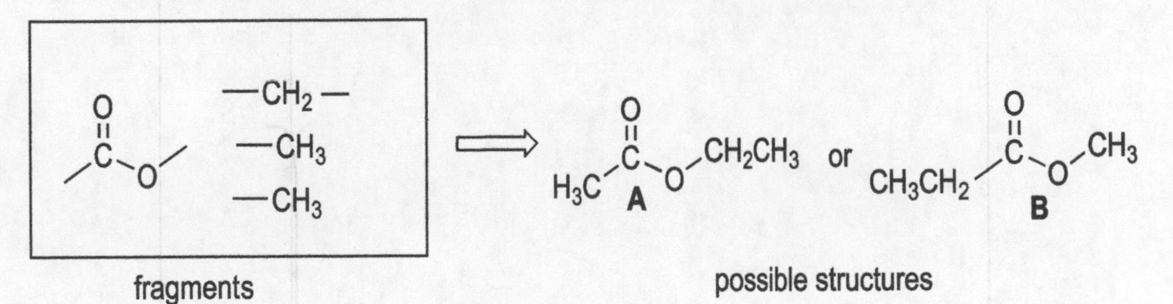

Finally, double check your structure to make sure everything makes sense. The peak with the highest chemical shift should be a quartet (because it's adjacent to three other hydrogens), and it should integrate for 2H. Check! You should see a peak at a lower chemical shift that integrates for 3H and is a triplet. Check! Finally, there should be another peak that integrates for 3H and is a singlet (because it has no neighboring hydrogens). Check! Looks like this structure's in the bag.

11. Give the structure of the molecule of formula C_4H_9Cl that gives the following ^1H NMR spectrum.

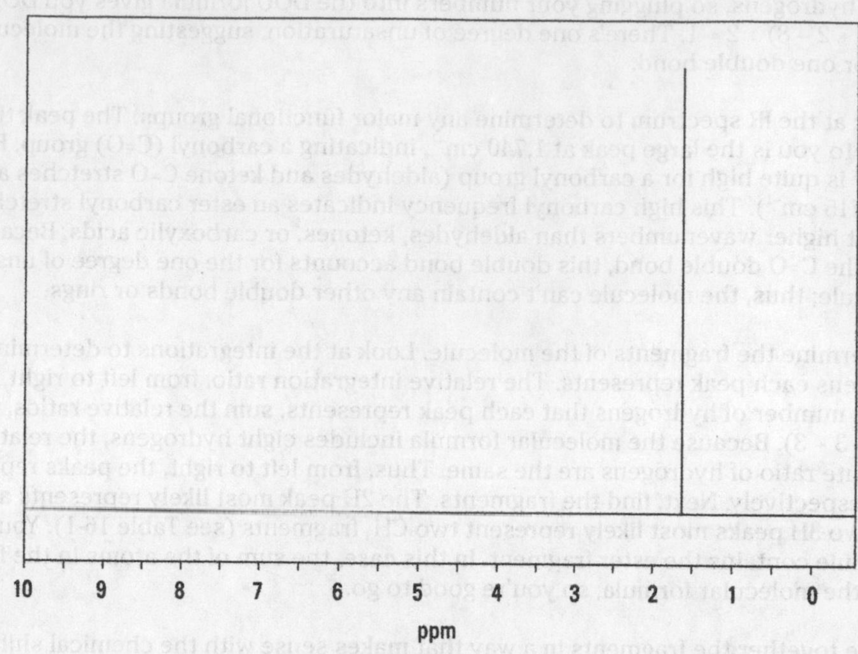

Solve It

12. Give the structure of the molecule of formula $C_{10}H_{14}$ that gives the following ^1H NMR spectrum.

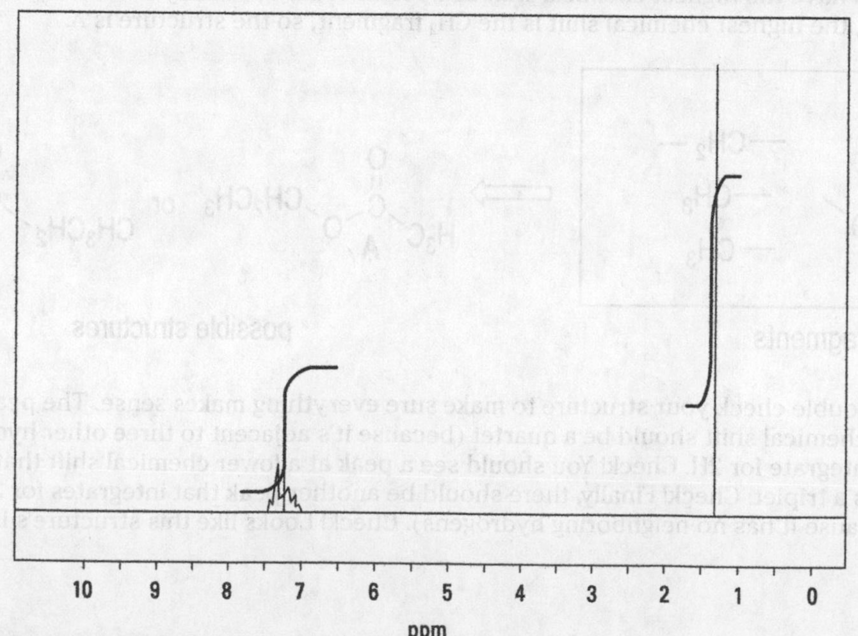

Solve It

13. Give the structure of the molecule of formula $C_4H_{10}O$ that gives the following ^1H NMR and IR spectra.

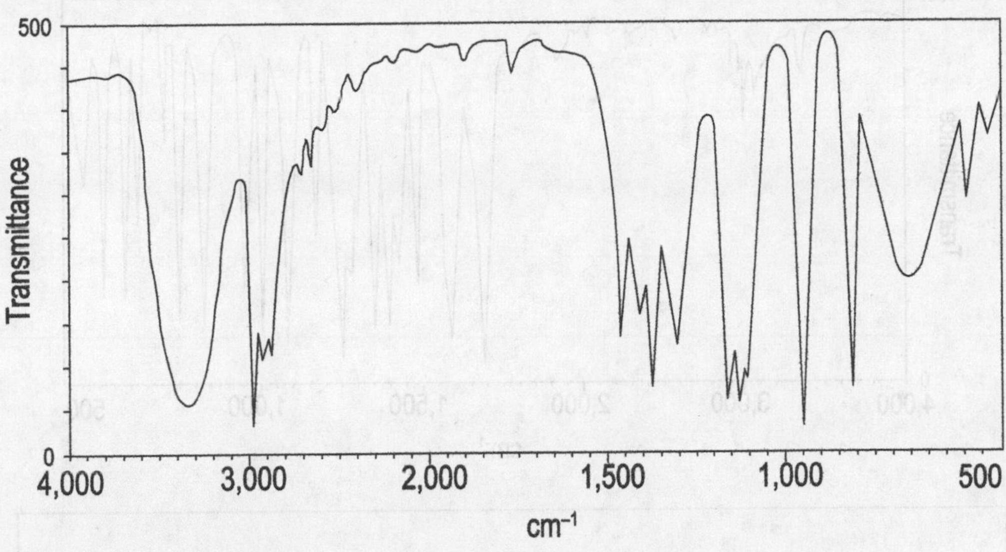

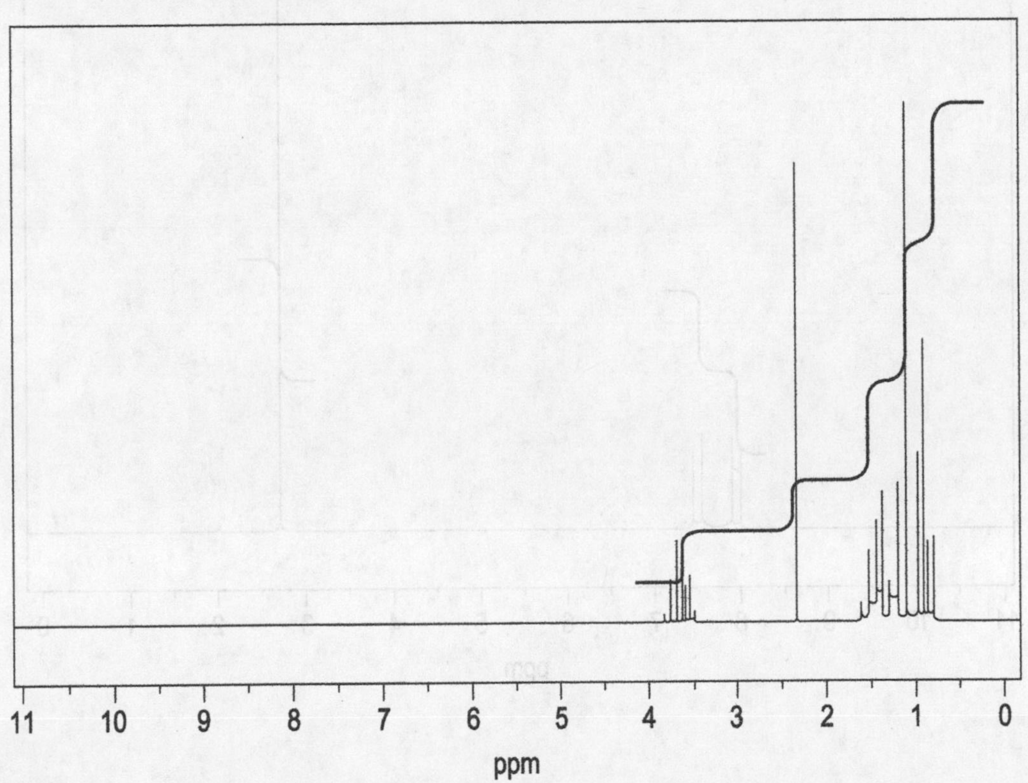

Solve It

14. Give the structure of the molecule of formula C_8H_7OCl that gives the following ^1H NMR and IR spectra.

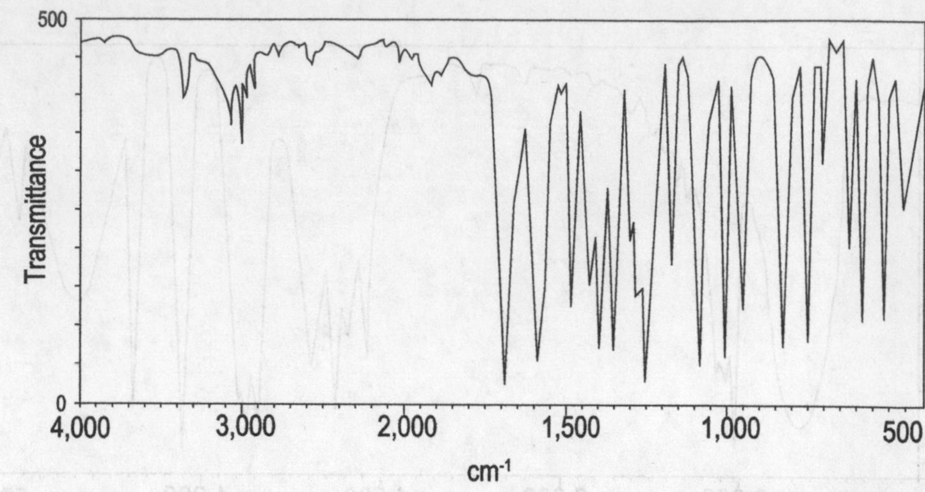

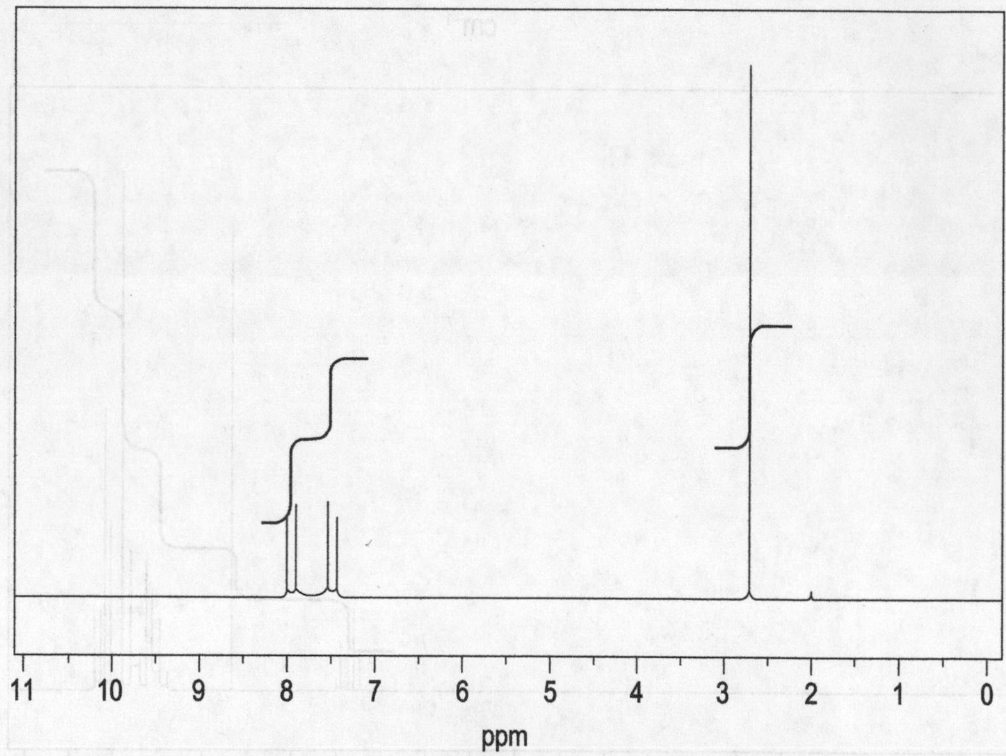

Solve It

15. Give the structure of the molecule of formula $C_9H_{10}O$ that gives the following ^1H NMR and IR spectra.

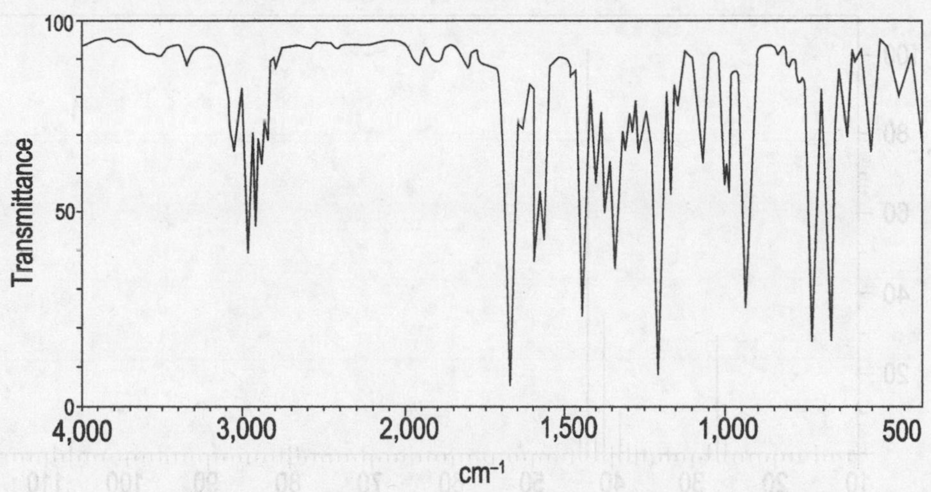

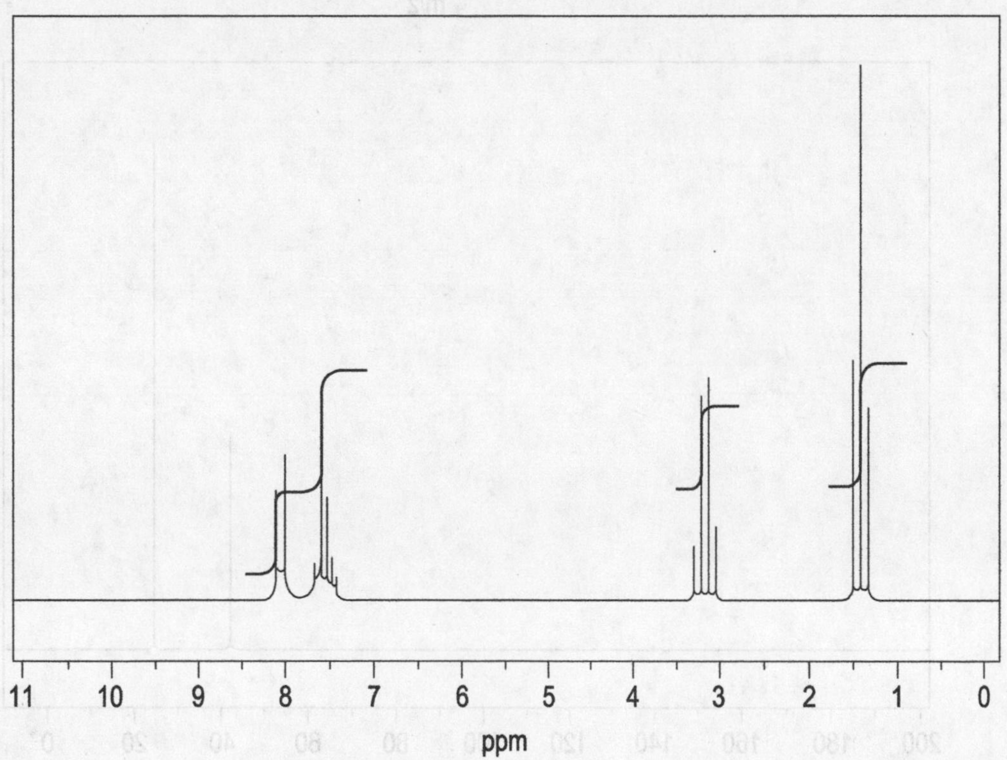

Solve It

16. Give the structure of the molecule that gives the following ^{13}C NMR and mass spectra.

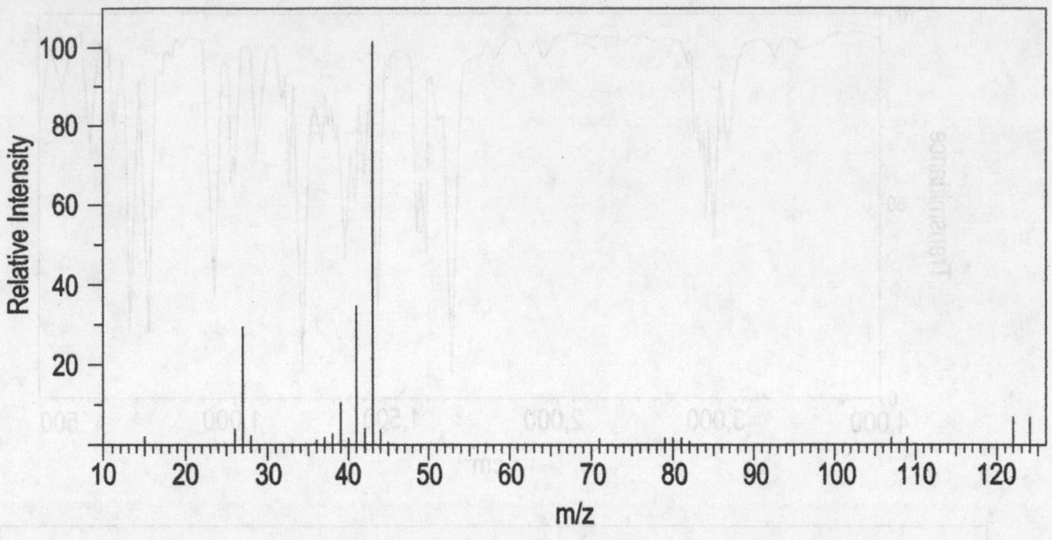

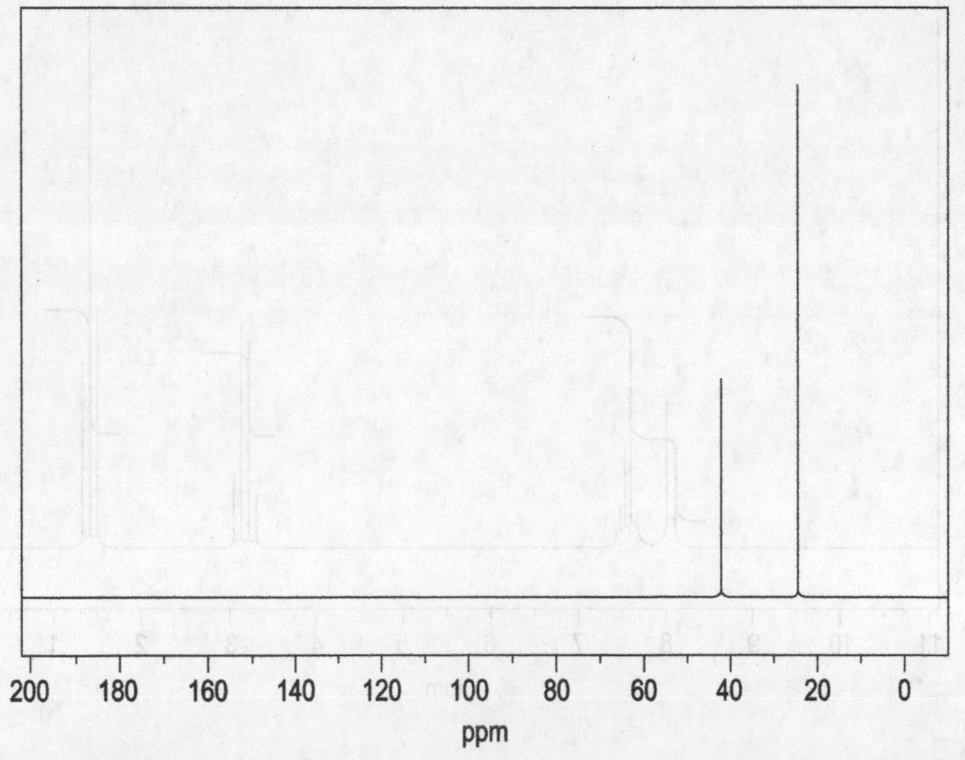

Solve It

17. Draw what you think the ^{1}H NMR spectrum of the following molecule would look like. Be sure to include coupling, integration curves, and reasonable approximations of the chemical shift values.

$CH_3CH_2CH_2Cl$

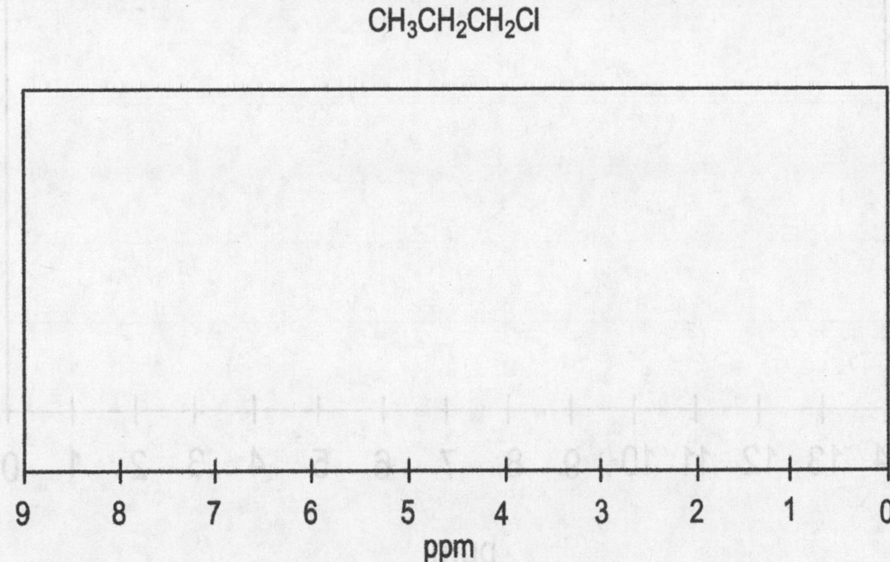

Solve It

18. Draw what you think the ^{1}H NMR spectrum of the following molecule would look like. Be sure to include coupling, integration curves, and reasonable approximations of the chemical shift values.

COOH

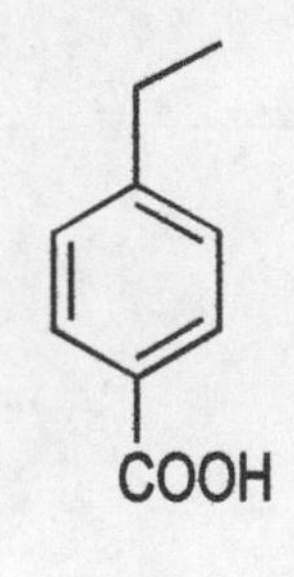

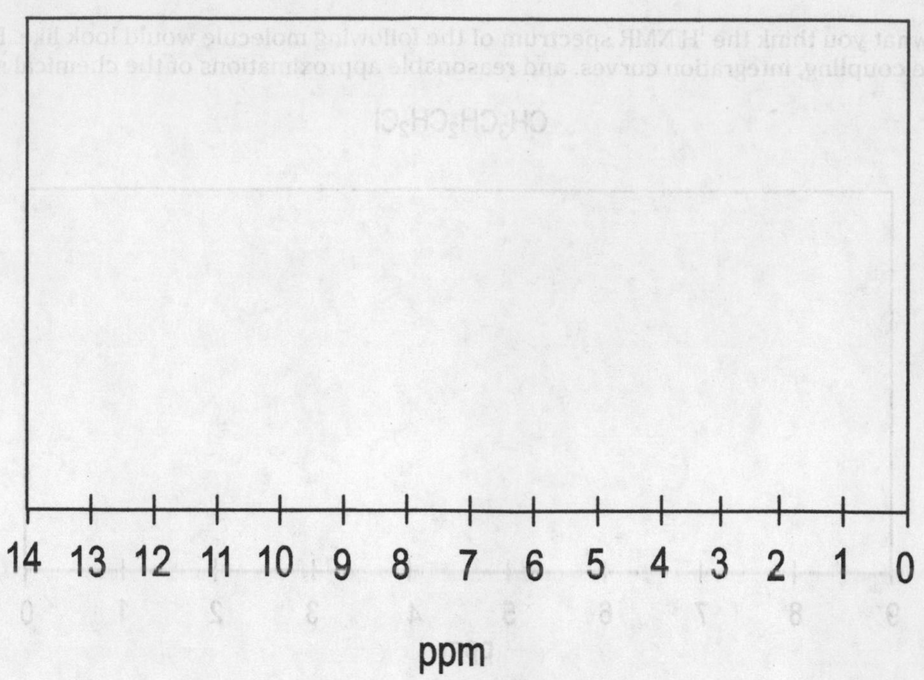

Solve It

Answer Key

The following are the answers to the practice questions presented in this chapter.

1 **The molecule has two planes of symmetry. As a result, there are two signals in the ^{1}H NMR and three signals in the ^{13}C NMR.**

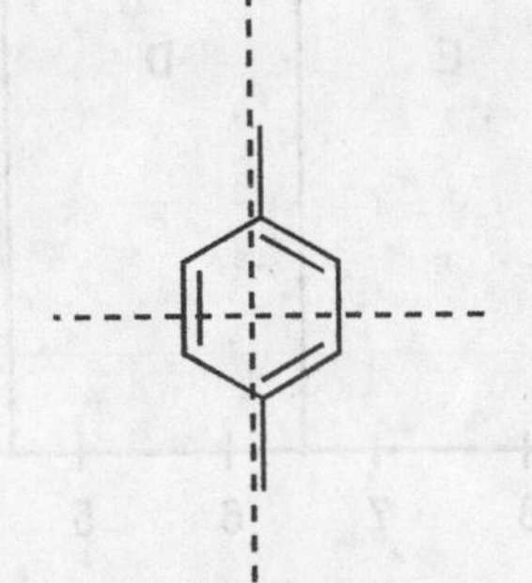

2 **This molecule has one plane of symmetry. Therefore, there are two signals in the ^{1}H NMR spectrum and three signals in the ^{13}C NMR spectrum.**

O

3 **This molecule has no planes of symmetry. Therefore, you see one signal for each carbon (five of them) and each set of hydrogens (four of them).**

4 **Cyclohexane has three vertical planes of symmetry. Therefore, every carbon and hydrogen is chemically equivalent. One peak appears in both the ^{13}C NMR and ^{1}H NMR spectra of this molecule.**

5 **The relative ratio is 1:2:2:3, which adds up to 8. Because 24 hydrogens are in the molecule, you multiply this relative ratio by 3. From left to right, the peaks thus represent 3H, 6H, 6H, and 9H.**

6

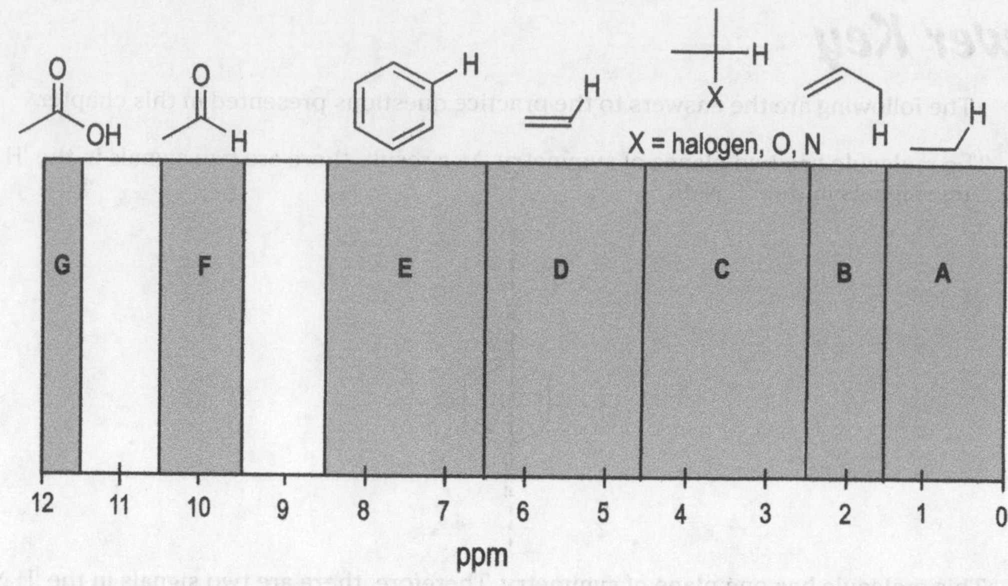

7

1 2
H H
F—C—C—Cl
F Cl

Fluorine is a more electronegative atom than chlorine (electronegativity increases as you go up and to the right on the periodic table). Thus, the proton on the carbon containing the fluorines has a higher chemical shift than the proton attached to the carbon containing the chlorine atoms. Similarly, in the ^{13}C NMR, the carbon with the attached fluorines would have a higher chemical shift than the carbon with the less electronegative chlorines attached.

8 Protons *a* are adjacent to two other protons, so the signal for these protons is split into three lines following the $n + 1$ rule (called a triplet). Protons *b* are adjacent to five hydrogens, so these split into approximately a six lines (called a sextet). Protons *c* are adjacent to two other protons, so these split into three lines (another triplet).

9 Protons *a* are adjacent to no other protons, so they don't split at all, following the $n + 1$ rule (where $n = 0$). A peak that isn't split is called a singlet. Protons *b* are adjacent to three proton neighbors, so the peak for these protons splits into four lines (called a quartet). Protons *c* are adjacent to two other protons, so this peak splits into three lines (a triplet).

10 Benzene rings are a special case. The pi electrons in benzene are induced to circulate around the ring when placed in a magnetic field (like a big magnet used in an NMR experiment). Anytime you have a moving charge (such as moving electrons), it creates a magnetic field of its own. In this case, the small magnetic field created by the moving pi electrons in the benzene ring reinforces the external magnetic field for protons attached to the benzene ring. The net result is that the protons attached to the ring are deshielded and come at a higher chemical shift as a result (this effect is termed *diamagnetic anisotropy*).

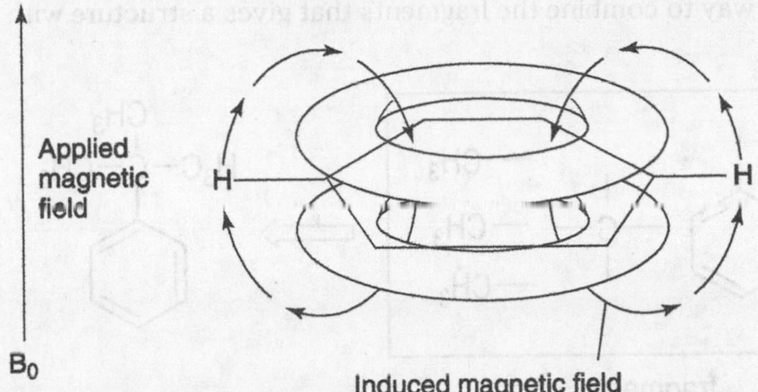

11 Insertion of the formula C_4H_9Cl into the formula for degrees of unsaturation shows that the structure has zero degrees of unsaturation, indicating that there are no double bonds or rings. Recall that with halides, you add hydrogen to the molecular formula, so you pretend the formula is C_4H_{10} when determining the degrees of unsaturation.

This problem is interesting because you see just a single peak at ~1.5 ppm. Thus, you can assume that this peak represents all nine hydrogens in the formula. From Table 16-1, you see that a peak with nine hydrogens generally indicates three chemically equivalent methyl groups (CH_3's), and it often indicates a tertiary butyl group. Indeed, summing up the atoms in the three CH_3 fragments and the chlorine leaves one carbon remaining (which can't have a proton on it because there's no signal for it in the ^{1}H NMR). There's only one way to combine these fragments, which gives you tertiary butyl chloride.

—Cl —CH_3 —CH_3 —CH_3 —C— ⟹ $H_3C-C(CH_3)_2-Cl$

fragments

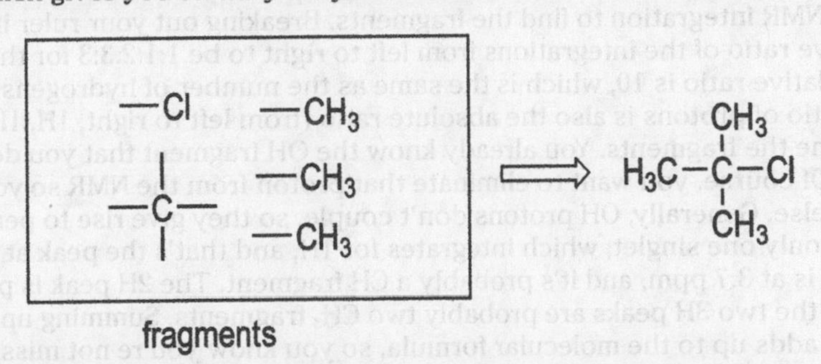

12 Plugging the formula $C_{10}H_{14}$ into the DOU equation gives you four degrees of unsaturation. Thus, there must be some combination of rings and double bonds that equals four. Typically, four or more degrees of unsaturation is a big tipoff for a benzene ring in the molecule because benzene rings have four degrees of unsaturation (three for the three double bonds and one for the ring). You can confirm this by looking at the ^{1}H NMR for peaks in the aromatic region (6.5–8.5 ppm). Indeed, you see a big haystack between 7–7.5 ppm, strongly suggesting a benzene ring.

Now use the NMR integration to find the fragments. After you whip out your ruler and get the ratios (or eyeball it if you're really, really good), you see that the ratio is 5:9 for the two peaks (you may have been tempted to go with 1:2, but the sum of these, 3, is not a divisor of 14, the number of hydrogens you have in the formula — a big clue that this can't be right). The fact that the aromatic protons integrate for five tells you that the benzene ring is substituted one time. Finally, the peak that integrates for 9H is probably three symmetrical CH_3 groups (and very probably, a tertiary butyl group). Summing the ring atoms (C_6H_5) and the three methyl groups (C_3H_9) leaves one carbon missing.

There's only one way to combine the fragments that gives a structure with carbons that all have four bonds:

$-CH_3$ $-C-$ $-CH_3$ $-CH_3$ ⟹ CH_3 $H_3C-C-CH_3$

fragments

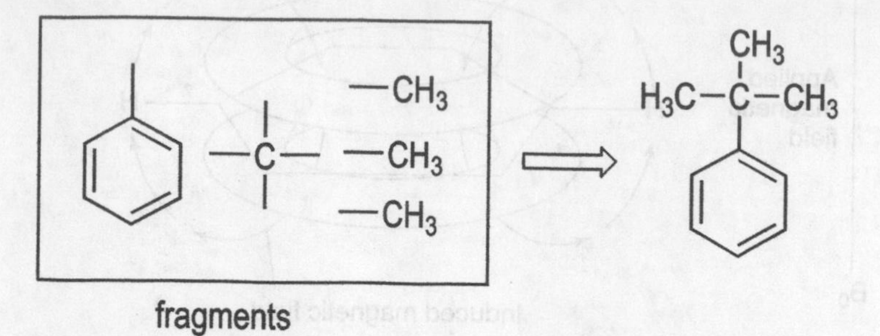

Finally, double check the structure. You should see peaks for aromatic protons that integrate for 5H, as well as a singlet that integrates for 9H. Check!

13 Plugging $C_4H_{10}O$ into the DOU formula gives you zero degrees of unsaturation (you ignore oxygen). Thus, this molecule can't contain any rings or double bonds.

Now look at the IR spectrum to determine the functional groups. The most glaring peak in the IR spectrum is the distinct OH peak at 3,400 cm^{-1}. You see a couple of small peaks that look like they may be double bond or aromatic protons — but of course, with no degrees of unsaturation, the molecule can't contain any rings or double bonds (part of the art of interpreting these spectra is to extract the important information and ignore the red herrings).

Next, use the NMR integration to find the fragments. Breaking out your ruler if you need to, you find the relative ratio of the integrations from left to right to be 1:1:2:3:3 for the five peaks. The sum of this relative ratio is 10, which is the same as the number of hydrogens in the formula, so the relative ratio of protons is also the absolute ratio (from left to right, 1H, 1H, 2H, 3H, and 3H). Next, determine the fragments. You already know the OH fragment that you deduced from the IR spectrum. Of course, you want to eliminate that proton from the NMR so you don't assign it to something else. Generally, OH protons don't couple, so they give rise to peaks that are singlets. You see only one singlet, which integrates for 1H, and that's the peak at 2.5 ppm. The other 1H peak is at 3.7 ppm, and it's probably a CH fragment. The 2H peak is probably a CH_2 fragment, and the two 3H peaks are probably two CH_3 fragments. Summing up all the atoms in the fragments adds up to the molecular formula, so you know you're not missing any atoms.

You have two options for putting the fragments together so that all the carbons are bonded four times. You can eliminate one of the possibilities (A) immediately because this structure has two chemically equivalent CH_3 groups, which should give rise to only one signal. You see two CH_3 peaks in the NMR spectrum, so that can't be the right structure.

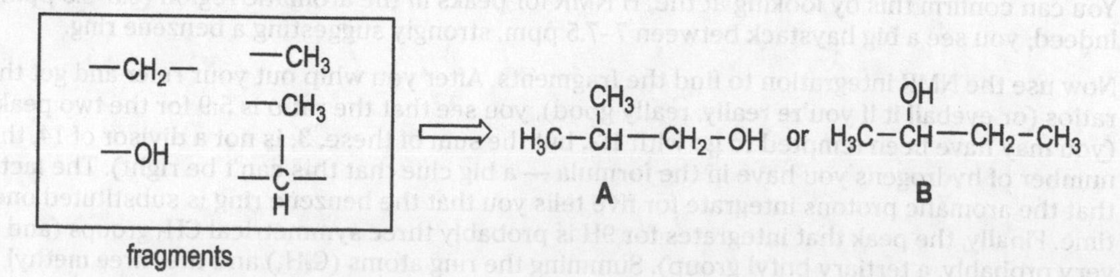

$-CH_2-$ $-CH_3$ $-CH_3$ $-OH$ $-C(H)-$ ⟹ $H_3C-CH(CH_3)-CH_2-OH$ or $H_3C-CH(OH)-CH_2-CH_3$

A B

fragments

Some of the couplings are difficult to read from the spectra. This happens on exams, so get used to having the coupling not always 100 percent clear (although in real life, you can zoom in on the coupling using computer tools). From left to right, the left methyl group (of B) should give a signal that's a doublet that integrates for 3H. Next, you should get a peak (which would give a *multiplet*) that integrates for 1H. Then the OH peak would give a singlet that integrates for 1H. The CH_2 group should give a multiplet (pentet) that integrates for 2H. And finally, the right-most CH_3 should be a triplet that integrates for 3H. As best you can tell with the fuzzy coupling data, everything checks out. And that's a wrap!

14 The DOU formula indicates that you have five degrees of unsaturation in this molecule. This is a big tip-off that you have an aromatic ring in the molecule, which would eat up four of those degrees of unsaturation.

The IR spectrum contains a big fat carbonyl peak at ~1,700 cm^{-1}. You possibly have some aromatic stretches (hard to tell because the spectrum's so noisy), but you can easily confirm aromatics by looking at the NMR for stretches between 6.5–8.5 ppm. Sure enough, by Jove, you have yourself an aromatic ring. The carbonyl (C=O) and the aromatic ring add up to five degrees of unsaturation, which takes care of all the double bonds and rings in the molecule.

The relative integration is 2:2:3. The sum of this relative ratio (2 + 2 + 3 = 7) matches the number of hydrogens in the molecular formula, so the peaks represent 2H, 2H, and 3H from left to right on the NMR spectrum. You've already identified the peaks between 7–8 ppm as hydrogens on an aromatic ring. The presence of four aromatic hydrogens tells you that the ring is substituted twice (because you've substituted two of the six benzene ring hydrogens with a substituent). When an aromatic ring is substituted twice, the substituents can be *ortho* (adjacent to each other), *meta* (two carbons away), or *para* (on opposite sides of the ring). Only in the para-substituted form would you get just two peaks as a result of the symmetry that bisects the ring (it looks like a mini field goal in the aromatic region).

The 3H peak most likely represents a CH_3 fragment. Adding up all the atoms in the fragments shows that you're missing a C, an O, and a Cl; the C and O are the C=O fragment that you identified earlier in the IR spectrum. Putting the fragments together leads to the following proposed structure:

-CH3
Cl
O
fragments
O
CH3
Cl

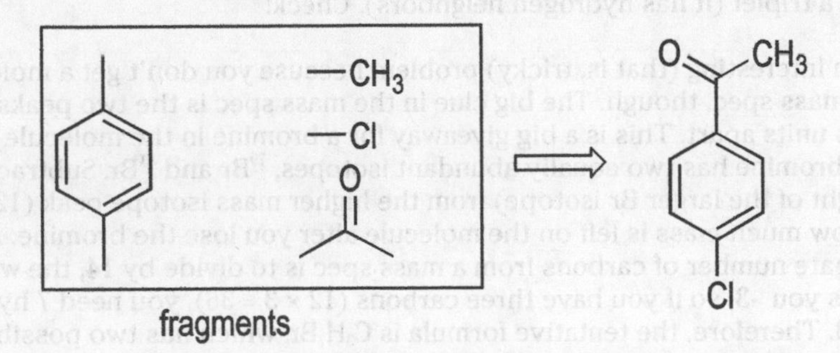

A quick check of the structure shows that everything is consistent with the spectral data. The CH_3 should be a singlet that's downfield (higher chemical shift) as a result of being adjacent to the C=O group, and it integrates for 3H. And there should be two peaks in the aromatic region of the spectrum, each of which should integrate for 2H. Check!

15 The formula $C_9H_{10}O$ has five degrees of unsaturation.

The IR shows a carbonyl stretch and aromatic stretches (little bumps at 1,800–2,000 cm^{-1}). The aromatics are confirmed by inspecting the NMR and observing peaks between 7–8 ppm.

The integration is 2:3:2:3, which adds up to the number of hydrogens in the molecular formula; thus, from left to right, the peaks represent 2H:3H:2H:3H. Assigning the fragments, there are five aromatic protons, indicating a monosubstituted benzene ring. The 2H peak probably represents a CH_2 fragment, and the 3H peak probably represents a CH_3 fragment.

Putting the fragments together leads to the following two possible structures, labeled A and B. So how do you determine which is the right structure? In the left-most structure, you'd expect the CH_2 group and the CH_3 group to give singlets in the NMR (because they have no hydrogen neighbors). However, the spectrum clearly shows that the peaks are coupled. Therefore, you can eliminate A as a possibility.

—CH_3
—CH_2—
C=O

fragments

H_2C–C(=O)–CH_3 (A) or O=C–CH_2CH_3 (B)

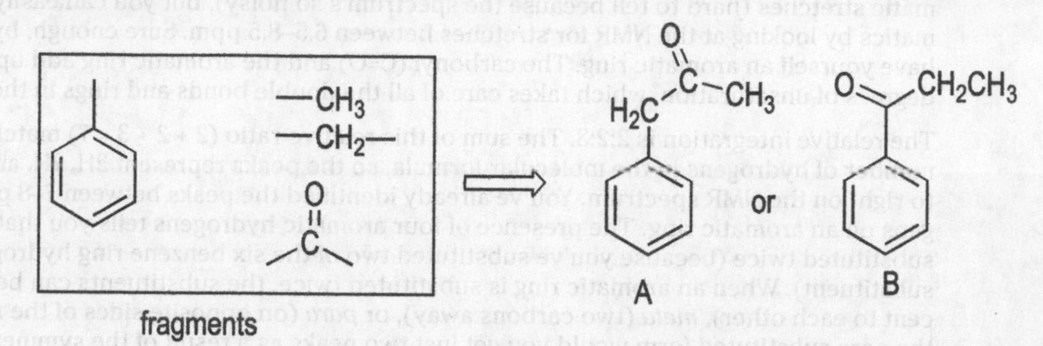

Now double check the proposed structure. The proposed structure B should give peaks for aromatic protons that integrate for 5H. Check! The CH_2 peak should be downfield because it's adjacent to the C=O group, should integrate for 2H, and should be split into a quartet (because it has three hydrogen neighbors). Check! And the CH_3 group should be further upfield (lower chemical shift) because it's farther from the C=O group, should integrate for 3H, and should split into a triplet (it has hydrogen neighbors). Check!

16 This is an interesting (that is, tricky) problem because you don't get a molecular formula. You do get a mass spec, though. The big clue in the mass spec is the two peaks of equal intensity two mass units apart. This is a big giveaway for a bromine in the molecule (see Chapter 15) because bromine has two equally abundant isotopes, ^{79}Br and ^{81}Br. Subtracting 81 mass units (the weight of the larger Br isotope) from the higher mass isotope peak (124) gives you 43. This is how much mass is left on the molecule after you lose the bromine. A good way to get an approximate number of carbons from a mass spec is to divide by 14, the weight of a CH_2 unit. This gives you ~3, so if you have three carbons (12 × 3 = 36), you need 7 hydrogens to make the weight 43. Therefore, the tentative formula is C_3H_7Br, which has two possible structures.

Br or Br

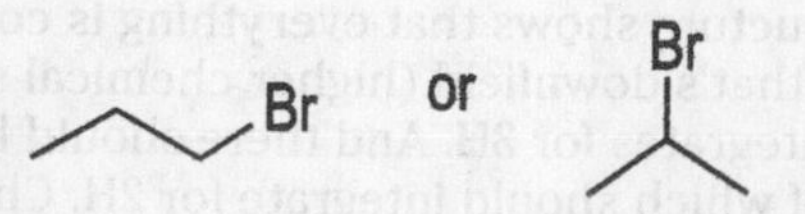

The left-most possibility has no symmetry, so you should see three signals in the ^{13}C NMR. In contrast, the right-most structure has a plane of symmetry, so you should see only two signals in the ^{13}C NMR. In fact, the spectrum shows only two signals, so the right-hand structure is the correct one.

17

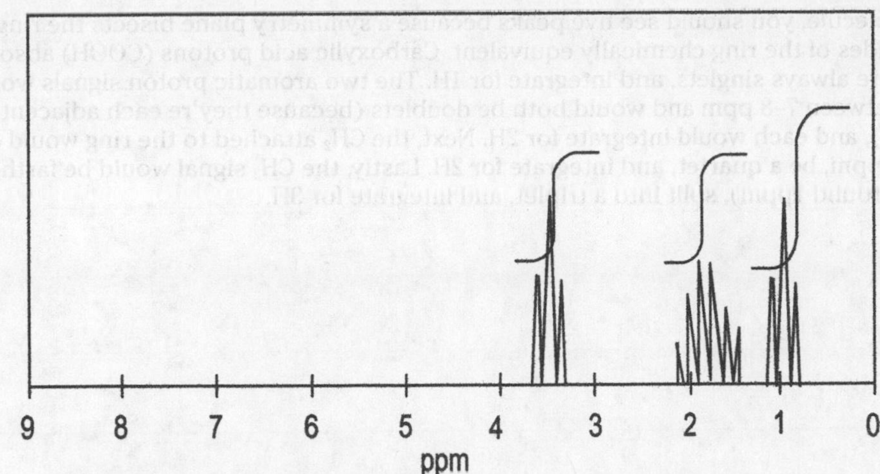

The spectrum of this molecule should show three peaks. The peak with the highest chemical shift is the one closest to the electronegative chlorine atom, followed by the second-closest CH_2 unit, with the CH_3 group at the lowest chemical shift.

Next, determine the coupling. The peak with the highest chemical shift integrates for 2H and splits into a triplet. The peak at a lower chemical shift (the adjacent CH_2) integrates for 2H and splits into a multiplet (approximately a sextet). Finally, the peak at lowest chemical shift is a triplet that integrates for 3H.

18

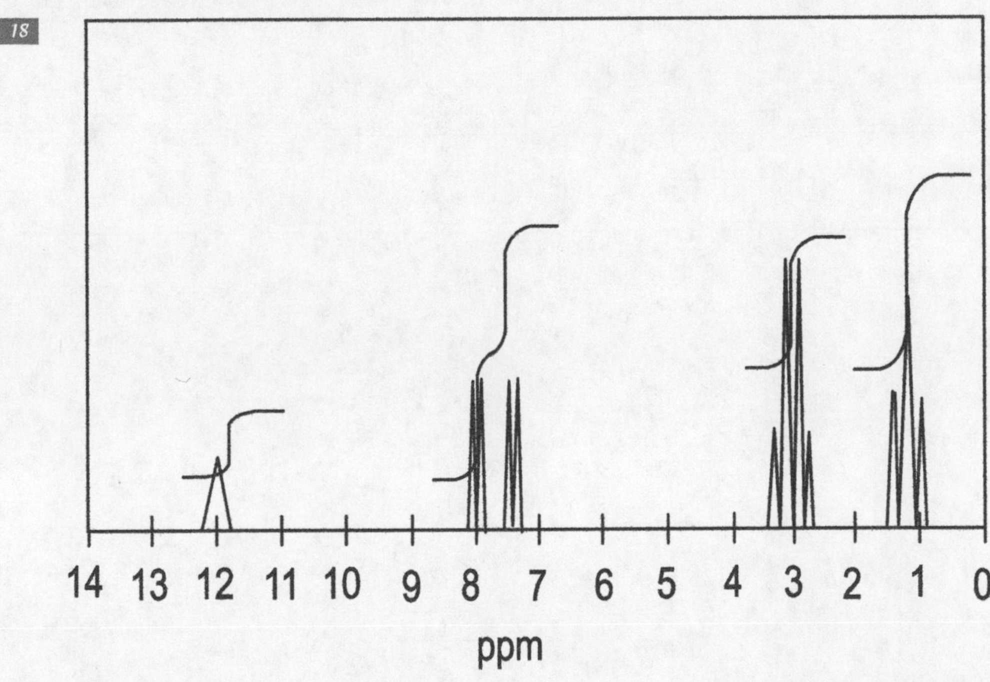

In this molecule, you should see five peaks because a symmetry plane bisects the ring, making the two sides of the ring chemically equivalent. Carboxylic acid protons (COOH) absorb around 12 ppm, are always singlets, and integrate for 1H. The two aromatic proton signals would absorb between 7–8 ppm and would both be doublets (because they're each adjacent to one hydrogen), and each would integrate for 2H. Next, the CH_2 attached to the ring would absorb around 3 ppm, be a quartet, and integrate for 2H. Lastly, the CH_3 signal would be farthest upfield (around 1ppm), split into a triplet, and integrate for 3H.

Part V
The Part of Tens

"Okay—now that the paramedic is here with the defibrillator and smelling salts, prepare to open your test booklets..."

In this part . . .

This part contains a few tidbits of wisdom about how to go about your business as a budding organic chemist. Included are Ten Commandments of Organic Chemistry, which can keep you on the straight and narrow, and ten tips for acing orgo exams.

Chapter 17

The Ten Commandments of Organic Chemistry

In This Chapter

- Seeing how to be successful in organic chemistry
- Avoiding the classic pitfalls

And lo, Friedrich Wohler, father of modern organic chemistry, came down from the mountainside and spoke over the throng of beginning organic students, saying, "If thou follow these commandments, blessed be your exams, and when thou get them returned unto you there will be no wailing or gnashing of teeth, but instead rejoicing. Or if not rejoicing, then certainly less wailing and gnashing of teeth than before. Grinding of teeth, perhaps, but not gnashing." If you follow Wohler's ways of organic chemistry, outlined in the commandments of this chapter, you may certainly feel less stress and anxiety toward your chemistry work. Break these commandments at your peril.

Thou Shalt Work the Practice Problems before Reading the Answers

Working the practice problems is the greatest commandment of all, so I put it first. Organic chem textbooks today weigh more than most small children, and practicing problems helps you remember the concepts for the long term. After you complete a problem and see where you made the mistakes and correct those mistakes, the ideas become engrained into your bones.

Reading the material is a good idea, but after a while, your brain reservoir fills and all the info that follows starts pouring over the dam. You soon forget any material you haven't practiced. That's a problem, because organic chemistry builds on itself, using information that's taught at the beginning of the course as a foundation for the new material. It's not like a history class, where if you didn't understand the culture of *Australopithecus,* you can still figure out the ancient Romans.

In organic chemistry, if you don't understand the basics of bonding, for example, then you're really sandbagged when you get to reactions, because the text and practice problems assume that you understand bonding. Of course, you can prevent all this trouble by working the problems.

Unfortunately, one of the most common ways that students "practice problems" is by reading the solutions manual. Similarly, I tried to "practice piano" by listening to concertos. That didn't work out for me, I'm afraid, and for those of you who work problems by reading the solutions guide — well, let me just suggest that it won't work out too well for you, either. Regrettably, there's no substitute for a few hours spent with the pencil, taking the time to honestly work the problems. In this class, lazy students lose, and the students who work hard (like you) win.

Thou Shalt Memorize Only What Thou Must

The rumors are true — organic chemistry involves some memorization, particularly at the beginning. For example, you have to commit rules of nomenclature and functional groups to memory. The problem comes when you try to memorize *everything.* This task is fruitless, impossible, and a waste of time. You can't memorize your way through organic chemistry. You have to understand the concepts and be able to apply those concepts to really succeed.

A professor can write any of millions of structures on an exam, and you can't possibly hope to memorize them all. You really need to understand the concepts to be able to apply what you know to structures you've never seen before. The concepts presented in the first couple chapters are particularly important because they lay the groundwork for the rest of organic chemistry. By working problems (honestly), you can see how to apply the concepts in organic chemistry to new systems and problem types.

Thou Shalt Understand Thy Mechanisms

This commandment is a specific extension of the preceding one. Memorizing what products come out of a reaction only gives you a limited understanding of that reaction. If you can draw the mechanism of that reaction, however, you can really understand how that reaction works and know exactly how the starting material is converted into product. I recommend two levels of knowledge for each reaction:

- Be able to quickly give the products of the reaction.
- Know the mechanism for each reaction so you understand how that reaction happens.

Thou Shalt Sleep at Night and Not in Class

Many students exercise their freedom as college students to sleep in class. This isn't your typical sleeping in class, though (although of course some students do physically fall asleep). Many students seem to sleep with their eyes open, like horses, sitting in a sort of trance-like hypnosis.

In truth, after a couple of weeks, many students in the classes I attend turn into a group of brain-dead zombies, mere scribes who occasionally scribble down notes off the board without giving the ideas in them a second thought. Their brains are powered off, they aren't learning, and they're wasting their time. Don't let this happen to you! (Following the next commandment may help with this.)

Thou Shalt Read Ahead Before Class

One of the secrets to success in organic chemistry is to read the material before you go to class. Admittedly, keeping up with even a very good professor in class is difficult, simply because the material comes so quickly. But if you've read the material beforehand, you may find that you understand much more in class than you ever did before (even if you didn't understand it all when you read). In doing so, you used your time efficiently because you didn't squander precious class time brainlessly copying down notes off the board without understanding a bit.

Working the problems after class and reading the material before class are the two single most important pieces of advice I can give you for doing well in organic chemistry.

Thou Shalt Not Fall Behind

Aside from not working problems in organic chemistry, the biggest source of trouble for students is falling behind. Winemaking has a term called the *Angel's Share,* the portion of wine that's mysteriously lost from the barrel each year the wine ages. In organic chemistry, there's a similar *Angel's Share* after each exam, as more and more students magically disappear from the classroom, never to be seen or heard from again (by their professors).

The biggest culprit is students' falling behind. Catching up is nearly impossible. The course is so fast-paced and the textbook is so large that after you fall behind, you have to devote twice the time studying to get back to speed. New material assumes you understand what was already covered. In other words, if you're behind, you won't be able to figure out what your professor is talking about in class *today* until you've gone back and relearned the old material.

When a student asks how much he or she should be studying, I make the following recommendation: Read the material that's going to be covered in the next class period the day before. After your professor finishes covering a chapter, start to work on the problems from that chapter. That means that every day, you should work problems from the previous chapter and begin reading the next chapter. If that seems like a lot of work, well, it is! The successful organic student studies, on average, every day for an hour. To make that commitment, you have to take charge of your own education.

Thou Shalt Know How Thou Learnest Best

People master concepts in different ways. Some people are visual learners, finding that they understand best by making diagrams and charts that connect the material. Others are auditory learners, gathering information best by listening quietly to a lecture. Some people study well in groups, whereas others are more productive working alone.

To make your studying as productive as possible, know how you learn the best. If you know that it's in groups, make it a priority at the beginning of the semester to organize a group studying session. Work best alone? Forget the groups and take a stroll to the bottom floor of the library. If you know that chemistry is just not your thing, find a tutor as early as possible in the semester so you don't commit the unforgivable sin of falling behind.

Thou Shalt Not Skip Class

This commandment of not skipping class is kind of like the Biblical commandment of not killing people — kind of a no-brainer, right? But apparently it's not for some students, because a good percentage of the class doesn't attend class after the first couple weeks. I know dragging yourself out of bed is difficult some days, but a good professor can explain the material better than a book ever can. You can ask questions in class, and some professors even slip hints about what'll be on exams. Even if they don't slip hints overtly, what they choose to cover in class is the biggest hint of all. But if you don't attend, you'll miss those hints.

Thou Shalt Ask Questions

Please, please, do ask questions. You may find that many students in the class have your exact question but were too afraid to pipe up about it for fear they'd look foolish. Keep in mind that some teachers have been giving lectures for many years (or decades), and occasionally they lose perspective on what students comprehend at that point.

Thou Shalt Keep a Positive Outlook

Keeping a positive outlook on the class can benefit you. It makes studying a lot easier, for one thing. Secondly, organic chemistry really does change the way you look at the world. After the class, you can see why carrots are orange and what plastics, polymers, and drugs look like on the molecular level. You'll have the tools to make compounds the world has never seen before, and you'll also have the foundation for understanding how biology *really* works, because biochemistry is essentially the organic chemistry that occurs in living systems. With the right attitude, organic chemistry can actually be quite fun. And you can brag to your friends that you mastered a difficult subject, which has to be at least as impressive as climbing a tall mountain (and less dangerous, too).

Chapter 18

Ten Tips for Acing Orgo Exams

In This Chapter

- Seeing the keys to smoking your next orgo exam
- Maximizing your studying success

For some people, taking exams isn't that big of a deal. However, for others, the thought of taking an exam makes their hands clammy and their brows sweat. And some individuals may even start to have a minor panic attack. Regardless of whether you worry about exams, this chapter gives some concrete tips on how you can improve your chances of success on organic chemistry exams.

Scan and Answer the Easy Questions First

To make the best use of your time during the exam when you first get your test, scan the exam quickly. You *don't* need to answer the exam in order, and by scanning, you can determine which problems are easiest. You can then tackle them first. Why?

- By answering all the easy questions first, you make sure you at least earn all the easy points. If you run out of time, you want to leave only really hard questions unanswered, the ones you may not have earned full credit on anyway, not some cakewalk problems you could've answered in your sleep.
- By starting on something you can do easily, you break the so-called *exam block,* the quite-frequent occurrence when students look at a tough first question on the exam and their minds freeze so they can't solve any of the problems. Getting in the flow of working problems, even if they're easy, lets your mind warm up so you can return to the state of mind that you were in when you worked problems before.

A common sight at the end of an organic exam is to see students scribbling answers furiously and the professor hollering for everyone to turn them in. That's because a lot of students don't finish their exams in the allotted time. Can there be any worse feeling than not doing as well as you liked, not because you didn't know the material but because you didn't finish quickly?

Read All of Every Question

As you work your way through the exam, make sure you take your time and read every question, all the way through. Doing so only takes a couple extra seconds. This tip may seem like a no-brainer, but there's always at least one exam during the course where a large percentage of the students lose points simply because they didn't read the whole question.

For example, a question may state, "Determine the products of the following reactions. Then label the nucleophile and electrophile." Half the class, seeing the reaction, may write the products and go on their merry way, never indicating what the nucleophile or the electrophile was because they failed to read the whole question. They then lose half the credit, even though they probably knew the answer. Don't let this happen to you.

Set Aside Time Each Day to Study

Each student is different and each class is different, but 45 minutes to an hour is probably the amount of time you should expect to study organic chemistry each night if you plan on getting an A or a B. That's the bad news. The good news is that anyone who's smart enough to pass a class in general chemistry is smart enough to get an A or at worst a B in organic chemistry, as long as they're willing to put in the work.

My favorite motto for the class is "it's not so much hard as it is hard work." In this class, the lazy students (no matter how smart) get blown away by those students who work really hard.

Form a Study Group

To improve how you do on your next exam, ask some classmates to form a study group. A lot of people work better in groups. Study groups are good for a couple reasons:

- **A study group allows you and other students to discuss concepts, which can help clarify them.** Sometimes a fellow student can explain a concept to you much better than a textbook (or even your professor) can. Or you can explain a concept to the others in your study group, and there's no better way of reinforcing your own knowledge than by teaching what you know to someone else.
- **A study group is a good way to force you to study.** If you make a commitment to meeting with a group every day, you're less likely to blow it off than if you were studying by yourself. It's kind of like getting an exercise partner.

You can make a study group from lab partners or studious students sitting next to you in class. In rare cases, a professor can help arrange study sections with sign-up sheets. To ensure everyone can contribute, I suggest keeping the study group on the small side, in the range of three to six people.

Get Old Exams

One of the easiest ways to study comes from practicing with old exams your professor has given (if they're available). You can get insight into the sort of questions that your professor likes to ask, and you can also get a feel for the depth of knowledge and level of detail that you're expected to know. Some professors like to ask a lot of straightforward problems; others prefer to give fewer, more challenging problems. The best way to find out is to check out what they've asked before.

Make Your Answers Clear by Using Structures

The words of the organic chemist are *structures*. Don't forget this mantra on your exams, and don't write long-winded explanations when a simple structure suffices. When solving the problems on the exam, follow the motto "show, don't tell." Keep your answers clear and concise by using structures.

For example, if you're asked to explain why one molecule is more stable than another, you can show how one compound has more resonance structures than another much more clearly than you can with a paragraph of text. Not only that, but using structures saves time — drawing a structure is a lot faster than writing a paragraph.

Don't Try to Memorize Your Way Through

Although organic chemistry does have some memorization involved, such as the rules for naming molecules and the steps in determining R and S stereochemistry, don't make the common mistake of trying to memorize your way through the course, which is a futile and impossible venture. If you've memorized a reaction, for example, but don't really understand it, it's very easy to get tripped up on an exam if a question is asked in a different way or on a different structure than the one you committed to memory. Also, concepts that you master at the beginning of the course are built on as the course goes on, so you start digging a hole quite quickly if you've memorized but don't really understand the material.

Work a Lot of Problems

There's an old saying that the best way to master organic chemistry is with a pencil. You're most likely to remember concepts when you've practiced problems that apply those concepts. Also, the more problems you practice, the faster you get at solving problems, making it less likely that you'll run out of time on an exam. This advice may be the most important that I can give you. Practice as many problems as you can get your hands on. (Working the problems in this book is a good start.)

Get Some Sleep the Night Before

In order to do well on your next orgo exam, you need to make sure you're fresh and not fatigued, stressed, and strung out on caffeine. Go to bed at a decent time the night before. Pulling all-nighters before orgo exams seems to be very popular, but I strongly advise against this practice.

A lot of times, information that's "learned" in cram sessions is soon forgotten. Even if you're fortunate enough not to forget it before you take the exam, you'll have to relearn the material again for the final. It's better to study a little every day and work the concepts into your bones so you never forget it.

Recognize Red Herrings

Many professors don't put red herrings into exams, but I've met enough who do, so you need to be able to identify them when you see them. *Red herrings* are problems that have many details that are unimportant to determining the answer. These questions are often designed simply to test your confidence. A lot of times, this distracter gives you a straightforward reaction on a complicated-looking structure. You can often identify these problems because they usually contain large paragraphs of text in the question.

For example, a question may ask the following:

> **Question 3 (15 pts).** Angosterolongmoleculename is a fascinating and novel steroid that was recently isolated from the udder of a rare breed of cow. Ten thousand pounds of the bovine udder was ground up and extracted with chloroform to obtain just a few milligrams of the material (what an udder waste!). Samuel J. Snodgrass characterized the molecule by NMR and IR spectroscopy. He converted the molecule into a derivative by reacting the molecule with BH_3 in THF, followed by the addition of peroxides. Give the structure of the derivative.

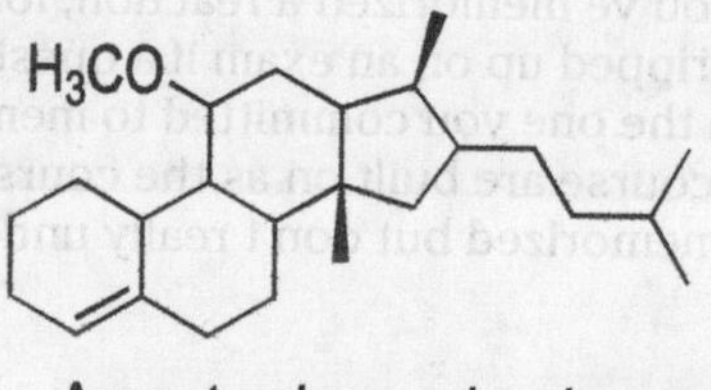

H_3CO

Angosterolongmoleculename

A lot of the information in the problem is irrelevant, but it makes the problem look intimidating. If you get right down to it, though, this problem is just a simple reaction of an alkene, provided you focus on the part of the molecule that's going to change in the reaction.

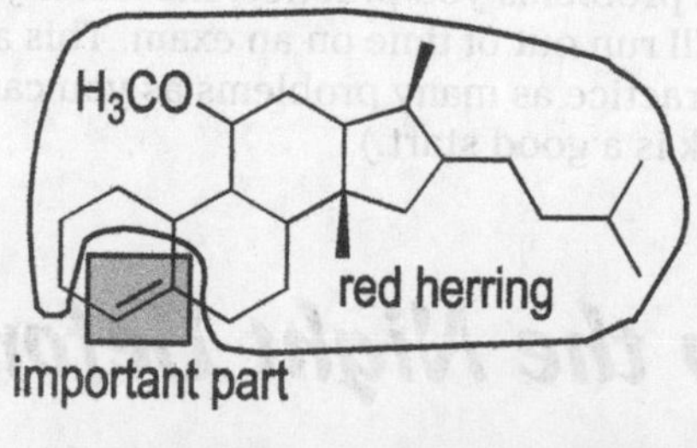

H_3CO

red herring

important part

Angosterolongmoleculename

Index

• Numerics •

• A •

• B •

• H •

• I •

• J •

• K •

• L •

• M •

• N •

• O •

• P •

• T •

• U •

• V •

• W •

• Y •

• Z •

SPORTS, FITNESS, PARENTING, RELIGION & SPIRITUALITY

0-471-76871-5

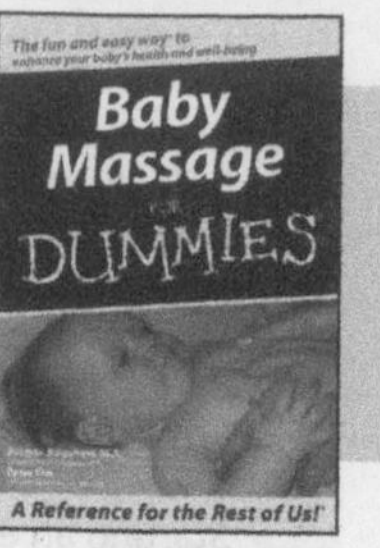

0-7645-7841-3

Also available:

- ✓ Catholicism For Dummies 0-7645-5391-7
- ✓ Exercise Balls For Dummies 0-7645-5623-1
- ✓ Fitness For Dummies 0-7645-7851-0
- ✓ Football For Dummies 0-7645-3936-1
- ✓ Judaism For Dummies 0-7645-5299-6
- ✓ Potty Training For Dummies 0-7645-5417-4
- ✓ Buddhism For Dummies 0-7645-5359-3
- ✓ Pregnancy For Dummies 0-7645-4483-7 †
- ✓ Ten Minute Tone-Ups For Dummies 0-7645-7207-5
- ✓ NASCAR For Dummies 0-7645-7681-X
- ✓ Religion For Dummies 0-7645-5264-3
- ✓ Soccer For Dummies 0-7645-5229-5
- ✓ Women in the Bible For Dummies 0-7645-8475-8

TRAVEL

0-7645-7749-2

0-7645-6945-7

Also available:

- ✓ Alaska For Dummies 0-7645-7746-8
- ✓ Cruise Vacations For Dummies 0-7645-6941-4
- ✓ England For Dummies 0-7645-4276-1
- ✓ Europe For Dummies 0-7645-7529-5
- ✓ Germany For Dummies 0-7645-7823-5
- ✓ Hawaii For Dummies 0-7645-7402-7
- ✓ Italy For Dummies 0-7645-7386-1
- ✓ Las Vegas For Dummies 0-7645-7382-9
- ✓ London For Dummies 0-7645-4277-X
- ✓ Paris For Dummies 0-7645-7630-5
- ✓ RV Vacations For Dummies 0-7645-4442-X
- ✓ Walt Disney World & Orlando For Dummies 0-7645-9660-8

GRAPHICS, DESIGN & WEB DEVELOPMENT

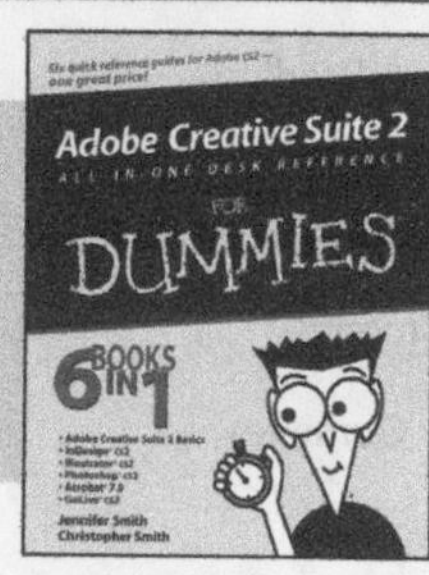

0-7645-8815-X

0-7645-9571-7

Also available:

- ✓ 3D Game Animation For Dummies 0-7645-8789-7
- ✓ AutoCAD 2006 For Dummies 0-7645-8925-3
- ✓ Building a Web Site For Dummies 0-7645-7144-3
- ✓ Creating Web Pages For Dummies 0-470-08030-2
- ✓ Creating Web Pages All-in-One Desk Reference For Dummies 0-7645-4345-8
- ✓ Dreamweaver 8 For Dummies 0-7645-9649-7
- ✓ InDesign CS2 For Dummies 0-7645-9572-5
- ✓ Macromedia Flash 8 For Dummies 0-7645-9691-8
- ✓ Photoshop CS2 and Digital Photography For Dummies 0-7645-9580-6
- ✓ Photoshop Elements 4 For Dummies 0-471-77483-9
- ✓ Syndicating Web Sites with RSS Feeds For Dummies 0-7645-8848-6
- ✓ Yahoo! SiteBuilder For Dummies 0-7645-9800-7

NETWORKING, SECURITY, PROGRAMMING & DATABASES

0-7645-7728-X

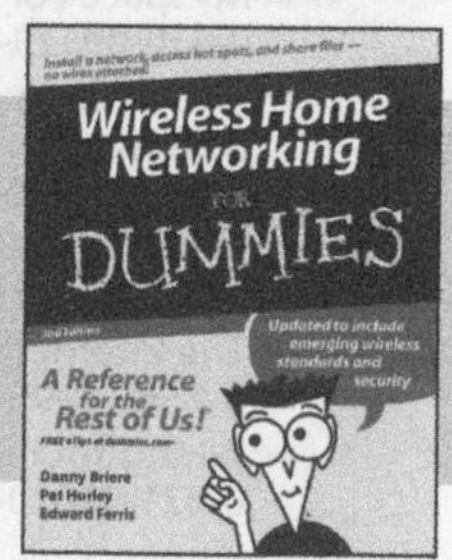

0-471-74940-0

Also available:

- ✓ Access 2007 For Dummies 0-470-04612-0
- ✓ ASP.NET 2 For Dummies 0-7645-7907-X
- ✓ C# 2005 For Dummies 0-7645-9704-3
- ✓ Hacking For Dummies 0-470-05235-X
- ✓ Hacking Wireless Networks For Dummies 0-7645-9730-2
- ✓ Java For Dummies 0-470-08716-1
- ✓ Microsoft SQL Server 2005 For Dummies 0-7645-7755-7
- ✓ Networking All-in-One Desk Reference For Dummies 0-7645-9939-9
- ✓ Preventing Identity Theft For Dummies 0-7645-7336-5
- ✓ Telecom For Dummies 0-471-77085-X
- ✓ Visual Studio 2005 All-in-One Desk Reference For Dummies 0-7645-9775-2
- ✓ XML For Dummies 0-7645-8845-1